McGraw-Hill Dictionary of
ELECTRICAL AND ELECTRONIC ENGINEERING

McGraw-Hill Dictionary of
ELECTRICAL AND ELECTRONIC ENGINEERING

Sybil P. Parker
EDITOR IN CHIEF

McGraw-Hill Book Company

New York   St. Louis   San Francisco

Auckland   Bogotá   Guatemala   Hamburg
Johannesburg   Lisbon   London   Madrid   Mexico
Montreal   New Delhi   Panama   Paris   San Juan
São Paulo   Singapore   Sydney   Tokyo   Toronto

McGRAW-HILL DICTIONARY OF ELECTRICAL AND
ELECTRONIC ENGINEERING
The material in this Dictionary has been published previously
in the McGRAW-HILL DICTIONARY OF SCIENTIFIC
AND TECHNICAL TERMS, Third Edition, copyright
© 1984 by McGraw-Hill, Inc. All rights reserved. Philippines
copyright 1984 by McGraw-Hill, Inc. Printed in the United
States of America. Except as permitted under the United
States Copyright Act of 1976, no part of this publication may
be reproduced or distributed in any form or by any means, or
stored in a data base or retrieval system, without the prior
written permission of the publisher.

1 2 3 4 5 6 7 8 9 0   FGFG   8 9 1 0 9 8 7 6 5

ISBN 0-07-045413-2

Library of Congress Cataloging in Publication Data

McGraw-Hill dictionary of electrical and electronic
engineering.

  1. Electric engineering—Dictionaries. 2. Electronics
—Dictionaries. I. Parker, Sybil P. II. McGraw-Hill
Company. III. Title: Dictionary of electrical and
electronic engineering.
TK9.M34   1985   621.3′03′21   84-28837
ISBN 0-07-045413-2 (pbk.)

Editorial Staff

Sybil P. Parker, Editor in Chief

Jonathan Weil, Editor
Betty Richman, Editor
Edward J. Fox, Art director
Joe Faulk, Editing manager
Frank Kotowski, Jr., Editing supervisor

Consulting and Contributing Editors
from the McGraw-Hill Dictionary of Scientific and Technical Terms

Prof. Roland H. Good, Jr.—Department of Physics, Pennsylvania State University. PHYSICS.

John Markus—Author and Consultant. ELECTRONICS.

How to Use the Dictionary

ALPHABETIZATION

The terms in the *McGraw-Hill Dictionary of Electrical and Electronic Engineering* are alphabetized on a letter-by-letter basis; word spacing, hyphen, comma, solidus, and apostrophe in a term are ignored in the sequencing. For example, an ordering of terms would be:

> **alloy junction**
> **all-wave receiver**
> **AM field signature**
> **AND/NOR gate**
> **AND-OR circuit**
> **A positive**

CROSS-REFERENCING

A cross-reference entry directs the user to the defining entry. For example, the user looking up "arc" finds:

> **arc** *See* electric arc.

The user then turns to the "E" terms for the definition.

Cross-references are also made from variant spellings, acronyms, abbreviations, and symbols.

> **Alice** *See* Alaska Integrated Communications Exchange.
> **i-f** *See* intermediate frequency.
> **mG** *See* milligauss.

The user turning directly to a defining entry will find the above type of information included, introduced by "Also known as . . . ," "Also spelled . . . ," "Abbreviated . . . ," "Symbolized . . . ," "Derived from . . . ."

McGraw-Hill Dictionary of
ELECTRICAL AND ELECTRONIC ENGINEERING

a *See* ampere.
aΩ *See* abohm.
(aΩ)⁻¹ *See* abmho.
A *See* ampere.
aA *See* abampere.
aAcm² *See* abampere centimeter squared.
aA/cm² *See* abampere per square centimeter.
A AND NOT B gate *See* AND NOT gate.
ab- A prefix used to identify centimeter-gram-second electromagnetic units, as in abampere, abcoulomb, abfarad, abhenry, abmho, abohm, and abvolt.
abampere The unit of electric current in the electromagnetic centimeter-gram-second system; 1 abampere equals 10 amperes in the absolute meter-kilogram-second-ampere system. Abbreviated aA. Also known as Bi; biot.
abampere centimeter squared The unit of magnetic moment in the electromagnetic centimeter-gram-second system. Abbreviated aAcm².
abampere per square centimeter The unit of current density in the electromagnetic centimeter-gram-second system. Abbreviated aA/cm².
A battery The battery that supplies power for filaments or heaters of electron tubes in battery-operated equipment.
ABC *See* automatic brightness control.
abcoulomb The unit of electric charge in the electromagnetic centimeter-gram-second system, equal to 10 coulombs. Abbreviated aC.
abcoulomb centimeter In the electromagnetic centimeter-gram-second system of units, the unit of electric dipole moment. Abbreviated aCcm.
abcoulomb per cubic centimeter The electromagnetic centimeter-gram-second unit of volume density of charge. Abbreviated aC/cm³.
abcoulomb per square centimeter The electromagnetic centimeter-gram-second unit of surface density of charge, electric polarization, and displacement. Abbreviated aC/cm².
abfarad A unit of capacitance in the electromagnetic centimeter-gram-second system equal to 10^9 farads. Abbreviated aF.
abhenry A unit of inductance in the electromagnetic centimeter-gram-second system of units which is equal to 10^{-9} henry. Abbreviated aH. Also known as centimeter.
abmho A unit of conductance in the electromagnetic centimeter-gram-second system of units equal to 10^9 mhos. Abbreviated (aΩ)⁻¹. Also known as absiemens (aS).

2 abnormal glow discharge

abnormal glow discharge A discharge of electricity in a gas tube at currents somewhat higher than those of an ordinary glow discharge, at which point the glow covers the entire cathode and the voltage drop decreases with increasing current.

abnormal reflections Sharply defined reflections of substantial intensity at frequencies greater than the critical frequency of the ionized layer of the ionosphere.

abohm The unit of electrical resistance in the centimeter-gram-second system; 1 abohm equals 10^{-9} ohm in the meter-kilogram-second system. Abbreviated aΩ.

abohm centimeter The centimeter-gram-second unit of resistivity. Abbreviated aΩcm.

AB power pack 1. Assembly in a single unit of the A battery and B battery for a battery-operated vacuum-tube circuit. 2. Unit that supplies the necessary A and B direct-current voltages from an alternating-current source of power.

abrupt junction A *pn* junction in which the concentration of impurities changes suddenly from acceptors to donors.

absiemens *See* abmho.

absolute electrometer A very precise type of attracted disk electrometer in which the attraction between two disks is balanced against the force of gravity.

absolute gain of an antenna Gain in a given direction when the reference antenna is an isotropic antenna isolated in space. Also known as isotropic gain of an antenna.

absolute permeability The ratio of the magnetic flux density to the intensity of the magnetic field in a medium; measurement is in webers per square meter in the meter-kilogram-second system. Also known as induced capacity.

absolute wavemeter A type of wavemeter in which the frequency of an injected radio-frequency voltage is determined by measuring the length of a resonant line.

absorb To take up energy from radiation.

absorber A material or device that takes up and dissipates radiated energy; may be used to shield an object from the energy, prevent reflection of the energy, determine the nature of the radiation, or selectively transmit one or more components of the radiation.

absorption 1. The property of a dielectric in a capacitor which causes a small charging current to flow after the plates have been brought up to the final potential, and a small discharging current to flow after the plates have been short-circuited, allowed to stand for a few minutes, and short-circuited again. Also known as dielectric soak. 2. The taking up of energy from radiation by the medium through which the radiation is passing.

absorption circuit A series-resonant circuit used to absorb power at an unwanted signal frequency by providing a low impedance to ground at this frequency.

absorption control *See* absorption modulation.

absorption cross section In radar, the ratio of the amount of power removed from a beam by absorption of radio energy by a target to the power in the beam incident upon the target.

absorption current The component of a dielectric current that is proportional to the rate of accumulation of electric charges within the dielectric.

absorption modulation A system of amplitude modulation in which a variable-impedance device is inserted in or coupled to the output circuit of the transmitter. Also known as absorption control; loss modulation.

absorption wavemeter A frequency- or wavelength-measuring instrument consisting of a calibrated tunable circuit and a resonance indicator.

abvolt The unit of electromotive force in the electromagnetic centimeter-gram-second system; 1 abvolt equals 10^{-8} volt in the absolute meter-kilogram-second system. Abbreviated aV.

abvolt per centimeter In the electromagnetic centimeter-gram-second system of units, the unit of electric field strength. Abbreviated aV/cm.

abwatt The unit of electrical power in the centimeter-gram-second system; 1 abwatt equals 1 watt in the absolute meter-kilogram-second system.

abWb *See* maxwell.

abweber *See* maxwell.

ac *See* alternating current.

aC *See* abcoulomb.

accelerated test A test of the serviceability of an electric cable in use for some time by applying twice the voltage normally carried.

accelerating electrode An electrode used in cathode-ray tubes and other electron tubes to increase the velocity of the electrons that contribute the space current or form a beam.

accelerating potential The energy potential in electron-beam equipment that imparts additional speed and energy to the electrons.

acceleration voltage The voltage between a cathode and accelerating electrode of an electron tube.

accentuation The enhancement of signal amplitudes in selected frequency bands with respect to other signals.

accentuator A circuit that provides for the first part of a process for increasing the strength of certain audio frequencies with respect to others, to help these frequencies override noise or to reduce distortion. Also known as accentuator circuit.

accentuator circuit *See* accentuator.

acceptor An impurity element that increases the number of holes in a semiconductor crystal such as germanium or silicon; aluminum, gallium, and indium are examples. Also known as acceptor impurity; acceptor material.

acceptor atom An atom of a substance added to a semiconductor crystal to increase the number of holes in the conduction band.

acceptor circuit A series-resonant circuit that has a low impedance at the frequency to which it is tuned and a higher impedance at all other frequencies.

acceptor impurity *See* acceptor.

acceptor material *See* acceptor.

aCcm *See* abcoulomb centimeter.

aC/cm² *See* abcoulomb per square centimeter.

aC/cm³ *See* abcoulomb per cubic centimeter.

accordion cable A flat, multiconductor cable prefolded into a zigzag shape and used to make connections to movable equipment such as a chassis mounted on pullout slides.

accumulator *See* storage battery.

accumulator battery *See* storage battery.

ac/dc motor *See* universal motor.

4 ac/dc receiver

ac/dc receiver A radio receiver designed to operate from either an alternating- or direct-current power line. Also known as universal receiver.

aΩcm *See* abohm centimeter.

acorn tube An ultra-high-frequency electron tube resembling an acorn in shape and size.

acoustic amplifier A device that amplifies mechanical vibrations directly at audio and ultrasonic frequencies. Also known as acoustoelectric amplifier.

acoustic branch One of the parts of the dispersion relation, frequency as a function of wave number, for crystal lattice vibrations, representing vibration at low (acoustic) frequencies.

acoustic bridge A device, based on the principle of the electrical Wheatstone bridge, used for analysis of deafness.

acoustic convolver *See* convolver.

acoustic delay line A device in which acoustic signals are propagated in a medium to make use of the sonic propagation time to obtain a time delay for the signals. Also known as sonic delay line.

acoustic detector The stage in a receiver at which demodulation of a modulated radio wave into its audio component takes place.

acoustic mode The type of crystal lattice vibrations which for long wavelengths act like an acoustic wave in a continuous medium, but which for shorter wavelengths approach the Debye frequency, showing a dispersive decrease in phase velocity.

acoustic phonon A quantum of excitation of an acoustic mode of vibration.

acoustic receiver The complete equipment required for receiving modulated radio waves and converting them into sound.

acoustic-wave amplifier An amplifier in which the charge carriers in a semiconductor are coupled to an acoustic wave that is propagated in a piezoelectric material, to produce amplification.

acoustoelectric amplifier *See* acoustic amplifier.

acoustoelectric effect The development of a direct-current voltage in a semiconductor or metal by an acoustic wave traveling parallel to the surface of the material. Also known as electroacoustic effect.

acoustooptical cell An electric-to-optical transducer in which an acoustic or ultrasonic electric input signal modulates or otherwise acts on a beam of light.

acquire 1. Of acquisition radars, the process of detecting the presence and location of a target in sufficient detail to permit identification. 2. Of tracking radars, the process of positioning a radar beam so that a target is in that beam to permit the effective employment of weapons. Also known as target acquisition.

ACSR *See* aluminum cable steel-reinforced.

actinodielectric Of a substance, exhibiting an increase in electrical conductivity when electromagnetic radiation is incident upon it.

actinoelectricity The electromotive force produced in a substance by electromagnetic radiation incident upon it.

action period The period of time during which data in a Williams tube storage device can be read or new data can be written into this storage.

activate 1. To make a cell or battery operative by addition of a liquid. 2. To treat the filament, cathode, or target of a vacuum tube to increase electron emission.

acyclic machine 5

activated cathode A thermionic cathode consisting of a tungsten filament to which thorium has been added, and then brought to the surface, by a process such as heating in the absence of an electric field in order to increase thermionic emission.

activation 1. The process of adding liquid to a manufactured cell or battery to make it operative. 2. The process of treating the cathode or target of an electron tube to increase its emission. Also known as sensitization.

active area The area of a metallic rectifier that acts as the rectifying junction and conducts current in the forward direction.

active component 1. In the phasor representation of quantities in an alternating-current circuit, the component of current, voltage, or apparent power which contributes power, namely, the active current, active voltage, or active power. Also known as power component. 2. See active element.

active current The component of an electric current in a branch of an alternating-current circuit that is in phase with the voltage. Also known as watt current.

active device A component, such as an electron tube or transistor, that is capable of amplifying the current or voltage in a circuit.

active electric network Electric network containing one or more sources of energy.

active electronic countermeasures The major subdivision of electronic countermeasures concerning electronic jamming and electronic deceptions.

active element Any generator of voltage or current in an impedance network. Also known as active component.

active filter A filter that uses an amplifier with conventional passive filter elements to provide a desired fixed or tunable pass or rejection characteristic.

active jamming *See* jamming.

active leg An electrical element within a transducer which changes its electrical characteristics as a function of the application of a stimulus.

active logic Logic that incorporates active components which provide such functions as level restoration, pulse shaping, pulse inversion, and power gain.

active material 1. A fluorescent material used in screens for cathode-ray tubes. 2. An energy-storing material, such as lead oxide, used in the plates of a storage battery. 3. A material, such as the iron of a core or the copper of a winding, that is involved in energy conversion in a circuit. 4. The material of the cathode of an electron tube that emits electrons when heated.

active power The product of the voltage across a branch of an alternating-current circuit and the component of the electric current that is in phase with the voltage.

active region The region in which amplifying, rectifying, light emitting, or other dynamic action occurs in a semiconductor device.

active substrate A semiconductor or ferrite material in which active elements are formed; also a mechanical support for the other elements of a semiconductor device or integrated circuit.

active transducer A transducer whose output is dependent upon sources of power, apart from that supplied by any of the actuating signals, which power is controlled by one or more of these signals.

active voltage In an alternating-current circuit, the component of voltage which is in phase with the current.

actual height Highest altitude at which refraction of radio waves actually occurs.

acyclic machine *See* homopolar generator.

6 adapter transformer

adapter transformer A transformer designed to supply a single electric lamp; its primary terminals are designed to fit into an ordinary lampholder, its secondary terminals into a lampholder of a low-voltage lamp.

Adcock antenna A pair of vertical antennas separated by a distance of one-half wavelength or less and connected in phase opposition to produce a radiation pattern having the shape of a figure eight.

adconductor cathode A cathode in which adsorbed alkali metal atoms provide electron emission in a glow or arc discharge.

adder A circuit in which two or more signals are combined to give an output-signal amplitude that is proportional to the sum of the input-signal amplitudes. Also known as adder circuit.

adder circuit *See* adder.

adding circuit A circuit that performs the mathematical operation of addition.

ADF *See* automatic direction finder.

adhesion Any mutually attractive force holding together two magnetic bodies, or two oppositely charged nonconducting bodies.

A display A radar oscilloscope display in cartesian coordinates; the targets appear as vertical deflection lines; their Y coordinates are proportional to signal intensity; their X coordinates are proportional to distance to targets.

adjacent-channel selectivity The ability of a radio receiver to respond to the desired signal and to reject signals in adjacent frequency channels.

adjustable resistor A resistor having one or more sliding contacts whose position may be changed.

adjustable transformer *See* variable transformer.

adjusted decibel A unit used to show the relationship between the interfering effect of a noise frequency, or band of noise frequencies, and a reference noise power level of -85 dBm. Abbreviated dBa. Also known as decibel adjusted.

admittance A measure of how readily alternating current will flow in a circuit; the reciprocal of impedance, it is expressed in mhos.

admittance matrix A matrix Y whose elements are the mutual admittances between the various meshes of an electrical network; it satisfies the matrix equation $I = YV$, where I and V are column vectors whose elements are the currents and voltages in the meshes.

advanced potential Any electromagnetic potential arising as a solution of the classical Maxwell field equations, analogous to a retarded potential solution, but lying on the future light cone of space-time; the potential appears, at present, to have no physical interpretation.

aerial *See* antenna.

aerodiscone antenna Electrically small antenna for airborne applications in the very-high-frequency and ultra-high-frequency bands; it is derived from, and preserves, the desirable electrical characteristics of the discone antenna and can be designed in various physical shapes.

aerogenerator A generator that is driven by the wind, designed to utilize wind power on a commercial scale.

aerospace electronics The field of electronics as applied to aircraft and spacecraft.

aF *See* abfarad.

AFC *See* automatic frequency control.

Alford loop   7

AGC *See* automatic gain control.

age coating The black deposit that is formed on the inner surface of an electric lamp by material evaporated from the filament.

aging 1. Allowing a permanent magnet, capacitor, meter, or other device to remain in storage for a period of time, sometimes with a voltage applied, until the characteristics of the device become essentially constant. 2. Change in the magnetic properties of iron with passage of time, for example, increase in the hysteresis.

aH *See* abhenry.

Ah *See* ampere-hour.

A/in.^{2} *See* ampere per square inch.

A indicator *See* A scope.

air battery A connected group of two or more air cells; also, a single air cell.

airblast circuit breaker An electric switch which, on opening, utilizes a high-pressure gas blast (air or sulfur hexafluoride) to break the arc.

air capacitor A capacitor having only air as the dielectric material between its plates. Also known as air condenser.

air cell A cell in which depolarization at the positive electrode is accomplished chemically by reduction of the oxygen in the air.

air condenser *See* air capacitor.

air-core coil An inductor without a magnetic core.

air-core transformer Transformer (usually radio-frequency) having a nonmetallic core.

aircraft antenna An airborne device used to detect or radiate electromagnetic waves.

aircraft decibel rating The ratio of the radar reflectivity of a specific type of aircraft to that of a selected reference aircraft, measured in decibels.

air depolarized battery A primary battery which is kept depolarized by atmospheric oxygen rather than chemical compounds. Also known as metal-air battery.

air gap 1. A gap or an equivalent filler of nonmagnetic material across the core of a choke, transformer, or other magnetic device. 2. A spark gap consisting of two electrodes separated by air.

air-insulated substation An electric power substation that has the busbars and equipment terminations generally open to air and utilizes insulation properties of ambient air for insulation to ground.

air-spaced coax Coaxial cable in which air is basically the dielectric material; the conductor may be centered by means of a spirally wound synthetic filament, beads, or braided filaments.

air switch A switch in which the breaking of the electric circuit takes place in air. Also known as air-break switch.

airwave A radio wave used in radio and television broadcasting.

alarm signal The international radiotelegraph alarm signal transmitted to actuate automatic devices that sound an alarm indicating that a distress message is about to be broadcast.

Alaska Integrated Communications Exchange A network of radio stations, generally using scatter-propagation equipment, that links early-warning radar stations. Also known as Alice; White Alice.

Alford loop An antenna utilizing multielements which usually are contained in the same horizontal plane and adjusted so that the antenna has approximately equal and in-

8 Alice

phase currents uniformly distributed along each of its peripheral elements and produces a substantially circular radiation pattern in the plane of polarization; it is known for its purity of polarization.

Alice *See* Alaska Integrated Communications Exchange.

alignment The process of adjusting components of a system for proper interrelationship, including the adjustment of tuned circuits for proper frequency response and the time synchronization of the components of a system.

alive *See* energized.

alkaline cell A primary cell that uses an alkaline electrolyte, usually potassium hydroxide, and delivers about 1.5 volts at much higher current rates than the common carbon-zinc cell. Also known as alkaline-manganese cell.

alkaline storage battery A storage battery in which the electrolyte consists of an alkaline solution, usually potassium hydroxide.

all-diffused monolithic integrated circuit Microcircuit consisting of a silicon substrate into which all of the circuit parts (both active and passive elements) are fabricated by diffusion and related processes.

alligator clip A long, narrow spring clip with meshing jaws; used with test leads to make temporary connections quickly. Also known as crocodile clip.

allowed energy bands The restricted regions of possible electron energy levels in a solid.

alloy junction A junction produced by alloying one or more impurity metals to a semiconductor to form a p or n region, depending on the impurity used. Also known as fused junction.

alloy junction diode A junction diode made by placing a pill of doped alloying material on a semiconductor material and heating until the molten alloy melts a portion of the semiconductor, resulting in a pn junction when the dissolved semiconductor recrystallizes. Also known as fused-junction diode.

alloy-junction transistor A junction transistor made by placing pellets of a p-type impurity such as indium above and below an n-type wafer of germanium, then heating until the impurity alloys with the germanium to give a pnp transistor. Also known as fused-junction transistor.

all-pass network A network designed to introduce a phase shift in a signal without introducing an appreciable reduction in energy of the signal at any frequency.

all-wave receiver A radio receiver capable of being tuned from about 535 kilohertz to at least 20 megahertz; some go above 100 megahertz and thus cover the FM band also.

alnico magnet A permanent magnet made of alnico.

alpha cutoff frequency The frequency at the high end of a transistor's range at which current amplification drops 3 decibels below its low-frequency value.

alphanumeric display device A device which visibly represents alphanumeric output information from some signal source.

alphanumeric reader A device capable of reading alphabetic, numeric, and special characters and punctuation marks.

alternating current Electric current that reverses direction periodically, usually many times per second. Abbreviated ac.

alternating-current circuit theory The mathematical description of conditions in an electric circuit driven by an alternating source or sources.

aluminum conductor 9

alternating-current coupling A coupling which passes alternating-current signals but blocks direct-current signals.

alternating-current/direct-current Pertaining to electronic equipment capable of operation from either an alternating-current or direct-current primary power source.

alternating-current dump The removal of all alternating-current power from a computer intentionally, accidentally, or conditionally.

alternating-current erase The use of an alternating current to energize a tape recorder erase head in order to remove previously recorded signals from a tape.

alternating-current erasing head In magnetic recording, an erasing head which uses alternating current to produce the magnetic field necessary for erasing.

alternating-current generator A machine, usually rotary, which converts mechanical power into alternating-current electric power.

alternating-current magnetic biasing Biasing with alternating current, usually well above the signal frequency range, in magnetic tape recording.

alternating-current motor A machine that converts alternating-current electrical energy into mechanical energy by utilizing forces exerted by magnetic fields produced by the current flow through conductors.

alternating-current power supply A power supply that provides one or more alternating-current output voltages, such as an ac generator, dynamotor, inverter, or transformer.

alternating-current resistance *See* high-frequency resistance.

alternating-current transmission In television, that form of transmission in which a fixed setting of the controls makes any instantaneous value of signal correspond to the same value of brightness for only a short time.

alternating gradient A magnetic field in which successive magnets have gradients of opposite sign, so that the field increases with radius in one magnet and decreases with radius in the next; used in synchrotrons and cyclotrons.

alternating voltage Periodic voltage, the average value of which over a period is zero.

alternator A mechanical, electrical, or electromechanical device which supplies alternating current.

altitude circle A bright circle which surrounds the central dark portion of a plan position indicator display or photograph, and which results from ground clutter.

altitude delay Synchronization delay introduced between the time of transmission of the radar pulse and the start of the trace on the indicator to eliminate the altitude/height hole on the plan position indicator–type display.

altitude hole The blank area in the center of a plan position indicator–type radarscope display caused by the time interval between transmission of a pulse and the receipt of the first ground return.

altitude signal The radio signals returned to an airborne electronics device by the ground or sea surface directly beneath the aircraft.

aluminum arrester *See* aluminum-cell arrester.

aluminum cable steel-reinforced A type of power transmission line made of an aluminum conductor provided with a core of steel. Abbreviated ACSR.

aluminum-cell arrester A lightning arrester consisting of a number of electrolytic cells in series formed from aluminum trays containing electrolyte. Also known as aluminum arrester; electrolytic arrester.

aluminum conductor Any of several aluminum alloys employed for conducting electric current; because its weight is one-half that of copper for the same conductance, it is used in high-voltage transmission lines.

10 AM

AM *See* amplitude modulation.

A/m *See* ampere per meter.

Am² *See* ampere meter squared.

A/m² *See* ampere per square meter.

amateur radio A radio used for two-way radio communications by private individuals as leisure-time activity. Also known as ham radio.

ambiguity The condition in which a synchro system or servosystem seeks more than one null position.

AM field signature The characteristic pattern of an alternating magnetic field, as displayed by detection and classification equipment.

A min *See* ampere-minute.

Am²/Js *See* ampere square meter per joule second.

amorphous semiconductor A semiconductor material which is not entirely crystalline, having only short-range order in its structure.

amortisseur winding *See* damper winding.

amp *See* amperage; ampere.

ampacity Current-carrying capacity in amperes; used as a rating for power cables.

amperage The amount of electric current in amperes. Abbreviated amp.

ampere The unit of electric current in the rationalized meter-kilogram-second system of units; defined in terms of the force of attraction between two parallel current-carrying conductors. Abbreviated a; A; amp.

Ampère currents Postulated "molecular-ring" currents to explain the phenomena of magnetism as well as the apparent nonexistence of isolated magnetic poles.

ampere-hour A unit for the quantity of electricity, obtained by integrating current flow in amperes over the time in hours for its flow; used as a measure of battery capacity. Abbreviated Ah; amp-hr.

ampere-hour capacity The charge, measured in ampere-hours, that can be delivered by a storage battery up to the limit to which the battery may be safely discharged.

Ampère law **1.** A law giving the magnetic induction at a point due to given currents in terms of the current elements and their positions relative to the point. Also known as Laplace law. **2.** A law giving the line integral over a closed path of the magnetic induction due to given currents in terms of the total current linking the path.

ampere meter squared The SI unit of electromagnetic moment. Abbreviated Am².

ampere-minute A unit of electrical charge, equal to the charge transported in 1 minute by a current of 1 ampere, or to 60 coulombs. Abbreviated A min.

ampere per meter The SI unit of magnetic field strength and magnetization. Abbreviated A/m.

ampere per square inch A unit of current density, equal to the uniform current density of a current of 1 ampere flowing through an area of 1 square inch. Abbreviated A/in².

ampere per square meter The SI unit of current density. Abbreviated A/m².

Ampère rule The rule which states that the direction of the magnetic field surrounding a conductor will be clockwise when viewed from the conductor if the direction of current flow is away from the observer.

amplitude versus frequency distortion 11

ampere square meter per joule second The SI unit of gyromagnetic ratio. Abbreviated Am^2/Js.

Ampère theorem The theorem which states that an electric current flowing in a circuit produces a magnetic field at external points equivalent to that due to a magnetic shell whose bounding edge is the conductor and whose strength is equal to the strength of the current.

ampere-turn A unit of magnetomotive force in the meter-kilogram-second system defined as the force of a closed loop of one turn when there is a current of 1 ampere flowing in the loop. Abbreviated amp-turn.

amp-hr *See* ampere-hour.

amplidyne A rotating magnetic amplifier having special windings and brush connections so that small changes in power input to the field coils produce large changes in power output.

amplification factor In a vacuum tube, the ratio of the incremental change in plate voltage to a given small change in grid voltage, under the conditions that the plate current and all other electrode voltages are held constant.

amplified back bias Degenerative voltage developed across a fast time-constant circuit within a stage of an amplifier and fed back into a preceding stage.

amplifying delay line Delay line used in pulse-compression systems to amplify delayed signals in the super-high-frequency region.

amplitron Crossed-field continuous cathode reentrant beam backward-wave amplifier for microwave frequencies.

amplitude discriminator *See* pulse-height discriminator.

amplitude distortion *See* frequency distortion.

amplitude-frequency distortion *See* frequency distortion.

amplitude gate A circuit which transmits only those portions of an input signal which lie between two amplitude boundary level values. Also known as slicer; slicer amplifier.

amplitude limiter *See* limiter.

amplitude-limiting circuit *See* limiter.

amplitude modulation Abbreviated AM. 1. Modulation in which the amplitude of a wave is the characteristic varied in accordance with the intelligence to be transmitted. 2. In telemetry, those systems of modulation in which each component frequency f of the transmitted intelligence produces a pair of sideband frequencies at carrier frequency plus f and carrier minus f.

amplitude noise Effect on radar accuracy of the fluctuations in the amplitude of the signal returned by the target; these fluctuations are caused by any change in aspect if the target is not a point source.

amplitude response The maximum output amplitude obtainable at various points over the frequency range of an instrument operating under rated conditions.

amplitude separator A circuit used to isolate the portion of a waveform with amplitudes above or below a given value or between two given values.

amplitude suppression ratio Ratio, in frequency modulation, of the undesired output to the desired output of a frequency-modulated receiver when the applied signal has simultaneous amplitude and frequency modulation.

amplitude versus frequency distortion The distortion caused by the nonuniform attenuation or gain of the system, with respect to frequency under specified terminal conditions.

amp-turn *See* ampere-turn.

analog A physical variable which remains similar to another variable insofar as the proportional relationships are the same over some specified range; for example, a temperature may be represented by a voltage which is its analog.

analog adder A device with one output voltage which is a weighted sum of two input voltages.

analog channel A channel on which the information transmitted can have any value between the channel limits, such as a voice channel.

analog comparator 1. A comparator that checks digital values to determine whether they are within predetermined upper and lower limits. 2. A comparator that produces high and low digital output signals when the sum of two analog voltages is positive and negative, respectively.

analog device A control device that operates with variables represented by continuously measured voltages or other quantities.

analog indicator A device in which the result of a measurement is indicated by a pointer deflection or other visual quantity.

analog multiplexer A multiplexer that provides switching of analog input signals to allow use of a common analog-to-digital converter.

analog multiplier A device that accepts two or more inputs in analog form and then produces an output proportional to the product of the input quantities.

analog network A circuit designed so that circuit variables such as voltages are proportional to the values of variables in a system under study.

analogous pole The pole of a crystal that acquires a positive charge when the crystal is heated.

analog recording Any method of recording in which some characteristic of the recording signal, such as amplitude or frequency, is continuously varied in a manner analogous to the time variations of the original signal.

analog signal A nominally continuous electrical signal that varies in amplitude or frequency in response to changes in sound, light, heat, position, or pressure.

analog switch 1. A device that either transmits an analog signal without distortion or completely blocks it. 2. Any solid-state device, with or without a driver, capable of bilaterally switching voltages or current.

analog-to-digital converter A device which translates continuous analog signals into proportional discrete digital signals.

analog-to-frequency converter A converter in which an analog input in some form other than frequency is converted to a proportional change in frequency.

analog voltage A voltage that varies in a continuous fashion in accordance with the magnitude of a measured variable.

analytical function generator An analog computer device in which the dependence of an output variable on one or more input variables is given by a function that also appears in a physical law. Also known as natural function generator; natural law function generator.

AND circuit *See* AND gate.

Anderson bridge A six-branch modification of the Maxwell-Wien bridge, used to measure self-inductance in terms of capacitance and resistance; bridge balance is independent of frequency.

AND gate A circuit which has two or more input signal ports and which delivers an output only if and when every input signal port is simultaneously energized. Also known as AND circuit; passive AND gate.

AND/NOR gate A single logic element whose operation is equivalent to that of two AND gates with outputs feeding into a NOR gate.

AND NOT gate A coincidence circuit that performs the logic operation AND NOT, under which a result is true only if statement A is true and statement B is not. Also known as A AND NOT B gate.

AND-OR circuit Gating circuit that produces a prescribed output condition when several possible combined input signals are applied; exhibits the characteristics of the AND gate and the OR gate.

AND-OR-INVERT gate A logic circuit with four inputs, a_1, a_2, b_1, and b_2, whose output is 0 only if either a_1 and a_2 or b_1 and b_2 are 1. Abbreviated A-O-I gate.

angle jamming An electronic countermeasure in which azimuth and elevation information, from a scanning fire control radar present in the modulation components on the returning echo pulse, is jammed by transmitting a pulse similar to the radar pulse but with modulation information out of phase with the returning target angle modulation information.

angle modulation The variation in the angle of a sine-wave carrier; particular forms are phase modulation and frequency modulation. Also known as sinusoidal angular modulation.

angle noise Tracking error introduced into radar by variations in the apparent angle of arrival of the echo from a target, because of finite target size.

angle of arrival A measure of the direction of propagation of electromagnetic radiation upon arrival at a receiver (the term is most commonly used in radio); it is the angle between the plane of the phase front and some plane of reference, usually the horizontal, at the receiving antenna.

angle of radiation Angle between the surface of the earth and the center of the beam of energy radiated upward into the sky from a transmitting antenna.

angle tracking noise Any deviation of the tracking axis from the center of reflectivity of a radar target; it is the resultant of servo noise, receiver noise, angle noise, and amplitude noise.

angular resolution A measure of the ability of a radar to distinguish between two targets solely by the measurement of angles.

angular resolver *See* resolver.

anisotropy constant In a ferromagnetic material, temperature-dependent parameters relating the magnetization in various directions to the anisotropy energy.

anisotropy energy Energy stored in a ferromagnetic crystal by virtue of the work done in rotating the magnetization of a domain away from the direction of easy magnetization.

annular conductor A number of wires stranded in three reversed concentric layers around a saturated hemp core.

annular transistor Mesa transistor in which the semiconductor regions are arranged in concentric circles about the emitter.

anode 1. The negative terminal of a primary cell or of a storage battery. 2. The collector of electrons in an electron tube. Also known as plate; positive electrode. 3. In a semiconductor diode, the terminal toward which forward current flows from the external circuit.

anode balancing coil A set of mutually coupled windings used to maintain approximately equal currents in anodes operating in parallel from the same transformer terminal.

anode circuit Complete external electrical circuit connected between the anode and the cathode of an electron tube. Also known as plate circuit.

14 anode-circuit detector

anode-circuit detector Detector functioning by virtue of a nonlinearity in its anode-circuit characteristic. Also known as plate-circuit detector.

anode current The electron current flowing through an electron tube from the cathode to the anode. Also known as plate current.

anode detector A detector in which rectification of radio-frequency signals takes place in the anode circuit of an electron tube. Also known as plate detector.

anode dissipation Power dissipated as heat in the anode of an electron tube because of bombardment by electrons and ions.

anode efficiency The ratio of the ac load circuit power to the dc anode power input for an electron tube. Also known as plate efficiency.

anode fall A very thin space-charge region in front of an anode surface, characterized by a steep potential gradient through the region.

anode impedance Total impedance between anode and cathode exclusive of the electron stream. Also known as plate impedance; plate-load impedance.

anode input power Direct-current power delivered to the plate (anode) of a vacuum tube by the source of supply. Also known as plate input power.

anode modulation Modulation produced by introducing the modulating signal into the anode circuit of any tube in which the carrier is present. Also known as plate modulation.

anode neutralization Method of neutralizing an amplifier in which the necessary 180° phase shift is obtained by an inverting network in the plate circuit. Also known as plate neutralization.

anode pulse modulation Modulation produced in an amplifier or oscillator by application of externally generated pulses to the plate circuit. Also known as plate pulse modulation.

anode rays Positive ions coming from the anode of an electron tube; generally due to impurities in the metal of the anode.

anode resistance The resistance value obtained when a small change in the anode voltage of an electron tube is divided by the resulting small change in anode current. Also known as plate resistance.

anode saturation The condition in which the anode current of an electron tube cannot be further increased by increasing the anode voltage; the electrons are then being drawn to the anode at the same rate as they are emitted from the cathode. Also known as current saturation; plate saturation; saturation; voltage saturation.

anode sheath The electron boundary which exists in a gas-discharge tube between the plasma and the anode when the current demanded by the anode circuit exceeds the random electron current at the anode surface.

anodized dielectric film An insulating film produced on a conducting surface by anodizing; used for producing thin-film capacitors, trimming resistor values, and passivation in the manufacture of integrated circuits.

anotron A cold-cathode glow-discharge diode having a copper anode and a large cathode of sodium or other material.

answering cord Cord nearest the face of the switchboard which is used for answering subscribers' calls and incoming trunks.

answering jack Jack on which a station calls in and is answered by an operator.

answer lamp Telephone switchboard lamp that lights when an answer cord is plugged into a line jack; the lamp goes out when the call is completed.

antenna scanner 15

antenna A device used for radiating or receiving radio waves. Also known as aerial; radio antenna.

antenna amplifier One or more stages of wide-band electronic amplification placed within or physically close to a receiving antenna to improve signal-to-noise ratio and mutually isolate various devices receiving their feed from the antenna.

antenna circuit A complete electric circuit which includes an antenna.

antenna coil Coil through which antenna current flows.

antenna coincidence That instance when two rotating, highly directional antennas are pointed toward each other.

antenna counterpoise *See* counterpoise.

antenna coupler A radio-frequency transformer, tuned line, or other device used to transfer energy efficiently from a transmitter to a transmission line or from a transmission line to a receiver.

antenna crosstalk The ratio or the logarithm of the ratio of the undesired power received by one antenna from another to the power transmitted by the other.

antenna detector Device consisting of an antenna and electronic equipment to warn aircraft crew members that they are being observed by radar sets.

antenna directivity diagram Curve representing, in polar or cartesian coordinates, a quantity proportional to the gain of an antenna in the various directions in a particular plane or cone.

antenna effect A distortion of the directional properties of a loop antenna caused by an input to the direction-finding receiver which is generated between the loop and ground, in contrast to that which is generated between the two terminals of the loop. Also known as electrostatic error; vertical component effect.

antenna effective area In any specified direction, the square of the wavelength multiplied by the power gain (or directive gain) in that direction, and divided by 4π.

antenna field A group of antennas placed in a geometric configuration.

antenna gain A measure of the effectiveness of a directional antenna as compared to a standard nondirectional antenna. Also known as gain.

antenna loading 1. The amount of inductance or capacitance in series with an antenna, which determines the antenna's electrical length. 2. The practice of loading an antenna in order to increase its electrical length.

antenna matching Process of adjusting impedances so that the impedance of an antenna equals the characteristic impedance of its transmission line.

antenna pair Two antennas located on a base line of accurately surveyed length, sometimes arranged so that the array may be rotated around an axis at the center of the base line; used to produce directional patterns and in direction finding.

antenna pattern *See* radiation pattern.

antenna power Radio-frequency power delivered to an antenna.

antenna power gain The power gain of an antenna in a given direction is 4π times the ratio of the radiation intensity in that direction to the total power delivered to the antenna.

antenna resistance The power supplied to an entire antenna divided by the square of the effective antenna current measured at the point where power is supplied to the antenna.

antenna scanner A microwave feed horn which moves in such a way as to illuminate sequentially different reflecting elements of an antenna array and thus produce the desired field pattern.

anticapacitance switch A switch designed to have low capacitance between its terminals when open.

anticathode The anode or target of an x-ray tube, on which the stream of electrons from the cathode is focused and from which x-rays are emitted.

anticlutter gain control Device which automatically and smoothly increases the gain of a radar receiver from a low level to the maximum, within a specified period after each transmitter pulse, so that short-range echoes producing clutter are amplified less than long-range echoes.

anticoincidence circuit Circuit that produces a specified output pulse when one (frequently predesignated) of two inputs receives a pulse and the other receives no pulse within an assigned time interval.

antifading antenna An antenna designed to confine radiation mainly to small angles of elevation to minimize the fading of radiation directed at larger angles of elevation.

antiferroelectric crystal A crystalline substance characterized by a state of lower symmetry consisting of two interpenetrating sublattices with equal but opposite electric polarization, and a state of higher symmetry in which the sublattices are unpolarized and indistinguishable.

antiferromagnetic domain A region in a solid within which equal groups of elementary atomic or molecular magnetic moments are aligned antiparallel.

antiferromagnetic resonance Magnetic resonance in antiferromagnetic materials which may be observed by rotating magnetic fields in either of two opposite directions.

antiferromagnetic substance A substance composed of antiferromagnetic domains.

antiferromagnetic susceptibility The magnetic response to an applied magnetic field of a substance whose atomic magnetic moments are aligned in antiparallel fashion.

antiferromagnetism A property possessed by some metals, alloys, and salts of transition elements by which the atomic magnetic moments form an ordered array which alternates or spirals so as to give no net total moment in zero applied magnetic field.

antihunt circuit A stabilizing circuit used in a closed-loop feedback system to prevent self-oscillations.

antijamming Any system or technique used to counteract the jamming of communications or of radar operation.

antilogous pole That crystal pole which becomes electrically negative when the crystal is heated or is expanded by decompression.

antiresonance *See* parallel resonance.

antiresonant circuit *See* parallel resonant circuit.

anti-sidetone circuit Telephone circuit which prevents sound, introduced in the local transmitter, from being reproduced in the local receiver.

Anti-Submarine Detection Investigation Committee *See* asdic.

anti-transmit-receive tube A switching tube that prevents the received echo signal from being dissipated in the transmitter.

A-O-I gate *See* AND-OR-INVERT gate.

APC *See* automatic phase control.

aperiodic antenna Antenna designed to have constant impedance over a wide range of frequencies because of the suppression of reflections within the antenna system; includes terminated wave and rhombic antennas.

aperiodic waves The transient current wave in a series circuit with resistance R, inductance L, and capacitance C when $R^2C = 4L$.

aperture An opening through which electrons, light, radio waves, or other radiation can pass.

aperture antenna Antenna in which the beam width is determined by the dimensions of a horn, lens, or reflector.

aperture grill picture tube An in-line gun-type picture tube in which the shadow mask is perforated by long, vertical stripes and the screen is painted with vertical phosphor stripes.

aperture illumination Field distribution in amplitude and phase over an aperture.

aperture plate A small part of a piece of perforated ferromagnetic material that forms a magnetic cell.

apodization A technique for modifying the response of a surface acoustic wave filter by varying the overlap between adjacent electrodes of the interdigital transducer.

A positive Also known as A+. 1. Positive terminal of an A battery or positive polarity of other sources of filament voltage. 2. Denoting the terminal to which the positive side of the filament voltage source should be connected.

A power supply *See* A supply.

apparent power The product of the root-mean-square voltage and the root-mean-square current delivered in an alternating-current circuit, no account being taken of the phase difference between voltage and current.

Applegate diagram A graph of the electron paths in a two-cavity klystron tube, showing how electron bunching occurs.

appliqué circuit Special circuit which is provided to modify existing equipment to allow for special usage; for example, some carrier telephone equipment designed for ringdown manual operation can be modified through the use of an appliqué circuit to allow for use between points having dial equipment.

APT system *See* automatic picture transmission system.

aquadag Graphite coating on the inside of certain cathode-ray tubes for collecting secondary electrons emitted by the face of the tube.

Arago's disk A device consisting of a horizontal disk of copper that can rotate about a vertical axis in an airtight box, and a horizontal bar magnet suspended above the disk but outside the box; upon rapid rotation of the disk, the bar magnet is deflected and eventually rotates in the same direction with smaller velocity.

arc *See* electric arc.

arcback The flow of a principal electron stream in the reverse direction in a mercury-vapor rectifier tube because of formation of a cathode spot on an anode; this results in failure of the rectifying action. Also known as backfire.

arc chute A collection of insulating barriers in a circuit breaker for confining the arc and preventing it from causing damage.

arc converter A form of oscillator using an electric arc as the generator of alternating or pulsating current.

arc discharge A direct-current electrical current between electrodes in a gas or vapor, having high current density and relatively low voltage drop.

arcing contacts Special contacts on which the arc is drawn after the main contacts of a switch or circuit breaker have opened.

arcing ring A metal ring attached to an insulator to protect it from damage by a power arc.

arcing time 1. Interval between the parting, in a switch or circuit breaker, of the arcing contacts and the extension of the arc. 2. Time elapsing, in a fuse, from the severance of the fuse link to the final interruption of the circuit under a specified condition.

18 arc lamp

arc lamp An electric lamp in which the light is produced by an arc made when current flows through ionized gas between two electrodes. Also known as electric-arc lamp.

arc-over An unwanted arc resulting from the opening of a switch or the breakdown of insulation.

arc-suppression coil A grounding reactor, used in alternating-current power transmission systems, which is designed to limit the current flowing to ground at the location of a fault almost to zero by setting up a reactive current to ground that balances the capacitive current to ground flowing from the lines. Also known as Petersen coil.

arc-through Of a gas tube, a loss of control resulting in the flow of a principal electron stream in the normal direction during a scheduled nonconducting period.

area effect In general, the condition of the dielectric strength of a liquid or vacuum separating two electrodes being higher for electrodes of smaller area.

arm *See* branch.

armature 1. That part of an electric rotating machine that includes the main current-carrying winding in which the electromotive force produced by magnetic flux rotation is induced; it may be rotating or stationary. 2. The movable part of an electromagnetic device, such as the movable iron part of a relay, or the spring-mounted iron part of a vibrator or buzzer.

armature chatter Vibration of the armature of a relay caused by pulsating coil current or by marginally low coil current.

armature contact *See* movable contact.

armature reactance The inductive reactance due to the flux produced by the armature current and enclosed by the conductors in the armature slots and the end connections.

armature reaction Interaction between the magnetic flux produced by armature current and that of the main magnetic field in an electric motor or generator.

armature resistance The ohmic resistance in the main current-carrying windings of an electric generator or motor.

armor Metal sheath enclosing a cable, primarily for mechanical protection.

armored cable An electrical cable provided with a sheath of metal primarily for mechanical protection.

Armstrong oscillator Inductive feedback oscillator that consists of a tuned-grid circuit and an untuned-tickler coil in the plate circuit; control of feedback is accomplished by varying the coupling between the tickler and the grid circuit.

array A group of components such as antennas, reflectors, or directors arranged to provide a desired variation of radiation transmission or reception with direction.

arrester *See* lightning arrester.

artificial antenna *See* dummy antenna.

artificial delay line *See* delay line.

artificial echo 1. Received reflections of a transmitted pulse from an artificial target, such as an echo box, corner reflector, or other metallic reflecting surface. 2. Delayed signal from a pulsed radio-frequency signal generator.

artificial line Circuit made up of lumped constants, which is used to simulate various characteristics of a transmission line.

artificial line duct Balancing network simulating the impedance of the real line and distant terminal apparatus, which is employed in a duplex circuit to make the receiving device unresponsive to outgoing signal currents.

atmospheric noise   19

artificial load Dissipative but essentially nonradiating device having the impedance characteristics of an antenna, transmission line, or other practical utilization circuit.

aS *See* abmho.

A scan *See* A scope.

A scope A radarscope on which the trace appears as a horizontal or vertical range scale and the signals appear as vertical or horizontal deflections. Also known as A indicator; A scan.

asdic Acronym for Anti-Submarine Detection Investigation Committee; British term for sonar and underwater listening devices.

asperomagnetic state The condition of a rare-earth glass in which the spins are oriented in fixed directions, with most nearest-neighbor spins parallel or nearly parallel, so that the spin directions are distributed in one hemisphere.

astable circuit A circuit that alternates automatically and continuously between two unstable states at a frequency dependent on circuit constants; for example, a blocking oscillator.

astable multivibrator A multivibrator in which each active device alternately conducts and is cut off for intervals of time determined by circuit constants, without use of external triggers. Also known as free-running multivibrator.

astatic pair A pair of parallel magnets, equal in strength and having polarities in opposite directions, and perpendicular to an axis which bisects both of them; there is no net force or torque on the pair in a uniform field.

astigmatism In an electron-beam tube, a focus defect in which electrons in different axial planes come to focus at different points.

Aston dark space A dark region in a glow-discharge tube which extends for a few millimeters from the cathode up to the cathode glow.

astrionics The science of adapting electronics to aerospace flight.

A supply Battery, transformer filament winding, or other voltage source that supplies power for heating filaments of vacuum tubes. Also known as A power supply.

asymmetrical cell A cell, such as a photoelectric cell, in which the impedance to the flow of current in one direction is greater than in the other direction.

asymmetrical deflection A type of electrostatic deflection in which one deflector plate is maintained at a fixed potential and the deflecting voltage is supplied to the other plate.

asynchronous inputs The terminals in a flip-flop circuit which affect the output state of the flip-flop independently of the clock.

asynchronous logic A logic network in which the speed of operation depends only on the signal propagation through the network.

asynchronous machine An ac machine whose speed is not proportional to the frequency of the power line.

asynchronous operation An operation that is started by a completion signal from a previous operation, proceeds at the maximum speed of the circuits until finished, and then generates its own completion signal.

asynchronous tie An installation at which power is transmitted between two alternating-current power systems, operating at the same nominal frequency but with different frequency controls, by a direct-current link.

athermancy Property of a substance which cannot transmit infrared radiation.

atmospheric noise Noise heard during radio reception due to atmospheric interference.

20 atmospheric radio wave

atmospheric radio wave Radio wave that is propagated by reflection in the atmosphere; may include either the ionospheric wave or the tropospheric wave, or both.

atomic paramagnetism The result of a permanent magnetic moment in an atom.

atomic susceptibility The magnetization of a material per atom per unit of applied field; measured in ergs/oersted/atom.

A trace The first trace of an oscilloscope, such as the upper trace of a loran indicator.

attachment plug A device having an attached flexible cord containing conductors, and capable of being inserted in a receptacle so as to form an electrical connection between the conductors in the cord and conductors permanently connected to the receptacle.

attenuation coefficient The space rate of attenuation of any transmitted electromagnetic radiation.

attenuation equalizer Corrective network which is designed to make the absolute value of the transfer impedance, with respect to two chosen pairs of terminals, substantially constant for all frequencies within a desired range.

attenuation network Arrangement of circuit elements, usually impedance elements, inserted in circuitry to introduce a known loss or to reduce the impedance level without reflections.

attenuator An adjustable or fixed transducer for reducing the amplitude of a wave without introducing appreciable distortion.

attracted-disk electrometer A type of electrometer in which the attraction between two oppositely charged disks is measured.

audio amplifier See audio-frequency amplifier.

audio-frequency amplifier An electronic circuit for amplification of signals within, and in some cases above, the audible range of frequencies in equipment used to record and reproduce sound. Also known as audio amplifier.

audio-frequency choke Choke used to impede the flow of audio-frequency currents; generally a coil wound on an iron core.

audio-frequency oscillator An oscillator circuit using an electron tube, transistor, or other nonrotating device to produce an audio-frequency alternating current. Also known as audio oscillator.

audio-frequency peak limiter A circuit used in an audio-frequency system to cut off signal peaks that exceed a predetermined value. Also known as audio peak limiter.

audio-frequency transformer An iron-core transformer used for coupling audio-frequency circuits. Also known as audio transformer.

audio oscillator See audio-frequency oscillator.

audio peak limiter See audio-frequency peak limiter.

aural radio range A radio-range station providing lines of position by virtue of aural identification or comparison of signals at the output of a receiver.

aurora See corona discharge.

aurora gating Operator-controlled gating to eliminate undesirable radar returns from aurora.

autoalarm See automatic alarm receiver.

autocorrelation A technique used to detect cyclic activity in a complex signal.

autocorrelator A correlator in which the input signal is delayed and multiplied by the undelayed signal, the product of which is then smoothed in a low-pass filter to give

automatic fine-tuning control 21

an approximate computation of the autocorrelation function; used to detect a nonperiodic signal or a weak periodic signal hidden in noise.

autodyne circuit A circuit in which the same tube elements serve as oscillator and detector simultaneously.

automatic alarm receiver A complete receiving, selecting, and warning device capable of being actuated automatically by intercepted radio-frequency waves forming the international automatic alarm signal. Also known as autoalarm.

automatic back bias Radar technique which consists of one or more automatic gain control loops to prevent overloading of a receiver by large signals, whether jamming or actual radar echoes.

automatic background control *See* automatic brightness control.

automatic bass compensation A circuit related to the volume control in some radio receivers and audio amplifiers to make bass notes sound properly balanced, in the audio spectrum, at low volume-control settings.

automatic brightness control A circuit used in a television receiver to keep the average brightness of the reproduced image essentially constant. Abbreviated ABC. Also known as automatic background control.

automatic C bias *See* self-bias.

automatic chroma control *See* automatic color control.

automatic chrominance control *See* automatic color control.

automatic color control A circuit used in a color television receiver to keep color intensity levels essentially constant despite variations in the strength of the received color signal; control is usually achieved by varying the gain of the chrominance bandpass amplifier. Also known as automatic chroma control; automatic chrominance control.

automatic connection Ability of electronic switching equipment to make a connection between users without human intervention.

automatic contrast control A circuit that varies the gain of the radio-frequency and video intermediate-frequency amplifiers in such a way that the contrast of the television picture is maintained at a constant average level.

automatic cutout A device, usually operated by centrifugal force or by an electromagnet, that automatically shorts part of a circuit at a particular time.

automatic degausser An arrangement of degaussing coils mounted around a color television picture tube, combined with a special circuit that energizes these coils only while the set is warming up; demagnetizes any parts of the receiver that have been affected by the magnetic field of the earth or of any nearby home appliance.

automatic dialer A device in which a telephone number up to a maximum of 14 digits can be stored in a memory and then activated, directly into the line, by the caller's pressing a button. Also known as mechanical dialer.

automatic direction finder A direction finder that without manual manipulation indicates the direction of arrival of a radio signal. Abbreviated ADF. Also known as radio compass.

automatic exchange A telephone, teletypewriter, or data-transmission exchange in which communication between subscribers is effected, without the intervention of an operator, by devices set in operation by the originating subscriber's instrument. Also known as automatic switching system; machine switching system.

automatic fine-tuning control A circuit used in a color television receiver to maintain the correct oscillator frequency in the tuner for best color picture by compensating for drift and incorrect tuning.

22 automatic frequency control

automatic frequency control Abbreviated AFC. **1.** A circuit used to maintain the frequency of an oscillator within specified limits, as in a transmitter. **2.** A circuit used to keep a superheterodyne receiver tuned accurately to a given frequency by controlling its local oscillator, as in an FM receiver. **3.** A circuit used in radar superheterodyne receivers to vary the local oscillator frequency so as to compensate for changes in the frequency of the received echo signal. **4.** A circuit used in television receivers to make the frequency of a sweep oscillator correspond to the frequency of the synchronizing pulses in the received signal.

automatic gain control A control circuit that automatically changes the gain (amplification) of a receiver or other piece of equipment so that the desired output signal remains essentially constant despite variations in input signal strength. Abbreviated AGC.

automatic grid bias *See* self-bias.

automatic light control Automatic adjustment of illumination reaching a film, television camera, or other imaging device as a function of scene brightness.

automatic peak limiter *See* limiter.

automatic phase control Abbreviated APC. **1.** A circuit used in color television receivers to reinsert a 3.58-megahertz carrier signal with exactly the correct phase and frequency by synchronizing it with the transmitted color-burst signal. **2.** An automatic frequency-control circuit in which the difference between two frequency sources is fed to a phase detector that produces the required control signal.

automatic picture control A multiple-contact switch used in some color television receivers to disconnect one or more of the regular controls and make connections to corresponding preset controls.

automatic picture transmission system A system in which a meteorological satellite continuously scans and transmits a view of a transverse swath directly beneath it; transmissions can be recorded by simple ground equipment to reconstruct an image of the cloud patterns within a thousand kilometers of the ground station. Abbreviated APT system.

automatic scanning receiver A receiver which can automatically and continuously sweep across a preselected frequency, either to stop when a signal is found or to plot signal occupancy within the frequency spectrum being swept.

automatic sensitivity control Circuit used for automatically maintaining receiver sensitivity at a predetermined level; it is similar to automatic gain control, but it affects the receiver constantly rather than during the brief interval selected by the range gate.

automatic short-circuiter Device designed to automatically short-circuit the commutator bars in some forms of single-phase commutator motors.

automatic switching system *See* automatic exchange.

automatic threshold variation Constant false-alarm rate scheme that is an open-loop of automatic gain control in which the decision threshold is varied continuously in proportion to the incoming intermediate frequency and video noise level.

automatic tint control A circuit used in color television receivers to maintain correct flesh tones when a station changes cameras or switches to commercials, by correcting phase errors before the chroma signal is demodulated.

automatic transfer equipment Equipment which automatically transfers a load so that a source of power may be selected from one of several incoming lines.

automatic video noise leveling Constant false-alarm rate scheme in which the video noise level at the output of the receiver is sampled at the end of each range sweep and the receiver gain is readjusted accordingly to maintain a constant video noise level at the output.

automatic voltage regulator *See* voltage regulator.

automatic volume control An automatic gain control that keeps the output volume of a radio receiver essentially constant despite variations in input-signal strength during fading or when tuning from station to station. Abbreviated AVC.

avalanche noise 23

automotive alternator An ac generator used in an automotive vehicle to provide current for the vehicle's electrical systems.

automotive voltage regulator A device in the automotive electrical system to prevent generator or alternator overvoltage.

autopolarity Automatic interchanging of connections to a digital meter when polarity is wrong; a minus sign appears ahead of the value on the digital display if the reading is negative.

autostarter 1. Automatic starting and switchover generating system consisting of a standby generator coupled to the station load through an automatic power transfer control unit. 2. See autotransformer starter.

autotransductor A saturable reactor in which the same windings carry the main and control currents.

autotransformer A power transformer having one continuous winding that is tapped; part of the winding serves as the primary and all of it serves as the secondary, or vice versa; small autotransformers are used to start motors.

autotransformer starter Motor starter having an autotransformer to furnish a reduced voltage for starting; includes the necessary switching mechanism. Also known as autostarter.

auxiliary contacts Contacts, in a switching device, in addition to the main circuit contacts, which function with the movement of the latter.

auxiliary relay Relay that operates in response to the opening or closing of its operating circuit to assist another relay or device in performing a function.

auxiliary switch A switch actuated by the main device (such as a circuit breaker) for signaling, interlocking, or other purposes.

aV See abvolt.

available line Portion of the length of the scanning line which can be used specifically for picture signals in a facsimile system.

available power The power which a linear source of energy is capable of delivering into its conjugate impedance.

available power gain Ratio, in an electronic transducer, of the available power from the output terminals of the transducer, under specified input termination conditions, to the available power from the driving generator.

avalanche 1. The cumulative process in which an electron or other charged particle accelerated by a strong electric field collides with and ionizes gas molecules, thereby releasing new electrons which in turn have more collisions, so that the discharge is thus self-maintained. Also known as avalanche effect; cascade; cumulative ionization; Townsend avalanche; Townsend ionization. 2. Cumulative multiplication of carriers in a semiconductor as a result of avalanche breakdown. Also known as avalanche effect.

avalanche breakdown Nondestructive breakdown in a semiconductor diode when the electric field across the barrier region is strong enough so that current carriers collide with valence electrons to produce ionization and cumulative multiplication of carriers.

avalanche diode A semiconductor breakdown diode, usually made of silicon, in which avalanche breakdown occurs across the entire pn junction and voltage drop is then essentially constant and independent of current; the two most important types are IMPATT and TRAPATT diodes.

avalanche effect See avalanche.

avalanche-induced migration A technique of forming interconnections in a field-programmable logic array by applying appropriate voltages for shorting selected base-emitter junctions.

avalanche noise 1. A junction phenomenon in a semiconductor in which carriers in a high-voltage gradient develop sufficient energy to dislodge additional carriers through physical impact; this agitation creates ragged current flows which are indicated by noise. 2. The noise produced when a junction diode is operated at the onset of avalanche breakdown.

avalanche oscillator An oscillator that uses an avalanche diode as a negative resistance to achieve one-step conversion from direct-current to microwave outputs in the gigahertz range.

avalanche photodiode A photodiode operated in the avalanche breakdown region to achieve internal photocurrent multiplication, thereby providing rapid light-controlled switching operation.

avalanche transistor A transistor that utilizes avalanche breakdown to produce chain generation of charge-carrying hole-electron pairs.

AVC *See* automatic volume control.

aV/cm *See* abvolt per centimeter.

average noise figure Ratio in a transducer of total output noise power to the portion thereof attributable to thermal noise in the input termination, the total noise being summed over frequencies from zero to infinity, and the noise temperature of the input termination being standard (290 K).

average power output Radio-frequency power, in an audio-modulation transmitter, delivered to the transmitter output terminals, averaged over a modulation cycle.

axial lead A wire lead extending from the end along the axis of a resistor, capacitor, or other component.

axial quadrupole *See* longitudinal quadrupole.

axial ratio The ratio of the major axis to the minor axis of the polarization ellipse of a waveguide. Also known as ellipticity.

Ayrton-Jones balance A type of balance with which force between current-carrying conductors is measured; uses single-layer solenoids as the fixed and movable coils.

Ayrton-Perry winding Winding of two wires in parallel but opposite directions to give better cancellation of magnetic fields than is obtained with a single winding.

Ayrton shunt A shunt used to increase the range of a galvanometer without changing the damping. Also known as universal shunt.

azel display Modified type of plan position indicator presentation showing two separate radar displays on one cathode-ray screen; one display presents bearing information and the other shows elevation.

azimuth blanking Blanking of the radar receiver as the scan traverses a selected azimuth region.

azimuth gain reduction Technique which allows control of the radar receiver system throughout any two azimuth sectors.

azimuth gating The practice of selectively brightening and enhancing the gain-desired sectors of a radar plan position indicator display, usually by applying a step waveform to the automatic gain control circuit.

azimuth marker *See* electronic azimuth marker.

azimuth resolution Angle or distance by which two targets must be separated in azimuth to be distinguished by a radar set, when the targets are at the same range.

azimuth versus amplitude Electronic counter-countermeasures receiver with a plan position indicator type of display attached to the main antenna and used to display strobes due to jamming aircraft; it is useful in making passive fixes when two or more radar sites can operate together.

B

B *See* intrinsic induction.

baby spot A small spotlight, usually equipped with a hood, used (as in the theater) to concentrate light on an area or an object a small distance from the spotlight.

back bias 1. Degenerative or regenerative voltage which is fed back to circuits before its originating point; usually applied to a control anode of a tube or other device. 2. Voltage applied to a grid of a tube (or tubes) or electrode of another device to reduce a condition which has been upset by some external cause.

back bond A chemical bond between an atom in the surface layer of a solid and an atom in the second layer.

back contact Normally closed stationary contact on a relay that is opened when the relay is energized.

back echo An echo signal produced on a radar screen by one of the minor back lobes of a search radar beam.

back echo reflection A radar echo produced by radiation reflected to the target by a large, fixed obstruction; that is, the ray path is from antenna to obstruction to target to antenna, instead of antenna to target to antenna.

back electromotive force *See* counterelectromotive force.

back-emission electron radiography A technique used in microradiography to visualize, among other things, the presence of material of different atomic numbers in the surface of the specimen being observed; the polished side of the specimen is facing and in close contact with the emulsion side of a fine-grain photographic plate; a light-tight cover holds the specimen and plate in place to be subjected to hardened x-rays.

backfire *See* arcback.

backfire antenna An antenna which exhibits significant gain in a direction 180° from its principal lobe.

background return *See* clutter.

backing Flexible material, usually cellulose acetate or polyester, used on magnetic tape as the carrier for the oxide coating.

backlash A small reverse current in a rectifier tube caused by the motion of positive ions produced in the gas by the impact of thermoelectrons.

back lobe The three-dimensional portion of the radiation pattern of a directional antenna that is directed away from the intended direction.

back porch The period of time in a television circuit immediately following a synchronizing pulse during which the signal is held at the instantaneous amplitude corresponding to a black area in the received picture.

back radiation *See* backscattering; counterradiation.

back resistance The resistance between the contacts opposing the inverse current of a metallic rectifier.

backscattering Also known as back radiation; backward scattering. 1. Radar echoes from a target. 2. Undesired radiation of energy to the rear by a directional antenna.

back-to-front ratio Ratio used in connection with an antenna, metal rectifier, or any device in which signal strength or resistance in one direction is compared with that in the opposite direction.

backup arrangement *See* cascade.

backup relay A relay designed to protect a power system in case a primary relay fails to operate as desired.

backward-acting regulator Transmission regulator in which the adjustment made by the regulator affects the quantity which caused the adjustment.

backward diode A semiconductor diode similar to a tunnel diode except that it has no forward tunnel current; used as a low-voltage rectifier.

backward scattering *See* backscattering.

backward wave An electromagnetic wave traveling opposite to the direction of motion of some other physical quantity in an electronic device such as a traveling-wave tube or mismatched transmission line.

backward-wave magnetron A magnetron in which the electron beam travels in a direction opposite to the flow of the radio-frequency energy.

backward-wave oscillator An electronic device which amplifies microwave signals simultaneously over a wide band of frequencies and in which the traveling wave produced is reflected backward so as to sustain the wave oscillations. Abbreviated BWO. Also known as carcinotron.

backward-wave tube A type of microwave traveling-wave electron tube in which electromagnetic energy on a slow-wave circuit flows opposite in direction to the travel of electrons in a beam.

baffle 1. Device for deflecting oil or gas in a circuit breaker. 2. An auxiliary member in a gas tube used, for example, to control the flow of mercury particles or deionize the mercury following conduction.

baffle plate Metal plate inserted in a waveguide to reduce the cross-sectional area for wave conversion purposes.

balance The state of an electrical network when it is adjusted so that voltage in one branch induces or causes no current in another branch.

balance coil An iron-core solenoid with adjustable taps near the center; used to convert a two-wire circuit to a three-wire circuit, the taps furnishing a neutral terminal for the latter.

balance control A control used in a stereo sound system to vary the volume of one loudspeaker system relative to the other while maintaining their combined volume essentially constant.

balanced amplifier An electronic amplifier in which there are two identical signal branches connected so as to operate with the inputs in phase opposition and with the output connections in phase, each balanced to ground.

balanced bridge Wheatstone bridge circuit which, when in a quiescent state, has an output voltage of zero.

balanced circuit A circuit whose two sides are electrically alike and symmetrical with respect to a common reference point, usually ground.

ballast factor 27

balanced converter *See* balun.

balanced currents Currents flowing in the two conductors of a balanced line which, at every point along the line, are equal in magnitude and opposite in direction. Also known as push-pull currents.

balanced detector A detector used in frequency-modulation receivers; in one form the audio output is the rectified difference between voltages produced across two resonant circuits, one being tuned slightly above the carrier frequency and one slightly below.

balanced input A symmetrical input circuit having equal impedance from both input terminals to reference.

balanced line A transmission line consisting of two conductors capable of being operated so that the voltages of the two conductors at any transverse plane are equal in magnitude and opposite in polarity with respect to ground.

balanced modulator A modulator in which the carrier and modulating signal are introduced in such a way that the output contains the two sidebands without the carrier.

balanced network Hybrid network in which the impedances of the opposite branches are equal.

balanced oscillator Any oscillator in which, at the oscillator frequency, the impedance centers of the tank circuits are at ground potential, and the voltages between either end and their centers are equal in magnitude and opposite in phase.

balanced output A three-conductor output (as from an amplifier) in which the signal voltage alternates above and below a third, neutral wire.

balanced ring modulator A modulator that uses tubes or diodes to suppress the carrier signal while providing double-sideband output.

balanced transmission line Transmission line having equal conductor resistances per unit length and equal impedances from each conductor to earth and to other electrical circuits.

balanced voltages Voltages that are equal in magnitude and opposite in polarity with respect to ground. Also known as push-pull voltages.

balanced wire circuit Circuit wherein the two sides are electrically alike and symmetrical with respect to ground and other conductors.

balancer A mechanism for equalizing the loads on the outer lines of a three-wire system for electric power distribution, consisting of two similar shunt or compound machines which are coupled together with the armatures connected in series across the outer lines.

balance-to-unbalance transformer Device for matching a pair of lines, balanced with respect to earth, to a pair of lines not balanced with respect to earth.

balancing capacitor A variable capacitor used to improve the accuracy of a radio direction finder. Also known as compensating capacitor.

balancing unit 1. Antenna-matching device used to permit efficient coupling of a transmitter or receiver having an unbalanced output circuit to an antenna having a balanced transmission line. 2. Device for converting balanced to unbalanced transmission lines, and vice versa, by placing suitable discontinuities at the junction between the lines instead of using lumped components.

ballast A circuit element that serves to limit an electric current or to provide a starting voltage, as in certain types of lamps, such as in fluorescent ceiling fixtures.

ballast factor The ratio of the luminous output of a lamp when operated on a ballast to its luminous output when operated under standardized rating conditions.

ballast lamp A light-producing electrical resistance device which maintains nearly constant current by increasing in resistance as the current increases.

ballast reactor A coil wound on an iron core and connected in series with a fluorescent lamp to compensate for the negative-resistance characteristics of the lamp by providing an increased voltage drop as the current through the lamp is increased.

ballast resistor A resistor that increases in resistance as current through it increases, and decreases in resistance as current decreases. Also known as barretter (British usage).

ballast tube A ballast resistor mounted in an evacuated glass or metal envelope, like that of a vacuum tube, to reduce radiation of heat from the resistance element and thereby improve the voltage-regulating action.

ballistic galvanometer A galvanometer having a long period of swing so that the deflection may measure the electric charge in a current pulse or the time integral of a voltage pulse.

balun A device used for matching an unbalanced coaxial transmission line or system to a balanced two-wire line or system. Also known as balanced converter; bazooka; line-balance converter.

banana jack A jack that fits a banana plug; generally designed for panel mounting.

banana plug A plug having a spring-metal tip shaped like a banana and used on test leads or as terminals for plug-in components.

band A restricted range in which the energies of electrons in solids lie, or from which they are excluded, as understood in quantum-mechanical terms

bandage Rubber ribbon about 4 inches (10 centimeters) wide for temporarily protecting a telephone or coaxial splice from moisture.

band-elimination filter *See* band-stop filter.

band gap An energy difference between two allowed bands of electron energy in a metal.

band-pass A range, in hertz or kilohertz, expressing the difference between the limiting frequencies at which a desired fraction (usually half power) of the maximum output is obtained.

band-pass amplifier An amplifier designed to pass a definite band of frequencies with essentially uniform response.

band-pass filter An electric filter which transmits more or less uniformly in a certain band, outside of which the frequency components are attenuated.

band-pass response Response characteristics in which a definite band of frequencies is transmitted uniformly. Also known as flat top response.

band-rejection filter *See* band-stop filter.

band scheme The identification of energy bands of a solid with the levels of independent atoms from which they arise as the atoms are brought together to form the solid, together with the width and spacing of the bands.

band selector A switch that selects any of the bands in which a receiver, signal generator, or transmitter is designed to operate and usually has two or more sections to make the required changes in all tuning circuits simultaneously. Also known as band switch.

band-spread tuning control A tuning control provided on some shortwave receivers to spread the stations in a single band of frequencies over an entire tuning dial.

band-stop filter An electric filter which transmits more or less uniformly at all frequencies of interest except for a band within which frequency components are largely attenuated. Also known as band-elimination filter; band-rejection filter.

band switch *See* band selector.

band theory of solids A quantum-mechanical theory of the motion of electrons in solids that predicts certain restricted ranges or bands for the energies of these electrons.

bang-bang circuit An operational amplifier with double feedback limiters that drive a high-speed relay (1–2 milliseconds) in an analog computer; involved in signal-controlled programming.

bank **1.** A number of similar electrical devices, such as resistors, connected together for use as a single device. **2.** An assemblage of fixed contacts over which one or more wipers or brushes move in order to establish electrical connections in automatic switching.

bank-and-wiper switch Switch in which electromagnetic ratchets or other mechanisms are used, first, to move the wipers to a desired group of terminals, and second, to move the wipers over the terminals of the group to the desired bank contacts.

banked winding A radio-frequency coil winding which proceeds from one end of the coil to the other without return by having, side by side, many flat spirals formed by winding single turns one over the other, thereby reducing the distributed capacitance of the coil.

bantam tube Vacuum tube having a standard octal base, but a considerably smaller glass tube than a standard glass tube.

Bardeen-Cooper-Schrieffer theory A theory of superconductivity that describes quantum-mechanically those states of the system in which conduction electrons cooperate in their motion so as to reduce the total energy appreciably below that of other states by exploiting their effective mutual attraction; these states predominate in a superconducting material. Abbreviated BCS theory.

bar generator Generator of pulses or repeating waves which are equally separated in time; these pulses are synchronized by the synchronizing pulses of a television system, so that they can produce a stationary bar pattern on a television screen.

BARITT diode *See* barrier injection transit-time diode.

barium fuel cell A fuel cell in which barium is used with either oxygen or chlorine to convert chemical energy into electrical energy.

Barkhausen effect The succession of abrupt changes in magnetization occurring when the magnetizing force acting on a piece of iron or other magnetic material is varied.

Barkhausen-Kurz oscillator An oscillator of the retarding-field type in which the frequency of oscillation depends solely on the transit time of electrons oscillating about a highly positive grid before reaching the less positive anode. Also known as Barkhausen oscillator; positive-grid oscillator.

Barkhausen oscillation Undesired oscillation in the horizontal output tube of a television receiver, causing one or more ragged dark vertical lines on the left side of the picture.

Barkhausen oscillator *See* Barkhausen-Kurz oscillator.

bar magnet A bar of hard steel that has been strongly magnetized and holds its magnetism, thereby serving as a permanent magnet.

Barnett effect The development of a slight magnetization in an initially unmagnetized iron rod when it is rotated at high speed about its axis.

Barnett method Use of the Barnett effect to determine the gyromagnetic moment of ferromagnetic material.

bar pattern Pattern of repeating lines or bars on a television screen.

barretter **1.** Bolometer that consists of a fine wire or metal film having a positive temperature coefficient of resistivity, so that resistance increases with temperature; used for making power measurements in microwave devices. **2.** *See* ballast resistor.

30 barrier-grid storage tube

barrier-grid storage tube See radechon.

barrier injection transit-time diode A microwave diode in which the carriers that traverse the drift region are generated by minority carrier injection from a forward-biased junction instead of being extracted from the plasma of an avalanche region. Abbreviated BARITT diode.

barrier layer See depletion layer.

barrier-layer cell See photovoltaic cell.

barrier-layer photocell See photovoltaic cell.

barrier-layer rectification See depletion-layer rectification.

barrier voltage The voltage necessary to cause electrical conduction in a junction of two dissimilar materials, such as pn junction diode.

bar winding An armature winding made up of a series of metallic bars connected at their ends.

base 1. The region that lies between an emitter and a collector of a transistor and into which minority carriers are injected. 2. The part of an electron tube that has the pins, leads, or other terminals to which external connections are made either directly or through a socket. 3. The plastic, ceramic, or other insulating board that supports a printed wiring pattern. 4. A plastic film that supports the magnetic powder of magnetic tape or the emulsion of photographic film.

base bias The direct voltage that is applied to the majority-carrier contact (base) of a transistor.

base electrode An ohmic or majority carrier contact to the base region of a transistor.

base insulator Heavy-duty insulator used to support the weight of an antenna mast and insulate the mast from the ground or some other surface.

base line The line traced on amplitude-modulated indicators which corresponds to the power level of the weakest echo detected by the radar; it is retraced with every pulse transmitted by the radar but appears as a nearly continuous display on the scope.

base-line break Technique in radar which uses the characteristic break in the base line on an A-scope display due to a pulse signal of significant strength in noise jamming.

baseload Minimum load of a power generator over a given period of time.

base-loaded antenna Vertical antenna having an impedance in series at the base for loading the antenna to secure a desired electrical length.

base modulation Amplitude modulation produced by applying the modulating voltage to the base of a transistor amplifier.

base pin See pin.

base-spreading resistance Resistance which is found in the base of any transistor and acts in series with it, generally a few ohms in value.

basic Q See nonloaded Q.

basket coil See basket winding.

basket winding A crisscross coil winding in which successive turns are far apart except at points of crossing, giving low distributed capacitance. Also known as basket coil.

bass boost A circuit that emphasizes the lower audio frequencies, generally by attenuating higher audio frequencies.

bass compensation A circuit that emphasizes the low-frequency response of an audio amplifier at low volume levels to offset the lower sensitivity of the human ear to weak low frequencies.

bass control A manual tone control that attenuates higher audio frequencies in an audio amplifier and thereby emphasizes bass frequencies.

bass response A measure of the output of an electronic device or system as a function of an input of low audio frequencies.

bat-handle switch A toggle switch having an actuating lever shaped like a baseball bat.

bathtub capacitor A capacitor enclosed in a metal housing having broadly rounded corners like those on a bathtub.

battery A direct-current voltage source made up of one or more units that convert chemical, thermal, nuclear, or solar energy into electrical energy.

battery charger A rectifier unit used to change alternating to direct power for charging a storage battery. Also known as charger.

battery clip A terminal of a connecting wire having spring jaws that can be quickly snapped on a terminal of a device, such as a battery, to which a temporary wire connection is desired.

battery eliminator A device which supplies electron tubes with voltage from electric power supply mains.

battery separator An insulating plate inserted between the positive and negative plates of a battery to prevent them from touching.

battle short Switch for short-circuiting safety interlocks and lighting a red warning light.

bay One segment of an antenna array.

Bayard-Alpert ionization gage A type of ionization vacuum gage using a tube with an electrode structure designed to minimize x-ray-induced electron emission from the ion collector.

bayonet base A tube base or lamp base having two projecting pins on opposite sides of a smooth cylindrical surface to engage in corresponding slots in a bayonet socket, and hold the base firmly in the socket.

bazooka *See* balun.

B battery The battery that furnishes required direct-current voltages to the plate and screen-grid electrodes of the electron tubes in a battery-operated circuit.

BBD *See* bucket brigade device.

B display The presentation of radar output data in rectangular coordinates in which range and azimuth are plotted on the coordinate axes. Also known as range-bearing display.

beacon delay The amount of transponding delay within a beacon, that is, the time between the arrival of a signal and the response of the beacon.

beacon presentation The radarscope presentation resulting from radio-frequency waves sent out by a radar beacon.

beacon skipping A condition where transponder return pulses from a beacon are missing at the interrogating radar.

beacon stealing Loss of beacon tracking by one radar due to stronger signals from an interfering radar.

bead A glass, ceramic, or plastic insulator through which passes the inner conductor of a coaxial transmission line and by means of which the inner conductor is supported in a position coaxial with the outer conductor.

beaded transmission line Line using beads to support the inner conductor in coaxial transmission lines.

32   bead thermistor

bead thermistor A thermistor made by applying the semiconducting material to two wire leads as a viscous droplet, which cements the leads upon firing.

beam angle *See* beam width.

beam antenna An antenna that concentrates its radiation into a narrow beam in a definite direction.

beam blank *See* blank.

beam coupling The production of an alternating current in a circuit connected between two electrodes that are close to, or in the path of, a density-modulated electron beam.

beam current The electric current determined by the number and velocity of electrons in an electron beam.

beam-deflection tube An electron-beam tube in which the current to an output electrode is controlled by transversely moving the electron beam.

beam drop Distortion of the normal rectilinear fan pattern of a detection radar in which a portion of the fan is at a lower elevation than the rest of the fan.

beam-forming electrode Electron-beam focusing elements in power tetrodes and cathode-ray tubes.

beam holding Use of a diffused beam of electrons to regenerate the charges stored on the screen of a cathode-ray storage tube.

beam-indexing tube A single-beam color television picture tube in which the color phosphor strips are arranged in groups of red, green, and blue.

beam lead A flat thick-film lead, sometimes of gold, deposited on a semiconductor chip chemically or by evaporation, as a connecting lead for a semiconductor device or integrated circuit.

beam lobe switching Method of determining the direction of a remote object by comparison of the signals corresponding to two or more successive beam angles, differing slightly from the direction of the object.

beam magnet *See* convergence magnet.

beam parametric amplifier Parametric amplifier that uses a modulated electron beam to provide a variable reactance.

beam power tube An electron-beam tube which uses directed electron beams to provide most of its power-handling capability and in which the control grid and screen grid are essentially aligned. Also known as beam tetrode.

beam recording A method of using an electron beam to write data generated by a computer directly on microfilm.

beam splitting Process for increasing accuracy in locating targets by radar; by noting the azimuths at which one radar scan first discloses a target and at which radar data from it ceases, beam splitting calculates the mean azimuth for the target.

beam steering Changing the direction of the major lobe of a radiation pattern, usually by switching antenna elements.

beam switching Method of obtaining more accurately the bearing or elevation of an object by comparing the signals received when the beam is in directions differing slightly in bearing or elevation; when these signals are equal, the object lies midway between the beam axes. Also known as lobe switching.

beam-switching tube An electron tube which has a series of electrodes arranged around a central cathode and in which an electron beam is switched from one electrode to another. Also known as cyclophon.

beam tetrode *See* beam-power tube.

beam width The angle, measured in a horizontal plane, between the directions at which the intensity of an electromagnetic beam, such as a radar or radio beam, is one-half its maximum value. Also known as beam angle.

bearing resolution Minimum angular separation in a horizontal plane between two targets at the same range that will allow an operator to obtain data on either target.

beat frequency The frequency of a signal equal to the difference in frequencies of two signals which produce the signal when they are combined in a nonlinear circuit.

beat-frequency oscillator An oscillator in which a desired signal frequency, such as an audio frequency, is obtained as the beat frequency produced by combining two different signal frequencies, such as two different radio frequencies. Abbreviated BFO. Also known as heterodyne oscillator.

beating-in Interconnecting two transmitter oscillators and adjusting one until no beat frequency is heard in a connected receiver; the oscillators are then at the same frequency.

beat note The beat frequency whose signal is produced by two signals having waves that are sinusoidal.

beat reception *See* heterodyne reception.

beavertail Fan-shaped radar beam, wide in the horizontal plane and narrow in the vertical plane, which is swept up and down for height finding.

Becquerel effect The phenomenon of a current flowing between two unequally illuminated electrodes of a certain type when they are immersed in an electrolyte.

bedspring array *See* billboard array.

B eliminator Power pack that changes the alternating-current powerline voltage to the direct-current source required by plant circuits of vacuum tubes or semiconductor devices.

bend A smooth change in the direction of the longitudinal axis of a waveguide.

bender element A combination of two thin strips of different piezoelectric materials bonded together so that when a voltage is applied, one strip increases in length and the other becomes shorter, causing the combination to bend.

beta The current gain of a transistor that is connected as a grounded-emitter amplifier, expressed as the ratio of change in collector current to resulting change in base current, the collector voltage being constant.

beyond-the-horizon communication *See* scatter propagation.

BFO *See* beat-frequency oscillator.

B-H curve A graphical curve showing the relation between magnetic induction B and magnetizing force H for a magnetic material. Also known as magnetization curve.

Bi *See* abampere.

bias 1. A direct-current voltage used on signaling or telegraph relays or electromagnets to secure desired time spacing of transitions from marking to spacing. **2.** The restraint of a relay armature by spring tension to secure a desired time spacing of transitions from marking to spacing. **3.** The effect on teleprinter signals produced by the electrical characteristics of the line and equipment. **4.** The force applied to a relay to hold it in a given position. **5.** A direct-current voltage applied to a transistor control electrode to establish the desired operating point. **6.** *See* grid bias.

bias cell A small dry cell used singly or in series to provide the required negative bias for the grid circuit of an electron tube. Also known as grid-bias cell.

bias current 1. An alternating electric current above about 40,000 hertz added to the audio current being recorded on magnetic tape to reduce distortion. **2.** An electric current flowing through the base-emitter junction of a transistor and adjusted to set the operating point of the transistor.

bias distortion Distortion resulting from the operation on a nonlinear portion of the characteristic curve of a vacuum tube or other device, due to improper biasing.

biased automatic gain control *See* delayed automatic gain control.

biased relay *See* percentage differential relay.

bias oscillator An oscillator used in a magnetic recorder to generate the alternating-current signal that is added to the audio current being recorded on magnetic tape to reduce distortion.

bias resistor A resistor used in the cathode or grid circuit of an electron tube to provide a voltage drop that serves as the bias.

bias voltage A voltage applied or developed between two electrodes as a bias.

bias winding A control winding that carries a steady direct current which serves to establish desired operating conditions in a magnetic amplifier or other magnetic device.

biconical antenna An antenna consisting of two metal cones having a common axis with their vertices coinciding or adjacent and with coaxial-cable or waveguide feed to the vertices.

bidirectional antenna An antenna that radiates or receives most of its energy in only two directions.

bidirectional clamping circuit A clamping circuit that functions at the prescribed time irrespective of the polarity of the signal source at the time the pulses used to actuate the clamping action are applied.

bidirectional clipping circuit An electronic circuit that prevents transmission of the portion of an electrical signal that exceeds a prescribed maximum or minimum voltage value.

bidirectional transducer A transducer capable of measuring in both positive and negative directions from a reference position. Also known as bilateral transducer.

bidirectional transistor A transistor that provides switching action in either direction of signal flow through a circuit; widely used in telephone switching circuits.

bidirectional triode thyristor A gate-controlled semiconductor switch designed for alternating-current power control.

bifilar electromagnetic oscillograph A writing low-frequency light-beam oscillograph usually using a moving coil with a single U-shaped turn (bifilar type).

bifilar resistor A resistor wound with a wire doubled back on itself to reduce the inductance.

bifilar transformer A transformer in which wires for the two windings are wound side by side to give extremely tight coupling.

bifilar winding A winding consisting of two insulated wires, side by side, with currents traveling through them in opposite directions.

bifurcated contact A contact having a forked shape such that it can slide over and interlock with an identical mating contact.

bilateral Having a voltage current characteristic curve that is symmetrical with respect to the origin.

bilateral amplifier An amplifier capable of receiving as well as transmitting signals; used primarily in transceivers.

bipolar integrated circuit 35

bilateral antenna An antenna having maximum response in exactly opposite directions, 180° apart, such as a loop.

bilateral circuit Circuit wherein equipment at opposite ends is managed, operated, and maintained by different services.

bilateral transducer See bidirectional transducer.

billboard array A broadside antenna array consisting of stacked dipoles spaced 1/4 to 3/4 wavelength apart in front of a large sheet-metal reflector. Also known as bedspring array; mattress array.

bimag core See bistable magnetic core.

bimorph cell Two piezoelectric plates cemented together in such a way that an applied voltage causes one to expand and the other to contract so that the cell bends in proportion to the applied voltage; conversely, applied pressure generates double the voltage of a single cell; used in phonograph pickups and microphones.

binary counter See binary scaler.

binary encoder An encoder that changes angular, linear, or other forms of input data into binary-coded output characters.

binary logic An assembly of digital logic elements which operate with two distinct states.

binary magnetic core A ferromagnetic core that can be made to take either of two stable magnetic states.

binary scaler A scaler that produces one output pulse for every two input pulses. Also known as binary counter; scale-of-two circuit.

binary signal A voltage or current which carries information by varying between two possible values, corresponding to 0 and 1 in the binary system.

B indicator See B scope.

binding post A manually turned screw terminal used for making electrical connections.

binistor A silicon *npn* tetrode that serves as a bistable negative-resistance device.

binode An electron tube with two anodes and one cathode used as a full-wave rectifier. Also known as double diode.

binomial array antenna Directional antenna array for reducing minor lobes and providing maximum response in two opposite directions.

biochemical fuel cell An electrochemical power generator in which the fuel source is bioorganic matter; air is the oxidant at the cathode, and microorganisms catalyze the oxidation of the bioorganic matter at the anode.

biot See abampere.

Biot-Savart law A law that gives the intensity of the magnetic field due to a wire carrying a constant electric current.

bipolar amplifier An amplifier capable of supplying a pair of output signals corresponding to the positive or negative polarity of the input signal.

bipolar circuit A logic circuit in which zeros and ones are treated in a symmetric or bipolar manner, rather than by the presence or absence of a signal; for example, a balanced arrangement in a square-loop-ferrite magnetic circuit.

bipolar electrode Electrode, without metallic connection with the current supply, one face of which acts as anode surface and the opposite face as a cathode surface when an electric current is passed through a cell.

bipolar integrated circuit An integrated circuit in which the principal element is the bipolar junction transistor.

36 bipolar power supply

bipolar power supply A high-precision, regulated, direct-current power supply that can be set to provide any desired voltage between positive and negative design limits, with a smooth transition from one polarity to the other.

bipolar transistor A transistor that uses both positive and negative charge carriers.

bipotential electrostatic lens An electron lens in which image and object space are field-free, but at different potentials; examples are the lenses formed between apertures of cylinders at different potentials. Also known as immersion electrostatic lens.

biquartic filter An active filter that uses operational amplifiers in combination with resistors and capacitors to provide infinite values of Q and simple adjustments for band-pass and center frequency.

bistable circuit A circuit with two stable states such that the transition between the states cannot be accomplished by self-triggering.

bistable magnetic core A magnetic core that can be in either of two possible states of magnetization. Also known as bimag core.

bistable multivibrator A multivibrator in which either of the two active devices may remain conducting, with the other nonconducting, until the application of an external pulse. Also known as Eccles-Jordan circuit; Eccles-Jordan multivibrator; flip-flop circuit; trigger circuit.

Bitter pattern A pattern produced when a drop of a colloidal suspension of ferromagnetic particles is placed on the surface of a ferromagnetic crystal; the particles collect along domain boundaries at the surface.

black-and-white groups *See* Shubnikov groups.

black level The level of the television picture signal corresponding to the maximum limit of black peaks.

blackout The shutting off of power in an electrical power transmission system, either deliberately or through failure of the system.

black scope Cathode-ray tube operating at the threshold of luminescence when no video signals are being applied.

black-surface field A layer of p^+ material which is applied to the back surface of a solar cell to reduce hole-electron recombinations there and thereby increase the cell's efficiency.

blade A flat moving conductor in a switch.

blank To cut off the electron beam of a television picture tube, camera tube, or cathode-ray oscilloscope tube during the process of retrace by applying a rectangular pulse voltage to the grid or cathode during each retrace interval. Also known as beam blank.

blanking circuit A circuit preventing the transmission of brightness variations during the horizontal and vertical retrace intervals in television scanning.

blanking level The level that separates picture information from synchronizing information in a composite television picture signal; coincides with the level of the base of the synchronizing pulses. Also known as pedestal; pedestal level.

blanking pulse A positive or negative square-wave pulse used to switch off a part of a television or radar set electronically for a predetermined length of time.

blanking signal The signal rendering the return trace invisible on the picture tube of a television receiver.

blanking time The length of time that the electron beam of a cathode-ray tube is shut off.

bleeder A high resistance connected across the dc output of a high-voltage power supply which serves to discharge the filter capacitors after the power supply has been turned off, and to provide a stabilizing load.

bleeder current Current drawn continuously from a voltage source to lessen the effect of load changes or to provide a voltage drop across a resistor.

bleeder resistor A resistor connected across a power pack or other voltage source to improve voltage regulation by drawing a fixed current value continuously; also used to dissipate the charge remaining in filter capacitors when equipment is turned off.

blinking Electronic-countermeasures technique employed by two aircraft separated by a short distance and within the same azimuth resolution so as to appear as one target to a tracking radar; the two aircraft alternately spot-jam, causing the radar system to oscillate from one place to another, making an accurate solution of a fire control problem impossible.

blip 1. The display of a received pulse on the screen of a cathode-ray tube. Also known as pip. 2. An ideal infrared radiation detector that detects with unit quantum efficiency all of the radiation in the signal for which the detector was designed, and responds only to the background radiation noise that comes from the field of view of the detector.

blip-scan ratio The ratio of the number of times a target appears on a radarscope to the number of times it could have been seen.

Bloch equations Approximate equations for the rate of change of magnetization of a solid in a magnetic field due to spin relaxation and gyroscopic precession.

Bloch function A wave function for an electron in a periodic lattice, of the form $u(\mathbf{r})$ exp $[i\mathbf{k}\cdot\mathbf{r}]$ where $u(\mathbf{r})$ has the periodicity of the lattice.

Bloch theorem The theorem that, in a periodic structure, every electronic wave function can be represented by a Bloch function.

Bloch wall A transition layer, with a finite thickness of a few hundred lattice constants, between adjacent ferromagnetic domains. Also known as domain wall.

blocked impedance The impedance at the input of a transducer when the impedance of the output system is made infinite, as by blocking or clamping the mechanical system.

blocking 1. Applying a high negative bias to the grid of an electron tube to reduce its anode current to zero. 2. Overloading a receiver by an unwanted signal so that the automatic gain control reduces the response to a desired signal. 3. Distortion occurring in a resistance-capacitance-coupled electron tube amplifier stage when grid current flows in the following tube. 4. The hindering of motion of dislocations in a solid substance by small particles of a second substance included in the solid; this results in hardening of the substance.

blocking capacitor *See* coupling capacitor.

blocking layer *See* depletion layer.

blocking oscillator A relaxation oscillator that generates a short-time-duration pulse by using a single transistor or electron tube and associated circuitry. Also known as squegger; squegging oscillator.

blocking oscillator driver Circuit which develops a square pulse used to drive the modulator tubes, and which usually contains a line-controlled blocking oscillator that shapes the pulse into the square wave.

block protector Rectangular piece of carbon, bakelite with a metal insert, or porcelain with a carbon insert which, in combination with each other, make one element of a protector; they form a gap which will break down and provide a path to ground for excessive voltages.

38 blooming

blooming 1. Defocusing of television picture areas where excessive brightness results in enlargement of spot size and halation of the fluorescent screen. 2. An increase in radarscope spot size due to an increase in signal intensity.

blow Opening of a circuit because of excess current, particularly when the current is heavy and a melting or breakdown point is reached.

blown-fuse indicator A neon warning light connected across a fuse so that it lights when the fuse is blown.

blowout 1. The melting of an electric fuse because of excessive current. 2. The extinguishing of an electric arc by deflection in a magnetic field. Also known as magnetic blowout.

blowout coil A coil that produces a magnetic field in an electrical switching device for the purpose of lengthening and extinguishing an electric arc formed as the contacts of the switching device part to interrupt the current.

blowout magnet An electromagnet or permanent magnet used to deflect and extinguish the arc formed when a high-current circuit breaker or switch is opened.

blue glow A glow normally seen in electron tubes containing mercury vapor, due to ionization of the mercury molecules.

Board of Trade unit *See* kilowatt-hour.

bobbin An insulated spool serving as a support for a coil.

bobbin core A magnetic core having a form or bobbin on which the ferromagnetic tape is wrapped for support of the tape.

bobbing Fluctuation of the strength of a radar echo, or its indication on a radarscope, due to alternate interference and reinforcement of returning reflected waves.

Bode diagram A diagram in which the phase shift or the gain of an amplifier, a servomechanism, or other device is plotted against frequency to show frequency response; logarithmic scales are customarily used for gain and frequency.

body capacitance Capacitance existing between the human hand or body and a circuit.

bombardment The use of induction heating to heat electrodes of electron tubes to drive out gases during evacuation.

bond The connection made by bonding electrically.

bonded NR diode An n^+ junction semiconductor device in which the negative resistance arises from a combination of avalanche breakdown and conductivity modulation which is due to the current flow through the junction.

bonding The use of low-resistance material to connect electrically a chassis, metal shield cans, cable shielding braid, and other supposedly equipotential points to eliminate undesirable electrical interaction resulting from high-impedance paths between them.

bonding pad A metallized area on the surface of a semiconductor device, to which connections can be made.

bonding wire Wire used to connect metal objects so they have the same potential (usually ground potential).

book capacitor A trimmer capacitor consisting of two plates which are hinged at one end; capacitance is varied by changing the angle between them.

boost To augment in relative intensity, as to boost the bass response in an audio system.

boost charge Partial charge of a storage battery, usually at a high current rate for a short period.

booster 1. A small generator inserted in series or parallel with a larger generator to maintain normal voltage output under heavy loads. 2. A separate radio-frequency amplifier connected between an antenna and a television receiver to amplify weak signals. 3. A radio-frequency amplifier that amplifies and rebroadcasts a received television or communication radio carrier frequency for reception by the general public.

booster battery A battery which increases the sensitivity of a crystal detector by maintaining a certain voltage across it and thereby adjusting conditions to increase the response to a given input.

booster voltage The additional voltage supplied by the damper tube to the horizontal output, horizontal oscillator, and vertical output tubes of a television receiver to give greater sawtooth sweep output.

boot A protective covering over any portion of a cable, wire, or connector.

bootstrap circuit A single-stage amplifier in which the output load is connected between the negative end of the anode supply and the cathode, while signal voltage is applied between grid and cathode; a change in grid voltage changes the input signal voltage with respect to ground by an amount equal to the output signal voltage.

bootstrap driver Electronic circuit used to produce a square pulse to drive the modulator tube; the duration of the square pulse is determined by a pulse-forming line.

bootstrap integrator A bootstrap sawtooth generator in which an integrating amplifier is used in the circuit. Also known as Miller generator.

bootstrapping A technique for lifting a generator circuit above ground by a voltage value derived from its own output signal.

bootstrap sawtooth generator A circuit capable of generating a highly linear positive sawtooth waveform through the use of bootstrapping.

Born-Haber cycle A sequence of chemical and physical processes by means of which the cohesive energy of an ionic crystal can be deduced from experimental quantities; it leads from an initial state in which a crystal is at zero pressure and 0 K to a final state which is an infinitely dilute gas of its constituent ions, also at zero pressure and 0 K.

Born-Madelung model A classical theory of cohesive energy, lattice spacing, and compressibility of ionic crystals.

Born-Mayer equation An equation for the cohesive energy of an ionic crystal which is deduced by assuming that this energy is the sum of terms arising from the Coulomb interaction and a repulsive interaction between nearest neighbors.

Born–von Kármán theory A theory of specific heat which considers an acoustical spectrum for the vibrations of a system of point particles distributed like the atoms in a crystal lattice.

Bosanquet's law The statement that, in analogy to Ohm's law for the resistance of an electric circuit, in a magnetic circuit the ratio of the magnetomotive force to the magnetic flux is a constant known as the reluctance.

boundary An interface between p- and n-type semiconductor materials, at which donor and acceptor concentrations are equal.

boundary-layer photocell *See* photovoltaic cell.

bound charge Electric charge which is confined to atoms or molecules, in contrast to free charge, such as metallic conduction electrons, which is not. Also known as polarization charge.

boxcar circuit A circuit used in radar for sampling voltage waveforms and storing the latest value sampled; the term is derived from the flat, steplike segments of the output voltage waveform.

B power supply *See* B supply.

Bragg angle One of the characteristic angles at which x-rays reflect specularly from planes of atoms in a crystal.

Bragg diffraction *See* Bragg scattering.

Bragg's equation *See* Bragg's law.

Bragg's law A statement of the conditions under which a crystal will reflect a beam of x-rays with maximum intensity. Also known as Bragg's equation; Bravais' law.

Bragg reflection *See* Bragg scattering.

Bragg scattering Scattering of x-rays or neutrons by the regularly spaced atoms in a crystal, for which constructive interference occurs only at definite angles called Bragg angles. Also known as Bragg diffraction; Bragg reflection.

braided wire A tube of fine wires woven around a conductor or cable for shielding purposes or used alone in flattened form as a grounding strap.

branch A portion of a network consisting of one or more two-terminal elements in series. Also known as arm.

branch circuit A portion of a wiring system in the interior of a structure that extends from a final overload protective device to a plug receptable or a load such as a lighting fixture, motor, or heater.

branch joint Joint used for connecting a branch conductor or cable, where the latter continues beyond the branch.

branch point A terminal in an electrical network that is common to more than two elements or parts of elements of the network. Also known as junction point; node.

Braun tube *See* cathode-ray tube.

breadboarding Assembling an electronic circuit in the most convenient manner, without regard for final locations of components, to prove the feasibility of the circuit and to facilitate changes when necessary.

break 1. A fault in a circuit. 2. The minimum distance in a circuit-opening device between the stationary and movable contacts when these contacts are in the open position. 3. A reflected radar pulse which appears on a radarscope as a line perpendicular to the base line.

break-before-make contact One of a pair of contacts that interrupt one circuit before establishing another.

break contact The contact of a switching device which opens a circuit upon the operation of the device.

breakdown A large, usually abrupt rise in electric current in the presence of a small increase in voltage; can occur in a confined gas between two electrodes, a gas tube, the atmosphere (as lightning), an electrical insulator, and a reverse-biased semiconductor diode.

breakdown diode A semiconductor diode in which the reverse-voltage breakdown mechanism is based either on the Zener effect or the avalanche effect.

breakdown impedance Of a semiconductor, the small-signal impedance at a specified direct current in the breakdown region.

breakdown potential *See* breakdown voltage.

breakdown region Of a semiconductor diode, the entire region of the volt-ampere characteristic beyond the initiation of breakdown for increasing magnitude of bias.

breakdown voltage 1. The voltage measured at a specified current in the electrical breakdown region of a semiconductor diode. Also known as Zener voltage. 2. The voltage at which an electrical breakdown occurs in a dielectric. 3. The voltage at

bridging connection   **41**

which an electrical breakdown occurs in a gas. Also known as breakdown potential; sparking potential; sparking voltage.

breaker-and-a-half A substation switching arrangement that involves two buses between which three breaker bays are installed.

breaker-and-a-third A substation switching arrangement having four breakers and three connections per bay.

breaker points Low-voltage contacts used to interrupt the current in the primary circuit of a gasoline engine's ignition system.

break-in device A device in a radiotelegraph communication system allowing an operator to receive signals in intervals between his own transmission signals.

breakout A joint at which one or more conductors are brought out from a multiconductor cable.

breakover In a silicon controlled rectifier or related device, a transition into forward conduction caused by the application of an excessively high anode voltage.

breakover voltage The positive anode voltage at which a silicon controlled rectifier switches into the conductive state with gate circuit open.

bremsstrahlung Radiation that is emitted by an electron accelerated in its collision with the nucleus of an atom.

bridge 1. An electrical instrument having four or more branches, by means of which one or more of the electrical constants of an unknown component may be measured. 2. An electrical shunt path.

bridge circuit An electrical network consisting basically of four impedances connected in series to form a rectangle, with one pair of diagonally opposite corners connected to an input device and the other pair to an output device.

bridged tap Portion of a cable pair connected to a circuit which is not a part of the useful path.

bridged-T network A T network with a fourth branch connected between an input and an output terminal and across two branches of the network.

bridge hybrid *See* hybrid junction.

bridge limiter A device employed in analog computers to keep the value of a variable within specified limits.

bridge magnetic amplifier A magnetic amplifier in which each of the gate windings is connected in series with an arm of a bridge rectifier; the rectifiers provide self-saturation and direct-current output.

bridge oscillator An oscillator using a balanced bridge circuit as the feedback network.

bridge rectifier A full-wave rectifier with four elements connected as a bridge circuit with direct voltage obtained from one pair of opposite junctions when alternating voltage is applied to the other pair.

bridge transformer *See* hybrid transformer.

bridging 1. Connecting one electric circuit in parallel with another. 2. The action of a selector switch whose movable contact is wide enough to touch two adjacent contacts so that the circuit is not broken during contact transfer.

bridging amplifier Amplifier with an input impedance sufficiently high so that its input may be bridged across a circuit without substantially affecting the signal level of the circuit across which it is bridged.

bridging connection Parallel connection by means of which some of the signal energy in a circuit may be withdrawn frequently, with imperceptible effect on the normal operation of the circuit.

bridging contacts A contact form in which the moving contact touches two stationary contacts simultaneously during transfer.

bridging loss Loss resulting from bridging an impedance across a transmission system; quantitatively, the ratio of the signal power delivered to that part of the system following the bridging point, and measured before the bridging, to the signal power delivered to the same part after the bridging.

Bridgman effect The phenomenon that when an electric current passes through an anisotropic crystal, there is an absorption or liberation of heat due to the nonuniformity in current distribution.

Bridgman relation $P = QT\sigma$ in a metal or semiconductor, where P is the Ettingshausen coefficient, Q the Nernst-Ettingshausen coefficient, T the temperature, and σ the thermal conductivity in a transverse magnetic field.

Bridgman technique A method of growing single crystals in which a vertical cylinder that tapers conically to a point at the bottom and contains the substance to be crystallized in molten form is slowly lowered into a cold zone, resulting in crystallization beginning at the tip.

brightness control A control that varies the luminance of the fluorescent screen of a cathode-ray tube, for a given input signal, by changing the grid bias of the tube and hence the beam current. Also known as brilliance control; intensity control.

brilliance 1. The degree of brightness and clarity of the display of a cathode-ray tube. 2. The degree to which the higher audio frequencies of an input sound are reproduced by a radio receiver, by a public address amplifier, or by a sound-recording playback system.

brilliance control *See* brightness control.

Brillouin function A function of x with index (or parameter) n that appears in the quantum-mechanical theories of paramagnetism and ferromagnetism and is expressed as $[(2n+1)/2n]$ coth $[(2n+1)x/2n] - (1/2n)\cdot$ coth $(x/2n)$.

Brillouin scattering Light scattering by acoustic phonons.

Brillouin zone A fundamental region of wave vectors in the theory of the propagation of waves through a crystal lattice; any wave vector outside this region is equivalent to some vector inside it.

broad-band amplifier An amplifier having essentially flat response over a wide range of frequencies.

broad-band antenna An antenna that functions satisfactorily over a wide range of frequencies, such as for all 12 very-high-frequency television channels.

broad-band klystron Klystron having three or more resonant cavities that are externally loaded and stagger-tuned to broaden the bandwidth.

broadcast transmitter A transmitter designed for use in a commercial amplitude-modulation, frequency-modulation, or television broadcast channel.

broadside Perpendicular to an axis or plane.

broadside array An antenna array whose direction of maximum radiation is perpendicular to the line or plane of the array.

broad tuning Poor selectivity in a radio receiver, causing reception of two or more stations at a single setting of the tuning dial.

Brooks variable inductometer An inductometer providing a nearly linear scale and consisting of two movable coils, side by side in a plane, sandwiched between two pairs of fixed coils.

brownout 1. A restriction of electrical power usage during a power shortage, especially for advertising and display purposes. 2. An extinguishing of some of the lights in a city as a defensive measure against enemy bombardment.

brush A conductive metal or carbon block used to make sliding electrical contact with a moving part.

brush discharge A luminous electric discharge that starts from a conductor when its potential exceeds a certain value but remains too low for the formation of an actual spark.

brush encoder An encoder in which brushes that make contact with conductive segments on a rotating or linearly moving surface convert positional information to digitally encoded data.

brush holder A structure in which a brush can slide in a direction perpendicular to the moving surface of a motor, generator, or other device.

brush rocker A yoke to which the brush holders in an electrical machine are attached, and which can be moved to adjust the positions of the brushes. Also known as brush rocker ring.

brush rocker ring *See* brush rocker.

brute-force filter Type of powerpack filter depending on large values of capacitance and inductance to smooth out pulsations rather than on resonant effects of tuned filters.

brute supply A type of power supply that is completely unregulated, employing no circuitry to maintain output voltage constant with changing input line or load variations.

B scan *See* B scope.

B scope A cathode-ray scope on which signals appear as spots, with bearing angle as the horizontal coordinate and range as the vertical coordinate. Also known as B indicator; B scan.

B supply Anode high voltage and screen-grid power source in vacuum tube circuits. Also known as B power supply.

B trace In loran the second trace of an oscilloscope which corresponds to the signal from the B station.

bubble *See* magnetic bubble.

bubble raft A visual demonstration for the structure of dislocations in metal lattices, showing slip propagation; it consists of many identical bubbles floating on a liquid surface in something like a crystalline array.

Buchholz protective device A protective relay which is attached to an oil-filled tank containing a transformer and which is activated either by gas produced by faults or by oil surges produced by explosive faults in the transformer. Also known as gas bubble protective device.

bucket brigade device A semiconductor device in which majority carriers store charges that represent information, and minority carriers transfer charges from point to point in sequence. Abbreviated BBD.

bucking coil A coil connected and positioned in such a way that its magnetic field opposes the magnetic field of another coil; for example, the hum-bucking coil of an excited-field loudspeaker.

bucking transformer A transformer whose voltage opposes that of a second transformer.

bucking voltage A voltage having a polarity opposite to that of another voltage against which it acts.

buffer 1. An electric circuit or component that prevents undesirable electrical interaction between two circuits or components. 2. An isolating circuit in an electronic computer used to prevent the action of a driven circuit from affecting the corresponding driving circuit. 3. *See* buffer amplifier.

buffer amplifier An amplifier used after an oscillator or other critical stage to isolate it from the effects of load impedance variations in subsequent stages. Also known as buffer; buffer stage.

buffer capacitor A capacitor connected across the secondary of a vibrator transformer or between the anode and cathode of a cold-cathode rectifier tube to suppress voltage surges that might otherwise damage other parts in the circuit.

buffer element A low-impedance inverting driver circuit.

buffer stage *See* buffer amplifier.

bug 1. A semiautomatic code-sending telegraph key in which movement of a lever to one side produces a series of correctly spaced dots and movement to the other side produces a single dash. **2.** An electronic listening device, generally concealed, used for commercial or military espionage.

build To increase in received signal strength.

building-out circuit Short section of transmission line, or a network which is shunted across a transmission line, for the purpose of impedance matching.

building-out network Network designed to be connected to a based network so that the combination will simulate the sending-end impedance, neglecting dissipation, of a line having a termination other than that for which the basic network was designed.

building-out section Short section of transmission line, either open or short-circuited at the far end, shunted across another transmission line for use on an impedance-matching transformer.

built-in antenna An antenna located inside the cabinet of a radio or television receiver.

bulk-acoustic-wave delay line A delay line in which the delay is determined by the distance traveled by a bulk acoustic wave between input and output transducers mounted on a piezoelectric block.

bulk diode A semiconductor microwave diode that uses the bulk effect, such as Gunn diodes and diodes operating in limited space-charge-accumulation modes.

bulk effect An effect that occurs within the entire bulk of a semiconductor material rather than in a localized region or junction.

bulk-effect device A semiconductor device that depends on a bulk effect, as in Gunn and avalanche devices.

bulk eraser A device used to erase an entire reel of recorded magnetic tape at once without running it through a recorder.

bulk lifetime The average time that elapses between the formation and recombination of minority charge carriers in the bulk material of a semiconductor.

bulk photoconductor A photoconductor having high power-handling capability and other unique properties that depend on the semiconductor and doping materials used.

bulk resistor An integrated-circuit resistor in which the n-type epitaxial layer of a semiconducting substrate is used as a noncritical high-value resistor; the spacing between the attached terminals and the sheet resistivity of the material together determine the resistance value.

bump contact A large-area contact used for alloying directly to the substrate of a transistor for mounting or interconnecting purposes.

bunched pair Group of pairs tied together or otherwise associated for identification.

butt joint   45

buncher resonator   The first or input cavity resonator in a velocity-modulated tube, next to the cathode; here the faster electrons catch up with the slower ones to produce bunches of electrons. Also known as buncher; input resonator.

bunching   The flow of electrons from cathode to anode of a velocity-modulated tube as a succession of electron groups rather than as a continuous stream.

bunching voltage   Radio-frequency voltage between the grids of the buncher resonator in a velocity-modulated tube such as a klystron; generally, the term implies the peak value of this oscillating voltage.

burden   The amount of power drawn from the circuit connecting the secondary terminals of an instrument transformer, usually expressed in volt-amperes.

burn-in   Operation of electronic components before they are applied in order to stabilize their characteristics and reveal defects.

burnout   Failure of a device due to excessive heat produced by excessive current.

burn-through   See jammer finder.

burst   1. An exceptionally large electric pulse in the circuit of an ionization chamber due to the simultaneous arrival of several ionizing particles. 2. A radar term for a single pulse of radio energy.

burst amplifier   An amplifier stage in a color television receiver that is keyed into conduction and amplification by a horizontal pulse at the instant of each arrival of the color burst. Also known as chroma band-pass amplifier.

burst separator   The circuit in a color television receiver that separates the color burst from the composite video signal.

bus   1. A set of two or more electric conductors that serve as common connections between load circuits and each of the polarities (in direct-current systems) or phases (in alternating-current systems) of the source of electric power. 2. One or more conductors in a computer along which information is transmitted from any of several sources to any of several destinations. 3. See busbar.

busbar   A heavy, rigid metallic conductor, usually uninsulated, used to carry a large current or to make a common connection between several circuits. Also known as bus.

bus reactor   An air-core inductor connected between two buses or two sections of the same bus in order to limit the effects of voltage transients on either bus.

busway   A prefabricated assembly of standard lengths of busbars rigidly supported by solid insulation and enclosed in a sheet-metal housing.

Butler oscillator   Oscillator in which a piezoelectric crystal is connected between the cathode of two tubes, one functioning as a cathode follower, and the other as a grounded-grid amplifier.

butt contact   A hemispherically shaped contact designed to mate against a similarly shaped contact.

butterfly capacitor   A variable capacitor having stator and rotor plates shaped like butterfly wings, with the stator plates having an outer ring to provide an inductance so that both capacitance and inductance may be varied, thereby giving a wide tuning range.

Butterworth filter   An electric filter whose pass band (graph of transmission versus frequency) has a maximally flat shape.

butt joint   1. A connection formed by placing the ends of two conductors together and joining them by welding, brazing, or soldering. 2. A connection giving physical contact between the ends of two waveguides to maintain electrical continuity.

button 1. A small, round piece of metal alloyed to the base wafer of an alloy-junction transistor. Also known as dot. **2.** The container that holds the carbon granules of a carbon microphone. Also known as carbon button.

buttonhook contact A curved, hooklike contact often used on feed-through terminals of headers to facilitate soldering or unsoldering of leads.

buzz The condition of a combinatorial circuit with feedback that has undergone a transition, caused by the inputs, from an unstable state to a new state that is also unstable.

buzzer An electromagnetic device having an armature that vibrates rapidly, producing a buzzing sound.

BWO *See* backward-wave oscillator.

BX cable Insulated wires in flexible metal tubing used for bringing electric power to electronic equipment.

bypass A shunt path around some element or elements of a circuit.

bypass capacitor A capacitor connected to provide a low-impedance path for radio-frequency or audio-frequency currents around a circuit element. Also known as bypass condenser.

bypass condenser *See* bypass capacitor.

bypass filter Filter which provides a low-attenuation path around some other equipment, such as a carrier frequency filter used to bypass a physical telephone repeater station.

C

C *See* capacitor; coulomb.

cable Strands of insulated electrical conductors laid together, usually around a central core, and surrounded by a heavy insulation.

cable-and-trunk schematic A drawing which shows, in block form, the interconnection between all major electric circuits in an office.

cable armor One or more layers of extra-strength material, such as steel wire or tape, to reinforce the usual lead wall in cable construction.

cable complement Group of wire pairs in a cable having some common distinguishing characteristic.

cable fill Ratio of the number of wire pairs in use to the total number of pairs in a cable.

cable messenger Stranded group of wires supported above the ground at intervals by poles or other structures and employed to furnish, within these intervals, frequent points of support for conductors or cables.

cable run Path occupied by a cable on cable racks or other support from one termination to another.

cable running list Drawing showing the code of cable, terminations, circuit names, and numbering of cables appearing in an office.

cable shield A metallic layer applied over insulation covering a cable, composed of woven or braided wires, foil wrap, or metal tube, which acts to prevent electromagnetic or electrostatic interference from affecting conductors within.

cable vault Vault in which the outside plant cables are spliced to the tipping cables.

cadmium cell A standard cell used as a voltage reference; at 20°C its voltage is 1.0186 volts.

cadmium lamp A lamp containing cadmium vapor; wavelength (6438.4696 international angstroms, or 643.84696 nanometers) of light emitted is a standard of length.

cadmium-nickel storage cell *See* nickel-cadmium battery.

cadmium selenide cell A photoconductive cell that uses cadmium selenide as the semiconductor material and has a fast response time and high sensitivity to longer wavelengths of light.

cadmium silver oxide cell An alkaline-electrolyte cell that may be used without recharging in primary batteries or that may be recharged for secondary-battery use.

cadmium sulfide cell A photoconductive cell in which a small wafer of cadmium sulfide provides an extremely high dark-light resistance ratio.

cadmium telluride detector A photoconductive cell capable of operating continuously at ambient temperatures up to 750°F (400°C); used in solar cells and infrared, nuclear-radiation, and gamma-ray detectors.

48   cage antenna

cage antenna Broad-band dipole antenna in which each pole consists of a cage of wires whose overall shape resembles that of a cylinder or a cone.

call announcer Device for receiving pulses from an automatic telephone office and audibly reproducing the corresponding number in words, so that it may be heard by a manual operator.

call circuit Communications circuit between switching points used by traffic forces for transmitting switching instructions.

call indicator Device for receiving pulses from an automatic switching system and displaying the corresponding called number before an operator at a manual switchboard.

calling device Apparatus which generates the pulses required for establishing connections in an automatic telephone switching system.

Calzecchi-Onesti effect A change in the conductivity of a loosely aggregated metallic powder caused by an applied electric field.

camera *See* television camera.

camera cable Cable or group of wires that carries the picture from the television camera to the control room.

camera tube An electron-beam tube used in a television camera to convert an optical image into a corresponding charge-density electric image and to scan the resulting electric image in a predetermined sequence to provide an equivalent electric signal. Also known as pickup tube; television camera tube.

Campbell bridge 1. A bridge designed for comparison of mutual inductances. 2. A circuit for measuring frequencies by adjusting a mutual inductance, until the current across a detector is zero.

Campbell's formula A formula which relates the propagation constant of a loaded transmission line to the propagation constant and characteristic impedance of an unloaded line and the impedance of each loading coil.

cancellation circuit A circuit used in providing moving-target indication on a plan position indicator scope; cancels constant-amplitude fixed-target pulses by subtraction of successive pulse trains.

capacitance The ratio of the charge on one of the conductors of a capacitor (there being an equal and opposite charge on the other conductor) to the potential difference between the conductors. Symbolized C. Formerly known as capacity.

capacitance box An assembly of capacitors and switches which permits adjustment of the capacitance existing at the terminals in nominally uniform steps, from a minimum value near zero to the maximum which exists when all the capacitors are connected in parallel.

capacitance bridge A bridge for comparing two capacitances, such as a Schering bridge.

capacitance relay An electronic relay that responds to a small change in capacitance, such as that created by bringing a hand near a pickup wire or plate.

capacitance standard *See* standard capacitor.

capacitive coupling Use of a capacitor to transfer energy from one circuit to another.

capacitive diaphragm A resonant window used in a waveguide to provide the equivalent of capacitive reactance at the frequency being transmitted.

capacitive-discharge ignition An automotive ignition system in which energy is stored in a capacitor and discharged across the gap of a spark plug through a step-up pulse transformer and distributor each time a silicon controlled rectifier is triggered.

capture effect 49

capacitive-discharge pilot light An electronic ignition system, operating off an alternating-current power line or battery power supply, that produces a spark for lighting a gas flame.

capacitive divider Two or more capacitors placed in series across a source, making available a portion of the source voltage across each capacitor; the voltage across each capacitor will be inversely proportional to its capacitance.

capacitive feedback Process of returning part of the energy in the plate (or output) circuit of a vacuum tube (or other device) to the grid (or input) circuit by means of a capacitance common to both circuits.

capacitive load A load in which the capacitive reactance exceeds the inductive reactance; the load draws a leading current.

capacitive post Metal post or screw extending across a waveguide at right angles to the E field, to provide capacitive susceptance in parallel with the waveguide for tuning or matching purposes.

capacitive reactance Reactance due to the capacitance of a capacitor or circuit, equal to the inverse of the product of the capacitance and the angular frequency.

capacitive tuning Tuning involving use of a variable capacitor.

capacitive window Conducting diaphragm extending into a waveguide from one or both sidewalls, producing the effect of a capacitive susceptance in parallel with the waveguide.

capacitor A device which consists essentially of two conductors (such as parallel metal plates) insulated from each other by a dielectric and which introduces capacitance into a circuit, stores electrical energy, blocks the flow of direct current, and permits the flow of alternating current to a degree dependent on the capacitor's capacitance and the current frequency. Symbolized C. Also known as condenser; electrical condenser.

capacitor antenna Antenna consisting of two conductors or systems of conductors, the essential characteristic of which is its capacitance. Also known as condenser antenna.

capacitor bank A number of capacitors connected in series or in parallel.

capacitor box A box-shaped structure in which a capacitor is submerged in a heat-absorbing medium, usually water. Also known as condenser box.

capacitor color code A method of marking the value on a capacitor by means of dots or bands of colors as specified in the Electronic Industry Association color code.

capacitor-input filter A power-supply filter in which a shunt capacitor is the first element after the rectifier.

capacitor motor 1. A single-phase induction motor having a main winding connected directly to a source of alternating-current power and an auxiliary winding connected in series with a capacitor to the source of ac power. 2. *See* capacitor-start motor.

capacitor-start motor A capacitor motor in which the capacitor is in the circuit only during the starting period; the capacitor and its auxiliary winding are disconnected automatically by a centrifugal switch or other device when the motor reaches a predetermined speed. Also known as capacitor motor.

capacitor start-run motor *See* permanent-split capacitor motor.

capacity *See* capacitance.

capacity cell 1. Capacitance-type device used to measure the dielectric constants of gases, liquids, or solids. 2. Capacitance-type device used to monitor certain composition changes in flowing streams.

capture effect The effect wherein a strong frequency-modulation signal in an FM receiver completely suppresses a weaker signal on the same or nearly the same frequency.

50 carbon arc

carbon arc An electric arc between two electrodes, at least one of which is made of carbon; used in welding and high-intensity lamps, such as in searchlights and photography lamps.

carbon arc lamp An arc lamp in which an electric current flows between two electrodes of pure carbon, with incandescence at one or both electrodes and some light from the luminescence of the arc.

carbon brush A rod made of carbon that bears against a commutator, collector ring, or slip ring to provide passage for the electric current from a dynamo through an outside circuit or for an external current through a motor.

carbon button *See* button.

carbon-film hygrometer element An electrical hygrometer element constructed of a plastic strip coated with a film of carbon black dispersed in a hygroscopic binder; variations in atmospheric moisture content vary the volume of the binder and thus change the resistance of the carbon coating.

carbon-film resistor A resistor made by depositing a thin carbon film on a ceramic form.

carbon lamp An arc lamp with carbon electrodes.

carbon pile A variable resistor consisting of a stack of carbon disks mounted between a fixed metal plate and a movable one that serve as the terminals of the resistor; the resistance value is reduced by applying pressure to the movable plate.

carbon resistor A resistor consisting of carbon particles mixed with a binder, molded into a cylindrical shape, and baked; terminal leads are attached to opposite ends. Also known as composition resistor.

carcinotron *See* backward-wave oscillator.

card-edge connector A connector that mates with printed-wiring leads running to the edge of a printed circuit board on one or both sides. Also known as edgeboard connector.

cardinal point effect The increased intensity of a line or group of returns on the radarscope occurring when the radar beam is perpendicular to the rectangular surface of a line or group of similarly aligned features in the ground pattern.

carrier *See* charge carrier.

carrier amplifier A direct-current amplifier in which the dc input signal is filtered by a low-pass filter, then used to modulate a carrier so it can be amplified conventionally as an alternating-current signal; the amplified dc output is obtained by rectifying and filtering the rectified carrier signal.

carrier density The density of electrons and holes in a semiconductor.

carrier isolating choke coil Inductor inserted in series with a line on which carrier energy is applied to impede the flow of carrier energy beyond that point.

carrier line Any transmission line used for multiple-channel carrier communication.

carrier loading The addition of lumped inductances to the cable section of a transmission line specifically designed for carrier transmission; it serves to minimize impedance mismatch between cable and open wire and to reduce the cable attenuation.

carrier mobility The average drift velocity of carriers per unit electric field in a homogeneous semiconductor; the mobility of electrons is usually different from that of holes.

carrier repeater Equipment designed to raise carrier signal levels to such a value that they may traverse a succeeding line section at such amplitude as to preserve an adequate signal-to-noise ratio; while the heart of a repeater is the amplifier, necessary adjuncts are filters, equalizers, level controls, and so on, depending upon the operating methods.

carrier terminal Apparatus at one end of a carrier transmission system, whereby the processes of modulation, demodulation, filtering, amplification, and associated functions are effected.

carrier transfer filters Filters arranged as a carrier-frequency crossover or bridge between two transmission circuits.

Carter chart An Argand diagram of the complex reflection coefficient of a waveguide junction on which are drawn lines of constant magnitude and phase of the impedance.

artridge fuse A type of electric fuse in which the fusible element is connected between metal ferrules at either end of an insulating tube.

cartridge lamp A pilot or dial lamp that has a tubular glass envelope with metal-ferrule terminals at each end.

cascade 1. An electric-power circuit arrangement in which circuit breakers of reduced interrupting ratings are used in the branches, the circuit breakers being assisted in their protection function by other circuit breakers which operate almost instantaneously. Also known as backup arrangement. 2. *See* avalanche.

cascade amplifier A vacuum-tube amplifier containing two or more stages arranged in the conventional series manner. Also known as multistage amplifier.

cascade-amplifier klystron A klystron having three resonant cavities to provide increased power amplification and output; the extra resonator, located between the input and output resonators, is excited by the bunched beam emerging from the first resonator gap and produces further bunching of the beam.

cascade connection A series connection of amplifier stages, networks, or tuning circuits in which the output of one feeds the input of the next. Also known as tandem connection.

cascaded feedback canceler Sophisticated moving-target-indicator canceler which provides clutter and chaff rejection. Also known as velocity shaped canceler.

cascade image tube An image tube having a number of sections stacked together, the output image of one section serving as the input for the next section; used for light detection at very low levels.

cascade junction Two *pn* semiconductor junctions in tandem such that the condition of the first governs that of the second.

cascade limiter A limiter circuit that uses two vacuum tubes in series to give improved limiter operation for both weak and strong signals in a frequency-modulation receiver. Also known as double limiter.

cascade networks Two networks in tandem such that the output of the first feeds the input of the second.

cascade noise The noise in a communications receiver after an input signal has been subjected to two tandem stages of amplification.

cascade transformer A source of high voltage that is made up of a collection of step-up transformers; secondary windings are in series, and primary windings, except the first, are supplied from a pair of taps on the secondary winding of the preceding transformer.

cascading An effect in which a failure of an electrical power system causes this system to draw excessive amounts of power from power systems which are interconnected with it, causing them to fail, and these systems cause adjacent systems to fail in a similar manner, and so forth.

cascode amplifier An amplifier consisting of a grounded-cathode input stage that drives a grounded-grid output stage; advantages include high gain and low noise; widely used in television tuners.

Casimir–du Pré theory A theory of spin-lattice relaxation which treats the lattice and spin systems as distinct thermodynamic systems in thermal contact with one another.

Cassegrain antenna A microwave antenna in which the feed radiator is mounted at or near the surface of the main reflector and aimed at a mirror at the focus; energy from the feed first illuminates the mirror, then spreads outward to illuminate the main reflector.

catalog-order device A logic circuit element that is readily obtainable from a manufacturer, and can be combined wih other such elements to provide a wide variety of logic circuits.

catcher Electrode in a velocity-modulated vacuum tube on which the spaced electron groups induce a signal; the output of the tube is taken from this element.

catching diode Diode connected to act as a short circuit when its anode becomes positive; the diode then prevents the voltage of a circuit terminal from rising above the diode cathode voltage.

cathode 1. The positively charged pole of a primary cell or a storage battery. 2. The primary source of electrons in an electron tube; in directly heated tubes the filament is the cathode, and in indirectly heated tubes a coated metal cathode surrounds a heater. Designated K. Also known as negative electrode. 3. The terminal of a semiconductor diode that is negative with respect to the other terminal when the diode is biased in the forward direction.

cathode bias Bias obtained by placing a resistor in the common cathode return circuit, between cathode and ground; flow of electrode currents through this resistor produces a voltage drop that serves to make the control grid negative with respect to the cathode.

cathode-coupled amplifier A cascade amplifier in which the coupling between two stages is provided by a common cathode resistor.

cathode coupling Use of an input or output element in the cathode circuit for coupling energy to another stage.

cathode dark space The relatively nonluminous region between the cathode glow and the negative flow in a glow-discharge cold-cathode tube. Also known as Crookes dark space; Hittorf dark space.

cathode disintegration The destruction of the active area of a cathode by positive-ion bombardment.

cathode drop The voltage between the arc stream and the cathode of a glow-discharge tube. Also known as cathode fall.

cathode emission A process whereby electrons are emitted from the cathode structure.

cathode fall *See* cathode drop.

cathode follower A vacuum-tube circuit in which the input signal is applied between the control grid and ground, and the load is connected between the cathode and ground. Also known as grounded-anode amplifier; grounded-plate amplifier.

cathode glow The luminous glow that covers all or part of the cathode in a glow-discharge cold-cathode tube.

cathode interface capacitance A capacitance which, when connected in parallel with an appropriate resistance, forms an impedance approximately equal to the cathode interface impedance. Also known as layer capacitance.

cavity 53

cathode interface impedance The impedance between the cathode base and coating in an electron tube, due to a high-resistivity layer or a poor mechanical bond. Also known as layer impedance.

cathode keying Transmitter keying by means of a key in the cathode lead of the keyed vacuum-tube stage, opening the direct-current circuits for the grid and anode simultaneously.

cathode modulation Amplitude modulation accomplished by applying the modulating voltage to the cathode circuit of an electron tube in which the carrier is present.

cathode ray A stream of electrons, such as that emitted by a heated filament in a tube, or that emitted by the cathode of a gas-discharge tube when the cathode is bombarded by positive ions.

cathode-ray oscillograph A cathode-ray oscilloscope in which a photographic or other permanent record is produced by the electron beam of the cathode-ray tube.

cathode-ray oscilloscope A test instrument that uses a cathode-ray tube to make visible on a fluorescent screen the instantaneous values and waveforms of electrical quantities that are rapidly varying as a function of time or another quantity. Abbreviated CRO. Also known as oscilloscope; scope.

cathode-ray storage tube A storage tube in which the information is written by means of a cathode-ray beam.

cathode-ray tube An electron tube in which a beam of electrons can be focused to a small area and varied in position and intensity on a surface. Abbreviated CRT. Originally known as Braun tube; also known as electron-ray tube.

cathode-ray tuning indicator A small cathode-ray tube having a fluorescent pattern whose size varies with the voltage applied to the grid; used in radio receivers to indicate accuracy of tuning and as a modulation indicator in some tape recorders. Also known as electric eye; electron-ray indicator; magic eye; tuning eye.

cathode-ray voltmeter An instrument consisting of a cathode-ray tube of known sensitivity, whose deflection can be used to measure voltages.

cathode resistor A resistor used in the cathode circuit of a vacuum tube, having a resistance value such that the voltage drop across it due to tube current provides the correct negative grid bias for the tube.

cathode spot The small cathode area from which an arc appears to originate in a discharge tube.

cathodoluminescence Luminescence produced when high-velocity electrons bombard a metal in vacuum, thus vaporizing small amounts of the metal in an excited state, which amounts emit radiation characteristic of the metal. Also known as electroluminescence.

cathodophosphorescence Phosphorescence produced when high-velocity electrons bombard a metal in a vacuum.

CATT *See* controlled avalanche transit-time triode.

catwhisker A sharply pointed, flexible wire used to make contact with the surface of a semiconductor crystal at a point that provides rectification.

Cauchy relations A set of six relations between the compliance constants of a solid which should be satisfied provided the forces between atoms in the solid depend only on the distances between them and act along the lines joining them, and provided that each atom is a center of symmetry in the lattice.

Cauer form A continued fraction expansion of the impedance used in the network synthesis for a driving point function resulting in a ladder network.

cavity *See* cavity resonator.

cavity coupling The extraction of electromagnetic energy from a resonant cavity, either waveguide or coaxial, using loops, probes, or apertures.

cavity filter A microwave filter that uses quarter-wavelength-coupled cavities inserted in waveguides or coaxial lines to provide band-pass or other response characteristics at frequencies in the gigahertz range.

cavity impedance The impedance of the cavity of a microwave tube which appears across the gap between the cathode and the anode.

cavity magnetron A magnetron having a number of resonant cavities forming the anode; used as a microwave oscillator.

cavity oscillator An ultra-high-frequency oscillator whose frequency is controlled by a cavity resonator.

cavity resonance The resonant vibration of a cavity.

cavity resonator A space totally enclosed by a metallic conductor and excited in such a way that it becomes a source of electromagnetic oscillations. Also known as cavity; microwave cavity; microwave resonance cavity; resonant cavity; resonant chamber; resonant element; rhumbatron; tuned cavity; waveguide resonator.

cavity tuning Use of an adjustable cavity resonator as a tuned circuit in an oscillator or amplifier, with tuning usually achieved by moving a metal plunger in or out of the cavity to change the volume, and hence the resonant frequency of the cavity.

C-band waveguide A rectangular waveguide, with dimensions 3.48 by 1.58 centimeters, which is used to excite only the dominant mode (TE_{01}) for wavelengths in the range 3.7–5.1 centimeters.

C battery The battery that supplies the steady bias voltage required by the control-grid electrodes of electron tubes in battery-operated equipment. Also known as grid battery.

C bias *See* grid bias.

C core A spirally wound magnetic core that is formed to a desired rectangular shape before being cut into two C-shaped pieces and placed around a transformer or magnetic amplifier coil.

C display In radar, a rectangular display in which targets appear as blips with bearing indicated by the horizontal coordinate, and angles of elevation by the vertical coordinate.

cell-type tube Gas-filled radio-frequency switching tube which operates in an external resonant circuit; a tuning mechanism may be incorporated in either the external resonant circuit or the tube.

center-coupled loop Coupling loop in the center of one of the resonant cavities of a multicavity magnetron.

centering control One of the two controls used for positioning the image on the screen of a cathode-ray tube; either the horizontal centering control or the vertical centering control.

center tap A terminal at the electrical midpoint of a resistor, coil, or other device. Abbreviated CT.

centimeter *See* abhenry; statfarad.

central office line *See* subscriber line.

ceramagnet A ferrimagnet composed of the hard magnetic material $BaO \cdot 6Fe_2O_3$.

ceramic amplifier An amplifier that utilizes the piezoelectric properties of semiconductors such as silicon.

ceramic-based microcircuit A microminiature circuit printed on a ceramic substrate.

ceramic capacitor A capacitor whose dielectric is a ceramic material such as steatite or barium titanate, the composition of which can be varied to give a wide range of temperature coefficients.

ceramic magnet A permanent magnet made from pressed and sintered mixtures of ceramic and magnetic powders. Also known as ferromagnetic ceramic.

ceramic tube An electron tube having a ceramic envelope capable of withstanding operating temperatures over 500°C, as required during reentry of guided missiles.

Cerenkov radiation Light emitted by a high-speed charged particle when the particle passes through a transparent, nonconducting material at a speed greater than the speed of light in the material.

Cerenkov rebatron radiator Device in which a tightly bunched, velocity-modulated electron beam is passed through a hole in a dielectric; the reaction between the higher velocity of the electrons passing through the hole and the slower velocity of the electromagnetic energy passing through the dielectric results in radiation at some frequency higher than the frequency of modulation of the electron beam.

cermet resistor A metal-glaze resistor, consisting of a mixture of finely powdered precious metals and insulating materials fired onto a ceramic substrate.

cesium-antimonide photocathode A photocathode obtained by exposing a thin layer of antimony to cesium vapor at elevated temperatures; has a maximum sensitivity in the blue and ultraviolet regions of the spectrum.

cesium-beam sputter source A source of negative ions in which a beam of positive cesium ions, accelerated through a potential difference on 20–30 kilovolts, sputters the cesium-coated inner surface of a hollow cone fabricated from or containing the element whose negative ion is required, and an appreciable fraction of the negative ions leaving the surface are extracted from the rear hole of the sputter cone.

cesium beam tube *See* cesium electron tube.

cesium electron tube An electronic device used as an atomic clock, producing electromagnetic energy that is accurate and stable in frequency. Also known as cesium beam tube.

cesium hollow cathode A cathode in which cesium is heated at the bottom of a cylinder serving as the cathode of an electron tube, to give current densities that can be as high as 800 amperes per square centimeter.

cesium phototube A phototube having a cesium-coated cathode; maximum sensitivity in the infrared portion of the spectrum.

cesium thermionic converter A thermionic diode in which cesium vapor is stored between the plates to neutralize space charge and to lower the work function of the emitter.

cesium-vapor lamp A lamp in which light is produced by the passage of current between two electrodes in ionized cesium vapor.

cesium-vapor Penning source A conventional Penning source modified for negative-ion generation through the introduction or a third, sputter cathode, made from or containing the element of interest, which is the source of negative ions, and through the introduction of cesium vapor into the arc chamber.

cesium-vapor rectifier A gas tube in which cesium vapor serves as the conducting gas and a condensed monatomic layer of cesium serves as the cathode coating.

challenger *See* interrogator.

changeover switch A means of moving a circuit from one set of connections to another.

channel 1. A path for a signal, as an audio amplifier may have several input channels. 2. The main current path between the source and drain electrodes in a field-effect transistor or other semiconductor device.

channel bank Part of a carrier-multiplex terminal that performs the first step of modulation of the transmitting voice frequencies into a higher-frequency band, and the final step in the demodulation of the received higher-frequency band into the received voice frequencies.

channel effect A leakage current flowing over a surface path between the collector and emitter in some types of transistors.

channel selector A control used to tune in the desired channel in a radio or television receiver.

channel shifter Radiotelephone carrier circuit that shifts one or two voice-frequency channels from normal channels to higher voice-frequency channels to reduce cross talk between channels; the channels are shifted back by a similar circuit at the receiving end.

channel synchronizer An electronic device providing the proper interface between the central processing unit and the peripheral devices.

characteristic temperature *See* Debye temperature.

character-writing tube A cathode-ray tube that forms alphanumeric and symbolic characters on its screen for viewing or recording purposes.

charge 1. A basic property of elementary particles of matter; the charge of an object may be a positive or negative number or zero; only integral multiples of the proton charge occur, and the charge of a body is the algebraic sum of the charges of its constituents; the value of the charge may be inferred from the Coulomb force between charged objects. Also known as electric charge. 2. To convert electrical energy to chemical energy in a secondary battery. 3. To feed electrical energy to a capacitor or other device that can store it.

charge carrier A mobile conduction electron or mobile hole in a semiconductor. Also known as carrier.

charge conservation *See* conservation of charge.

charge-coupled devices Semiconductor devices arrayed so that the electric charge at the output of one provides the input stimulus to the next.

charge-coupled image sensor A device in which charges are introduced when light from a scene is focused on the surface of the device; image points are accessed sequentially to produce a television-type output signal. Also known as solid-state image sensor.

charge density The charge per unit area on a surface or per unit volume in space.

charge-density wave The ground state of a metal in which the conduction–electron charge density is sinusoidally modulated in space.

charge exchange source A source of negative ions, generally negative helium ions, in which positive ions generated in a duoplasmatron are directed through a donor canal, usually containing lithium vapor, where they pick up sequentially two electrons to form negative ions.

charge-injection device A charge-transfer device used as an image sensor in which the image points are accessed by reference to their horizontal and vertical coordinates. Abbreviated CID.

charge-mass ratio The ratio of the electric charge of a particle to its mass.

charge neutrality The condition in which electrons and holes are present in equal numbers in a semiconductor.

chip circuit 57

charge quantization The principle that the electric charge of an object must equal an integral multiple of a universal basic charge.

charger *See* battery charger.

charger-eliminator A battery charger with a low-noise, low-impedance output which can either charge a storage battery or supply a dc load directly, without a storage battery in parallel.

charge-state process A process involving the motion of preexisting crystal defects in a solid, following a change in the charges of the defects.

charge-storage transistor A transistor in which the collector-base junction will charge when forward bias is applied with the base at a high level and the collector at a low level.

charge-storage tube A storage tube in which information is retained on a surface in the form of electric charges.

charge-storage varactor A varactor that uses semiconductor techniques to achieve power outputs above 50 watts at ultra-high and microwave frequencies.

charge-transfer device A semiconductor device that depends upon movements of stored charges between predetermined locations, as in charge-coupled and charge-injection devices.

charging current The current that flows into a capacitor when a voltage is first applied.

chassis ground A connection to the metal chassis on which the components of a circuit are mounted, to serve as a common return path to the power source.

chatter Prolonged undesirable opening and closing of electric contacts, as on a relay. Also known as contact chatter.

Chebyshev filter A filter in which the transmission frequency curve has an equal-ripple shape, with very small peaks and valleys.

cheese antenna An antenna having a parabolic reflector between two metal plates, dimensioned to permit propagation of more than one mode in the desired direction of polarization.

chemical film dielectric An extremely thin layer of material on one or both electrodes of an electrolytic capacitor, which conducts electricity in only one direction and thereby constitutes the insulating element of the capacitor.

Child-Langmuir equation *See* Child's law.

Child-Langmuir-Schottky equation *See* Child's law.

Child's law A law stating that the current in a thermionic diode varies directly with the three-halves power of anode voltage and inversely with the square of the distance between the electrodes, provided the operating conditions are such that the current is limited only by the space charge. Also known as Child-Langmuir equation; Child-Langmuir-Schottky equation; Langmuir-Child equation.

chimney A pipelike enclosure that is placed over a heat sink to improve natural upward convection of heat and thereby increase the dissipating ability of the sink.

chip 1. The shaped and processed semiconductor die that is mounted on a substrate to form a transistor, diode, or other semiconductor device. 2. An integrated microcircuit performing a significant number of functions and constituting a subsystem.

chip capacitor A single-layer or multilayer monolithic capacitor constructed in chip form, with metallized terminations to facilitate direct bonding on hybrid integrated circuits.

chip circuit *See* large-scale integrated circuit.

58　chip resistor

chip resistor　A thick-film resistor constructed in chip form, with metallized terminations to facilitate direct bonding on hybrid integrated circuits.

choke　**1.** An inductance used in a circuit to present a high impedance to frequencies above a specified frequency range without appreciably limiting the flow of direct current. Also known as choke coil. **2.** A groove or other discontinuity in a waveguide surface so shaped and dimensioned as to impede the passage of guided waves within a limited frequency range.

choke coil　*See* choke.

choke coupling　Coupling between two parts of a waveguide system that are not in direct mechanical contact with each other.

choke filter　*See* choke input filter.

choke flange　A waveguide flange having in its mating surface a slot (choke) so shaped and dimensioned as to restrict leakage of microwave energy within a limited frequency range.

choke input filter　A power-supply filter in which the first filter element is a series choke. Also known as choke filter.

choke joint　A connection between two waveguides that uses two mating choke flanges to provide effective electrical continuity without metallic continuity at the inner walls of the waveguide.

choke piston　A piston in which there is no metallic contact with the walls of the waveguide at the edges of the reflecting surface; the short circuit for high-frequency currents is achieved by a choke system. Also known as noncontacting piston; noncontacting plunger.

chopper amplifier　A carrier amplifier in which the direct-current input is filtered by a low-pass filter, then converted into a square-wave alternating-current signal by either one or two choppers.

chopper-stabilized amplifier　A direct-current amplifier in which a direct-coupled amplifier is in parallel with a chopper amplifier.

chopper transistor　A bipolar or field-effect transistor operated as a repetitive "on/off" switch to produce square-wave modulation of an input signal.

chopping　The removal, by electronic means, of one or both extremities of a wave at a predetermined level.

chroma band-pass amplifier　*See* burst amplifier.

chroma control　The control that adjusts the amplitude of the carrier chrominance signal fed to the chrominance demodulators in a color television receiver, so as to change the saturation or vividness of the hues in the color picture. Also known as color control; color-saturation control.

chroma oscillator　A crystal oscillator used in color television receivers to generate a 3.579545-megahertz signal for comparison with the incoming 3.579545-megahertz chrominance subcarrier signal being transmitted. Also known as chrominance-subcarrier oscillator; color oscillator; color-subcarrier oscillator.

chromatic aberration　An electron-gun defect causing enlargement and blurring of the spot on the screen of a cathode-ray tube, because electrons leave the cathode with different initial velocities and are deflected differently by the electron lenses and deflection coils.

chromatron　A single-gun color picture tube having color phosphors deposited on the screen in strips instead of dots. Also known as Lawrence tube.

chrominance demodulator　A demodulator used in a color television receiver for deriving the I and Q components of the chrominance signal from the chrominance

circuit conditioning 59

signal and the chrominance-subcarrier frequency. Also known as chrominance-subcarrier demodulator.

chrominance gain control Variable resistors in red, green, and blue matrix channels that individually adjust primary signal levels in color television.

chrominance modulator A modulator used in a color television transmitter to generate the chrominance signal from the video-frequency chrominance components and the chrominance subcarrier. Also known as chrominance-subcarrier modulator.

chrominance-subcarrier demodulator *See* chrominance demodulator.

chrominance-subcarrier modulator *See* chrominance modulator.

chrominance-subcarrier oscillator *See* chroma oscillator.

chrominance video signal Voltage output from the red, green, or blue section of a color television camera or receiver matrix.

chromium dioxide tape A magnetic recording tape developed primarily to improve quality and brilliance of reproduction when used in cassettes operated at 1⅞ inches per second (4.76 centimeters per second); requires special recorders that provide high bias.

chromium-gold metallizing A metal film used on a silicon or silicon oxide surface in semiconductor devices because it is not susceptible to purple plague deterioration; a layer of chromium is applied first for adherence to silicon, then a layer of chromium-gold mixture, and finally a layer of gold to which bonding contacts can be applied.

chronistor A subminiature elapsed-time indicator that uses electroplating principles to totalize operating time of equipment up to several thousand hours.

chronometric encoder An encoder that uses an electronic counter to time or count electrical events and deliver in digital form a number equivalent to the input magnitude.

chronopher Instrument for emitting standard time signal impulses from a standard clock or timing device.

chronotron A device that measures millimicrosecond time intervals between pulses on a transmission line to determine the time between the events which initiated the pulses.

CID *See* charge-injection device.

C indicator *See* C scope.

ciphony equipment Any equipment attached to a radio transmitter, radio receiver, or telephone for scrambling or unscrambling voice messages.

circle diagram A diagram which gives a graphical solution of equations for a transmission line, giving the input impedance of the line as a function of load impedance and electrical length of the line.

circle-dot mode Mode of cathode-ray storage of binary digits in which one kind of digit is represented by a small circle of excitation of the screen, and the other kind by a similar circle with a concentric dot.

circuit 1. A complete wire, radio, or carrier communications channel. 2. *See* electric circuit.

circuit breaker An electromagnetic device that opens a circuit automatically when the current exceeds a predetermined value.

circuit conditioning Test, analysis, engineering, and installation actions to upgrade a communications circuit to meet an operational requirement; includes the reduction of noise, the equalization of phase and level stability and frequency response, and

60 circuit design

the correction of impedance discontinuities, but does not include normal maintenance and repair activities.

circuit design The art of specifying the components and interconnections of an electrical network.

circuit diagram A drawing, using standardized symbols, of the arrangement and interconnections of the conductors and components of an electrical or electronic device or installation. Also known as schematic circuit diagram; wiring diagram.

circuit efficiency Of an electron tube, the power delivered to a load at the output terminals of the output circuit at a desired frequency divided by the power delivered by the electron stream to the output circuit at that frequency.

circuit element *See* component.

circuit interrupter A device in a circuit breaker to remove energy from an arc in order to extinguish it.

circuit loading Power drawn from a circuit by an electric measuring instrument, which may alter appreciably the quantity being measured.

circuit protection Provision for automatically preventing excess or dangerous temperatures in a conductor and limiting the amount of energy liberated when an electrical failure occurs.

circuitron Combination of active and passive components mounted in a single envelope like that used for tubes, to serve as one or more complete operating stages.

circuitry The complete combination of circuits used in an electrical or electronic system or piece of equipment.

circuit testing The testing of electric circuits to determine and locate an open circuit, or a short circuit or leakage.

circuit theory The mathematical analysis of conditions and relationships in an electric circuit. Also known as electric circuit theory.

circular antenna A folded dipole that is bent into a circle, so the transmission line and the abutting folded ends are at opposite ends of a diameter.

circular coil In eddy-current nondestructive tests, a type of test coil which surrounds an object.

circular current An electric current moving in a circular path.

circular electric wave A transverse electric wave for which the lines of electric force form concentric circles.

circular horn A circular-waveguide section that flares outward into the shape of a horn, to serve as a feed for a microwave reflector or lens.

circular magnetic wave A transverse magnetic wave for which the lines of magnetic force form concentric circles.

circular polarized loop vee Airborne communications antenna with an omnidirectional radiation pattern to provide optimum near-horizon communications coverage.

circular sweep generation The use of electronic circuits to provide voltage or current which causes an electron beam in a device such as a cathode-ray tube to move in a circular deflection path at constant speed.

circular waveguide A waveguide whose cross-sectional area is circular.

circulating memory A digital computer device that uses a delay line to store information in the form of a pattern of pulses in a train; the output pulses are detected electrically, amplified, reshaped, and reinserted in the delay line at the beginning. Also known as delay-line memory; delay-line storage; circulating storage.

circulating storage *See* circulating memory.

circulator A waveguide component having a number of terminals so arranged that energy entering one terminal is transmitted to the next adjacent terminal in a particular direction. Also known as microwave circulator.

clamp *See* clamping circuit.

clamper *See* direct-current restorer.

clamping The introduction of a reference level that has some desired relation to a pulsed waveform, as at the negative or positive peaks. Also known as direct-current reinsertion; direct-current restoration.

clamping circuit A circuit that reestablishes the direct-current level of a waveform; used in the dc restorer stage of a television receiver to restore the dc component to the video signal after its loss in capacitance-coupled alternating-current amplifiers, to reestablish the average light value of the reproduced image. Also known as clamp.

clamping diode A diode used to clamp a voltage at some point in a circuit.

clamp-on ammeter *See* snap-on ammeter.

clapper A hinged or pivoted relay armature.

Clapp oscillator A series-tuned Colpitts oscillator, having low drift.

Clark cell An early form of standard cell, having 1.433 volts at 15°C, now largely replaced by the Weston standard cell as a voltage standard.

class A amplifier 1. An amplifier in which the grid bias and alternating grid voltages are such that anode current in a specific tube flows at all times. 2. A transistor amplifier in which each transistor is in its active region for the entire signal cycle.

class AB amplifier 1. An amplifier in which the grid bias and alternating grid voltages are such that anode current in a specific tube flows for appreciably more than half but less than the entire electric cycle. 2. A transistor amplifier whose operation is class A for small signals and class B for large signals.

class A modulator A class A amplifier used to supply the necessary signal power to modulate a carrier.

class B amplifier 1. An amplifier in which the grid bias is approximately equal to the cutoff value, so that anode current is approximately zero when no exciting grid voltage is applied, and flows for approximately half of each cycle when an alternating grid voltage is applied. 2. A transistor amplifier in which each transistor is in its active region for approximately half the signal cycle.

class B auxiliary power Standby power plant to cover extended outages (days) of primary power.

class B modulator A class B amplifier used to supply the necessary signal power to modulate a carrier; usually connected in push-pull.

class C amplifier 1. An amplifier in which the bias on the control element is appreciably greater than the cutoff valve, so that the output current in each device is zero when no alternating control signal is applied, and flows for appreciably less than half of each cycle when an alternating control signal is applied. 2. A transistor amplifier in which each transistor is in its active region for significantly less than half the signal cycle.

class C auxiliary power Quick start (10–60 seconds) power unit to cover short-term outages (hours) of primary power.

class D auxiliary power Uninterruptible (no-break) power unit using stored energy to provide continuous power within specified voltage and frequency tolerances.

classical electron radius The quantity expressed as $e^2/m_e c^2$, where e is the electron's charge in electrostatic units, m its mass, and c the speed of light; equal to approximately 2.82×10^{-13} centimeter.

Clausius-Mossotti equation An expression for the polarizability γ of an individual molecule in a medium which has the relative dielectric constant ϵ and has N molecules per unit volume: $\gamma = (3/4\pi N) [(\epsilon - 1)/(\epsilon + 2)]$ (Gaussian units).

Clausius-Mossotti-Lorentz-Lorenz equation The equation that results from replacing the real relative dielectric constant in the Clausius-Mossotti equation, or the real index of refraction in the Lorentz-Lorenz equation, with its complex counterpart.

cleanup Gradual disappearance of gases from an electron tube during operation, due to absorption by getter material or the tube structure.

clipper *See* limiter.

clipper diode A bidirectional breakdown diode that clips signal voltage peaks of either polarity when they exceed a predetermined amplitude.

clipper-limiter A device whose output is a function of the instantaneous input amplitude for a range of values lying between two predetermined limits but is approximately constant, at another level, for input values above the range.

clipping *See* limiting.

clipping circuit *See* limiter.

clipping level The level at which a clipping circuit is adjusted; for example, the magnitude of the clipped wave shape.

clock A source of accurately timed pulses, used for synchronization in a digital computer or as a time base in a transmission system.

clocked flip-flop A flip-flop circuit that is set and reset at specific times by adding clock pulses to the input so that the circuit is triggered only if both trigger and clock pulses are present simultaneously.

clocked logic A logic circuit in which the switching action is controlled by repetitive pulses from a clock.

clock frequency The master frequency of the periodic pulses that schedule the operation of a digital computer.

clock motor *See* timing motor.

clock oscillator An oscillator that controls an electronic clock.

clock rate The rate at which bits or words are transferred from one internal element of a computer to another.

close coupling 1. The coupling obtained when the primary and secondary windings of a radio-frequency or intermediate-frequency transformer are close together. 2. A degree of coupling that is greater than critical coupling. Also known as tight coupling.

closed circuit A complete path for current.

closed-cycle fuel cell A fuel cell in which the reactants are regenerated by an auxiliary process, such as electrolysis.

closed-loop voltage gain The voltage gain of an amplifier with feedback.

closed magnetic circuit A complete circulating path for magnetic flux around a core of ferromagnetic material.

cloud attenuation The attenuation of microwave radiation by clouds (for the centimeter-wavelength band, clouds produce Rayleigh scattering); due largely to scattering, rather than absorption, for both ice and water clouds.

coaxial diode   63

cloud pulse The output resulting from space charge effects produced by turning the electron beam on or off in a charge-storage tube.

cloverleaf antenna Antenna having radiating units shaped like a four-leaf clover.

clutter Unwanted echoes on a radar screen, such as those caused by the ground, sea, rain, stationary objects, chaff, enemy jamming transmissions, and grass. Also known as background return; radar clutter.

clutter gating A technique which provides switching between moving-target-indicator and normal videos; this results in normal video being displayed in regions with no clutter and moving-target-indicator video being switched in only for the clutter areas.

CMOS device A device formed by the combination of a PMOS (p-type-channel metal oxide semiconductor device) with an NMOS (n-type-channel metal oxide semiconductor device). Derived from complementary metal oxide semiconductor device.

C network Network composed of three impedance branches in series, the free ends being connected to one pair of terminals, and the junction points being connected to another pair of terminals.

coastal refraction An apparent change in the direction of travel of a radio wave when it crosses a shoreline obliquely. Also known as land effect.

coated cathode A cathode that has been coated with compounds to increase electron emission.

coated filament A vacuum-tube filament coated with metal oxides to provide increased electron emission.

coax *See* coaxial cable.

coaxial antenna An antenna consisting of a quarter-wave extension of the inner conductor of a coaxial line and a radiating sleeve that is in effect formed by folding back the outer conductor of the coaxial line for a length of approximately a quarter wavelength.

coaxial attenuator An attenuator that has a coaxial construction and terminations suitable for use with coaxial cable.

coaxial bolometer A bolometer in which the desired square-law detection characteristic is provided by a fine Wollaston wire element that has been thoroughly cleaned before being axially located and soldered in position in its cylinder.

coaxial cable A transmission line in which one conductor is centered inside and insulated from an outer metal tube that serves as the second conductor. Also known as coax; coaxial line; coaxial transmission line; concentric cable; concentric line; concentric transmission line.

coaxial capacitor *See* cylindrical capacitor.

coaxial cavity A cylindrical resonating cavity having a central conductor in contact with its pistons or other reflecting devices.

coaxial cavity magnetron A magnetron which achieves mode separation, high efficiency, stability, and ease of mechanical tuning by coupling a coaxial high Q cavity to a normal set of quarter-wavelength vane cavities.

coaxial connector An electric connector between a coaxial cable and an equipment circuit, so constructed as to maintain the conductor configuration, through the separable connection, and the characteristic impedance of the coaxial cable.

coaxial cylinder magnetron A magnetron in which the cathode and anode consist of coaxial cylinders.

coaxial diode A diode having the same outer diameter and terminations as a coaxial cable, or otherwise designed to be inserted in a coaxial cable.

64   coaxial filter

coaxial filter A section of coaxial line having reentrant elements that provide the inductance and capacitance of a filter section.

coaxial hybrid A hybrid junction of coaxial transmission lines.

coaxial isolator An isolator used in a coaxial cable to provide a higher loss for energy flow in one direction than in the opposite direction; all types use a permanent magnetic field in combination with ferrite and dielectric materials.

coaxial line *See* coaxial cable.

coaxial line resonator A resonator consisting of a length of coaxial line short-circuited at one or both ends.

coaxially fed linear array A beacon antenna having a uniform azimuth pattern.

coaxial relay A relay designed for opening or closing a coaxial cable circuit without introducing a mismatch that would cause wave reflections.

coaxial stub A length of nondissipative cylindrical waveguide or coaxial cable branched from the side of a waveguide to produce some desired change in its characteristics.

coaxial switch A switch that changes connections between coaxial cables going to antennas, transmitters, receivers, or other high-frequency devices without introducing impedance mismatch.

coaxial transistor A point-contact transistor in which the emitter and collector are point electrodes making pressure contact at the centers of opposite sides of a thin disk of semiconductor material serving as base.

coaxial transmission line *See* coaxial cable.

cobinotron The combination of a corbino disk and a coil arranged to produce a magnetic field perpendicular to the disk.

codan A device that silences a receiver except when a modulated carrier signal is being received.

coded passive reflector antenna An object intended to reflect Hertzian waves and having variable reflecting properties according to a predetermined code for the purpose of producing an indication on a radar receiver.

code practice oscillator An oscillator used with a key and either headphones or a loudspeaker to practice sending and receiving Morse code.

codistor A multijunction semiconductor device which provides noise rejection and voltage regulation functions.

coercive force The magnetic field H which must be applied to a magnetic material in a symmetrical, cyclicly magnetized fashion, to make the magnetic induction B vanish. Also known as magnetic coercive force.

coercivity The coercive force of a magnetic material in a hysteresis loop whose maximum induction approximates the saturation induction.

cogging Variations in torque and speed of an electric motor due to variations in magnetic flux as rotor poles move past stator poles.

cohered video The video detector output signal in a coherent moving-target indicator radar system.

coherence distance *See* coherence length.

coherence length A measure of the distance through which the effect of any local disturbance is spread out in a superconducting material. Also known as coherence distance.

coherent detector A detector used in moving-target indicator radar to give an output-signal amplitude that depends on the phase of the echo signal instead of on its strength, as required for a display that shows only moving targets.

cold cathode   65

coherent echo   A radar echo whose phase and amplitude at a given range remain relatively constant.

coherent oscillator   An oscillator used in moving-target indicator radar to serve as a reference by which changes in the radio-frequency phase of successively received pulses may be recognized. Abbreviated coho.

coherent-pulse radar   A radar in which the radio-frequency oscillations of recurrent pulses bear a constant phase relation to those of a continuous oscillation.

coherent pulses   Characterizing pulses in which the phase of the radio-frequency waves is maintained through successive pulses.

coherent reference   A reference signal, usually of stable frequency, to which other signals are phase-locked to establish coherence throughout a system.

coherent signal   In a pulsed radar system, a signal having a constant phase; it is mixed with the echo signal, whose phase depends upon the range of the target, in order to detect the phase shift and measure the target's range.

coherent transponder   A transponder in which a fixed relation between frequency and phase of input and output signals is maintained.

coherent video   The video signal produced in a moving-target indicator system by combining a radar echo signal with the output of a continuous-wave oscillator; after delay, this signal is detected, amplified, and subtracted from the next pulse train to give a signal representing only moving targets.

coherer   A cell containing a granular conductor between two electrodes; the cell becomes highly conducting when it is subjected to an electric field, and conduction can then be stopped only by jarring the granules.

cohesive energy   The difference between the energy per atom of a system of free atoms at rest far apart from each other, and the energy of the solid.

coho   *See* coherent oscillator.

coil   A number of turns of wire used to introduce inductance into an electric circuit, to produce magnetic flux, or to react mechanically to a changing magnetic flux; in high-frequency circuits a coil may be only a fraction of a turn. Also known as electric coil; inductance; inductance coil; inductor.

coil antenna   An antenna that consists of one or more complete turns of wire.

coil form   The tubing or spool of insulating material on which a coil is wound.

coil neutralization   *See* inductive neutralization.

coincidence amplifier   An electronic circuit that amplifies only that portion of a signal present when an enabling or controlling signal is simultaneously applied.

coincidence circuit   A circuit that produces a specified output pulse only when a specified number or combination of two or more input terminals receives pulses within an assigned time interval. Also known as coincidence counter; coincidence gate.

coincidence counter   *See* coincidence circuit.

coincidence gate   *See* coincidence circuit.

coincident-current selection   The selection of a particular magnetic cell, for reading or writing in computer storage, by simultaneously applying two or more currents.

cold   Pertaining to electrical circuits that are disconnected from voltage supplies and at ground potential; opposed to hot, pertaining to carrying an electrical charge.

cold cathode   A cathode whose operation does not depend on its temperature being above the ambient temperature.

66 cold-cathode counter tube

cold-cathode counter tube A counter tube having one anode and three sets of 10 cathodes; two sets of cathodes serve as guides that direct the flow discharge to each of the 10 output cathodes in correct sequence in response to driving pulses.

cold-cathode discharge See glow discharge.

cold-cathode ionization gage See Philips ionization gage.

cold-cathode rectifier A cold-cathode gas tube in which the electrodes differ greatly in size so electron flow is much greater in one direction than in the other. Also known as gas-filled rectifier.

cold-cathode tube An electron tube containing a cold cathode, such as a cold-cathode rectifier, mercury-pool rectifier, neon tube, phototube, or voltage regulator.

cold junction The reference junction of thermocouple wires leading to the measuring instrument; normally at room temperature.

cold neutron A very-low-energy neutron in a reactor, used for research into solid-state physics because it has a wavelength of the order of crystal lattice spacings and can therefore be diffracted by crystals.

collector 1. A semiconductive region through which a primary flow of charge carriers leaves the base of a transistor; the electrode or terminal connected to this region is also called the collector. 2. An electrode that collects electrons or ions which have completed their functions within an electron tube; a collector receives electrons after they have done useful work, whereas an anode receives electrons whose useful work is to be done outside the tube. Also known as electron collector.

collector capacitance The depletion-layer capacitance associated with the collector junction of a transistor.

collector cutoff The reverse saturation current of the collector-base junction.

collector junction A semiconductor junction located between the base and collector electrodes of a transistor.

collector modulation Amplitude modulation in which the modulator varies the collector voltage of a transistor.

collector plate One of several metal inserts that are sometimes embedded in the lining of an electrolyte cell to make the resistance between the cell lining and the current leads as small as possible.

collector resistance The back resistance of the collector-base diode of a transistor.

collector voltage The direct-current voltage, obtained from a power supply, that is applied between the base and collector of a transistor.

colliding-beam source A device for generating beams of polarized negative hydrogen or deuterium ions, in which polarized negative hydrogen or deuterium atoms are converted to negative ions through charge exchange during collisions with cesium atoms.

collinear array See linear array.

collinear heterodyning An optical processing system in which the correlation function is developed from an ultrasonic light modulator; the output signal is derived from a reference beam in such a way that the two beams are collinear until they enter the detection aperture; variations in optical path length then modulate the phase of both signal and reference beams simultaneously, and phase differences cancel out in the heterodyning process.

color balance Adjustment of the circuits feeding the three electron guns of a television color picture tube to compensate for differences in light-emitting efficiencies of the three color phosphors on the screen of the tube.

color-bar generator A signal generator that delivers to the input of a color television receiver the signal needed to produce a color-bar test pattern on one or more channels.

color burst The portion of the composite color television signal consisting of a few cycles of a sine wave of chrominance subcarrier frequency. Also known as burst; reference burst.

color center A point lattice defect which produces optical absorption bands in an otherwise transparent crystal.

color code A system of colors used to indicate the electrical value of a component or to identify terminals and leads.

color coder *See* matrix.

color control *See* chroma control.

color decoder *See* matrix.

color-difference signal A signal that is added to the monochrome signal in a color television receiver to obtain a signal representative of one of the three tristimulus values needed by the color picture tube.

color encoder *See* matrix.

color fringing Spurious chromaticity at boundaries of objects in a television picture.

color killer circuit The circuit in a color television receiver that biases chrominance amplifier tubes to cutoff during reception of monochrome programs. Also known as killer stage.

color kinescope *See* color picture tube.

color oscillator *See* chroma oscillator.

color-phase detector The color television receiver circuit that compares the frequency and phase of the incoming burst signal with those of the locally generated 3.579545-megahertz chroma oscillator and delivers a correction voltage to a reactance tube to ensure that the color portions of the picture will be in exact register with the black-and-white portions on the screen.

color picture tube A cathode-ray tube having three different colors of phosphors, so that when these are appropriately scanned and excited in a color television receiver, a color picture is obtained. Also known as color kinescope; color television picture tube; tricolor picture tube.

color purity Absence of undesired colors in the spot produced on the screen by each beam of a television color picture tube.

color-saturation control *See* chroma control.

color-subcarrier oscillator *See* chroma oscillator.

color television picture tube *See* color picture tube.

Colpitts oscillator An oscillator in which a parallel-tuned tank circuit has two voltage-dividing capacitors in series, with their common connection going to the cathode in the electron-tube version and the emitter circuit in the transistor version.

coma A cathode-ray tube image defect that makes the spot on the screen appear comet-shaped when away from the center of the screen.

coma lobe Side lobe that occurs in the radiation pattern of a microwave antenna when the reflector alone is tilted back and forth to sweep the beam through space because the feed is no longer always at the center of the reflector; used to eliminate the need for a rotary joint in the feed waveguide.

comb antenna A broad-band antenna for vertically polarized signals, in which half of a fishbone antenna is erected vertically and fed against ground by a coaxial line.

comb filter A wave filter whose frequency spectrum consists of a number of equispaced elements resembling the teeth of a comb.

combinational circuit A switching circuit whose outputs are determined only by the concurrent inputs.

combination cable A cable having conductors grouped in both quads and pairs.

combination distributing frame Frame which combines the functions of a main distributing frame and an intermediate distributing frame.

combiner circuit The circuit that combines the luminance and chrominance signals with the synchronizing signals in a color television camera chain.

command pulses The electrical representations of bit values of 1 or 0 which control input/output devices.

common-base connection *See* grounded-base connection.

common-base feedback oscillator A bipolar transistor amplifier with a common-base connection and a positive feedback network between the collector (output) and the emitter (input).

common branch A branch of an electrical network which is common to two or more meshes. Also known as mutual branch.

common-collector connection *See* grounded-collector connection.

common-drain amplifier An amplifier using a field-effect transistor so that the input signal is injected between gate and drain, while the output is taken between the source and drain. Also known as source-follower amplifier.

common-emitter connection *See* grounded-emitter connection.

common-gate amplifier An amplifier using a field-effect transistor in which the gate is common to both the input circuit and the output circuit.

common impedance coupling The interaction of two circuits by means of an inductance or capacitance in a branch which is common to both circuits.

common mode Having signals that are identical in amplitude and phase at both inputs, as in a differential operational amplifier.

common-mode error The error voltage that exists at the output terminals of an operational amplifier due to the common-mode voltage at the input.

common-mode gain The ratio of the output voltage of a differential amplifier to the common-mode input voltage.

common-mode input capacitance The equivalent capacitance of both inverting and noninverting inputs of an operational amplifier with respect to ground.

common-mode input impedance The open-loop input impedance of both inverting and noninverting inputs of an operational amplifier with respect to ground.

common-mode input resistance The equivalent resistance of both inverting and noninverting inputs of an operational amplifier with respect to ground or reference.

common-mode rejection The ability of an amplifier to cancel a common-mode signal while responding to an out-of-phase signal. Also known as in-phase rejection.

common-mode signal A signal applied equally to both ungrounded inputs of a balanced amplifier stage or other differential device. Also known as in-phase signal.

common-mode voltage A voltage that appears in common at both input terminals of a device with respect to the output reference (usually ground).

common return A return conductor that serves two or more circuits.

common-source amplifier An amplifier stage using a field-effect transistor in which the input signal is applied between gate and source and the output signal is taken between drain and source.

common user circuit A circuit designated to furnish a communications service to a number of users.

communication receiver A receiver designed especially for reception of voice or code messages transmitted by radio communication systems.

communications zone indicator Device to indicate whether or not long-distance high-frequency broadcasts are successfully reaching their destinations.

commutating capacitor A capacitor used in gas-tube rectifier circuits to prevent the anode from going highly negative immediately after extinction.

commutating pole One of several small poles between the main poles of a direct-current generator or motor, which serves to neutralize the flux distortion in the neutral plane caused by armature reaction. Also known as compole; interpole.

commutating reactance An inductive reactance placed in the cathode lead of a three-phase mercury-arc rectifier to ensure that tube current holds over during transfer of conduction from one anode to the next.

commutating reactor A reactor found primarily in silicon controlled rectifier (SCR) converters where it is connected in series with a commutation capacitor to form a highly efficient resonant circuit used to cause a current oscillation which turns off (commutates) the conducting SCR.

commutating zone The part of the armature of an electric machine that contains the windings which are short-circuited by the brush on the commutator at a particular instant.

commutation 1. The transfer of current from one channel to another in a gas tube. 2. The process of current reversal in the armature windings of a direct-current rotating machine to provide direct current at the brushes.

commutator That part of a direct-current motor or generator which serves the dual function, in combination with brushes, of providing an electrical connection between the rotating armature winding and the stationary terminals, and of permitting reversal of the current in the armature windings.

commutator head The butt end of a commutator.

commutator motor An electric motor having a commutator.

commutator switch A switch, usually rotary and mechanically driven, that performs a set of switching operations in repeated sequential order, such as is required for telemetering many quantities. Also known as sampling switch; scanning switch.

companding A process in which compression is followed by expansion; often used for noise reduction in equipment, in which case compression is applied before noise exposure and expansion after exposure.

compandor A system for improving the signal-to-noise ratio by compressing the volume range of the signal at a transmitter or recorder by means of a compressor and restoring the normal range at the receiving or reproducing apparatus with an expander.

comparator circuit An electronic circuit that produces an output voltage or current whenever two input levels simultaneously satisfy predetermined amplitude requirements; may be linear (continuous) or digital (discrete).

comparing unit An electromechanical device which compares two groups of timed pulses and signals to establish either identity or nonidentity.

comparison bridge A bridge circuit in which any change in the output voltage with respect to a reference voltage creates a corresponding error signal, which, by means of negative feedback, is used to correct the output voltage and thereby restore bridge balance.

compatible monolithic integrated circuit Device in which passive components are deposited by thin-film techniques on top of a basic silicon-substrate circuit containing the active components and some passive parts.

compensated amplifier A broad-band amplifier in which the frequency range is extended by choice of circuit constants.

compensated-loop direction finder A direction finder employing a loop antenna and a second antenna system to compensate for polarization error.

compensated semiconductor Semiconductor in which one type of impurity or imperfection (for example, donor) partially cancels the electrical effects on the other type of impurity or imperfection (for example, acceptor).

compensated winding *See* pole-face winding.

compensating capacitor *See* balancing capacitor.

compensation The modification of the amplitude-frequency response of an amplifier to broaden the bandwidth or to make the response more nearly uniform over the existing bandwidth. Also known as frequency compensation.

compensator A component that offsets an error or other undesired effect.

complementary Having *pnp* and *npn* or *p*- and *n*-channel semiconductor elements on or within the same integrated-circuit substrate or working together in the same functional amplifier state.

complementary logic switch A complementary transistor pair which has a common input and interconnections such that one transistor is on when the other is off, and vice versa.

complementary metal oxide semiconductor device *See* CMOS device.

complementary symmetry A circuit using both *pnp* and *npn* transistors in a symmetrical arrangement that permits push-pull operation without an input transformer or other form of phase inverter.

complementary transistors Two transistors of opposite conductivity (*pnp* and *npn*) in the same functional unit.

complementary wave Wave brought into existence at the ends of a coaxial cable, or two-conductor transmission lines, or any discontinuity along the line.

complex impedance *See* electrical impedance.

complex permeability A property, designated by μ^*, of a magnetic material, equal to $\mu_0 \, (L/L_0)$, where L is the complex inductance of an inductance coil in which the magnetic material forms the core when the coil is connected to a sinusoidal voltage source, and L_0 is the vacuum inductance of the coil.

complex permittivity A property of a dielectric, equal to $\epsilon_0(C/C_0)$, where C is the complex capacitance of a capacitor in which the dielectric is the insulating material when the capacitor is connected to a sinusoidal voltage source, and C_0 is the vacuum capacitance of the capacitor.

complex relative attenuation The ratio of the peak output voltage, in complex notation, of an electric filter to the output voltage at the frequency being considered.

compole *See* commutating pole.

component Any electric device, such as a coil, resistor, capacitor, generator, line, or electron tube, having distinct electrical characteristics and having terminals at which it may be connected to other components to form a circuit. Also known as circuit element; element.

component symbol A graphical design used to represent a component in a circuit diagram.

composite balance An electric balance made by modifying the Kelvin balance to measure amperage, voltage, or wattage.

composite cable Cable in which conductors of different gages or types are combined under one sheath.

composite circuit A circuit used simultaneously for voice communication and telegraphy, with frequency-discriminating networks serving to separate the two types of signals.

composite filter A filter constructed by linking filters of different kinds in series.

composite pulse A pulse composed of a series of overlapping pulses received from the same source over several paths in a pulse navigation system.

composite set Assembly of apparatus designed to provide one end of a composite circuit.

composite wave filter A combination of two or more low-pass, high-pass, band-pass, or band-elimination filters.

composition resistor *See* carbon resistor.

compound cryosar A cryosar consisting of two normal cryosars with different electrical characteristics in series.

compound generator A direct-current generator which has both a series field winding and a shunt field winding, both on the main poles with the shunt field winding on the outside.

compound motor A direct-current motor with two separate field windings, one connected in parallel with the armature circuit, the other connected in series with the armature circuit.

compound winding A winding that is a combination of series and shunt winding.

compression 1. Reduction of the effective gain of a device at one level of signal with respect to the gain at a lower level of signal, so that weak signal components will not be lost in background and strong signals will not overload the system. 2. *See* compression ratio.

compression ratio The ratio of the gain of a device at a low power level to the gain at some higher level, usually expressed in decibels. Also known as compression.

compressive intercept receiver An electromagnetic surveillance receiver that instantaneously analyzes and sorts all signals within a broad radio-frequency spectrum by using pulse compression techniques which perform a complete analysis up to 10,000 times faster than a superheterodyne receiver or spectrum analyzer.

compressor The part of a compandor that is used to compress the intensity range of signals at the transmitting or recording end of a circuit.

compromise network 1. Network employed in conjunction with a hybrid coil to balance a subscriber's loop; adjusted for an average loop length or an average subscriber's set, or both, to secure compromise (not precision) isolation between the two directional paths of the hybrid. 2. Hybrid balancing network which is designed to balance the average of the impedances that may be connected to the switchboard side of a hybrid arrangement of a repeater.

concatenation A method of speed control of induction motors in which the rotors of two wound-rotor motors are mechanically coupled together and the stator of the second motor is supplied with power from the rotor slip rings of the first motor.

72 concentrator

concentrator Buffer switch (analog or digital) which to reduces the number of trunks required.

concentric cable *See* coaxial cable.

concentric line *See* coaxial cable.

concentric slip ring A large slip-ring assembly consisting of concentrically arranged insulators and conducting materials.

concentric transmission line *See* coaxial cable.

concentric windings Transformer windings in which the low-voltage winding is in the form of a cylinder next to the core, and the high-voltage winding, also cylindrical, surrounds the low-voltage winding.

condensation An increase of electric charge on a capacitor conductor.

condenser *See* capacitor.

condenser antenna *See* capacitor antenna.

condenser box *See* capacitor box.

condenser bushing An insulation made up of alternate layers of insulating material and metal foil placed between the conductor and outer casing in terminals of transformers and other high-voltage equipment such as switchgears.

conditionally stable circuit A circuit which is stable for certain values of input signal and gain, and unstable for other values.

conditioning Equipment modifications or adjustments necessary to match transmission levels and impedances or to provide equalization between facilities.

conductance The real part of the admittance of a circuit; when the impedance contains no reactance, as in a direct-current circuit, it is the reciprocal of resistance, and is thus a measure of the ability of the circuit to conduct electricity. Also known as electrical conductance. Designated G.

conductance-variation method A technique for measuring low admittances; measurements in a parallel-resonance circuit with the terminals open-circuited, with the unknown admittance connected, and then with the unknown admittance replaced by a known conductance standard are made; from them the unknown can be calculated.

conduction The passage of electric charge, which can occur by a variety of processes, such as the passage of electrons or ionized atoms. Also known as electrical conduction.

conduction band An energy band in which electrons can move freely in a solid, producing net transport of charge.

conduction current A current due to a flow of conduction electrons through a body.

conduction electron An electron in the conduction band of a solid, where it is free to move under the influence of an electric field. Also known as outer-shell electron; valence electron.

conduction field Energy surrounding a conductor when an electric current is passed through the conductor, which, because of the difference in phase between the electrical field and magnetic field set up in the conductor, cannot be detached from the conductor.

conductive coupling Electric connection of two electric circuits by their sharing the same resistor.

conductive gasket A flexible metallic gasket used to reduce radio-frequency leakage at joints in shielding.

conductivity The ratio of the electric current density to the electric field in a material. Also known as electrical conductivity; specific conductance.

conductivity bridge A modified Kelvin bridge for measuring very low resistances.

conductivity cell A glass vessel with two electrodes at a definite distance apart and filled with a solution whose conductivity is to be measured.

conductivity modulation Of a semiconductor, the variation of the conductivity of a semiconductor through variation of the charge carrier density.

conductivity modulation transistor Transistor in which the active properties are derived from minority carrier modulation of the bulk resistivity of the semiconductor.

conductivity tensor A tensor which, when multiplied by the electric field vector according to the rules of matrix multiplication, gives the current density vector.

conductor A wire, cable, or other body or medium that is suitable for carrying electric current.

conduit Solid or flexible metal or other tubing through which insulated electric wires are run.

cone of nulls In antenna practice, a conical surface formed by directions of negligible radiation.

configuration A group of components interconnected to perform a desired circuit function.

confocal resonator A wavemeter for millimeter wavelengths, consisting of two spherical mirrors facing each other; changing the spacing between the mirrors affects propagation of electromagnetic energy between them, permitting direct measurement of free-space wavelength.

conformable optical mask An optical mask made on a flexible glass substrate so that it can be pulled down under vacuum into intimate contact with the substrate for accurate circuit fabrication.

conformal array A circular, cylindrical, hemispherical, or other shaped array of electronically switched antennas; provides the special radiation patterns required for Tacan, IFF, and other air navigation, radar, and missile control applications.

conformal reflection chart An Argand diagram for plotting the complex reflection coefficient of a waveguide junction and its image, the two being related by a conformal transformation.

confusion jamming An electronic countermeasure technique in which the signal from an enemy tracking radar is amplified and retransmitted with distortion to create a false echo that affects accuracy of target range, azimuth, and velocity data.

conical antenna A wide-band antenna in which the driven element is conical in shape. Also known as cone antenna.

conical beam The radar beam produced by conical scanning methods.

conical helimagnet A helimagnet in which the directions of atomic magnetic moments all make the same angle (greater than 0° and less than 90°) with a specified axis of the crystal; moments of atoms in successive basal planes are separated by equal azimuthal angles, and all moments have the same magnitude.

conical-horn antenna A horn antenna having a circular cross section and straight sides.

conical scanning Scanning in radar in which the direction of maximum radiation generates a cone, the vertex angle of which is of the order of the beam width; may be either rotating or nutating, according to whether the direction of polarization rotates or remains unchanged.

conjugate branches Any two branches of an electrical network such that a change in the electromotive force in either branch does not result in a change in current in the other. Also known as conjugate conductors.

conjugate bridge A bridge in which the detector circuit and the supply circuits are interchanged, as compared with a normal bridge of the given type.

conjugate conductors See conjugate branches.

conjugate impedances Impedances having resistance components that are equal, and reactance components that are equal in magnitude but opposite in sign.

connected load The sum of the continuous power ratings of all load-consuming apparatus connected to an electric power distribution system or any part thereof.

connecting circuit A functional switching circuit which directly couples other functional circuit units to each other to exchange information as dictated by the momentary needs of the switching system.

connector A switch, or relay group system, which finds the telephone line being called as a result of digits being dialed; it also causes interrupted ringing voltage to be placed on the called line or of returning a busy tone to the calling party if the line is busy.

consequent poles Pairs of magnetic poles in a magnetized body that are in excess of the usual single pair.

conservation of charge A law which states that the total charge of an isolated system is constant; no violation of this law has been discovered. Also known as charge conservation.

console receiver A television or radio receiver in a console.

constant-conductance network See constant-resistance network.

constant-current characteristic The relation between the voltages of two electrodes in an electron tube when the current to one of them is maintained constant and all other electrode voltages are constant.

constant-current dc potentiometer A potentiometer in which the unknown electromotive force is balanced by a constant current times the resistance of a calibrated resistor or slide-wire. Also known as Poggendorff's first method.

constant-current filter A filter network intended to be connected to a source whose internal impedance is so high it can be assumed as infinite.

constant-current generator A vacuum-tube circuit, generally containing a pentode, in which the alternating-current anode resistance is so high that anode current remains essentially constant despite variations in load resistance.

constant-current source A circuit which produces a specified current, independent of the load resistance or applied voltage.

constant-current supply The power supply for repeatered submarine telephone cables; the voltage is varied automatically to maintain a constant current through the use of variable-voltage rectifiers and constant-current regulators at each shore station.

constant-current transformer A transformer that automatically maintains a constant current in its secondary circuit under varying loads, when supplied from a constant-voltage source.

constant-false-alarm rate Radar system devices used to prevent receiver saturation and overload so as to present clean video information to the display, and to present a constant noise level to an automatic detector.

constant-k filter A filter in which the product of the series and shunt impedances is a constant that is independent of frequency.

contact resistance 75

constant-k lens A microwave lens that is constructed as a solid dielectric sphere; a plane electromagnetic wave brought to a focus at one point on the sphere emerges from the opposite side of the sphere as a parallel beam.

constant-k network A ladder network in which the product of the series and shunt impedances is independent of frequency within the operating frequency range.

constant-resistance dc potentiometer A potentiometer in which the ratio of an unknown and a known potential are set equal to the ratio of two known constant resistances. Also known as Poggendorff's second method.

constant-resistance network A network having at least one driving-point impedance that is a positive constant. Also known as constant-conductance network.

constant-voltage generator An axle generator that is equipped with a regulator which keeps voltage constant.

constant-voltage transformer A power transformer which will supply a constant voltage to an unvarying load, even with changes in the primary voltage.

constitutive equations The equations $D = \epsilon E$ and $B = \mu H$, which relate the electric displacement D with the electric field intensity E, and the magnetic induction B with the magnetic field intensity H.

contact See electric contact.

contact block A block of conducting material such as carbon, used in a relay.

contact bounce The uncontrolled making and breaking of contact one or more times, but not continuously, when relay contacts are moved to the closed position.

contact chatter See chatter.

contact clip The clip which the blade of a knife switch is clamped to in the closed condition.

contact electricity An electric charge at the surface of contact of two different materials.

contact electromotive force See contact potential difference.

contact follow The distance two contacts travel together after just touching. Also known as contact overtravel.

contact force The force exerted by the moving contact of a switch or relay on a stationary contact.

contact-making meter See instrument-type relay.

contactor A heavy-duty relay used to control electric power circuits. Also known as electric contactor.

contact piston A waveguide piston that makes contact with the walls of the waveguide. Also known as contact plunger.

contact plunger See contact piston.

contact potential difference The potential difference that exists across the space between two electrically connected materials. Also known as contact electromotive force; contact potential; Volta effect.

contact pressure The amount of pressure holding a set of contacts together.

contact protection Any method for suppressing the surge which results when an inductive circuit is suddenly interrupted; the break would otherwise produce arcing at the contacts, leading to their deterioration.

contact rectifier See metallic rectifier.

contact resistance The resistance in ohms between the contacts of a relay, switch, or other device when the contacts are touching each other.

continuity Continuous effective contact of all components of an electric circuit to give it high conductance by providing low resistance.

continuity test An electrical test used to determine the presence and location of a broken connection.

continuous comparator *See* linear comparator.

continuous-duty rating The rating that defines the load which can be carried for an indefinite time without exceeding a specified temperature rise.

continuous film scanner A television film scanner in which the motion picture film moves continuously while being scanned by a flying-spot kinescope.

continuous loading Loading in which the added inductance is distributed uniformly along a line by wrapping magnetic material around each conductor.

continuously adjustable transformer *See* variable transformer.

continuous-tone squelch Squelch in which a continuous subaudible tone, generally below 200 hertz, is transmitted by frequency-modulation equipment along with a desired voice signal.

continuous wave A radio or radar wave whose successive sinusoidal oscillations are identical under steady-state conditions. Abbreviated CW. Also known as type A wave.

continuous-wave jammer An electronic jammer that emits a single frequency which gives the appearance of a picket or rail fence on an enemy's radarscope. Also known as rail-fence jammer.

continuous-wave tracking system Tracking system which operates by keeping a continuous radio beam on a target and determining its behavior from changes in the antenna necessary to keep the beam on the target.

continuous x-rays The electromagnetic radiation, having a continuous spectral distribution, that is produced when high-velocity electrons strike a target.

contourograph 1. Device using a cathode-ray oscilloscope to produce imagery that has a three-dimensional appearance. 2. Device using a cathode-ray oscilloscope to produce imagery that has a three-dimensional appearance.

contrast control A manual control that adjusts the range of brightness between highlights and shadows on the reproduced image in a television receiver.

contrast ratio The ratio of the maximum to the minimum luminance values in a television picture.

control An input element of a cryotron.

control board A panel at which one can make circuit changes, as in lighting a theater.

control characteristic 1. The relation, usually shown by a graph, between critical grid voltage and anode voltage of a gas tube. 2. The relation between control ampere-turns and output current of a magnetic amplifier.

control circuit 1. A circuit that controls some function of a machine, device, or piece of equipment. 2. The circuit that feeds the control winding of a magnetic amplifier.

control electrode An electrode used to initiate or vary the current between two or more electrodes in an electron tube.

control grid A grid, ordinarily placed between the cathode and an anode, that serves to control the anode current of an electron tube.

control-grid bias Average direct-current voltage between the control grid and cathode of a vacuum tube.

convergence control 77

control-grid plate transconductance Ratio of the amplification factor of a vacuum tube to its plate resistance, combining the effects of both into one term.

control inductor *See* control winding.

controlled avalanche device A semiconductor device that has rigidly specified maximum and minimum avalanche voltage characteristics and is able to operate and absorb momentary power surges in this avalanche region indefinitely without damage.

controlled avalanche rectifier A silicon rectifier in which carefully controlled, nondestructive internal avalanche breakdown across the entire junction area protects the junction surface, thereby eliminating local heating that would impair or destroy the reverse blocking ability of the rectifier.

controlled avalanche transit-time triode A solid-state microwave device that uses a combination of IMPATT diode and *npn* bipolar transistor technologies; avalanche and drift zones are located between the base and collector regions. Abbreviated CATT.

controlled mercury-arc rectifier A mercury-arc rectifier in which one or more electrodes control the start of the discharge in each cycle and thereby control output current.

controlled rectifier A rectifier that has provisions for regulating output current, such as with thyratrons, ignitrons, or silicon controlled rectifiers.

control limits In radar evaluation, upper and lower control limits are established at those performance figures within which it is expected that 95% of quality-control samples will fall when the radar is performing normally.

control synchro *See* control transformer.

control transformer A synchro in which the electrical output of the rotor is dependent on both the shaft position and the electric input to the stator. Also known as control synchro.

control winding A winding used on a magnetic amplifier or saturable reactor to apply control magnetomotive forces to the core. Also known as control inductor.

convection current The time rate at which the electric charges of an electron stream are transported through a given surface.

convective discharge The movement of a visible or invisible stream of charged particles away from a body that has been charged to a sufficiently high voltage. Also known as electric wind; static breeze.

convenience receptacle *See* outlet.

conventional current The concept of current as the transfer of positive charge, so that its direction of flow is opposite to that of electrons which are negatively charged.

convergence A condition in which the electron beams of a multibeam cathode-ray tube intersect at a specified point, such as at an opening in the shadow mask of a three-gun color television picture tube; both static convergence and dynamic convergence are required.

convergence circuit An auxiliary deflection system in a color television receiver which maintains convergence, having separate convergence coils for electromagnetic controls of the positions of the three beams in a convergence yoke around the neck of the kinescope.

convergence coil One of the coils used to obtain convergence of electron beams in a three-gun color television picture tube.

convergence control A control used in a color television receiver to adjust the potential on the convergence electrode of the three-gun color picture tube to achieve convergence.

convergence electrode An electrode whose electric field converges two or more electron beams.

convergence magnet A magnet assembly whose magnetic field converges two or more electron beams; used in three-gun color picture tubes. Also known as beam magnet.

conversion gain **1.** Ratio of the intermediate-frequency output voltage to the input signal voltage of the first detector of a superheterodyne receiver. **2.** Ratio of the available intermediate-frequency power output of a converter or mixer to the available radio-frequency power input.

converter **1.** Any device for changing alternating current to direct current, or direct current to alternating current. **2.** The section of a superheterodyne radio receiver that converts the desired incoming radio-frequency signal to the intermediate-frequency value; the converter section includes the oscillator and the mixer-first detector. Also known as heterodyne conversion transducer; oscillator-mixer-first detector. **3.** An auxiliary unit used with a television or radio receiver to permit reception of channels or frequencies for which the receiver was not originally designed. **4.** In facsimile, a device that changes the type of modulation delivered by the scanner. **5.** Unit of a radar system in which the mixer of a superheterodyne receiver and usually two stages of intermediate-frequency amplification are located; performs a preamplifying operation. **6.** *See* remodulator; synchronous converter.

converter substation An electric power substation whose main function is the conversion of power from ac to dc, and vice versa.

converter tube An electron tube that combines the mixer and local-oscillator functions of a heterodyne conversion transducer.

convolver A surface acoustic-wave device in which signal processing is performed by a nonlinear interaction between two waves traveling in opposite directions. Also known as acoustic convolver.

Conwell-Weisskopf equation An equation for the mobility of electrons in a semiconductor in the presence of donor or acceptor impurities, in terms of the dielectric constant of the medium, the temperature, the concentration of ionized donors (or acceptors), and the average distance between them.

cooled infrared detector An infrared detector that must be operated at cryogenic temperatures, such as at the temperature of liquid nitrogen, to obtain the desired infrared sensitivity.

Coolidge tube An x-ray tube in which the needed electrons are produced by a hot cathode.

cooperative phenomenon A process that involves a simultaneous collective interaction among many atoms or electrons in a crystal, such as ferromagnetism, superconductivity, and order-disorder transformations.

Cooper pairs Pairs of bound electrons which occur in a superconducting medium according to the Bardeen-Cooper-Schrieffer theory.

coordinate data receiver A receiver specifically designed to accept the signal of a coordinate data transmitter and reconvert this signal into a form suitable for input to associated equipment such as a plotting board, computer, or radar set.

coordinate data transmitter A transmitter that accepts two or more coordinates, such as those representing a target position, and converts them into a form suitable for transmission.

coordinated transpositions Transpositions which are installed in either electric supply or communications circuits or in both, for the purpose of reducing inductive coupling, and which are located effectively with respect to the discontinuities in both the electric supply and communications circuits.

corona discharge 79

coplanar electrodes Electrodes mounted in the same plane.

copper cable A mechanically assembled group of copper wires, used in place of a single, large wire for increased flexibility.

copper loss Power loss in a winding due to current flow through the resistance of the copper conductors. Also known as I^2R loss.

copper oxide photovoltaic cell A photovoltaic cell in which light acting on the surface of contact between layers of copper and cuprous oxide causes a voltage to be produced.

copper oxide rectifier A metallic rectifier in which the rectifying barrier is the junction between metallic copper and cuprous oxide.

copper sulfide rectifier A semiconductor rectifier in which the rectifying barrier is the junction between magnesium and copper sulfide.

corbino disk A variable-resistance device utilizing the effect of a magnetic field on the flow of carriers from the center to the circumference of a disk made of semiconducting or conducting material.

cord A small, very flexible insulated cable.

cord circuit Connecting circuit terminating in a plug at one or both ends and used at switchboard positions in establishing telephone connections.

cordwood module High-density circuit module in which discrete components are mounted between and perpendicular to two small, parallel printed circuit boards to which their terminals are attached.

core See magnetic core.

core array A rectangular grid arrangement of magnetic cores.

core bank A stack of core arrays and associated electronics, the stack containing a specific number of core arrays.

core hitch Attachment to a cable core to permit pulling it into a duct without damaging the sheath.

core logic Logic performed in ferrite cores that serve as inputs to diode and transistor circuits.

core loss The rate of energy conversion into heat in a magnetic material due to the presence of an alternating or pulsating magnetic field. Also known as excitation loss; iron loss.

core stack A number of core arrays, next to one another and treated as a unit.

core-type induction heater A device in which a charge is heated by induction, with a magnetic core being used to link the induction coil to the charge.

corkscrew rule The rule that the direction of the current and that of the resulting magnetic field are related to each other as the forward travel of a corkscrew and the direction in which it is rotated.

corner reflector An antenna consisting of two conducting surfaces intersecting at an angle that is usually 90°, with a dipole or other antenna located on the bisector of the angle.

corona See corona discharge.

corona current The current of electricity equivalent to the rate of charge transferred to the air from an object experiencing corona discharge.

corona discharge A discharge of electricity appearing as a bluish-purple glow on the surface of and adjacent to a conductor when the voltage gradient exceeds a certain critical value; due to ionization of the surrounding air by the high voltage. Also known as aurora; corona; electric corona.

corona failure High-voltage failure initiated by corona discharge at areas of high-voltage stress such as metal inserts or terminals.

corona resistance Ability of a conductor to resist destruction when a high-voltage electrostatic field ionizes within insulation voids.

corona shield A shield placed about a point of high potential to redistribute electrostatic lines of force.

corona stabilization The increase in the breakdown voltage of a gas separating two electrodes, where the electric field is very high at one pointed electrode and low at the other, due to the reduction of electric field around the pointed electrode by corona discharge.

corona start voltage The voltage difference at which corona discharge is initiated in a given system.

corona tube A gas-discharge voltage-reference tube employing a corona discharge.

corona voltmeter A voltmeter in which the crest value of a voltage is indicated by the inception of corona at a known electrode spacing.

corrective network An electric network inserted in a circuit to improve its transmission properties, impedance properties, or both. Also known as shaping circuit; shaping network.

correed relay Hermetically sealed reed capsule surrounded by a coil winding, used as a switching device with telephone equipment.

correlator A device that detects weak signals in noise by performing an electronic operation approximating the computation of a correlation function. Also known as correlation-type receiver.

cosecant antenna An antenna that gives a beam whose amplitude varies as the cosecant of the angle of depression below the horizontal; used in navigation radar.

cosecant-squared antenna An antenna having a cosecant-squared pattern.

cosecant-squared pattern A ground radar-antenna radiation pattern that sends less power to nearby objects than to those farther away in the same sector; the field intensity varies as the square of the cosecant of the elevation angle.

cosine winding A winding used in the deflection yoke of a cathode-ray tube to prevent changes in focus as the beam is deflected over the entire area of the screen.

Cottrell hardening Hardening of a material caused by locking of its dislocations when impurity atoms whose size differs from that of the solvent cluster around them.

coul *See* coulomb.

coulomb A unit of electric charge, defined as the amount of electric charge that crosses a surface in 1 second when a steady current of 1 absolute ampere is flowing across the surface; this is the absolute coulomb and has been the legal standard of quantity of electricity since 1950; the previous standard was the international coulomb, equal to 0.999835 absolute coulomb. Abbreviated coul. Symbolized C.

Coulomb attraction The electrostatic force of attraction exerted by one charged particle on another charged particle of opposite sign. Also known as electrostatic attraction.

Coulomb field The electric field created by a stationary charged particle.

Coulomb force The electrostatic force of attraction or repulsion exerted by one charged particle on another, in accordance with Coulomb's law.

Coulomb gage A gage in which the divergence of the magnetic vector potential is equal to 0.

Coulomb interactions Interactions of charged particles associated with the Coulomb forces they exert on one another. Also known as electrostatic interactions.

Coulomb's law The law that the attraction or repulsion between two electric charges acts along the line between them, is proportional to the product of their magnitudes, and is inversely proportional to the square of the distance between them. Also known as law of electrostatic attraction.

Coulomb potential A scalar point function equal to the work per unit charge done against the Coulomb force in transferring a particle bearing an infinitesimal positive charge from infinity to a point in the field of a specific charge distribution.

Coulomb repulsion The electrostatic force of repulsion exerted by one charged particle on another charged particle of the same sign. Also known as electrostatic repulsion.

counter *See* scaler.

counter circuit *See* counting circuit.

counter decade *See* decade scaler.

counterelectromotive cell Cell of practically no ampere-hour capacity, used to oppose the line voltage.

counterelectromotive force The voltage developed in an inductive circuit by a changing current; the polarity of the induced voltage is at each instant opposite that of the applied voltage. Also known as back electromotive force.

countermeasures set A complete electronic set specifically designed to provide facilities for intercepting and analyzing electromagnetic energy propagated by transmitter and to provide a source of radio-frequency signals which deprive the enemy of effective use of his electronic equipment.

counterpoise A system of wires or other conductors that is elevated above and insulated from the ground to form a lower system of conductors for an antenna. Also known as antenna counterpoise.

counter tube An electron tube having one signal-input electrode and 10 or more output electrodes, with each input pulse serving to transfer conduction sequentially to the next output electrode; beam-switching tubes and cold-cathode counter tubes are examples.

counting circuit A circuit that counts pulses by frequency-dividing techniques, by charging a capacitor in such a way as to produce a voltage proportional to the pulse count, or by other means. Also known as counter circuit.

counting-down circuit *See* frequency divider.

couple 1. To connect two circuits so signals are transferred from one to the other. 2. Two metals placed in contact, as in a thermocouple.

coupled antenna An antenna electromagnetically coupled to another.

coupled circuits Two or more electric circuits so arranged that energy can transfer electrically or magnetically from one to another.

coupled field vectors The electric- and magnetic-field vectors, which depend upon each other according to Maxwell's field equations.

coupled oscillators A set of alternating-current circuits which interact with each other, for example, through mutual inductances or capacitances.

coupled transistors Transistors connected in series by transformers or resistance-capacitance networks, in much the same manner as electron tubes.

coupler 1. A component used to transfer energy from one circuit to another. 2. A passage which joins two cavities or waveguides, allowing them to exchange energy.

82 coupling

3. A passage which joins the ends of two waveguides, whose cross section changes continuously from that of one to that of the other.

coupling 1. A mutual relation between two circuits that permits energy transfer from one to another, through a wire, resistor, transformer, capacitor, or other device. 2. A hardware device used to make a temporary connection between two wires.

coupling aperture An aperture in the wall of a waveguide or cavity resonator, designed to transfer energy to or from an external circuit. Also known as coupling hole; coupling slot.

coupling capacitor A capacitor used to block the flow of direct current while allowing alternating or signal current to pass; widely used for joining two circuits or stages. Also known as blocking capacitor; stopping capacitor.

coupling coefficient The ratio of the maximum change in energy of an electron traversing an interaction space to the product of the peak alternating gap voltage and the electronic charge.

coupling hole See coupling aperture.

coupling loop A conducting loop projecting into a waveguide or cavity resonator, designed to transfer energy to or from an external circuit.

coupling probe A probe projecting into a waveguide or cavity resonator, designed to transfer energy to or from an external circuit.

coupling slot See coupling aperture.

C power supply A device connected in the circuit between the cathode and grid of a vacuum tube to apply grid bias.

crater lamp A glow-discharge tube used as a point source of light whose brightness is proportional to the signal current sent through the tube; used for photographic recording of facsimile signals.

credence In radar, a measure of confidence in a target detection, generally proportional to target return amplitude.

creep A slow change in a characteristic with time or usage.

creepage The conduction of electricity across the surface of a dielectric.

crest value See peak value.

crest voltmeter A voltmeter reading the peak value of the voltage applied to its terminals.

crimp contact A contact whose back portion is a hollow cylinder that will accept a wire; after a bared wire is inserted, a swaging tool is applied to crimp the contact metal firmly against the wire. Also known as solderless contact.

critical anode voltage The anode voltage at which breakdown occurs in a gas tube.

critical area See picture element.

critical coupling The degree of coupling that provides maximum transfer of signal energy from one radio-frequency resonant circuit to another when both are tuned to the same frequency. Also known as optimum coupling.

critical current The current in a superconductive material above which the material is normal and below which the material is superconducting, at a specified temperature and in the absence of external magnetic fields.

critical field The smallest theoretical value of steady magnetic flux density that would prevent an electron emitted from the cathode of a magnetron at zero velocity from reaching the anode. Also known as cutoff field.

cross flux 83

critical frequency 1. The limiting frequency below which a radio wave will be reflected by an ionospheric layer at vertical incidence at a given time. 2. *See* cutoff frequency.

critical grid current Instantaneous value of grid current when the anode current starts to flow in a gas-filled vacuum tube.

critical grid voltage The grid voltage at which anode current starts to flow in a gas tube. Also known as firing point.

critical magnetic field The field below which a superconductive material is superconducting and above which the material is normal, at a specified temperature and in the absence of current.

critical potential A potential which results in sudden change in magnitude of the current.

critical voltage The highest theoretical value of steady anode voltage, at a given steady magnetic flux density, at which electrons emitted from the cathode of a magnetron at zero velocity would fail to reach the anode. Also known as cutoff voltage.

CR law A law which states that when a constant electromotive force is applied to a circuit consisting of a resistor and capacitor connected in series, the time taken for the potential on the plates of the capacitor to rise to any given fraction of its final value depends only on the product of capacitance and resistance.

CRO *See* cathode-ray oscilloscope.

crocodile A unit of potential difference or electromotive force, equal to 10^6 volts; used informally at some nuclear physics laboratories.

crocodile clip *See* alligator clip.

Crookes dark space *See* cathode dark space.

Crookes tube An early form of low-pressure discharge tube whose cathode was a flat aluminum disk at one end of the tube, and whose anode was a wire at one side of the tube, outside the electron stream; used to study cathode rays.

crossbar switch A switch having a three-dimensional arrangement of contacts and a magnet system that selects individual contacts according to their coordinates in the matrix.

cross-color In color television, the interference in the receiver chrominance channel caused by cross talk from monochrome signals.

cross-correlator A correlator in which a locally generated reference signal is multiplied by the incoming signal and the result is smoothed in a low-pass filter to give an approximate computation of the cross-correlation function. Also known as synchronous detector.

crossed-field amplifier A forward-wave, beam-type microwave amplifier that uses crossed-field interaction to achieve good phase stability, high efficiency, high gain, and wide bandwidth for most of the microwave spectrum.

crossed-field backward-wave oscillator One of several types of backward-wave oscillators that utilize a crossed field, such as the amplitron and carcinotron.

crossed-field device Any instrument which uses the motion of electrons in perpendicular electric and magnetic fields to generate microwave radiation, either as an amplifier or oscillator.

crossed-field multiplier phototube A multiplier phototube in which repeated secondary emission is obtained from a single active electrode by the combined effects of a strong radio-frequency electric field and a perpendicular direct-current magnetic field.

cross flux A component of magnetic flux perpendicular to that produced by the field magnets in an electrical rotating machine.

crosshatch generator A signal generator that generates a crosshatch pattern for adjusting color television receiver circuits.

cross-magnetizing effect The distortion in the flux-density distribution in the air gap of an electric rotating machine caused by armature reaction.

cross-neutralization Method of neutralization used in push-pull amplifiers, whereby a portion of the plate-cathode alternating-current voltage of each vacuum tube is applied to the grid-cathode circuit of the other vacuum tube through a neutralizing capacitor.

crossover A point at which two conductors cross, with appropriate insulation between them to prevent contact.

crossover distortion Amplitude distortion in a class B transistor power amplifier which occurs at low values of current, when input impedance becomes appreciable compared with driver impedance.

crossover voltage In a cathode-ray storage tube, the voltage of a secondary writing surface, with respect to cathode voltage, on which the secondary emission is unity.

cross-polarization The component of the electric field vector normal to the desired polarization component.

crosstalk *See* magnetic printing.

crowbar A device or action that in effect places a high overload on the actuating element of a circuit breaker or other protective device, thus triggering it.

crowbar voltage protector A separate circuit which monitors the output of a regulated power supply and instantaneously throws a short circuit (or crowbar) across the output terminals of the power supply whenever a preset voltage limit is exceeded.

crown cell The generic name for alkaline zinc—manganese dioxide dry-cell battery; manganese dioxide—graphite cathode mix is pressed into a steel can onto which a steel cap is spot-welded to contain the amalgamated powdered-zinc anode.

CRT *See* cathode-ray tube.

cruciform core A transformer core in which all windings are on one center leg, and four additional legs arranged in the form of a cross serve as return paths for magnetic flux.

cryoelectronics *See* cryolectronics.

cryogenic conductor *See* superconductor.

cryogenic transformer A transformer designed to operate in digital cryogenic circuits, such as a controlled-coupling transformer.

cryolectronics Technology concerning the characteristics of electronic components at cryogenic temperatures. Also known as cryoelectronics.

cryoresistive transmission line An electric power transmission line whose conducting cables are cooled to the temperature of liquid nitrogen, 77 K ($-196°C$), resulting in a reduction of the resistance of the conductor by a factor of approximately 10, leading to increased transmission capacity.

cryosar A cryogenic, two-terminal, negative-resistance semiconductor device, consisting essentially of two contacts on a germanium wafer operating in liquid helium.

cryosistor A cryogenic semiconductor device in which a reverse-biased pn junction is used to control the ionization between two ohmic contacts.

cryotron A switch that operates at very low temperatures at which its components are superconducting; when current is sent through a control element to produce a magnetic field, a gate element changes from a superconductive zero-resistance state to its normal resistive state.

crystal operation 85

cryotronics The branch of electronics that deals with the design, construction, and use of cryogenic devices.

crystal A natural or synthetic piezoelectric or semiconductor material whose atoms are arranged with some degree of geometric regularity.

crystal activity A measure of the amplitude of vibration of a piezoelectric crystal plate under specified conditions.

crystal-audio receiver Similar to the crystal-video receiver, except for the path detection bandwidth which is audio rather than video.

crystal blank The result of the final cutting operation on a piezoelectric or semiconductor crystal.

crystal calibrator A crystal-controlled oscillator used as a reference standard to check frequencies.

crystal control Control of the frequency of an oscillator by means of a quartz crystal unit.

crystal-controlled oscillator An oscillator whose frequency of operation is controlled by a crystal unit.

crystal-controlled transmitter A transmitter whose carrier frequency is directly controlled by the electromechanical characteristics of a quartz crystal unit.

crystal current The actual alternating current flowing through a crystal unit.

crystal detector 1. A crystal diode, or an equivalent earlier crystal-catwhisker combination, used to rectify a modulated radio-frequency signal to obtain the audio or video signal directly. 2. A crystal diode used in a microwave receiver to combine an incoming radio-frequency signal with a local oscillator signal to produce an intermediate-frequency signal.

crystal diffraction Diffraction by a crystal of beams of x-rays, neutrons, or electrons whose wavelengths (or de Broglie wavelengths) are comparable with the interatomic spacing of the crystal.

crystal diode *See* semiconductor diode.

crystal dynamics *See* lattice dynamics.

crystal filter A highly selective tuned circuit employing one or more quartz crystals; sometimes used in intermediate-frequency amplifiers of communication receivers to improve the selectivity.

crystal harmonic generator A type of crystal-controlled oscillator which produces an output rich in harmonics (overtones or multiples) of its fundamental frequency.

crystal lattice filter A crystal filter that uses two matched pairs of series crystals and a higher-frequency matched pair of shunt or lattice crystals.

crystalline anisotropy The tendency of crystals to have different properties in different directions; for example, a ferromagnet will spontaneously magnetize along certain crystallographic axes.

crystalline field The internal electric field in a solid due to localized charges, especially ions, inside.

crystallomagnetic Pertaining to magnetic properties of crystals.

crystal mixer A mixer that uses the nonlinear characteristic of a crystal diode to mix two frequencies; widely used in radar receivers to convert the received radar signal to a lower intermediate-frequency value by mixing it with a local oscillator signal.

crystal operation Operation using crystal-controlled oscillators.

86 crystal oscillator

crystal oscillator An oscillator in which the frequency of the alternating-current output is determined by the mechanical properties of a piezoelectric crystal. Also known as piezoelectric oscillator.

crystal plate A precisely cut slab of quartz crystal that has been lapped to final dimensions, etched to improve stability and efficiency, and coated with metal on its major surfaces for connecting purposes. Also known as quartz plate.

crystal rectifier *See* semiconductor diode.

crystal resonator A precisely cut piezoelectric crystal whose natural frequency of vibration is used to control or stabilize the frequency of an oscillator. Also known as piezoelectric resonator.

crystal set A radio receiver having a crystal detector stage for demodulation of the received signals, but no amplifier stages.

crystal shutter Mechanical waveguide or coaxial-cable shorting switch that, when closed, prevents undesired radio-frequency energy from reaching and damaging a crystal detector.

crystal-stabilized transmitter A transmitter employing automatic frequency control, in which the reference frequency is that of a crystal oscillator.

crystal transducer A transducer in which a piezoelectric crystal serves as the sensing element.

crystal unit A complete assembly of one or more quartz plates in a crystal holder.

crystal video receiver A broad-tuning radar or other microwave receiver consisting only of a crystal detector and a video or audio amplifier.

crystal video rectifier A crystal rectifier transforming a high-frequency signal directly into a video-frequency signal.

C scan *See* C scope.

C scope A cathode-ray scope on which signals appear as spots, with bearing angle as the horizontal coordinate and elevation angle as the vertical coordinate. Also known as C indicator; C scan.

CT *See* center tap.

cube-surface coil A system of five equally spaced square coils that produces a region of uniform magnetic field over a large volume which is easily accessible from outside the coils.

cubical antenna An antenna array, the elements of which are positioned to form a cube.

Cuccia coupler *See* electron coupler.

cue circuit A one-way communication circuit used to convey program control information.

cumulative compound generator A compound generator in which the series field is connected to aid the shunt field magnetomotive force.

cumulative ionization *See* avalanche.

cup core A core that encloses a coil to provide magnetic shielding; usually has a powdered iron center post through the coil.

Curie constant The electric or magnetic susceptibility at some temperature times the difference of the temperature and the Curie temperature, which is a constant at temperatures above the Curie temperature according to the Curie-Weiss law.

Curie's law The law that the magnetic susceptibilities of most paramagnetic substances are inversely proportional to their absolute temperatures.

Curie temperature The temperature marking the transition between ferromagnetism and paramagnetism, or between the ferroelectric phase and paraelectric phase. Also known as Curie point.

Curie-Weiss law A relation between magnetic or electric susceptibilities and the absolute temperatures which is followed by ferromagnets, antiferromagnets, nonpolar ferroelectrics, antiferroelectrics, and some paramagnets.

current The net transfer of electric charge per unit time; a specialization of the physics definition. Also known as electric current.

current amplification The ratio of output-signal current to input-signal current for an electron tube, transistor, or magnetic amplifier, the multiplier section of a multiplier phototube, or any other amplifying device; often expressed in decibels by multiplying the common logarithm of the ratio by 20.

current amplifier An amplifier capable of delivering considerably more signal current than is fed in.

current antinode A point at which current is a maximum along a transmission line, antenna, or other circuit element having standing waves. Also known as current loop.

current attenuation The ratio of input-signal current for a transducer to the current in a specified load impedance connected to the transducer; often expressed in decibels.

current balance An apparatus with which force is measured between current-carrying conductors, with the purpose of assigning the value of the ampere.

current-carrying capacity The maximum current that can be continuously carried without causing permanent deterioration of electrical or mechanical properties of a device or conductor.

current-controlled switch A semiconductor device in which the controlling bias sets the resistance at either a very high or very low value, corresponding to the "off" and "on" conditions of a switch.

current density The current per unit cross-sectional area of a conductor; a specialization of the physics definition. Also known as electric current density.

current divider A device used to deliver a desired fraction of a total current to a circuit.

current drain The current taken from a voltage source by a load. Also known as drain.

current-equalizing reactor A reactor that is used to achieve a desired division of current between several circuits operating in parallel.

current feed Feed to a point where current is a maximum, as at the center of a half-wave antenna.

current feedback Feedback introduced in series with the input circuit of an amplifier.

current feedback circuit A circuit used to eliminate effects of amplifier gain instability in an indirect-acting recording instrument, in which the voltage input (error signal) to an amplifier is the difference between the measured quantity and the voltage drop across a resistor.

current gain The fraction of the current flowing into the emitter of a transistor which flows through the base region and out the collector.

current generator A two-terminal circuit element whose terminal current is independent of the voltage between its terminals.

current hogging A condition in which the largest fraction of a current passes through one of several parallel logic circuits because it has a lower resistance than the others.

current intensity The magnitude of an electric current. Also known as current strength.

current interrupter Mechanism connected into a current-carrying line to periodically interrupt current flow to allow no-current tests of system components.

current limiter A device that restricts the flow of current to a certain amount, regardless of applied voltage. Also known as demand limiter.

current-limiting resistor A resistor inserted in an electric circuit to limit the flow of current to some predetermined value; used chiefly to protect tubes and other components during warm-up.

current loop *See* current antinode.

current measurement The measurement of the flow of electric current.

current-mode logic Integrated-circuit logic in which transistors are paralleled so as to eliminate current hogging.

current node A point at which current is zero along a transmission line, antenna, or other circuit element having standing waves.

current noise Electrical noise of uncertain origin which is observed in certain resistances when a direct current is present, and which increases with the square of this current.

current phasor A line referenced to a point, whose length and angle represent the magnitude and phase of a current.

current ratio In a waveguide, the ratio of maximum to minimum current.

current regulator A device that maintains the output current of a voltage source at a predetermined, essentially constant value despite changes in load impedance.

current relay A relay that operates at a specified current value rather than at a specified voltage value.

current saturation *See* anode saturation.

current strength *See* current intensity.

current tap *See* multiple lamp holder; plug adapter lamp holder.

current transformer An instrument transformer intended to have its primary winding connected in series with a circuit carrying the current to be measured or controlled; the current is measured across the secondary winding.

current transformer phase angle Angle between the primary current vector and the secondary current vector reversed; it is conveniently considered as positive when the reversed secondary current vector leads the primary current vector.

current-voltage dual A circuit which is equivalent to a specified circuit when one replaces quantities with dual quantities; current and voltage impedance and admittance, and meshes and nodes are examples of dual quantities.

curtain array An antenna array consisting of vertical wire elements stretched between two suspension cables.

curtain rhombic antenna A multiple-wire rhombic antenna having a constant input impedance over a wide frequency range; two or more conductors join at the feed and terminating ends but are spaced apart vertically from 1 to 5 feet (30 to 150 centimeters) at the side poles.

curvature effect Generally, the condition in which the dielectric strength of a liquid or vacuum separating two electrodes is higher for electrodes of smaller radius of curvature.

cusped magnetic field A magnetic field created by adjacent parallel coils that carry current in opposite directions; used in fusion research, to contain a plasma of high-energy deuterium ions.

cyclophon 89

custom-designed device An integrated logic circuit element that is generated by a series of steps resembling photographic development from highly complicated artwork patterns.

customer substation A distribution substation located on the premises of a larger customer, such as a shopping center, commercial building, or industrial plant.

Cutler feed A resonant cavity that transfers radio-frequency energy from the end of a waveguide to the reflector of a radar spinner assembly.

cutoff 1. The minimum value of negative grid bias that will prevent the flow of anode current in an electron tube. 2. *See* cutoff frequency.

cutoff attenuator Variable length of waveguide used below its cutoff frequency to introduce variable nondissipative attenuation.

cutoff bias The direct-current bias voltage that must be applied to the grid of an electron tube to stop the flow of anode current.

cutoff field *See* critical field.

cutoff frequency A frequency at which the attenuation of a device begins to increase sharply, such as the limiting frequency below which a traveling wave in a given mode cannot be maintained in a waveguide, or the frequency above which an electron tube loses efficiency rapidly. Also known as critical frequency; cutoff.

cutoff limiting Limiting the maximum output voltage of a vacuum tube circuit by driving the grid beyond cutoff.

cutoff voltage 1. The electrode voltage value that reduces the dependent variable of an electron-tube characteristic to a specified low value. 2. *See* critical voltage.

cutoff wavelength 1. The ratio of the velocity of electromagnetic waves in free space to the cutoff frequency in a uniconductor waveguide. 2. The wavelength corresponding to the cutoff frequency.

cutout 1. Pairs brought out of a cable and terminated at some place other than at the end of the cable. 2. An electrical device that is used to interrupt the flow of current through any particular apparatus or instrument, either automatically or manually. Also known as electric cutout.

cutout box A fireproof cabinet or box with one or more hinged doors that contains fuses and switches for various leads in an electrical wiring system. Also known as fuse box.

cut-set A set of branches of a network such that the cutting of all the branches of the set increases the number of separate parts of the network, but the cutting of all the branches except one does not.

cut-signal-branch operation In systems where radio reception continues without cutting off the carrier, the cut-signal-branch operation technique disables a signal branch in one direction when it is enabled in the other to preclude unwanted signal reflections.

CW *See* continuous wave.

cycle timer A timer that opens or closes circuits according to a predetermined schedule.

cyclic magnetization A magnetizing force varying between two specific limits long enough so that the magnetic induction has the same value for corresponding points in successive cycles.

cycloconverter A device that produces an alternating current of constant or precisely controllable frequency from a variable-frequency alternating-current input, with the output frequency usually one-third or less of the input frequency.

cyclophon *See* beam-switching tube.

cyclotron emission *See* cyclotron radiation.

cyclotron frequency The frequency at which an electron traverses an orbit when moving subject to a uniform magnetic field, at right angles to the field. Also known as gyrofrequency.

cyclotron-frequency magnetron A magnetron whose frequency of operation depends on synchronism between the alternating-current electric field and the electrons oscillating in a direction parallel to this field.

cyclotron radiation The electromagnetic radiation emitted by charged particles as they orbit in a magnetic field, at a speed which is not close to the speed of light. Also known as cyclotron emission.

cyclotron resonance maser *See* gyrotron.

cyclotron wave A wave associated with the electron beam of a traveling-wave tube.

cylindrical antenna An antenna in which hollow cylinders serve as radiating elements.

cylindrical array An electronic scanning antenna that may consist of several hundred columns of vertical dipoles mounted in cylindrical radomes arranged in a circle.

cylindrical capacitor A capacitor made of two concentric metal cylinders of the same length, with dielectric filling the space between the cylinders. Also known as coaxial capacitor.

cylindrical cavity A cavity resonator in the shape of a right circular cylinder.

cylindrical-film storage A computer storage in which each storage element consists of a short length of glass tubing having a thin film of nickel-iron alloy on its outer surface.

cylindrical pinch *See* pinch effect.

cylindrical reflector A reflector that is a portion of a cylinder; this cylinder is usually parabolic.

cylindrical wave A wave whose equiphase surfaces form a family of coaxial cylinders.

cylindrical winding The current-carrying element of a core-type transformer, consisting of a single coil of one or more layers wound concentrically with the iron core.

D

dac *See* digital-to-analog converter.

damper A diode used in the horizontal deflection circuit of a television receiver to make the sawtooth deflection current decrease smoothly to zero instead of oscillating at zero; the diode conducts each time the polarity is reversed by a current swing below zero.

damper winding A winding consisting of several conducting bars on the field poles of a synchronous machine, short-circuited by conducting rings or plates at their ends, and used to prevent pulsating variations of the position or magnitude of the magnetic field linking the poles. Also known as amortisseur winding.

damping magnet A permanent magnet used in conjunction with a disk or other moving conductor to produce a force that opposes motion of the conductor and thereby provides damping.

dangling bond A chemical bond associated with an atom in the surface layer of a solid that does not join the atom with a second atom but extends in the direction of the solid's exterior.

daraf The unit of elastance, equal to the reciprocal of 1 farad.

dark conduction Residual conduction in a photosensitive substance that is not illuminated.

dark current *See* electrode dark current.

dark-current pulse A phototube dark-current excursion that can be resolved by the system employing the phototube.

dark discharge An invisible electrical discharge in a gas.

dark resistance The resistance of a selenium cell or other photoelectric device in total darkness.

dark space A region in a glow discharge that produces little or no light.

dark spot A spot on a television receiver tube that results from a spurious signal generated in the television camera tube during rescan, generally from the redistribution of secondary electrons over the mosaic in the tube.

dark-trace tube A cathode-ray tube with a bright face that does not necessarily luminesce, on which signals are displayed as dark traces or dark blips where the potassium chloride screen is hit by the electron beam. Also known as skiatron.

Darlington amplifier A current amplifier consisting essentially of two separate transistors and often mounted in a single transistor housing.

d'Arsonval current A current consisting of isolated trains of heavily damped high-frequency oscillations of high voltage and relatively low current, used in diathermy.

data circuit A telephone facility allowing transmission of digital data pulses with minimum distortion.

92 data concentrator

data concentrator A device, such as a microprocessor, that takes data from several different teletypewriter or other slow-speed lines and feeds them to a single higher-speed line.

dataset *See* modem.

data stabilization Stabilization of the display of radar signals with respect to a selected reference, regardless of changes in radar-carrying vehicle attitude, as in azimuth-stabilized plan-position indicator.

data transmission line A system of electrical conductors, such as a coaxial cable or pair of wires, used to send information from one place to another or one part of a system to another.

daylight lamp An incandescent or fluorescent lamp that emits light whose spectral distribution is approximately that of daylight.

dBf *See* decibels above 1 femtowatt.

dBk *See* decibels above 1 kilowatt.

dBm *See* decibels above 1 milliwatt.

dBp *See* decibels above 1 picowatt.

dBrn *See* decibels above reference noise.

dBV *See* decibels above 1 volt.

dBW *See* decibels above 1 watt.

dBx *See* decibels above reference coupling.

dc *See* direct current.

D cable Two-conductor cable, each conductor having the shape of the letter D, with insulation between the conductors and between the conductors and the sheath.

DCTL *See* direct-coupled transistor logic.

dc-to-ac converter *See* inverter.

dc-to-ac inverter *See* inverter.

dc-to-dc converter An electronic circuit which converts one direct-current voltage into another, consisting of an inverter followed by a step-up or step-down transformer and rectifier.

dcwv *See* direct-current working volts.

D display In radar, a C display in which the blips extend vertically to give a rough estimate of distance.

deaccentuator A circuit used in a frequency-modulation receiver to offset the preemphasis of higher audio frequencies introduced at the transmitter.

dead Free from any electric connection to a source of potential difference from electric charge; not having a potential different from that of earth; the term is used only with reference to current-carrying parts which are sometimes alive or charged.

dead band The portion of a potentiometer element that is shortened by a tap; when the wiper traverses this area, there is no change in output.

dead-center position Position in which a brush would be placed on the commutator of a direct-current motor or generator if the field flux were not distorted by armature reaction.

dead earth A connection between a line conductor and earth by means of a path of low resistance.

dead end The portion of a tapped coil through which no current is flowing at a particular switch position.

dead-end effect Absorption of energy by unused portions of a tapped coil.

dead-end switch A switch used to short-circuit unused portions of a tapped coil to prevent dead-end effects.

dead ground A low-resistance connection between the ground and an electric circuit.

dead short A short-circuit path that has extremely low resistance.

dead spot A portion of the tuning range of a receiver in which stations are heard poorly or not at all, due to improper design of tuning circuits.

debug To detect and remove secretly installed listening devices popularly known as bugs.

debunching A tendency for electrons in a beam to spread out both longitudinally and transversely due to mutual repulsion; the effect is a drawback in velocity modulation tubes.

debye A unit of electric dipole moment, equal to 10^{-18} Franklin centimeter.

Debye effect Selective absorption of electromagnetic waves by a dielectric, due to molecular dipoles.

Debye equation The equation for the Debye specific heat, which satisfies the Dulong and Petit law at high temperatures and the Debye T^3 law at low temperatures.

Debye frequency The maximum allowable frequency in the computation of the Debye specific heat.

Debye-Jauncey scattering Incoherent background scattering of x-rays from a crystal in directions between those of the Bragg reflections.

Debye potentials Two scalar potentials, designated Π_e and Π_m, in terms of which one can express the electric and magnetic fields resulting from radiation or scattering of electromagnetic waves by a distribution of localized sources in a homogeneous isotropic medium.

Debye-Scherrer method An x-ray diffraction method in which the sample, consisting of a powder stuck to a thin fiber or contained in a thin-walled silica tube, is rotated in a monochromatic beam of x-rays, and the diffraction pattern is recorded on a cylindrical film whose axis is parallel to the axis of rotation of the sample.

Debye specific heat The specific heat of a solid under the assumption that the energy of the lattice arises entirely from acoustic lattice vibration modes which all have the same sound velocity, and that frequencies are cut off at a maximum such that the total number of modes equals the number of degrees of freedom of the solid.

Debye temperature The temperature Θ arising in the computation of the Debye specific heat, defined by $k\Theta = h\nu$, where k is the Boltzmann constant, h is Planck's constant, and ν is the Debye frequency. Also known as characteristic temperature.

Debye T^3 law The law that the specific heat of a solid at constant volume varies as the cube of the absolute temperature T at temperatures which are small with respect to the Debye temperature.

Debye-Waller factor A reduction factor for the intensity of coherent (Bragg) scattering of x-rays, neutrons, or electrons by a crystal, arising from thermal motion of the atoms in the lattice.

decade A group or assembly of 10 units; for example, a decade counter counts 10 in one column, and a decade box inserts resistance quantities in multiples of powers of 10.

decade box An assembly of precision resistors, coils, or capacitors whose individual values vary in submultiples and multiples of 10; by appropriately setting a 10-position selector switch for each section, the decade box can be set to any desired value within its range.

decade bridge Electronic apparatus for measurement of unknown values of resistances or capacitances by comparison with known values (bridge); one secondary section of the oscillator-driven transformer is tapped in decade steps, the other in 10 uniform steps.

decade counter See decade scaler.

decade scaler A scaler that produces one output pulse for every 10 input pulses. Also known as counter decade; decade counter; scale-of-ten circuit.

decametric wave British term for a radio wave ranging from 10 to 100 meters long.

decelerating electrode Of an electron-beam tube, an electrode to which a potential is applied to decrease the velocity of the electrons in the beam.

deception The deliberate radiation, reradiation, alteration, absorption, or reflection of electromagnetic energy in a manner intended to mislead an enemy in the interpretation of information received by his electronic systems.

decibel adjusted See adjusted decibel.

decibels above 1 femtowatt A power level equal to 10 times the common logarithm of the ratio of the given power in watts to 1 femtowatt (10^{-15} watt). Abbreviated dBf.

decibels above 1 kilowatt A measure of power equal to 10 times the common logarithm of the ratio of a given power to 1000 watts. Abbreviated dBk.

decibels above 1 milliwatt A measure of power equal to 10 times the common logarithm of the ratio of a given power to 0.001 watt; a negative value, such as -2.7 dBm, means decibels below 1 milliwatt. Abbreviated dBm.

decibels above 1 picowatt A measure of power equal to 10 times the common logarithm of the ratio of a given power to 1 picowatt. Abbreviated dBp.

decibels above reference coupling A measure of the coupling between two circuits, expressed in relation to a reference value of coupling that gives a specified reading on a specified noise-measuring set when a test tone of 90 dBa is impressed on one circuit. Abbreviated dBx.

decibels above reference noise Units used to show the relationship between the interfering effect of a noise frequency, or band of noise frequencies, and a fixed amount of noise power commonly called reference noise; a 1000-hertz tone having a power level of -90 dBm was selected as the reference noise power; superseded by the adjusted decibel unit. Abbreviated dBrn.

decibels above 1 volt A measure of voltage equal to 20 times the common logarithm of the ratio of a given voltage to 1 volt. Abbreviated dBV.

decibels above 1 watt A measure of power equal to 10 times the common logarithm of the ratio of a given power to 1 watt. Abbreviated dBW.

decimal attenuator System of attenuators arranged so that a voltage or current can be reduced decimally.

decimal-binary switch A switch that connects a single input lead to appropriate combinations of four output leads (representing 1, 2, 4, and 8) for each of the decimal-numbered settings of its control knob; thus, for position 7, output leads 1, 2, and 4 would be connected to the input.

decimetric wave An electromagnetic wave having a wavelength between 0.1 and 1 meter, corresponding to a frequency between 300 and 3000 megahertz.

deflection defocusing   95

decision element A circuit that performs a logical operation such as "and," "or," "not," or "except" on one or more binary digits of input information representing "yes" or "no" and that expresses the result in its output. Also known as decision gate.

decision gate *See* decision element.

deck switch *See* gang switch.

decoder 1. A matrix of logic elements that selects one or more output channels, depending on the combination of input signals present. 2. *See* decoder circuit; matrix; tree.

decoder circuit A circuit that responds to a particular coded signal while rejecting others. Also known as decoder.

decometer An adding-type phasemeter which rotates continuously and adds up the total number of degrees of phase shift between two signals, such as those received from two transmitters in the Decca navigation system.

decommutation The process of recovering a signal from the composite signal previously created by a commutation process.

decommutator The section of a telemetering system that extracts analog data from a time-serial train of samples representing a multiplicity of data sources transmitted over a single radio-frequency link.

decoupling Preventing transfer or feedback of energy from one circuit to another.

decoupling filter One of a number of low-pass filters placed between each of several amplifier stages and a common power supply.

decoupling network Any combination of resistors, coils, and capacitors placed in power supply leads or other leads that are common to two or more circuits, to prevent unwanted interstage coupling.

decoy transponder A transponder that returns a strong signal when triggered directly by a radar pulse, to produce large and misleading target signals on enemy radar screens.

decrypt To convert a cryptogram or series of electronic pulses into plain text by electronic means.

deenergize To disconnect from the source of power.

deerhorn antenna A dipole antenna whose ends are swept back to reduce wind resistance when mounted on an airplane.

defect chemistry The study of the dynamic properties of crystal defects under particular conditions, such as raising of the temperature or exposure to electromagnetic particle radiation.

defect conduction Electric conduction in a semiconductor by holes in the valence band.

definition The extent to which the fine-line details of a printed circuit correspond to the master drawing.

deflection The displacement of an electron beam from its straight-line path by an electrostatic or electromagnetic field.

deflection circuit A circuit which controls the deflection of an electron beam in a cathode-ray tube.

deflection coil One of the coils in a deflection yoke.

deflection defocusing Defocusing that becomes greater as deflection is increased in a cathode-ray tube, because the beam hits the screen at a greater slant and the beam spot becomes more elliptical as it approaches the edges of the screen.

96 deflection electrode

deflection electrode An electrode whose potential provides an electric field that deflects an electron beam. Also known as deflection plate.

deflection factor The reciprocal of the deflection sensitivity in a cathode-ray tube.

deflection plate See deflection electrode.

deflection polarity Relationship between the direction of a displacement of the cathode beam and the polarity of the applied signal wave.

deflection sensitivity The displacement of the electron beam at the target or screen of a cathode-ray tube per unit of change in the deflection field; usually expressed in inches per volt applied between deflection electrodes or inches per ampere in a deflection coil.

deflection voltage The voltage applied between a pair of deflection electrodes to produce an electric field.

deflection yoke An assembly of one or more electromagnets that is placed around the neck of an electron-beam tube to produce a magnetic field for deflection of one or more electron beams. Also known as scanning yoke; yoke.

defocus-dash mode A mode of cathode-ray tube storage of binary digits in which the writing beam is initially defocused so as to excite a small circular area on the screen; for one kind of binary digit it remains defocused, and for the other kind it is suddenly focused to a concentric dot and drawn out into a dash.

defocus-focus mode A variation of the defocus-dash mode in which the focused dot is drawn out into a dash.

deformation potential The effective electric potential experienced by free electrons in a semiconductor or metal resulting from a local deformation in the crystal lattice.

defruit To remove random asynchronous replies from the video input of a display unit in a radar beacon system by comparing the video signals on successive sweeps.

degas To drive out and exhaust the gases occluded in the internal parts of an electron tube or other gastight apparatus, generally by heating during evacuation.

degauss 1. To remove, erase, or clear information from a magnetic tape, disk, drum, or core. 2. To neutralize (demagnetize) a magnetic field of, for example, a ship hull or television tube; a direct current of the correct value is sent through a cable around the ship hull; a current-carrying coil is brought up to and then removed from the television tube. Also known as deperm.

degaussing cable A single-conductor or multiple-conductor cable used on ships for degaussing.

degaussing coil A plastic-encased coil, about 1 foot (0.3 meter) in diameter, that can be plugged into a 120-volt alternating-current wall outlet and moved slowly toward and away from a color television picture tube to demagnetize adjacent parts.

degaussing control A control that automatically varies the current in the degaussing cable as a ship changes heading or rolls and pitches.

degenerate amplifier Parametric amplifier with a pump frequency exactly twice the signal frequency, producing an idler frequency equal to that of the signal input; it is considered as a single-frequency device.

degenerate conduction band A band in which two or more orthogonal quantum states exist that have the same energy, the same spin, and zero mean velocity.

degenerate semiconductor A semiconductor in which the number of electrons in the conduction band approaches that of a metal.

degeneration The loss or gain in an amplifier through unintentional negative feedback.

deglitcher A nonlinear filter or other special circuit used to limit the duration of switching transients in digital converters.

degree of current rectification Ratio between the average unidirectional current output and the root mean square value of the alternating current input from which it was derived.

degree of voltage rectification Ratio between the average unidirectional voltage and the root mean square value of the alternating voltage from which it was derived.

de Haas–Van Alphen effect An effect occurring in many complex metals at low temperatures, consisting of a periodic variation in the diamagnetic susceptibility of conduction electrons with changes in the component of the applied magnetic field at right angles to the principal axis of the crystal.

deion circuit breaker Circuit breaker built so that the arc that forms when the circuit is broken is magnetically blown into a stack of insulated copper plates, giving the effect of a large number of short arcs in series; each arc becomes almost instantly deionized when the current drops to zero in the alternating current cycle, and the arc cannot reform.

deionization The return of an ionized gas to its neutral state after all sources of ionization have been removed, involving diffusion of ions to the container walls and volume recombination of negative and positive ions.

deionization potential The potential at which ionization of the gas in a gas-filled tube ceases and conduction stops.

deionization time The time required for a gas tube to regain its preconduction characteristics after interruption of anode current, so that the grid regains control. Also called recontrol time.

de la Rue and Miller's law The law that in a field between two parallel plates, the sparking potential of a gas is a function of the product of gas pressure and sparking distance only.

delay circuit *See* time-delay circuit.

delay distortion Phase distortion in which the rate of change of phase shift with frequency of a circuit or system is not constant over the frequency range required for transmission. Also called envelope delay distortion.

delayed automatic gain control An automatic gain control system that does not operate until the signal exceeds a predetermined magnitude; weaker signals thus receive maximum amplification. Also known as biased automatic gain control; delayed automatic volume control; quiet automatic volume control.

delayed automatic volume control *See* delayed automatic gain control.

delayed PPI A plan position indicator in which initiation of the time base is delayed a fixed time after each transmitted pulse, to give expansion of the range scale for distant targets so that they show more clearly on the screen.

delayed sweep A sweep whose beginning is delayed for a definite time after the pulse that initiates the sweep.

delay equalizer A corrective network used to make the phase delay or envelope delay of a circuit or system substantially constant over a desired frequency range.

delay flip-flop *See* D flip-flop.

delay line A transmission line (as dissipationless as possible), or an electric network approximation of it, which, if terminated in its characteristic impedance, will reproduce at its output a waveform applied to its input terminals with little distortion, but at a time delayed by an amount dependent upon the electrical length of the line. Also known as artificial delay line.

98 delay-line memory

delay-line memory See circulating memory.

delay-line storage See circulating memory.

delay multivibrator A monostable multivibrator that generates an output pulse a predetermined time after it is triggered by an input pulse.

delay relay A relay having predetermined delay between energization and closing of contacts or between deenergization and dropout.

delay time The time taken for collector current to start flowing in a transistor that is being turned on from the cutoff condition.

delay unit Unit of a radar system in which pulses may be delayed a controllable amount.

Delco Remy distributor A distributor in which two breaker arms are connected in parallel, one coil and one capacitor are used, and one set of contacts closes a few degrees after the other is broken.

delta The difference between a partial-select output of a magnetic cell in a one state and a partial-select output of the same cell in a zero state.

delta connection A combination of three components connected in series to form a triangle like the Greek letter delta. Also known as mesh connection.

delta current Electricity going through a delta connection.

delta-gun tube A color television picture tube in which three electron guns, arranged in a triangle, provide electron beams that fall on phosphor dots on the screen, causing them to emit light in three primary colors; a shadow mask located just behind the screen ensures that each beam excites only dots of one color.

delta matching transformer Impedance device used to match the impedance of an open-wire transmission line to an antenna; the two ends of the transmission line are fanned out so that the impedance of the line gradually increases; the ends of the transmission line are attached to the antenna at points of equal impedance, symmetrically located with respect to the center of the antenna.

delta modulation A pulse-modulation technique in which a continuous signal is converted into a binary pulse pattern, for transmission through low-quality channels.

delta network A set of three branches connected in series to form a mesh.

delta pulse code modulation A modulation system that converts audio signals into corresponding trains of digital pulses to give greater freedom from interference during transmission over wire or radio channels.

delta-Y transformation See Y-delta transformation.

deltic method A method of sampling incoming radar, sonar, seismic, speech, or other waveforms along with reference signals, compressing the samples in time, and comparing them by autocorrelation.

demagnetization 1. The process of reducing or removing the magnetism of a ferromagnetic material. 2. The reduction of magnetic induction by the internal field of a magnet.

demagnetizer A device for removing undesired magnetism, as from the playback head of a tape recorder or from a recorded reel of magnetic tape that is to be erased.

demand See demand factor.

demand factor The ratio of the maximum demand of a building for electric power to the total connected load. Also known as demand.

demand limiter See current limiter.

demand rate The maximum amount of electric power that must be kept available to a customer.

Destriau effect 99

Dember effect Creation of a voltage in a conductor or semiconductor by illumination of one surface. Also known as photodiffusion effect.

demodulator See detector.

demountable tube High-power radio tube having a metal envelope with porcelain insulation; can be taken apart for inspection and for renewal of electrodes.

demultiplexer A device used to separate two or more signals that were previously combined by a compatible multiplexer and transmitted over a single channel.

demultiplexing circuit A circuit used to separate the signals that were combined for transmission by multiplex.

density modulation Modulation of an electron beam by making the density of the electrons in the beam vary with time.

density of states A function of energy E equal to the number of quantum states in the energy range between E and $E + dE$ divided by the product of dE and the volume of the substance.

depletion Reduction of the charge-carrier density in a semiconductor below the normal value for a given temperature and doping level.

depletion layer An electric double layer formed at the surface of contact between a metal and a semiconductor having different work functions, because the mobile carrier charge density is insufficient to neutralize the fixed charge density of donors and acceptors. Also known as barrier layer (deprecated); blocking layer (deprecated); space-charge layer.

depletion-layer rectification Rectification at the junction between dissimilar materials, such as a pn junction or a junction between a metal and a semiconductor. Also known as barrier-layer rectification.

depletion-layer transistor A transistor that relies directly on motion of carriers through depletion layers, such as spacistor.

depletion-mode field-effect transistor See junction-gate field-effect transistor.

depletion region The portion of the channel in a metal oxide field-effect transistor in which there are no charge carriers.

depolarization The removal or prevention of polarization in a substance (for example, through the use of a depolarizer in an electric cell) or of polarization arising from the field due to the charges induced on the surface of a dielectric when an external field is applied.

depolarization factor The ratio of the internal electric field induced by the charges on the surface of a dielectric when an external field is applied to the polarization of the dielectric.

deposited carbon resistor A resistor in which the resistive element is a carbon film pyrolytically deposited on a ceramic substrate.

derating The reduction of the rating of a device to improve reliability or to permit operation at high ambient temperatures.

DeSauty's bridge A four-arm bridge used to compare two capacitances; two adjacent arms contain capacitors in series with resistors, while the other two arms contain resistors only. Also known as Wien–DeSauty bridge.

despun antenna Satellite directional antenna pointed continuously at earth by electrically or mechanically despinning the antenna at the same rate that the satellite is spinning for stabilization.

Destriau effect Sustained emission of light by suitable phosphor powders that are embedded in an insulator and subjected only to the action of an alternating electric field.

destructive breakdown

destructive breakdown Breakdown of the barrier between the gate and channel of a field-effect transistor, causing failure of the transistor.

detectivity The normalized radiation power required to give a signal from a photoconductor that is equal to the noise.

detector The stage in a receiver at which demodulation takes place; in a superheterodyne receiver this is called the second detector. Also known as demodulator; envelope detector.

detector balanced bias Controlling circuit used in radar systems for anticlutter purposes.

detune To change the inductance or capacitance of a tuned circuit so its resonant frequency is different from the incoming signal frequency.

detuning stub Quarter-wave stub used to match a coaxial line to a sleeve-stub antenna; the stub detunes the outside of the coaxial feed line while tuning the antenna itself.

deuterium discharge tube A tube similar to a hydrogen discharge lamp, but with deuterium replacing the hydrogen; source of high-intensity ultraviolet radiation for spectroscopic microanalysis.

device An electronic element that cannot be divided without destroying its stated function; commonly applied to active elements such as transistors and transducers.

D flip-flop A flip-flop whose output is a function of the input which appeared one pulse earlier. Also known as delay flip-flop.

DG synchro amplifier Synchro differential generator driven by servosystem.

diac *See* trigger diode.

diactor Direct-acting automatic regulator for control of shunt generator voltage output.

diagonal horn antenna Horn antenna in which all cross sections are square and the electric vector is parallel to one of the diagonals; the radiation pattern in the far field has almost perfect circular symmetry.

dial jacks Strip of jacks associated with and bridged to a regular outgoing trunk jack circuit to provide a connection between the dial cords and the outgoing trunks.

dial key Key unit of the subscriber's cord circuit used to connect the dial into the line.

dial lamp A small lamp used to illuminate a dial.

dial leg Conductor in a circuit brought out for direct-current dial signaling.

dial pulse interpreter A device that converts the signaling pulses of a dial telephone to a form suitable for data entry to a computer.

diamagnet A substance which is diamagnetic, such as the alkali and alkaline earth metals, the halogens, and the noble gases.

diamagnetic Having a magnetic permeability less than 1; materials with this property are repelled by a magnet and tend to position themselves at right angles to magnetic lines of force.

diamagnetic susceptibility The susceptibility of a diamagnetic material, which is always negative and usually on the order of -10^{-5} cm^3/mole.

diamagnetism The property of a material which is repelled by magnets.

diamond antenna *See* rhombic antenna.

diamond circuit A gate circuit that provides isolation between input and output terminals in its off state, by operating transistors in their cutoff region; in the on state the output voltage follows the input voltage as required for gating both analog and

dielectric heating 101

digital signals, while the transistors provide current gain to supply output current on demand.

diaphragm See iris.

diathermy machine A radio-frequency oscillator, sometimes followed by rf amplifier stages, used to generate high-frequency currents that produce heat within some part of the body for therapeutic purposes.

di-cap storage Device capable of holding data in the form of an array of charged capacitors and using diodes for controlling information flow.

DICE See digital intercontinental conversion equipment.

dicing Sawing or otherwise machining a semiconductor wafer into small squares, or dice, from which transistors and diodes can be fabricated.

Dicke fix Technique designed to protect a receiver from fast sweep jamming.

Dicke radiometer A radiometer-type receiver that detects weak signals in noise by modulating or switching the incoming signal before it is processed by conventional receiver circuits.

die The tiny, sawed or otherwise machined piece of semiconductor material used in the construction of a transistor, diode, or other semiconductor device; plural is dice.

dielectric absorption 1. The persistence of electric polarization in certain dielectrics after removal of the electric field. 2. See dielectric loss.

dielectric amplifier An amplifier using a ferroelectric capacitor whose capacitance varies with applied voltage so as to give signal amplification.

dielectric antenna An antenna in which a dielectric is the major component used to produce a desired radiation pattern.

dielectric breakdown Breakdown which occurs in an alkali halide crystal at field strengths on the order of 10^6 volts per centimeter.

dielectric circuit Any electric circuit which has capacitors.

dielectric constant 1. For an isotropic medium, the ratio of the capacitance of a capacitor filled with a given dielectric to that of the same capacitor having only a vacuum as dielectric. 2. More generally, $1 + \gamma\chi$, where γ is 4π in Gaussian and cgs electrostatic units or 1 in rationalized mks units, and χ is the electric susceptibility tensor. Also known as relative dielectric constant; relative permittivity; specific inductive capacity (SIC).

dielectric crystal A crystal which is electrically nonconducting.

dielectric current The current flowing at any instant through a surface of a dielectric that is located in a changing electric field.

dielectric displacement See electric displacement.

dielectric ellipsoid For an anisotropic medium in which the dielectric constant is a tensor quantity $\mathbf{K}$, the locus of points $\mathbf{r}$ satisfying $\mathbf{r}\cdot\mathbf{K}\cdot\mathbf{r} = 1$.

dielectric fatigue The property of some dielectrics in which resistance to breakdown decreases after a voltage has been applied for a considerable time.

dielectric film A film possessing dielectric properties; used as the central layer of a capacitor.

dielectric flux density See electric displacement.

dielectric gas A gas having a high dielectric constant, such as sulfur hexafluoride.

dielectric heating Heating of a nominally electrical insulating material due to its own electrical (dielectric) losses, when the material is placed in a varying electrostatic field.

102 dielectric hysteresis

dielectric hysteresis *See* ferroelectric hysteresis.

dielectric lens A lens made of dielectric material so that it refracts radio waves in the same manner that an optical lens refracts light waves; used with microwave antennas.

dielectric-lens antenna An aperture antenna in which the beam width is determined by the dimensions of a dielectric lens through which the beam passes.

dielectric loss The electric energy that is converted into heat in a dielectric subjected to a varying electric field. Also known as dielectric absorption.

dielectric loss angle Difference between 90° and the dielectric phase angle.

dielectric loss factor Product of the dielectric constant of a material and the tangent of its dielectric loss angle.

dielectric matching plate In waveguide technique, a dielectric plate used as an impedance transformer for matching purposes.

dielectric phase angle Angular difference in phase between the sinusoidal alternating potential difference applied to a dielectric and the component of the resulting alternating current having the same period as the potential difference.

dielectric polarization *See* polarization.

dielectric power factor Cosine of the dielectric phase angle (or sine of the dielectric loss angle).

dielectric-rod antenna A surface-wave antenna in which an end-fire radiation pattern is produced by propagation of a surface wave on a tapered dielectric rod.

dielectric shielding The reduction of an electric field in some region by interposing a dielectric substance, such as polystyrene, glass, or mica.

dielectric soak *See* absorption.

dielectric strength The maximum electrical potential gradient that a material can withstand without rupture; usually specified in volts per millimeter of thickness. Also known as electric strength.

dielectric susceptibility *See* electric susceptibility.

dielectric test A test involving application of a voltage higher than the rated value for a specified time, to determine the margin of safety against later failure of insulating materials.

dielectric waveguide A waveguide consisting of a dielectric cylinder surrounded by air.

dielectric wedge A wedge-shaped piece of dielectric used in a waveguide to match its impedance to that of another waveguide.

dielectric wire A dielectric waveguide used to transmit ultra-high-frequency radio waves short distances between parts of a circuit.

difference amplifier *See* differential amplifier.

difference detector A detector circuit in which the output is a function of the difference between the amplitudes of the two input waveforms.

differential amplifier An amplifier whose output is proportional to the difference between the voltages applied to its two inputs. Also called difference amplifier.

differential capacitance The derivative with respect to voltage of a charge characteristic, such as an alternating charge characteristic or a mean charge characteristic, at a given point on the characteristic.

differential capacitor A two-section variable capacitor having one rotor and two stators so arranged that as capacitance is reduced in one section it is increased in the other.

differential operational amplifier 103

differential comparator A comparator having at least two high-gain differential-amplifier stages, followed by level-shifting and buffering stages, as required for converting a differential input to single-ended output for digital logic applications.

differential compound motor A direct-current motor whose speed may be made nearly constant or may be adjusted to increase with increasing load.

differential discriminator A discriminator that passes only pulses whose amplitudes are between two predetermined values, neither of which is zero.

differential duplex system System in which the sent currents divide through two mutually inductive sections of a receiving apparatus, connected respectively to the line and to a balancing artificial line in opposite directions, so that there is substantially no net effect on the receiving apparatus; the received currents pass mainly through one section, or through the two sections in the same direction, and operate the apparatus.

differential frequency circuit A circuit that provides a continuous output frequency equal to the absolute difference between two continuous input frequencies.

differential gain control Device for altering the gain of a radio receiver according to expected change of signal level, to reduce the amplitude differential between the signals at the output of the receiver. Also known as gain sensitivity control.

differential galvanometer A galvanometer having a magnetic needle which is free to rotate in the magnetic field produced by currents flowing in opposite directions through two separate identical coils, so that there is no deflection when the currents are equal.

differential input Amplifier input circuit that rejects voltages that are the same at both input terminals and amplifies the voltage difference between the two input terminals.

differential-input capacitance The capacitance between the inverting and noninverting input terminals of a differential amplifier.

differential-input impedance The impedance between the inverting and noninverting input terminals of a differential amplifier.

differential-input measurement A measurement in which the two inputs to a differential amplifier are connected to two points in a circuit under test and the amplifier displays the difference voltage between the points.

differential-input resistance The resistance between the inverting and noninverting input terminals of a differential amplifier.

differential-input voltage The maximum voltage that can be applied across the input terminals of a differential amplifier without causing damage to the amplifier.

differential keying Method for obtaining chirp-free break-in keying of continuous wave transmitters by using circuitry that arranges to have the oscillator turn on fast before the keyed amplifier stage can pass any signal, and turn off fast after the keyed amplifier stage has cut off.

differential-mode gain The ratio of the output voltage of a differential amplifier to the differential-mode input voltage.

differential-mode input The voltage difference between the two inputs of a differential amplifier.

differential-mode signal A signal that is applied between the two ungrounded terminals of a balanced three-terminal system.

differential operational amplifier An amplifier that has two input terminals, used with additional circuit elements to perform mathematical functions on the difference in voltage between the two input signals.

104 differential output voltage

differential output voltage The difference between the values of two ac voltages, 180° out of phase, present at the output terminals of an amplifier when a differential input voltage is applied to the input terminals of the amplifier.

differential permeability The slope of the magnetization curve for a magnetic material.

differential phase Difference in output phase of a small high-frequency sine-wave signal at two stated levels of a low-frequency signal on which it is superimposed in a video transmission system.

differential pressure pickup An instrument that measures the difference in pressure between two pressure sources and translates this difference into a change in inductance, resistance, voltage, or some other electrical quality.

differential relay A two-winding relay that operates when the difference between the currents in the two windings reaches a predetermined value.

differential selsyn Selsyn in which both rotor and stator have similar windings that are spread 120° apart; position of the rotor corresponds to the algebraic sum of the fields produced by the stator and rotor.

differential stage A symmetrical amplifier stage with two inputs balanced against each other so that with no input signal or equal input signals, no output signal exists, while a signal to either input, or an input signal unbalance, produces an output signal proportional to the difference.

differential synchro See synchro differential receiver; synchro differential transmitter.

differential transducer A transducer that simultaneously senses two separate sources and provides an output proportional to the difference between them.

differential transformer A transformer used to join two or more sources of signals to a common transmission line.

differential-transformer transducer A transducer in which movement of the iron core of a transformer varies the output voltage across two series-opposing secondary windings.

differential voltage gain Ratio of the change in output signal voltage at either terminal, or in a differential device, to the change in signal voltage applied to either input terminal, all voltages being measured to common reference.

differential voltmeter A voltmeter that measures only the difference between a known voltage and an unknown voltage.

differential winding A winding whose magnetic field opposes that of a nearby winding.

differential wound field Type of motor or generator field having both series and shunt coils that are connected to oppose each other.

differentiating circuit A circuit whose output voltage is proportional to the rate of change of the input voltage. Also known as differentiating network.

differentiating network See differentiating circuit.

differentiator A device whose output function is proportional to the derivative, or rate of change, of the input function with respect to one or more variables.

diffraction propagation Propagation of electromagnetic waves around objects or over the horizon by diffraction.

diffraction zone The portion of a radio propagation path which lies outside a line-of-sight path.

diffused-alloy transistor A transistor in which the semiconductor wafer is subjected to gaseous diffusion to produce a nonuniform base region, after which alloy junctions are formed in the same manner as for an alloy-junction transistor; it may also have an intrinsic region, to give a *pnip* unit. Also known as drift transistor.

digital filter 105

diffused-base transistor A transistor in which a nonuniform base region is produced by gaseous diffusion; the collector-base junction is also formed by gaseous diffusion, while the emitter-base junction is a conventional alloy junction.

diffused emitter-collector transistor A transistor in which both the emitter and collector are produced by diffusion.

diffused junction A semiconductor junction that has been formed by the diffusion of an impurity within a semiconductor crystal.

diffused-junction rectifier A semiconductor diode in which the pn junction is produced by diffusion.

diffused-junction transistor A transistor in which the emitter and collector electrodes have been formed by diffusion by an impurity metal into the semiconductor wafer without heating.

diffused-mesa transistor A diffused-junction transistor in which an n-type impurity is diffused into one side of a p-type wafer; a second pn junction, required for the emitter, is produced by alloying or diffusing a p-type impurity into the newly formed n-type surface; after contacts have been applied, undesired diffused areas are etched away to create a flat-topped peak called a mesa.

diffused resistor An integrated-circuit resistor produced by a diffusion process in a semiconductor substrate.

diffusion 1. A method of producing a junction by diffusing an impurity metal into a semiconductor at a high temperature. 2. The actual transport of mass, in the form of discrete atoms, through the lattice of a crystalline solid. 3. The movement of carriers in a semiconductor.

diffusion capacitance The rate of change of stored minority-carrier charge with the voltage across a semiconductor junction.

diffusion constant The diffusion current density in a homogeneous semiconductor divided by the charge carrier concentration gradient.

diffusion theory The theory that in semiconductors, where there is a variation of carrier concentration, a motion of the carriers is produced by diffusion in addition to the drift determined by the mobility and the electric field.

diffusion transistor A transistor in which current flow is a result of diffusion of carriers, donors, or acceptors, as in a junction transistor.

digit absorbing selector Dial switch arranged to set up and then fall back on the first one of two digits dialed; it then operates on the next digit dialed.

digital circuit A circuit designed to respond at input voltages at one of a finite number of levels and, similarly, to produce output voltages at one of a finite number of levels.

digital comparator A comparator circuit operating on input signals at discrete levels. Also known as discrete comparator.

digital converter A device that converts voltages to digital form; examples include analog-to-digital converters, pulse-code modulators, encoders, and quantizing encoders.

digital counter A discrete-state device (one with only a finite number of output conditions) that responds by advancing to its next output condition.

digital delay generator A high-precision adjustable time-delay generator in which delays may be selected in increments such as 1, 10, or 100 nanoseconds by means of panel switches and sometimes by remote programming.

digital filter An electrical filter that responds to an input which has been quantified, usually as pulses.

106 digital frequency meter

digital frequency meter A frequency meter in which the value of the frequency being measured is indicated on a digital display.

digital intercontinental conversion equipment Equipment which uses pulse-code modulation to convert a 525-line, 60-frame-per-second television signal used in the United States into a 625-line, 50-frame-per-second phase-alternation line signal used in Europe; the 525-line signal is sampled and quantized into a pulse-code modulation signal which is stored in shift registers from which the phase-alternation line signal is read out. Abbreviated DICE.

digital message entry system A system that encodes formatted messages in digital form; it enters the encoded digital information into a voice communications transceiver by frequency shift techniques.

digital multiplier A multiplier that accepts two numbers in digital form and gives their product in the same digital form, usually by making repeated additions; the multiplying process is simpler if the numbers are in binary form wherein digits are represented by a 0 or 1.

digital output An output signal consisting of a sequence of discrete quantities coded in an appropriate manner for driving a printer or digital display.

digital phase shifter Device which provides a signal phase shift by the application of a control pulse; a reversal or phase shift requires a control pulse of opposite polarity.

digital plotter A recorder that produces permanent hard copy in the form of a graph from digital input data.

digital recording Magnetic recording in which the information is first coded in a digital form, generally with a binary code that uses two discrete values of residual flux.

digital signal analyzer A signal analyzer in which one or more analog inputs are sampled at regular intervals, converted to digital form, and fed to a memory.

digital synchronometer A time comparator that provides a direct-reading digital display of time with high precision by making accurate comparisons between its own digital clock and high-accuracy time transmissions from radio station WWV or a loran C station.

digital television converter A converter used to convert television programs from one system to another, such as for converting 525-line 60-field United States broadcasts to 625-line 50-field European PAL (phase-alternation line) or SECAM (sequential couleur à memoire) standards; the video signal is digitized before conversion.

digital-to-analog converter A converter in which digital input signals are changed to essentially proportional analog signals. Abbreviated dac.

digital-to-synchro converter A converter that changes BCD or other digital input data to a three-wire synchro output signal representing corresponding angular data.

digital transducer A transducer that measures physical quantities and transmits the information as coded digital signals rather than as continuously varying currents or voltages.

digital voltmeter A voltmeter in which the unknown voltage is compared with an internally generated analog voltage, the result being indicated in digital form rather than by a pointer moving over a meter scale.

digit delay element A logic element that introduces a delay of one digit period in a series of signals or pulses.

digit period The time interval between successive pulses, usually representing binary digits, in a computer or in pulse modulation, determined by the pulse-repetition frequency. Also known as digit time.

diode function generator 107

digit pulse An electrical pulse which induces a magnetizing force in a number of magnetic cores in a computer storage, all corresponding to a particular digit position in a number of different words.

digit time *See* digit period.

diheptal base A tube base having 14 pins or 14 possible pin positions; used chiefly on television cathode-ray tubes.

DIM *See* nonthermal decimetric emission.

dimmer An electrical or electronic control for varying the intensity of a lamp or other light source.

dina An airborne radar-jamming transmitter operating in the band from 92 to 210 megahertz with an output of 30 watts, radiating noise in one side band for spot or barrage jamming; the carrier and the other side band are suppressed.

D indicator *See* D scope.

diode 1. A two-electrode electron tube containing an anode and a cathode. 2. *See* semiconductor diode.

diode alternating-current switch *See* trigger diode.

diode amplifier A microwave amplifier using an IMPATT, TRAPATT, or transferred-electron diode in a cavity, with a microwave circulator providing the input/output isolation required for amplification; center frequencies are in the gigahertz range, from about 1 to 100 gigahertz, and power outputs are up to 20 watts continuous-wave or more than 200 watts pulsed, depending on the diode used.

diode bridge A series-parallel configuration of four diodes, whose output polarity remains unchanged whatever the input polarity.

diode-capacitor transistor logic A circuit that uses diodes, capacitors, and transistors to provide logic functions.

diode characteristic The composite electrode characteristic of an electron tube when all electrodes except the cathode are connected together.

diode clamp *See* diode clamping circuit.

diode clamping circuit A clamping circuit in which a diode provides a very low resistance whenever the potential at a certain point rises above a certain value in some circuits or falls below a certain value in others. Also known as diode clamp.

diode clipping circuit A clipping circuit in which a diode is used as a switch to perform the clipping action.

diode-connected transistor A bipolar transistor in which two terminals are shorted to give diode action.

diode demodulator A demodulator using one or more crystal or electron tube diodes to provide a rectified output whose average value is proportional to the original modulation. Also known as diode detector.

diode detector *See* diode demodulator.

diode drop *See* diode forward voltage.

diode forward voltage The voltage across a semiconductor diode that is carrying current in the forward direction; it is usually approximately constant over the range of currents commonly used. Also known as diode drop; diode voltage; forward voltage drop.

diode function generator A function generator that uses the transfer characteristics of resistive networks containing biased diodes; the desired function is approximated by linear segments.

diode gate An AND gate that uses diodes as switching elements.

diode limiter A peak-limiting circuit employing a diode that becomes conductive when signal peaks exceed a predetermined value.

diode logic An electronic circuit using current-steering diodes, such that the relations between input and output voltages correspond to AND or OR logic functions.

diode matrix A two-dimensional array of diodes used for a variety of purposes such as decoding and read-only memory.

diode mixer A mixer that uses a crystal or electron tube diode; it is generally small enough to fit directly into a radio-frequency transmission line.

diode modulator A modulator using one or more diodes to combine a modulating signal with a carrier signal; used chiefly for low-level signaling because of inherently poor efficiency.

diode pack Combination of two or more diodes integrated into one solid block.

diode peak detector Diode used in a circuit to indicate when peaks exceed a predetermined value.

diode-pentode Vacuum tube having a diode and a pentode in the same envelope.

diode rectifier A half-wave rectifier of two elements between which current flows in only one direction.

diode rectifier-amplifier meter The most widely used vacuum tube voltmeter for measurement of alternating-current voltage; has separate tubes for rectification and direct-current amplification, permitting an optimum design for each.

diode-switch Diode which is made to act as a switch by the successive application of positive and negative biasing voltages to the anode (relative to the cathode), thereby allowing or preventing, respectively, the passage of other applied waveforms within certain limits of voltage.

diode theory The theory that in a semiconductor, when the barrier thickness is comparable to or smaller than the mean free path of the carriers, then the carriers cross the barrier without being scattered, much as in a vacuum tube diode.

diode transistor logic A circuit that uses diodes, transistors, and resistors to provide logic functions. Abbreviated DTL.

diode-triode Vacuum tube having a diode and a triode in the same envelope.

diode voltage *See* diode forward voltage.

diode voltage regulator A voltage regulator with a Zener diode, making use of its almost constant voltage over a range of currents. Also known as Zener diode voltage regulator.

DIP *See* dual in-line package.

diplexer A coupling system that allows two different transmitters to operate simultaneously or separately from the same antenna.

diplex reception Simultaneous reception of two signals which have some features in common, such as a single receiving antenna or a single carrier frequency.

dipole Any object or system that is oppositely charged at two points, or poles, such as a magnet or a polar molecule; more precisely, the limit as either charge goes to infinity, the separation distance to zero, while the product remains constant. Also known as doublet; electric doublet.

dipole antenna An antenna approximately one-half wavelength long, split at its electrical center for connection to a transmission line whose radiation pattern has a

direct-current motor control 109

maximum at right angles to the antenna. Also known as doublet antenna; half-wave dipole.

dipole disk feed Antenna, consisting of a dipole near a disk, used to reflect energy to the disk.

dipole moment See electric dipole moment; magnetic dipole moment.

dipole radiation The electromagnetic radiation generated by an oscillating electric or magnetic dipole.

dipole relaxation The process, occupying a certain period of time after a change in the applied electric field, in which the orientation polarization of a substance reaches equilibrium.

direct-aperture antenna An antenna whose conductor or dielectric is a surface or solid, such as a horn, mirror, or lens.

direct-coupled amplifier A direct-current amplifier in which a resistor or a direct connection provides the coupling between stages, so small changes in direct currents can be amplified.

direct-coupled transistor logic Integrated-circuit logic using only resistors and transistors, with direct conductive coupling between the transistors; speed can be up to 1 megahertz. Abbreviated DCTL.

direct coupling Coupling of two circuits by means of a non-frequency-sensitive device, such as a wire, resistor, or battery, so both direct and alternating current can flow through the coupling path.

direct current Electric current which flows in one direction only, as opposed to alternating current. Abbreviated dc.

direct-current amplifier An amplifier that is capable of amplifying dc voltages and slowly varying voltages.

direct-current circuit Any combination of dc voltage or current sources, such as generators and batteries, in conjunction with transmission lines, resistors, and power converters such as motors.

direct-current circuit theory An analysis of relationships within a dc circuit.

direct-current continuity Property of a circuit in which there is an established pathway for conduction of current from a direct-current source.

direct-current coupling That type of coupling in which the zero-frequency term of the Fourier series representing the input signal is transmitted.

direct-current discharge The passage of a direct current through a gas.

direct-current dump Removal of all direct-current power from a computer system or component intentionally, accidentally, or conditionally; in some types of storage, this results in loss of stored information.

direct-current erase Use of direct current to energize an erasing head of a tape recorder.

direct-current generator A rotating electric machine that converts mechanical power into dc power.

direct-current inserter A television transmitter stage that adds to the video signal a dc component known as the pedestal level.

direct-current motor An electric rotating machine energized by direct current and used to convert electric energy to mechanical energy.

direct-current motor control See electronic motor control.

direct-current offset A direct-current level that may be added to the input signal of an amplifier or other circuit.

direct-current plate resistance Value or characteristic used in vacuum-tube computations; it is equal to the direct-current plate voltage divided by the direct-current plate current.

direct-current power The power delivered by a dc power system, equal to the line voltage times the load current.

direct-current power supply A power supply that provides one or more dc output voltages, such as a dc generator, rectifier-type power supply, converter, or dynamotor.

direct-current receiver A radio receiver designed to operate directly from a 115-volt dc power line.

direct-current reinsertion See clamping.

direct-current restoration See clamping.

direct-current restorer A clamp circuit used to establish a dc reference level in a signal without modifying to any important degree the waveform of the signal itself. Also known as clamper; reinserter.

direct-current signaling A transmission method that uses direct current.

direct-current SQUID A type of superconducting quantum interference device (SQUID) which contains two Josephson junctions in a superconducting loop; its state is determined from direct-current measurements.

direct-current tachometer A dc generator operating with negligible load current and with constant field flux provided by a permanent magnet, so its dc output voltage is proportional to speed.

direct-current transducer A transducer that requires dc excitation and provides a dc output that varies with the parameter being sensed.

direct-current vacuum-tube voltmeter The amplifying and indicating portions of the diode rectifier-amplifier meter, which are usually designed so that the diode rectifier can be disconnected for dc measurements.

direct-current voltage See direct voltage.

direct-current working volts The maximum continuously applied dc voltage for which a capacitor is rated. Abbreviated dcwV.

direct electromotive force Unidirectional electromotive force in which the changes in values are either zero or so small that they may be neglected.

direct-gap semiconductor A semiconductor in which the minimum of the conduction band occurs at the same wave vector as the maximum of the valence band, and recombination radiation consequently occurs with relatively large intensity.

direct grid bias See grid bias.

direct interelectrode capacitance See interelectrode capacitance.

directional antenna An antenna that radiates or receives radio waves more effectively in some directions than others.

directional beam A radio or radar wave that is concentrated in a given direction.

directional coupler A device that couples a secondary system only to a wave traveling in a particular direction in a primary transmission system, while completely ignoring a wave traveling in the opposite direction. Also known as directive feed.

directional filter A low-pass, band-pass, or high-pass filter that separates the bands of frequencies used for transmission in opposite directions in a carrier system. Also known as directional separation filter.

discharge 111

directional pattern See radiation pattern.

directional phase shifter Passive phase shifter in which the phase change for transmission in one direction differs from that for transmission in the opposite direction.

directional relay Relay which functions in conformance with the direction of power, voltage, current, pulse, rotation, and so on.

directional separation filter See directional filter.

direct ionization See extrinsic photoemission.

direction rectifier A rectifier that supplies a direct-current voltage whose magnitude and polarity vary with the magnitude and relative polarity of an alternating-current synchro error voltage.

directive feed See directional coupler.

directive gain Of an antenna in a given direction, 4π times the ratio of the radiation intensity in that direction to the total power radiated by the antenna.

directivity 1. The ability of a logic circuit to ensure that the input signal is not affected by the output signal. 2. The value of the directive gain of an antenna in the direction of its maximum value. 3. The ratio of the power measured at the forward-wave sampling terminals of a directional coupler, with only a forward wave present in the transmission line, to the power measured at the same terminals when the direction of the forward wave in the line is reversed; the ratio is usually expressed in decibels.

directly heated cathode See filament.

director 1. Telephone switch which translates the digits dialed into the directing digits actually used to switch the call. 2. A parasitic element placed a fraction of a wavelength ahead of a dipole receiving antenna to increase the gain of the array in the direction of the major lobe.

direct piezoelectricity Name sometimes given to the piezoelectric effect in which an electric charge is developed on a crystal by the application of mechanical stress.

direct point repeater Telegraph repeater in which the receiving relay controlled by the signals received over a line repeats corresponding signals directly into another line or lines without the interposition of any other repeating or transmitting apparatus.

direct resistance-coupled amplifier Amplifier in which the plate of one stage is connected either directly or through a resistor to the control grid of the next stage, with the plate-load resistor being common to both stages; used to amplify small changes in direct current.

direct route In wire communications, the trunks that connect a pair of switching centers, regardless of the geographical direction the actual trunk facilities may follow.

direct stroke A lightning stroke that actually strikes some part of a power or communication system.

direct-view storage tube A cathode-ray tube in which secondary emission of electrons from a storage grid is used to provide an intensely bright display for long and controllable periods of time. Also known as display storage tube; viewing storage tube.

direct voltage A voltage that forces electrons to move through a circuit in the same direction continuously, thereby producing a direct current. Also known as direct-current voltage.

direct-wire circuit Supervised protective signaling circuit usually consisting of one metallic conductor and a ground return and having signal-receiving equipment responsive to either an increase or a decrease in current.

discharge 1. To remove a charge from a battery, capacitor, or other electric-energy storage device. 2. The passage of electricity through a gas, usually accompanied by a glow, arc, spark, or corona. Also known as electric discharge.

discharge key Device for switching a capacitor suddenly from a charging circuit to a load through which it can discharge.

discharge lamp A lamp in which light is produced by an electric discharge between electrodes in a gas (or vapor) at low or high pressure. Also known as electric-discharge lamp; gas-discharge lamp; vapor lamp.

discharger A silver-impregnated cotton wick encased in a flexible plastic tube with an aluminum mounting lug, used on aircraft to reduce precipitation static.

discharge tube An evacuated enclosure containing a gas at low pressure, through which current can flow when sufficient voltage is applied between metal electrodes in the tube. Also known as electric-discharge tube.

discone antenna A biconical antenna in which one of the cones is spread out to 180° to form a disk; the center conductor of the coaxial line terminates at the center of the disk, and the cable shield terminates at the vertex of the cone.

disconnect To open a circuit by removing wires or connections, as distinguished from opening a switch to stop current flow.

disconnect fitting An electrical connection that can be disconnected without tools.

disconnecting switch A switch that isolates a circuit or piece of electrical apparatus after interruption of the current. Also known as disconnector.

disconnector *See* disconnecting switch.

disconnector release Device which disengages the apparatus used in a telephone connection to restore it to its original condition when not in use.

discontinuity An abrupt change in the shape of a waveguide.

discontinuous amplifier Amplifier in which the input waveform is reproduced on some type of averaging basis.

discrete comparator *See* digital comparator.

discrete sampling Sampling in which the individual samples are of such long duration that the frequency response of the channel is not deteriorated by the sampling process.

discriminator A circuit in which magnitude and polarity of the output voltage depend on how an input signal differs from a standard or from another signal.

discriminator transformer A transformer designed to be used in a stage where frequency-modulated signals are converted directly to audio-frequency signals or in a stage where frequency changes are converted to corresponding voltage changes.

dish *See* parabolic reflector.

disintegration voltage The lowest anode voltage at which destructive positive-ion bombardment of the cathode occurs in a hot-cathode gas tube.

disk armature The armature in a motor that has a disk winding or is made up of a metal disk.

disk memory *See* disk storage.

disk-seal tube An electron tube having disk-shaped electrodes arranged in closely spaced parallel layers, to give low interelectrode capacitance along with high power output, up to 2500 megahertz. Also known as lighthouse tube; megatron.

disk storage An external computer storage device consisting of one or more disks spaced on a common shaft, and magnetic heads mounted on arms that reach between the disks to read and record information on them. Also known as disk memory; magnetic disk storage.

dissymmetrical transducer 113

disk thermistor A thermistor which is produced by pressing and sintering an oxide binder mixture into a disk, 0.2–0.6 inch (5–15 millimeters) in diameter and 0.04–0.5 inch (1.0–13 millimeters) thick, coating the major surfaces with conducting material, and attaching leads.

disordered crystalline alloy A mixture of two elements in which the atoms of the mixture are found at more or less random positions on a crystal lattice.

dispenser cathode An electron tube cathode having provisions for continuously replacing evaporated electron-emitting material.

dispersion Scattering of microwave radiation by an obstruction.

dispersive line A delay line that delays each frequency a different length of time.

dispersive medium A medium in which the phase velocity of an electromagnetic wave is a function of frequency.

displacement *See* electric displacement.

displacement angle The change in the phase of an alternator's terminal voltage when a load is applied.

displacement current The rate of change of the electric displacement vector, which must be added to the current density to extend Ampère's law to the case of time-varying fields (meter-kilogram-second units). Also known as Maxwell's displacement current.

display 1. A visible representation of information, in words, numbers, or drawings, as on the cathode-ray tube screen of a radar set, navigation system, or computer console. 2. The device on which the information is projected. Also known as display device. 3. The image of the information.

display device *See* display.

display loss *See* visibility factor.

display storage tube *See* direct-view storage tube.

display tube A cathode-ray tube used to provide a visual display. Also known as visual display unit.

disruptive discharge A sudden and large increase in current through an insulating medium due to complete failure of the medium under electrostatic stress.

dissector tube Camera tube having a continuous photo cathode on which is formed a photoelectric emission pattern which is scanned by moving its electron-optical image over an aperture.

dissipation factor The inverse of Q, the storage factor.

dissipation line A length of stainless steel or Nichrome wire used as a noninductive terminating impedance for a rhombic transmitting antenna when several kilowatts of power must be dissipated.

dissipation loss A measure of the power loss of a transducer in transmitting signals, expressed as the ratio of its input power to its output power.

dissipative tunneling Quantum-mechanical tunneling of individual electrons, rather than pairs, across a thin insulating layer separating two superconducting metals when there is a voltage across this layer, resulting in partial disruption of cooperative motion.

dissipator *See* heat sink.

dissymmetrical network *See* dissymmetrical transducer.

dissymmetrical transducer A transducer whose input and output image impedances are not equal. Also known as dissymmetrical network.

114 distance mark

distance mark A movable point produced on a radar display by a special signal generator, so that when the mark is moved to a target position on the screen the range to the target can be read on the calibrated dial of the signal generator; usually used for gun laying where highly accurate distance is important.

distance protection Effect of a device operative within a predetermined electrical distance on the protected circuit to cause and maintain an interruption of power in a faulty circuit.

distance relay Protective relay, the operation of which is a function of the distance between the relay and the point of fault.

distant field The electromagnetic field at a distance of five wavelengths or more from a transmitter, where the radial electric field becomes negligible.

distortion Any undesired change in the waveform of an electric signal passing through a circuit or other transmission medium.

distributed amplifier A wide-band amplifier in which tubes are distributed along artificial delay lines made up of coils acting with the input and output capacitances of the tubes.

distributed capacitance Capacitance that exists between the turns in a coil or choke, or between adjacent conductors or circuits, as distinguished from the capacitance concentrated in a capacitor.

distributed circuit A film circuit whose effective components cannot be easily recognized as discrete.

distributed constant A circuit parameter that exists along the entire length of a transmission line. Also known as distributed parameter.

distributed-emission photodiode A broad-band photodiode proposed for detection of modulated laser beams at millimeter wavelengths; incident light falls on a photocathode strip that generates a traveling wave of photocurrent having the same wave velocity as the transmission line which the photodiode feeds.

distributed inductance The inductance that exists along the entire length of a conductor, as distinguished from inductance concentrated in a coil.

distributed parameter *See* distributed constant.

distributed paramp Paramagnetic amplifier that consists essentially of a transmission line shunted by uniformly spaced, identical varactors; the applied pumping wave excites the varactors in sequence to give the desired traveling-wave effect.

distributing frame Structure for terminating permanent wires of a central office, private branch exchange, or private exchange, and for permitting the easy change of connections between them by means of cross-connecting wires.

distributing terminal assembly Frame situated between each pair of selector bays to provide terminal facilities for the selector bank wiring and facilities for cross-connection to trunks running to succeeding switches.

distribution amplifier A radio-frequency power amplifier used to feed television or radio signals to a number of receivers, as in an apartment house or a hotel.

distribution cable Cable extending from a feeder cable into a specific area for the purpose of providing service to that area.

distribution center In an alternating-current power system, the point at which control and routing equipment is installed.

distribution control *See* linearity control.

distribution substation An electric power substation associated with the distribution system and the primary feeders for supply to residential, commercial, and industrial loads.

domain theory 115

distribution switchboard Power switchboard used for the distribution of electrical energy at the voltage common for each distribution within a building.

distribution system Circuitry involving high-voltage switchgear, step-down transformers, voltage dividers, and related equipment used to receive high-voltage electricity from a primary source and redistribute it at lower voltages. Also known as electric distribution system.

distribution transformer An element of an electric distribution system located near consumers which changes primary distribution voltage to secondary distribution voltage.

distributor 1. Any device which allocates a telegraph line to each of a number of channels, or to each row of holes on a punched tape, in succession. 2. A rotary switch that directs the high-voltage ignition current in the proper firing sequence to the various cylinders of an internal combustion engine. 3. The electronic circuitry which acts as an intermediate link between the accumulator and drum storage.

distributor points Cam-operated contacts, the opening of which triggers the ignition pulse in an internal combustion engine.

disturbed-one output One output of a magnetic cell to which partial-read pulses have been applied since that cell was last selected for writing.

divergence The spreading of a cathode-ray stream due to repulsion of like charges (electrons).

diversity factor Ratio of the sum of the individual maximum demands to total maximum demand, as applied to an electrical distribution system.

diversity receiver A radio receiver designed for space or frequency diversity reception.

diverter A low resistance which is connected in parallel with the series or compole winding of a direct-current machine and diverts current from it, causing the magnetomotive force produced by the winding to vary.

diverter-pole generator Compound wound direct-current generator with the series winding of the diverter pole opposing the flux generated by the shunt wound main pole; provides a close voltage regulation.

Dobrowolsky generator Three-wire, direct-current generator with a balance coil connected across the armature; the coil's midpoint produces the midpoint voltage for the system.

doghouse Small enclosure placed at the base of a transmitting antenna tower to house antenna tuning equipment.

Doherty amplifier A linear radio-frequency power amplifier that is divided into two sections whose inputs and outputs are connected by quarter-wave networks; for all values of input signal voltage up to one-half maximum amplitude, section no. 1 delivers all the power to the load; above this level, section no. 2 comes into operation.

domain A region in a solid within which elementary atomic or molecular magnetic or electric moments are uniformly arrayed.

domain growth A stage in the process of magnetization in which there is a growth of those magnetic domains in a ferromagnet oriented most nearly in the direction of an applied magnetic field.

domain rotation The stage in the magnetization process in which there is rotation of the direction of magnetization of magnetic domains in a ferromagnet toward the direction of a magnetic applied field and against anisotropy forces.

domain theory A theory of the behavior of ferromagnetic and ferroelectric crystals according to which changes in the bulk magnetization and polarization arise from changes in size and orientation of domains that are each polarized to saturation but which point in different directions.

domain wall *See* Bloch wall.

dominant mode *See* fundamental mode.

dominant wave The electromagnetic wave that has the lowest cutoff frequency in a given uniconductor waveguide.

donor An impurity that is added to a pure semiconductor material to increase the number of free electrons. Also known as donor impurity; electron donor.

donor impurity *See* donor.

donor level An intermediate energy level close to the conduction band in the energy diagram of an extrinsic semiconductor.

doorknob capacitor A high-voltage, plastic-encased capacitor resembling a doorknob in size and shape.

dopant *See* doping agent.

dope *See* doping agent.

doped junction A junction produced by adding an impurity to the melt during growing of a semiconductor crystal.

doping The addition of impurities to a semiconductor to achieve a desired characteristic, as in producing an n-type or p-type material. Also known as semiconductor doping.

doping agent An impurity element added to semiconductor materials used in crystal diodes and transistors. Also known as dopant; dope.

doping compensation The addition of donor impurities to a p-type semiconductor or of acceptor impurities to an n-type semiconductor.

doroid A coil resembling half a toroid, using a removable core segment to simplify the winding process.

dot *See* button.

dot angel An angel that appears on the screens of vertically pointing radars, often on clear cloudless days, as a bright dot; believed to be produced by a vertical column of rising air passing through air layers having different indices of refraction.

dot generator A signal generator that produces a dot pattern on the screen of a three-gun color television picture tube, for use in convergence adjustments.

dot matrix A method of generating characters with a matrix of dots.

dot-sequential color television A color television system in which the red, blue, and green primary-color dots are formed in rapid succession along each scanning line.

dot system Manufacturing technique for producing microelectronic circuitry.

double-amplitude-modulation multiplier A multiplier in which one variable is amplitude-modulated by a carrier, and the modulated signal is again amplitude-modulated by the other variable; the resulting double-modulated signal is applied to a balanced demodulator to obtain the product of the two variables.

double-base diode *See* unijunction transistor.

double-base junction diode *See* unijunction transistor.

double-base junction transistor A tetrode transistor that is essentially a junction triode transistor having two base connections on opposite sides of the central region of the transistor. Also known as tetrode junction transistor.

double-stream amplifier   117

double-beam cathode-ray tube A cathode-ray tube having two beams and capable of producing two independent traces that may overlap; the beams may be produced by splitting the beam of one gun or by using two guns.

double-bounce calibration Method of radar calibration which is used to determine the zero set error by using round-trip echoes; the correct range is the difference between the first and second echoes.

double-break switch Switch which opens the connected circuit at two points.

double bridge *See* Kelvin bridge.

double bus-double breaker A substation switching arrangement having two common buses and two breakers per connection.

double bus-single breaker A substation switching arrangement that involves two common buses and only one breaker per connection.

double-current generator Machine which supplies both direct and alternating current from the same armature winding.

double-diffused transistor A transistor in which two pn junctions are formed in the semiconductor wafer by gaseous diffusion of both p-type and n-type impurities; an intrinsic region can also be formed.

double diode *See* binode; duodiode.

double-diode limiter Type of limiter which is used to remove all positive signals from a combination of positive and negative pulses, or to remove all the negative signals from such a combination of positive and negative pulses.

double-doped transistor The original grown-junction transistor, formed by successively adding p-type and n-type impurities to the melt during growing of the crystal.

double-doublet antenna Two half-wave doublet antennas criss-crossed at their center, one being shorter than the other to give broader frequency coverage.

double image A television picture consisting of two overlapping images due to reception of the signal over two paths of different length so that signals arrive at slightly different times.

double limiter *See* cascade limiter.

double moding Undesirable shifting of a magnetron from one frequency to another at irregular intervals.

double-pole double-throw switch A six-terminal switch or relay contact arrangement that simultaneously connects one pair of terminals to either of two other pairs of terminals. Abbreviated dpdt switch.

double-pole single-throw switch A four-terminal switch or relay contact arrangement that simultaneously opens or closes two separate circuits or both sides of the same circuit. Abbreviated dpst switch.

double-pole switch A switch that operates simultaneously in two separate electric circuits or in both lines of a single circuit.

doubler *See* frequency doubler; voltage doubler.

double screen Three-layer cathode-ray tube screen consisting of a two-layer screen with the addition of a second long-persistence coating having a different color and different persistence from the first.

double-shield enclosure Type of shielded enclosure or room in which the inner wall is partially isolated electrically from the outer wall.

double-stream amplifier Microwave traveling-wave amplifier in which amplification occurs through interaction of two electron beams having different average velocities.

double-stub tuner Impedance-matching device, consisting of two stubs, usually fixed three-eighths of a wavelength apart, in parallel with the main transmission lines.

doublet *See* dipole.

doublet antenna *See* dipole antenna.

double-throw circuit breaker Circuit breaker by means of which a change in the circuit connections can be obtained by closing either of two sets of contacts.

double-throw switch A switch that connects one set of two or more terminals to either of two other similar sets of terminals.

double triode An electron tube having two triodes in the same envelope. Also known as duotriode.

doublet trigger A trigger signal consisting of two pulses spaced a predetermined amount for coding purposes.

double-tuned amplifier Amplifier of one or more stages in which each stage uses coupled circuits having two frequencies of resonance, to obtain wider bands than those obtainable with single tuning.

double-tuned circuit A circuit that is resonant to two adjacent frequencies, so that there are two approximately equal values of peak response, with a dip between.

double-tuned detector A type of frequency-modulation discriminator in which the limiter output transformer has two secondaries, one tuned above the resting frequency and the other tuned an equal amount below.

double-winding synchronous generator Synchronous generator which has two similar windings, in phase with one another, mounted on the same magnetic structure but not connected electrically, designed to supply power to two independent external circuits.

down-lead *See* lead-in.

Dow oscillator *See* electron-coupled oscillator.

dpdt switch *See* double-pole double-throw switch.

dpst switch *See* double-pole single-throw switch.

drag-cup motor An induction motor having a cup-shaped rotor or conducting material, inside of which is a stationary magnetic core.

drain 1. One of the electrodes in a thin-film transistor. 2. *See* current drain.

drain wire Metallic conductor frequently used in contact with foil-type signal-cable shielding to provide a low-resistance ground return at any point along the shield.

dress The arrangement of connecting wires in a circuit to prevent undesirable coupling and feedback.

drift The movement of current carriers in a semiconductor under the influence of an applied voltage.

drift-corrected amplifier A type of amplifier that includes circuits designed to reduce gradual changes in output, used in analog computers.

drift mobility The average drift velocity of carriers per unit electric field in a homogeneous semiconductor. Also known as mobility.

drift space A space in an electron tube which is substantially free of externally applied alternating fields and in which repositioning of electrons takes place.

drift speed Average speed at which electrons or ions progress through a medium.

drift transistor 1. A transistor having two plane parallel junctions, with a resistivity gradient in the base region between the junctions to improve the high-frequency response. 2. *See* diffused-alloy transistor.

drift velocity The average velocity of a carrier that is moving under the influence of an electric field in a semiconductor, conductor, or electron tube.

drive *See* excitation.

drive control *See* horizontal drive control.

driven array An antenna array consisting of a number of driven elements, usually half-wave dipoles, fed in phase or out of phase from a common source.

driven blocking oscillator *See* monostable blocking oscillator.

driven element An antenna element that is directly connected to the transmission line.

drive pulse An electrical pulse which induces a magnetizing force in an element of a magnetic core storage, reversing the polarity of the core.

driver The amplifier stage preceding the output stage in a receiver or transmitter.

driver element Antenna array element that receives power directly from the transmitter.

driver sweep Sweep triggered only by an incoming signal or trigger.

driver transformer A transformer in the input circuit of an amplifier, especially in the transmitter.

drive winding A coil of wire that is inductively coupled to an element of a magnetic memory. Also known as drive wire.

driving point impedance The complex ratio of applied alternating voltage to the resulting alternating current in an electron tube, network, or other transducer.

driving signal Television signal that times the scanning at the pickup point.

drop bar Protective device used to ground a high-voltage capacitor when opening a door.

drop bracket transposition Reversal of the relative positions of two parallel wire conductors while depressing one, so that the crossover is in a vertical plane.

dropout 1. Of a relay, the maximum current, voltage, power, or such, at which it will release from its energized position. 2. A reduction in output signal level during reproduction of recorded data, sufficient to cause a processing error.

dropout current The maximum current at which a relay or other magnetically operated device will release to its deenergized position.

dropout error Loss of a recorded bit or any other error occurring in recorded magnetic tape due to foreign particles on or in the magnetic coating or to defects in the backing.

dropout fuse A fuse used on utility line poles which springs open when the fuse metal melts to provide rapid arc extinction, and which drops to an open-circuit position readily distinguishable from the ground. Also known as flip-open cutout fuse.

dropout voltage The maximum voltage at which a relay or other magnetically operated device will release to its deenergized position.

dropping resistor A resistor used in series with a load to decrease the voltage applied to the load.

drop relay Relay activated by incoming ringing current to call an operator's attention to a subscriber's line.

drop repeater Microwave repeater that is provided with the necessary equipment for local termination of one or more circuits.

drop wire Wire suitable for extending an open wire or cable pair from a pole or cable terminal to a building.

Drude's theory of conduction A theory which treats the electrons in a metal as a gas of classical particles.

drum A computer storage device consisting of a rapidly rotating cylinder with a magnetizable external surface on which data can be read or written by many read/write heads floating a few millionths of an inch off the surface. Also known as drum memory; drum storage; magnetic drum; magnetic drum storage.

drum armature An armature that has a drum winding.

drum controller An electric device that has a drum switch for its main switching element; used to govern the way electric power is delivered to a motor.

drum disk rectifier A mechanical rectifier using synchronous contacts and a copper oxide dry disk.

drum memory *See* drum.

drum recorder A facsimile recorder in which the record sheet is mounted on a rotating drum or cylinder.

drum storage *See* drum.

drum switch A switch in which the electrical contacts are made on pins, segments, or surfaces on the periphery of a rotating cylinder or sector, or by the operation of a rotating cam.

drum transmitter A facsimile transmitter in which the subject copy is mounted on a rotating drum or cylinder.

drum winding A type of winding in electric machines in which coils are housed in long, narrow gaps either in the outer surface of a cylindrical core or in the inner surface of a core with a cylindrical bore.

dry battery A battery made up of a series, parallel, or series-parallel arrangement of dry cells in a single housing to provide desired voltage and current values.

dry cell A voltage-generating cell having an immobilized electrolyte.

dry-charged battery A storage battery in which the electrolyte is drained from the battery for storage, and which is filled with electrolyte and charged for a few minutes to prepare for use.

dry circuit A relay circuit in which open-circuit voltages are very low and closed-circuit currents extremely small, so there is no arcing to roughen the contacts.

dry contact A contact that does not break or make current.

dry-disk rectifier *See* metallic rectifier.

dry electrolytic capacitor An electrolytic capacitor in which the electrolyte is a paste rather than a liquid; the dielectric is a thin film of gas formed on one of the plates by chemical action.

dry flashover voltage Voltage at which the air surrounding a clean dry insulator or shell completely breaks down between electrodes.

dry-plate rectifier *See* metallic rectifier.

dry-reed relay Reed-type relay which does not use mercury at the relay contacts.

dry-reed switch A switch having contacts mounted on magnetic reeds in a vacuum enclosure, designed for reliable operation in dry circuits.

dry-tape fuel cell A fuel cell in which the fuel is in the form of a dry tape, coated with fuel, oxidant, and electrolyte, which is fed into the cell at a rate corresponding to the demand for electric energy.

D scan *See* D scope.

duolateral coil   121

D scope A cathode-ray scope which combines the features of B and C scopes, the signal appearing as a spot with bearing angle as the horizontal coordinate and elevation angle as the vertical coordinate, but with each spot expanded slightly in a vertical direction to give a rough range indication. Also known as D indicator; D scan.

dual diversity receiver A diversity radio receiver in which the two antennas feed separate radio-frequency systems, with mixing occurring after the converter.

dual-emitter transistor A passivated *pnp* silicon planar epitaxial transistor having two emitters, for use in low-level choppers.

dual-gun cathode-ray tube A dual-trace oscilloscope in which beams from two electron guns are controlled by separate balanced vertical-deflection plates and also have separate brightness and focus controls.

dual in-line package Microcircuit package with two rows of seven vertical leads that are easily inserted into an etched circuit board. Abbreviated DIP.

duality principle 1. The principle that for any theorem in electrical circuit analysis there is a dual theorem in which one replaces quantities with dual quantities; current and voltage, impedance and admittance, and meshes and nodes are examples of dual quantities. 2. The principle that analogies may be drawn between a transistor circuit and the corresponding vacuum tube circuit. 3. The principle that one can obtain new solutions of Maxwell's equations from known solutions by replacing E with H, H with $-E$, ϵ with μ, and μ with ϵ.

dual network A network which has the same number of terminal pairs as a given network, and whose open-circuit impedance network is the same as the short-circuit admittance matrix of the given network, and vice versa.

dual-trace amplifier An oscilloscope amplifier that switches electronically between two signals under observation in the interval between sweeps, so that waveforms of both signals are displayed on the screen.

dual-trace oscilloscope An oscilloscope which can compare two waveforms on the face of a single cathode-ray tube, using any one of several methods.

Duddell oscillograph A moving-coil oscillograph; the current to be observed passes through a coil in a magnetic field and a mirror attached to the coil reveals its movement.

dull emitter An electron tube whose cathode is a filament that does not glow brightly.

dummy antenna A device that has the impedance characteristic and power-handling capacity of an antenna but does not radiate or receive radio waves; used chiefly for testing a transmitter. Also known as artificial antenna.

dummy load A dissipative device used at the end of a transmission line or waveguide to convert transmitted energy into heat, so that essentially no energy is radiated outward or reflected back to its source.

dump To withdraw all power from a system or component accidentally or intentionally.

dump power Electric power, generated by any source, which is in excess of the needs of the electric system and which cannot be stored or conserved.

duodiode An electron tube having two diodes in the same envelope, with either a common cathode or separate cathodes. Also known as double diode.

duodiode-pentode An electron tube having two diodes and a pentode in the same envelope, generally with a common cathode.

duodiode-triode An electron tube having two diodes and a triode in the same envelope, generally with a common cathode.

duolateral coil *See* honeycomb coil.

122 duoplasmatron

duoplasmatron An ion-beam source in which electrons from a hot filament are accelerated sufficiently to ionize a gas by impact; the resulting positive ions are drawn out by high-voltage electrons and focused into a beam by electrostatic lens action.

duotriode *See* double triode.

duplex artificial line A balancing network, simulating the impedance of the real line and distant terminal apparatus, which is employed in a duplex circuit for the purpose of making the receiving device unresponsive to outgoing signal currents.

duplex cable Two insulated stranded conductors twisted together; they may have a common insulating covering.

duplexer A switching device used in radar to permit alternate use of the same antenna for both transmitting and receiving; other forms of duplexers serve for two-way radio communication using a single antenna at lower frequencies. Also known as duplexing assembly.

duplexing assembly *See* duplexer.

duplex tube Combination of two vacuum tubes in one envelope.

duration control Control for adjusting the time duration of reduced gain in a sensitivity-time control circuit.

Dushman equation *See* Richardson-Dushman equation.

dust core *See* ferrite core.

duty classification of a relay Expression of the frequency with which the relay may be required to operate without exceeding prescribed limitations.

duty ratio In a pulse radar or similar system, the ratio of average to peak pulse power. Also known as duty cycle.

dynamic characteristic *See* load characteristic.

dynamic circuit An MOS circuit designed to make use of its high input impedance to store charge temporarily at certain nodes of the circuit and thereby increase the speed of the circuit.

dynamic condenser electrometer A sensitive voltage-measuring instrument in which an object carrying charge resulting from the voltage is moved back and forth in an electrostatic field and the resulting alternating-current signal is observed.

dynamic convergence The process whereby the locus of the point of convergence of electron beams in a color-television or other multibeam cathode-ray tube is made to fall on a specified surface during scanning.

dynamic error Error in a time-varying signal resulting from inadequate dynamic response of a transducer.

dynamic focusing The process of varying the focusing electrode voltage for a color picture tube automatically so the electron-beam spots remain in focus as they sweep over the flat surface of the screen.

dynamic impedance The impedance of a circuit having an inductance and a capacitance in parallel at the frequency at which this impedance has a maximum value. Also known as rejector impedance.

dynamic pickup A pickup in which the electric output is due to motion of a coil or conductor in a constant magnetic field. Also known as dynamic reproducer; moving-coil pickup.

dynamic plate impedance Internal resistance to the flow of alternating current between the cathode and plate of a tube.

dynamic plate resistance Opposition that the plate circuit of a vacuum tube offers to a small increment of plate voltage; it is the ratio of a small change in plate voltage to the resulting change in the plate current, other tube voltages remaining constant.

dynamic range The ratio of the specified maximum signal level capability of a system or component to its noise level; usually expressed in decibels.

dynamic regulator Transmission regulator in which the adjusting mechanism is in self-equilibrium at only one or a few settings and requires control power to maintain it at any other setting.

dynamic reproducer *See* dynamic pickup.

dynamic resistance A device's electrical resistance when it is in operation.

dynamo *See* generator.

dynamoelectric amplifier generator A generator that serves as a power amplifier at low frequencies or direct current; the input signal is applied to the stationary field to change the excitation, and the amplified output is taken from the rotating armature.

dynamometer multiplier A multiplier in which a fixed and a moving coil are arranged so that the deflection of the moving coil is proportional to the product of the currents flowing in the coils.

dynamostatic Pertaining to a machine that uses direct or alternating current to produce static electricity.

dynamotor A rotating electric machine having two or more windings on a single armature containing a commutator for direct-current operation and slip rings for alternating-current operation; when one type of power is fed in for motor operation, the other type is delivered by generator action. Also known as rotary converter; synchronous inverter.

dynatron A screen-grid tube in which secondary emission of electrons from the anode causes the anode current to decrease as anode voltage increases, resulting in a negative resistance characteristic. Also known as negatron.

dynatron oscillator An oscillator in which secondary emission of electrons from the anode of a screen-grid tube causes the anode current to decrease as anode voltage is increased, giving the negative resistance characteristic required for oscillation.

dynode An electrode whose primary function is secondary emission of electrons; used in multiplier phototubes and some types of television camera tubes. Also known as electron mirror.

E

E *See* electric field vector; emitter.

Earnshaw's theorem The theorem that a charge cannot be held in stable equilibrium by an electrostatic field.

earth *See* ground.

earth current Return, fault, leakage, or stray current passing through the earth from electrical equipment. Also known as ground current.

Easter-egging An undirected procedure for checking electronic equipment, which derives its name from the children's activity of searching for hidden eggs at Eastertime.

E bend A smooth change in the direction of the axis of a waveguide, throughout which the axis remains in a plane parallel to the direction of polarization. Also known as E-plane bend.

E cell A timing device that converts the current-time integral of an electrical function into an equivalent mass integral (or the converse operation) up to a maximum of several thousand microampere-hours.

echo 1. The signal reflected by a radar target, or the trace produced by this signal on the screen of the cathode-ray tube in a radar receiver. Also known as radar echo; return. 2. *See* ghost signal.

echo amplitude In radar, an empirical measure of the strength of a target signal as determined from the appearance of the echo; the amplitude of the echo waveform usually is measured by the deflection of the electron beam from the base line of an amplitude-modulated indicator.

echo area In radar, the area of a fictitious perfect reflector of electromagnetic waves that would reflect the same amount of energy back to the radar as the actual target. Also known as radar cross section; target cross section.

echo attenuation The power transmitted at an output terminal of a transmission line, divided by the power reflected back to the same output terminal.

echo box A calibrated high-Q resonant cavity that stores part of the transmitted radar pulse power and gradually feeds this energy into the receiving system after completion of the pulse transmission; used to provide an artificial target signal for test and tuning purposes. Also known as phantom target.

echo contour A trace of equal signal intensity of the radar echo displayed on a range height indicator or plan position indicator scope.

echo frequency The number of fluctuations, per unit time, in the power or amplitude of a radar target signal.

echo intensity The brightness or brilliance of a radar echo as displayed on an intensity-modulated indicator; echo intensity is, within certain limits, proportional to the voltage of the target signal or to the square root of its power.

echo power The electrical strength, or power, of a radar target signal, normally measured in watts or dBm (decibels referred to 1 milliwatt).

echo pulse A pulse of radio energy received at the radar after reflection from a target; that is, the target signal of a pulse radar.

echo suppressor 1. A circuit that desensitizes radar navigation equipment for a fixed period after the reception of one pulse, for the purpose of rejecting delayed pulses arriving from longer, indirect reflection paths. 2. A relay or other device used on a transmission line to prevent a reflected wave from returning to the sending end of the line.

ECL *See* emitter-coupled logic.

E core A transformer core made from E-shaped laminations and used in conjunction with I-shaped laminations.

eddy current An electric current induced within the body of a conductor when that conductor either moves through a nonuniform magnetic field or is in a region where there is a change in magnetic flux. Also known as Foucault current.

eddy-current loss Energy loss due to undesired eddy currents circulating in a magnetic core.

eddy-current test A nondestructive test in which the change of impedance of a test coil brought close to a conducting specimen indicates the eddy currents induced by the coil, and thereby indicates certain properties or defects of the specimen.

edge effect An outward-curving distortion of lines of force near the edges of two parallel metal plates that form a capacitor.

edge focusing Axial focusing of a stream of ions which occurs when it crosses a fringe magnetic field obliquely; used in mass spectrometers and cyclotrons.

Edison battery A storage battery composed of cells having nickel and iron in an alkaline solution. Also known as nickel-iron battery.

Edison distribution system Three-wire direct-current distribution system, usually 120 to 240 volts, for combined light and power service from a single set of mains.

E display A rectangular radar display in which targets appear as blips with distance indicated by the horizontal coordinate, and elevation by the vertical coordinate.

effective ampere The amount of alternating current flowing through a resistance that produces heat at the same average rate as 1 ampere of direct current flowing in the same resistance.

effective antenna length Electrical length of an antenna, as distinguished from its physical length.

effective area Of an antenna in any specified direction, the square of the wavelength multiplied by the power gain (or directive gain) in that direction and divided by 4π (12.57).

effective bandwidth The bandwidth of an assumed rectangular band-pass having the same transfer ratio at a reference frequency as a given actual band-pass filter, and passing the same mean-square value of a hypothetical current having even distribution of energy throughout that bandwidth.

effective capacitance Total capacitance existing between any two given points of an electric circuit.

effective current The value of alternating current that will give the same heating effect as the corresponding value of direct current. Also known as root-mean-square current.

effective field intensity Root-mean-square value of the inverse distance fields at a distance of 1 mile (1.6 kilometers) from the transmitting antenna in all directions in the horizontal plane.

effective height The height of the center of radiation of a transmitting antenna above the effective ground level.

effectively grounded Grounded through a connection of sufficiently low impedances (inherent or intentionally added) so that fault grounds which may occur cannot build up voltages dangerous to connected personnel or other equipment.

effective mass A parameter with the dimensions of mass that is assigned to electrons in a solid; in the presence of an external electromagnetic field the electrons behave in many respects as if they were free, but with a mass equal to this parameter rather than the true mass.

effective radiated power The product of antenna input power and antenna power gain, expressed in kilowatts. Abbreviated ERP.

effective thermal resistance Of a semiconductor device, the effective temperature rise per unit power dissipation of a designated junction above the temperature of a stated external reference point under conditions of thermal equilibrium. Also known as thermal resistance.

Ehrenhaft effect A helical motion of fine particles along the lines of force of a magnetic field during exposure to light, resulting from radiometer effects.

eht See extra-high tension.

E-H T junction In microwave waveguides, a combination of E- and H-plane T junctions forming a junction at a common point of intersection with the main waveguide.

E-H tuner Tunable E-H T junction having two arms terminated in adjustable plungers used for impedance transformation.

ehv See extra-high voltage.

Einstein–de Haas effect A freely suspended body consisting of a ferromagnetic material acquires a rotation when its magnetization changes.

Einstein–de Haas method Method of measuring the gyromagnetic ratio of a ferromagnetic substance; one measures the angular displacement induced in a ferromagnetic cylinder suspended from a torsion fiber when magnetization of the object is reversed, and the magnetization change is measured with a magnetometer.

Einstein frequency Single frequency with which each atom vibrates independently of other atoms, in a model of lattice vibrations; equal to the frequency observed in infrared absorption studies.

Einstein frequency condition The assumption that all vibrations of a crystal lattice are harmonic with the same characteristic frequency.

Einstein's equation for specific heat The earliest equation based on quantum mechanics for the specific heat of a solid; uses the assumption that each atom oscillates with the same frequency.

elastance The reciprocal of capacitance.

elastoresistance The change in a material's electrical resistance as it undergoes a stress within its elastic limit.

elbow In a waveguide, a bend of comparatively short radius, normally 90°, and sometimes for acute angles down to 15°.

electret A solid dielectric possessing persistent electric polarization, by virtue of a long time constant for decay of a charge instability.

electret transducer An electroacoustic or electromechanical transducer in which a foil electret, stretched out to form a diaphragm, is placed next to a metal or metal-coated plate, and motion of the diaphragm is converted to voltage between diaphragm and plate, or vice versa.

electric Containing, producing, arising from, or actuated by electricity; often used interchangeably with electrical.

electrical Related to or associated with electricity, but not containing it or having its properties or characteristics; often used interchangeably with electric.

electrical angle An angle that specifies a particular instant in an alternating-current cycle or expresses the phase difference between two alternating quantities; usually expressed in electrical degrees.

electrical axis The x axis in a quartz crystal; there are three such axes in a crystal, each parallel to one pair of opposite sides of the hexagon; all pass through and are perpendicular to the optical, or z, axis.

electrical center Point approximately midway between the ends of an inductor or resistor that divides the inductor or resistor into two equal electrical values.

electrical code A systematic body of rules governing the practical application and installation of electrically operated equipment and devices and electric wiring systems.

electrical conductivity analyzer Alternating-current, resistance-bridge device used to measure the electrical conductivity of solutions, slurries, or wet solids.

electrical degree A unit equal to $1/360$ cycle of an alternating quantity.

electrical distance The distance between two points, expressed in terms of the duration of travel of an electromagnetic wave in free space between the two points.

electrical drainage Diversion of electric currents from subterranean pipes to prevent electrolytic corrosion.

electrical equipment Apparatus, appliances, devices, wiring, fixtures, fittings, and material used as a part of or in connection with an electrical installation.

electrical impedance meter An instrument which measures the complex ratio of voltage to current in a given circuit at a given frequency. Also known as impedance meter.

electrical impedance Also known as impedance. 1. The total opposition that a circuit presents to an alternating current, equal to the complex ratio of the voltage to the current in complex notation. Also known as complex impedance. 2. The ratio of the maximum voltage in an alternating-current circuit to the maximum current; equal to the magnitude of the quantity in the first definition.

electrical instability A persistent condition of unwanted self-oscillation in an amplifier or other electric circuit.

electrical length The length of a conductor expressed in wavelengths, radians, or degrees.

electrically connected Connected by means of a conducting path, or through a capacitor, as distinguished from connection merely through electromagnetic induction.

electrical measurement The measurement of any one of the many quantities by which electricity is characterized.

electrical noise Noise generated by electrical devices, for example, motors, engine ignition, power lines, and so on, and propagated to the receiving antenna direct from the noise source.

electrical potential energy Energy possessed by electric charges by virtue of their position in an electrostatic field.

electrical properties Properties of a substance which determine its response to an electric field, such as its dielectric constant or conductivity.

electrical resistivity The electrical resistance offered by a material to the flow of current, times the cross-sectional area of current flow and per unit length of current path; the reciprocal of the conductivity. Also known as resistivity; specific resistance.

electrical symbol A simple geometrical symbol used to represent a component of a circuit in a schematic circuit diagram.

electrical system System of wiring, switches, relays, and other equipment associated with receiving and distributing electricity.

electrical unit A standard in terms of which some electrical quantity is evaluated.

electrical zero A standard reference position from which rotor angles are measured in synchros and other rotating devices.

electric arc A discharge of electricity through a gas, normally characterized by a voltage drop approximately equal to the ionization potential of the gas. Also known as arc.

electric cell 1. A single unit of a primary or secondary battery that converts chemical energy into electric energy. 2. A single unit of a device that converts radiant energy into electric energy, such as a nuclear, solar, or photovoltaic cell.

electric chopper A chopper in which an electromagnet driven by a source of alternating current sets into vibration a reed carrying a moving contact that alternately touches two fixed contacts in a signal circuit, thus periodically interrupting the signal.

electric circuit Also known as circuit. 1. A path or group of interconnected paths capable of carrying electric currents. 2. An arrangement of one or more complete, closed paths for electron flow.

electric comparator A comparator in which movement results in a change in some electrical quantity, which is then amplified by electrical means.

electric connection A direct wire path for current between two points in a circuit.

electric connector A device that joins electric conductors mechanically and electrically to other conductors and to the terminals of apparatus and equipment.

electric constant The permittivity of empty space, equal to 1 in centimeter-gram-second electrostatic units and to $10^7/4\pi c^2$ farads per meter or, numerically, to 8.854 × 10^{-12} farads per meter in International System units, where c is the speed of light in meters per second. Symbolized ϵ_0.

electric contact A physical contact that permits current flow between conducting parts. Also known as contact.

electric control The control of a machine or device by switches, relays, or rheostats, as contrasted with electronic control by electron tubes or by devices that do the work of electron tubes.

electric controller A device that governs in some predetermined manner the electric power delivered to apparatus.

electric delay line A delay line using properties of lumped or distributed capacitive and inductive elements; can be used for signal storage by recirculating information-carrying wave patterns.

electric dipole A localized distribution of positive and negative electricity, without net charge, whose mean positions of positive and negative charges do not coincide.

electric dipole moment A quantity characteristic of a charge distribution, equal to the vector sum over the electric charges of the product of the charge and the position vector of the charge.

electric displacement The electric field intensity multiplied by the permittivity. Symbolized D. Also known as dielectric displacement; dielectric flux density; displacement; electric displacement density; electric flux density; electric induction.

electric energy 1. Energy of electric charges by virtue of their position in an electric field. 2. Energy of electric currents by virtue of their position in a magnetic field.

130 electric energy measurement

electric energy measurement The measurement of the integral, with respect to time, of the power in an electric circuit.

electric energy meter A device which measures the integral, with respect to time, of the power in an electric circuit.

electric field 1. One of the fundamental fields in nature, causing a charged body to be attracted to or repelled by other charged bodies; associated with an electromagnetic wave or a changing magnetic field. 2. Specifically, the electric force per unit test charge.

electric field vector The force on a stationary positive charge per unit charge at a point in an electric field. Designated E. Also known as electric field intensity; electric field strength; electric vector.

electric filter 1. A network that transmits alternating currents of desired frequencies while substantially attenuating all other frequencies. Also known as frequency-selective device. 2. See filter.

electric flowmeter Fluid-flow measurement device relying on an inductance or impedance bridge or on electrical-resistance rod elements to sense flow-rate variations.

electric flux 1. The integral over a surface of the component of the electric displacement perpendicular to the surface; equal to the number of electric lines of force crossing the surface. 2. The electric lines of force in a region.

electric forming The process of applying electric energy to a semiconductor or other device to modify permanently its electrical characteristics.

electric image A fictitious charge used in finding the electric field set up by fixed electric charges in the neighborhood of a conductor; the conductor, with its distribution of induced surface charges, is replaced by one or more of these fictitious charges. Also known as image.

electric induction See electric displacement.

electric lamp A lamp in which light is produced by electricity, as the incandescent lamp, arc lamp, glow lamp, mercury-vapor lamp, and fluorescent lamp.

electric line of force An imaginary line drawn so that each segment of the line is parallel to the direction of the electric field or of the electric displacement at that point, and the density of the set of lines is proportional to the electric field or electrical displacement. Also known as electric flux line.

electric main See power transmission line.

electric moment One of a series of quantities characterizing an electric charge distribution; an l-th moment is given by integrating the product of the charge density, the l-th power of the distance from the origin, and a spherical harmonic $Y*_{lm}$ over the charge distribution.

electric monopole A distribution of electric charge which is concentrated at a point or is spherically symmetric.

electric motor See motor.

electric multipole One of a series of types of static or oscillating charge distributions; the multipole of order 1 is a point charge or a spherically symmetric distribution, and the electric and magnetic fields produced by an electric multipole of order 2^n are equivalent to those of two electric multipoles of order 2^{n-1} of equal strengths, but opposite sign, separated from each other by a short distance.

electric multipole field The electric and magnetic fields generated by a static or oscillating electric multipole.

electric network See network.

electric solenoid 131

electric octupole moment A quantity characterizing an electric charge distribution; obtained by integrating the product of the charge density, the third power of the distance from the origin, and a spherical harmonic Y^*_{3m} over the charge distribution.

electric outlet *See* outlet.

electric polarizability Induced dipole moment of an atom or molecule in a unit electric field.

electric polarization *See* polarization.

electric potential The work which must be done against electric forces to bring a unit charge from a reference point to the point in question; the reference point is located at an infinite distance, or, for practical purposes, at the surface of the earth or some other large conductor. Also known as electrostatic potential; potential.

electric power The rate at which electric energy is converted to other forms of energy, equal to the product of the current and the voltage drop.

electric power line *See* power line.

electric power station A generating station or an electric power substation.

electric power substation An assembly of equipment in an electric power system through which electric energy is passed for transmission, transformation, distribution, or switching. Also known as substation.

electric power transmission Process of transferring electric energy from one point to another in an electric power system.

electric pressure transducer *See* pressure transducer.

electric protective device A particular type of equipment used in electric power systems to detect abnormal conditions and to initiate appropriate corrective action. Also known as protective device.

electric quadrupole A charge distribution that produces an electric field equivalent to that produced by two electric dipoles whose dipole moments have the same magnitude but point in opposite directions and which are separated from each other by a small distance.

electric quadrupole lens A device for focusing beams of charged particles which has four electrodes with alternately positive and negative polarity; used in electron microscopes and particle accelerators.

electric quadrupole moment A quantity characterizing an electric charge distribution, obtained by integrating the product of the charge density, the second power of the distance from the origin, and a spherical harmonic Y^*_{2m} over the charge distribution.

electric raceway *See* raceway.

electric reactor *See* reactor.

electric rotating machinery Any form of apparatus which has a rotating member and generates, converts, transforms, or modifies electric power, such as a motor, generator, or synchronous converter.

electric scanning Scanning in which the required changes in radar beam direction are produced by variations in phase or amplitude of the currents fed to the various elements of the antenna array.

electric shielding Any means of avoiding pickup of undesired signals or noise, suppressing radiation of undesired signals, or confining wanted signals to desired paths or regions, such as electrostatic shielding or electromagnetic shielding. Also known as screening; shielding.

electric solenoid *See* solenoid.

electric spark

electric spark See spark.

electric strength See dielectric strength.

electric susceptibility A dimensionless parameter measuring the ease of polarization of a dielectric, equal (in meter-kilogram-second units) to the ratio of the polarization to the product of the electric field strength and the vacuum permittivity. Also known as dielectric susceptibility.

electric switch See switch.

electric switchboard See switchboard.

electric terminal See terminal.

electric transducer A transducer in which all of the waves are electric.

electric transient A temporary component of current and voltage in an electric circuit which has been disturbed.

electric tuning Tuning a receiver to a desired station by switching a set of preadjusted trimmer capacitors or coils into the tuning circuits.

electric twinning A defect occurring in natural quartz crystals, in which adjacent regions of quartz have their electric axes oppositely poled.

electric vector See electric field vector.

electric wave An electromagnetic wave, especially one whose wavelength is at least a few centimeters. Also known as Hertzian wave.

electric-wave filter See filter.

electric wind See convective discharge.

electric wire See wire.

electric wiring See wiring.

electrification 1. The process of establishing a charge in an object. 2. The generation, distribution, and utilization of electricity.

electrization The electric polarization divided by the permittivity of empty space.

electroacoustic effect See acoustoelectric effect.

electrochemical recording Recording by means of a chemical reaction brought about by the passage of signal-controlled current through the sensitized portion of the record sheet.

electrochemical valve Electric valve consisting of a metal in contact with a solution or compound, across the boundary of which current flows more readily in one direction than in the other direction, and in which the valve action is accompanied by chemical changes.

electrochromic display A solid-state passive display that uses organic or inorganic insulating solids which change color when injected with positive or negative charges.

electrode 1. An electric conductor through which an electric current enters or leaves a medium, whether it be an electrolytic solution, solid, molten mass, gas, or vacuum. 2. One of the terminals used in dielectric heating or diathermy for applying the high-frequency electric field to the material being heated.

electrode admittance Quotient of dividing the alternating component of the electrode current by the alternating component of the electrode voltage, all other electrode voltages being maintained constant.

electrode capacitance Capacitance between one electrode and all the other electrodes connected together.

electroluminescent lamp   133

electrode characteristic Relation between the electrode voltage and the current to an electrode, all other electrode voltages being maintained constant.

electrode conductance Quotient of the inphase component of the electrode alternating current by the electrode alternating voltage, all other electrode voltage being maintained constant; this is a variational and not a total conductance. Also known as grid conductance.

electrode couple The pair of electrodes in an electric cell, between which there is a potential difference.

electrode current Current passing to or from an electrode, through the interelectrode space within a vacuum tube.

electrode dark current The electrode current that flows when there is no radiant flux incident on the photocathode in a phototube or camera tube. Also known as dark current.

electrode dissipation Power dissipated in the form of heat by an electrode as a result of electron or ion bombardment.

electrode drop Voltage drop in the electrode due to its resistance.

electrode impedance Reciprocal of the electrode admittance.

electrode inverse current Current flowing through an electrode in the direction opposite to that for which the tube is designed.

electrode potential Also known as electrode voltage. The instantaneous voltage of an electrode with respect to the cathode of an electron tube.

electrode resistance Reciprocal of the electrode conductance; this is the effective parallel resistance and is not the real component of the electrode impedance.

electrode voltage *See* electrode potential.

electrodynamic machine An electric generator or motor in which the output load current is produced by magnetomotive currents generated in a rotating armature.

electrodynamics The study of the relations between electrical, magnetic, and mechanical phenomena.

electrodynamic shaker *See* shaker.

electrogram A record of an image of an object made by sparking, usually on paper.

electrographic pencil A pencil used to make a conductive mark on paper, for detection by a conductive-mark sensing device.

electrokinetics The study of the motion of electric charges, especially of steady currents in electric circuits, and of the motion of electrified particles in electric or magnetic fields.

electrokinetic transducer An instrument which converts dynamic physical forces, such as vibration and sound, into corresponding electric signals by measuring the streaming potential generated by passage of a polar fluid through a permeable refractory-ceramic or fritted-glass member between two chambers.

electroluminescence The emission of light, not due to heating effects alone, resulting from application of an electric field to a material, usually solid.

electroluminescent cell *See* electroluminescent panel.

electroluminescent display A display in which various combinations of electroluminescent segments may be activated by applying voltages to produce any desired numeral or other character.

electroluminescent lamp *See* electroluminescent panel.

electroluminescent panel A surface-area light source employing the principle of electroluminescence; consists of a suitable phosphor placed between sheet-metal electrodes, one of which is essentially transparent, with an alternating current applied between the electrodes. Also known as electroluminescent cell; electroluminescent lamp; light panel; luminescent cell.

electrolyte-activated battery A reserve battery in which an aqueous electrolyte is stored in a separate chamber, and a mechanism, which may be operated from a remote location, drives the electrolyte out of the reservoir and into the cells of the battery for activation.

electrolytic arrester See aluminum-cell arrester.

electrolytic capacitor A capacitor consisting of two electrodes separated by an electrolyte; a dielectric film, usually a thin layer of gas, is formed on the surface of one electrode. Also known as electrolytic condenser.

electrolytic condenser See electrolytic capacitor.

electrolytic interrupter An interrupter that consists of two electrodes in an electrolytic solution; bubbles formed in the solution continually interrupt the passage of current between the electrodes.

electrolytic recording Electrochemical recording in which the chemical change is made possible by the presence of an electrolyte.

electrolytic rectifier A rectifier consisting of metal electrodes in an electrolyte, in which rectification of alternating current is accompanied by electrolytic action; polarizing film formed on one electrode permits current flow in one direction but not the other.

electrolytic rheostat A rheostat that consists of a tank of conducting liquid in which electrodes are placed, and resistance is varied by changing the distance between the electrodes, the depth of immersion of the electrodes, or the resistivity of the solution. Also known as water rheostat.

electrolytic switch A switch having two electrodes projecting into a chamber partly filled with electrolyte, leaving an air bubble of predetermined width; the bubble shifts position and changes the amount of electrolyte in contact with the electrodes when the switch is tilted from true horizontal.

electromagnet A magnet consisting of a coil wound around a soft iron or steel core; the core is strongly magnetized when current flows through the coil, and is almost completely demagnetized when the current is interrupted.

electromagnetic amplifying lens Large numbers of waveguides symmetrically arranged with respect to an excitation medium in order to become excited with equal amplitude and phase to provide a net gain in energy.

electromagnetic cathode-ray tube A cathode-ray tube in which electromagnetic deflection is used on the electron beam.

electromagnetic compatibility The capability of electronic equipment or systems to be operated in the intended electromagnetic environment at design levels of efficiency.

electromagnetic complex Electromagnetic configuration of an installation, including all significant radiators of energy.

electromagnetic constant See speed of light.

electromagnetic countermeasure See electronic countermeasure.

electromagnetic coupling Coupling that exists between circuits when they are mutually affected by the same electromagnetic field.

electromagnetic current Motion of charged particles (for example, in the ionosphere) giving rise to electric and magnetic fields.

electromagnetic pulse   135

electromagnetic damping Retardation of motion that results from the reaction between eddy currents in a moving conductor and the magnetic field in which it is moving.

electromagnetic deflection Deflection of an electron stream by means of a magnetic field.

electromagnetic energy The energy associated with electric or magnetic fields.

electromagnetic field An electric or magnetic field, or a combination of the two, as in an electromagnetic wave.

electromagnetic field equations See Maxwell field equations.

electromagnetic field tensor An antisymmetric, second-rank Lorentz tensor, whose elements are proportional to the electric and magnetic fields; the Maxwell field equations can be expressed in a simple form in terms of this tensor.

electromagnetic focusing Focusing the electron beam in a telelvision picture tube by means of a magnetic field parallel to the beam; the field is produced by sending an adjustable value of direct current through a focusing coil mounted on the neck of the tube.

electromagnetic horn See horn antenna.

electromagnetic induction The production of an electromotive force either by motion of a conductor through a magnetic field so as to cut across the magnetic flux or by a change in the magnetic flux that threads a conductor. Also known as induction.

electromagnetic inertia 1. Characteristic delay of a current in an electric circuit in reaching its maximum value, or in returning to zero, after the source voltage has been removed or applied. 2. The property of a circuit whereby variation of the current in the circuit gives rise to a voltage in the circuit.

electromagnetic interference Interference, generally at radio frequencies, that is generated inside systems, as contrasted to radio-frequency interference coming from sources outside a system. Abbreviated emi.

electromagnetic lens An electron lens in which electron beams are focused by an electromagnetic field.

electromagnetic mass The contribution to the mass of an object from its electric and magnetic field energy.

electromagnetic mirror Surface or region capable of reflecting radio waves, such as one of the ionized layers in the upper atmosphere.

electromagnetic momentum The momentum transported by electromagnetic radiation; its volume density equals the Poynting vector divided by the square of the speed of light.

electromagnetic noise Noise in a communications system resulting from undesired electromagnetic radiation. Also known as radiation noise.

electromagnetic oscillograph An oscillograph in which the recording mechanism is controlled by a moving-coil galvanometer, such as a direct-writing recorder or a light-beam oscillograph.

electromagnetic potential Collective name for a scalar potential, which reduces to the electrostatic potential in a time-independent system, and the vector potential for the magnetic field; the electric and magnetic fields can be written in terms of these potentials.

electromagnetic properties The response of materials or equipment to electromagnetic fields, and their ability to produce such fields.

electromagnetic pulse The pulse of electromagnetic radiation generated by a large thermonuclear explosion.

electromagnetic pump A pump in which a conductive liquid is made to move through a pipe by sending a large current transversely through the liquid; this current reacts with a magnetic field that is at right angles to the pipe and to current flow, to move the current-carrying liquid conductor.

electromagnetic radiation Electromagnetic waves and, especially, the associated electromagnetic energy.

electromagnetic reconnaissance Reconnaissance for the purpose of locating and identifying potentially hostile transmitters of electromagnetic radiation, including radar, communication, missile-guidance, and navigation-aid equipment.

electromagnetic relay A relay in which current flow through a coil produces a magnetic field that results in contact actuation.

electromagnetic separator Device in which ions of varying mass are separated by a combination of electric and magnetic fields.

electromagnetic shielding Means, similar to electrostatic or magnetostatic shielding, for suppressing changing magnetic fields or electromagnetic radiation at a device.

electromagnetic shock wave Electromagnetic wave of great intensity which results when waves with different intensities propagate with different velocities in a nonlinear optical medium, and faster-traveling waves from a pulse of light catch up with preceding, slower traveling waves.

electromagnetic spectrum The total range of wavelengths or frequencies of electromagnetic radiation, extending from the longest radio waves to the shortest known cosmic rays.

electromagnetic susceptibility The tolerance of circuits and components to all sources of interfering electromagnetic energy.

electromagnetic system of units A centimeter-gram-second system of electric and magnetic units in which the unit of current is defined as the current which, if maintained in two straight parallel wires having infinite length and being 1 centimeter apart in vacuum, would produce between these conductors a force of 2 dynes per centimeter of length; other units are derived from this definition by assigning unit coefficients in equations relating electric and magnetic quantities. Also known as electromagnetic units (emu).

electromagnetic theory of light Theory according to which light is an electromagnetic wave whose electric and magnetic fields obey Maxwell's equations.

electromagnetic transducer *See* electromechanical transducer.

electromagnetic units *See* electromagnetic system of units.

electromagnetic wave A disturbance which propagates outward from any electric charge which oscillates or is accelerated; far from the charge it consists of vibrating electric and magnetic fields which move at the speed of light and are at right angles to each other and to the direction of motion.

electromagnetic-wave filter Any device to transmit electromagnetic waves of desired frequencies while substantially attenuating all other frequencies.

electromechanical circuit A circuit containing both electrical and mechanical parameters of consequence in its analysis.

electromechanical dialer Telephone dialer which activates one of a set of desired numbers, precoded into it, when the user selects and presses a start button.

electromechanical recording Recording by means of a signal-actuated mechanical device, such as a pen arm or mirror attached to the moving coil of a galvanometer.

electromechanical relay A protective relay operating on the principle of electromagnetic attraction, as a plunger relay, or of electromagnetic induction.

electron coupling 137

electromechanical transducer A transducer for receiving waves from an electric system and delivering waves to a mechanical system, or vice versa. Also known as electromagnetic transducer.

electrometer amplifier A low-noise amplifier having sufficiently low current drift and other characteristics required for measuring currents smaller than 10^{-12} ampere.

electrometer tube A high-vacuum electron tube having a high input impedance (low control-electrode conductance) to facilitate measurement of extremely small direct currents or voltages.

electron acceptor See acceptor.

electron beam A narrow stream of electrons moving in the same direction, all having about the same velocity.

electron-beam channeling The technique of transporting high-energy, high-current electron beams from an accelerator to a target through a region of high-pressure gas by creating a path through the gas where the gas density may be temporarily reduced; the gas may be ionized; or a current may flow whose magnetic field focuses the electron beam on the target.

electron-beam drilling Drilling of tiny holes in a ferrite, semiconductor, or other material by using a sharply focused electron beam to melt and evaporate or sublimate the material in a vacuum.

electron-beam generator Velocity-modulated generator, such as a klystron tube, used to generate extremely high frequencies.

electron-beam ion source A source of multiply charged heavy ions which uses an intense electron beam with energies of 5 to 10 kiloelectronvolts to successively ionize injected gas. Abbreviated EBIS.

electron-beam parametric amplifier A parametric amplifier in which energy is pumped from an electrostatic field into a beam of electrons traveling down the length of the tube, and electron couplers impress the input signal at one end of the tube and translate spiraling electron motion into electric output at the other.

electron-beam pumping The use of an electron beam to produce excitation for population inversion and lasing action in a semiconductor laser.

electron-beam recorder A recorder in which a moving electron beam is used to record signals or data on photographic or thermoplastic film in a vacuum chamber.

electron-beam tube An electron tube whose performance depends on the formation and control of one or more electron beams.

electron-bombardment-induced conductivity In a multimode display-storage tube, a process using an electron gun to erase the image on the cathode-ray tube interface.

electron collector See collector.

electron conduction Conduction of electricity resulting from motion of electrons, rather than from ions in a gas or solution, or holes in a solid.

electron-coupled oscillator An oscillator employing a multigrid tube in which the cathode and two grids operate as an oscillator; the anode-circuit load is coupled to the oscillator through the electron stream. Abbreviated eco. Also known as Dow oscillator.

electron coupler A microwave amplifier tube in which electron bunching is produced by an electron beam projected parallel to a magnetic field and, at the same time, subjected to a transverse electric field produced by a signal generator. Also known as Cuccia coupler.

electron coupling A method of coupling two circuits inside an electron tube, used principally with multigrid tubes; the electron stream passing between electrodes in one circuit transfers energy to electrodes in the other circuit. Also known as electronic coupling.

electron cyclotron resonance source A source of multiply charged heavy ions that uses microwave power to heat electrons to energies of tens of kilovolts in two magnetic mirror confinement chambers in series; ions formed in the first chamber drift into the second chamber, where they become highly charged. Abbreviated ECR source.

electron device A device in which conduction is principally by electrons moving through a vacuum, gas, or semiconductor, as in a crystal diode, electron tube, transistor, or selenium rectifier.

electron donor See donor.

electron efficiency The power which an electron stream delivers to the circuit of an oscillator or amplifier at a given frequency, divided by the direct power supplied to the stream. Also known as electronic efficiency.

electronegative 1. Carrying a negative electric charge. 2. Capable of acting as the negative electrode in an electric cell.

electron emission The liberation of electrons from an electrode into the surrounding space, usually under the influence of heat, light, or a high electric field.

electron emitter The electrode from which electrons are emitted.

electron flow A current produced by the movement of free electrons toward a positive terminal; the direction of electron flow is opposite to that of current.

electron gun An electrode structure that produces and may control, focus, deflect, and converge one or more electron beams in an electron tube.

electron-gun density multiplication Ratio of the average current density at any specified aperture through which the electron stream passes to the average current density at the cathode surface.

electron hole See hole.

electron hole droplets A form of electronic excitation observed in germanium and silicon at sufficiently low cryogenic temperatures; it is associated with a liquid-gas phase transition of the charge carriers, and consists of regions of conducting electron-hole Fermi liquid coexisting with regions of insulating exciton gas.

electronic Pertaining to electron devices or to circuits or systems utilizing electron devices, including electron tubes, magnetic amplifiers, transistors, and other devices that do the work of electron tubes.

electronic alternating-current voltmeter A voltmeter consisting of a direct-current milliammeter calibrated in volts and connected to an amplifier-rectifier circuit.

electronic azimuth marker On an airborne radar plan position indicator (PPI) a bright rotatable radial line used for bearing determination. Also known as azimuth marker.

electronic bearing cursor Of a marine radar set, the bright rotatable radial line on the plan position indicator used for bearing determination. Also known as electronic bearing marker.

electronic bearing marker See electronic bearing cursor.

electronic calculator A calculator in which integrated circuits perform calculations and show results on a digital display; the displays usually use either seven-segment light-emitting diodes or liquid crystals.

electronic camouflage Use of electronic means, or exploitation of electronic characteristics to reduce, submerge, or eliminate the radar echoing properties of a target.

electronic microradiography

electronic circuit An electric circuit in which the equilibrium of electrons in some of the components (such as electron tubes, transistors, or magnetic amplifiers) is upset by means other than an applied voltage.

electronic commutator An electron-tube or transistor circuit that switches one circuit connection rapidly and successively to many other circuits, without the wear and noise of mechanical switches.

electronic component A component which is able to amplify or control voltages or currents without mechanical or other nonelectrical command, or to switch currents or voltages without mechanical switches; examples include electron tubes, transistors, and other solid-state devices.

electronic computing units The sensing sections of tabulating equipment which enable the machine to handle the contents of punched cards in a prescribed manner.

electronic confusion area Amount of space that a target appears to occupy in a radar resolution cell, as it appears to that radar beam.

electronic control The control of a machine or process by circuits using electron tubes, transistors, magnetic amplifiers, or other devices having comparable functions.

electronic controller Electronic device incorporating vacuum tubes or solid-state devices and used to control the action or position of equipment; for example, a valve operator.

electronic counter A circuit using electron tubes or equivalent devices for counting electric pulses. Also known as electronic tachometer.

electronic countermeasure An offensive or defensive tactic or device using electronic and reflecting apparatus to reduce the military effectiveness of enemy equipment involving electromagnetic radiation, such as radar, communication, guidance, or other radio-wave devices. Abbreviated ECM. Also known as electromagnetic countermeasure.

electronic coupling See electron coupling.

electronic defense evaluation A mutual evaluation of radar and aircraft, with the aircraft trying to penetrate the radar's area of coverage in an electronic countermeasure environment.

electronic efficiency Ratio of the power at the desired frequency, delivered by the electron stream to the circuit in an oscillator or amplifier, to the average power supplied to the stream.

electronic interference Any electrical or electromagnetic disturbance that causes undesirable response in electronic equipment.

electronic jammer See jammer.

electronic jamming See jamming.

electronic line scanning Method which provides motion of the scanning spot along the scanning line by electronic means.

electronic listening device A device used to capture the sound waves of conversation originating in an ostensibly private setting in a form, usually as a magnetic tape recording, which can be used against the target by adverse interests.

electronic locator See metal detector.

electronic locking A technique for preventing the operation of a switch until a specific electrical signal (the unlocking signal) is introduced into circuitry associated with the switch; usually, but not necessarily, the unlocking signal is a binary sequence.

electronic microradiography Microradiography of very thin specimens in which the emission of electrons from an irradiated object, either the specimen or a lead screen behind it, is used to produce a photographic image of the specimen, which is then enlarged.

electronic motor control A control circuit used to vary the speed of a direct-current motor operated from an alternating-current power line. Also known as direct-current motor control; motor control.

electronic multimeter A multimeter that uses semiconductor or electron-tube circuits to drive a conventional multiscale meter.

electronic noise jammer An electronic jammer which emits a radio-frequency carrier modulated with a white noise signal usually derived from a gas tube; used against enemy radar.

electronic organ A musical instrument which uses electronic circuits to produce music similar to that of a pipe organ.

electronic phase-angle meter A phasemeter that makes use of electronic devices, such as amplifiers and limiters, that convert the alternating-current voltages being measured to square waves whose spacings are proportional to phase.

electronic piano A piano without a sounding board, in which vibrations of each string affect the capacitance of a capacitor microphone and thereby produce audio-frequency signals that are amplified and reproduced by a loudspeaker.

electronic polarization Polarization arising from the displacement of electrons with respect to the nuclei with which they are associated, upon application of an external electric field.

electronic power supply See power supply.

electronic radiography Radiography in which the image is detached by direct image converter tubes or by the use of television pickup or electronic scanning, and the resulant signals are amplified and presented for viewing on a kinescope.

electronic-raster scanning See electronic scanning.

electronic reconnaissance The detection, identification, evaluation, and location of foreign, electromagnetic radiations emanating from other than nuclear detonations or radioactive sources.

electronic recording The process of making a graphical record of a varying quantity or signal (or the result of such a process) by electronic means, involving control of an electron beam by electric or magnetic fields, as in a cathode-ray oscillograph, in contrast to light-beam recording.

electronic scanning Scanning in which an electron beam, controlled by electric or magnetic fields, is swept over the area under examination, in contrast to mechanical or electromechanical scanning. Also known as electronic-raster scanning.

electronic security Protection resulting from all measures designed to deny to unauthorized persons information of value which might be derived from the possession and study of electromagnetic radiations.

electronic sky screen equipment Electronic device that indicates the departure of a missile from a predetermined trajectory.

electronic specific heat Contribution to the specific heat of a metal from the motion of conduction electrons.

electronic surge arrester Device used to switch to ground high-energy surges, thereby reducing transient energy to a level safe for secondary protectors, for example, Zener diodes, silicon rectifiers and so on.

electronic switch 1. Vacuum tube, crystal diodes, or transistors used as an on and off switching device. 2. Test instrument used to present two wave shapes on a single gun cathode-ray tube.

electronic switching The use of electronic circuits to perform the functions of a high-speed switch.

electronic tachometer *See* electronic counter.

electronic tuning Tuning of a transmitter, receiver, or other tuned equipment by changing a control voltage rather than by adjusting or switching components by hand.

electronic video recording The recording of black and white or color television visual signals on a reel of photographic film as coded black and white images. Abbreviated EVR.

electronic voltage regulator A device which maintains the direct-current power supply voltage for electronic equipment nearly constant in spite of input alternating-current line voltage variations and output load variations.

electronic warfare Military action involving the use of electromagnetic energy to determine, exploit, reduce, or prevent hostile use of the electromagnetic spectrum, and action which retains friendly use of electromagnetic spectrum.

electronic warfare support measures That division of electronic warfare involving actions taken to search for, intercept, locate, record, and analyze radiated electromagnetic energy for the purpose of exploiting such radiations in support of military operations.

electronic work function The energy required to raise an electron with the Fermi energy in a solid to the energy level of an electron at rest in vacuum outside the solid.

electronic writing The use of electronic circuits and electron devices to reproduce symbols, such as an alphabet, in a prescribed order on an electronic display device for the purpose of transferring information from a source to a viewer of the display device.

electron image tube *See* image tube.

electron injection 1. The emission of electrons from one solid into another. 2. The process of injecting a beam of electrons with an electron gun into the vacuum chamber of a mass spectrometer, betatron, or other large electron accelerator.

electron lens An electric or magnetic field, or a combination thereof, which acts upon an electron beam in a manner analogous to that in which an optical lens acts upon a light beam. Also known as lens.

electron microscope A device for forming greatly magnified images of objects by means of electrons, usually focused by electron lenses.

electron mirror *See* dynode.

electron mobility The drift mobility of electrons in a semiconductor, being the electron velocity divided by the applied electric field.

electron multiplier An electron-tube structure which produces current amplification; an electron beam containing the desired signal is reflected in turn from the surfaces of each of a series of dynodes, and at each reflection an impinging electron releases two or more secondary electrons, so that the beam builds up in strength. Also known as multiplier.

electron-multiplier phototube *See* multiplier phototube.

electronographic tube An image tube used in astronomy in which the electron image formed by the tube is recorded directly upon film or plates.

electronography The use of image tubes to form intensified electron images of astronomical objects and record them directly on film or plates.

142 electronoluminescence

electronoluminescence *See* cathodoluminescence.

electron optics The study of the motion of free electrons under the influence of electric and magnetic fields.

electron-ray indicator *See* cathode-ray tuning indicator.

electron-ray tube *See* cathode-ray tube.

electron-stream potential At any point in an electron stream, the time average of the potential difference between that point and the electron-emitting surface.

electron-stream transmission efficiency At an electrode through which the electron stream (beam) passes, the ratio of the average stream current through the electrode to the stream current approaching the electrode.

electron telescope A telescope in which an infrared image of a distant object is focused on the photosensitive cathode of an image converter tube; the resulting electron image is enlarged by electron lenses and made visible by a fluorescent screen.

electron trap A defect or chemical impurity in a semiconductor or insulator which captures mobile electrons in a special way.

electron tube An electron device in which conduction of electricity is provided by electrons moving through a vacuum or gaseous medium within a gastight envelope. Also known as radio tube; tube; valve (British usage).

electron-tube amplifier An amplifier in which electron tubes provide the required increase in signal strength.

electron-tube generator A generator in which direct-current energy is converted to radio-frequency energy by an electron tube in an oscillator circuit.

electron-tube heater *See* heater.

electron tube static characteristic Relation between a pair of variables such as electrode voltage and electrode current with all other voltages maintained constant.

electroosmotic driver A type of solion for converting voltage into fluid pressure, which uses depolarizing electrodes sealed in an electrolyte and operates through the streaming potential effect. Also known as micropump.

electrophorus A device used to produce electric charges; it consists of a hard-rubber disk, which is negatively charged by rubbing with fur, and a metal plate, held by an insulating handle, which is placed on the disk; the plate is then touched with a grounded conductor, so that negative charge is removed and the plate has net positive charge.

electrophotoluminescence Emission of light resulting from application of an electric field to a phosphor which is concurrently, or has been previously, excited by other means.

electropositive 1. Carrying a positive electric charge. 2. Capable of acting as the positive electrode in an electric cell.

electroresistive effect The change in the resistivity of certain materials with changes in applied voltage.

electrosensitive recording Recording in which the image is produced by passing electric current through the record sheet.

electrostatic Pertaining to electricity at rest, such as an electric charge on an object.

electrostatic accelerator Any instrument which uses an electrostatic field to accelerate charged particles to high velocities in a vacuum.

electrostatic analyzer A device which filters an electron beam, permitting only electrons within a very narrow velocity range to pass through.

electrostatic storage 143

electrostatic attraction See Coulomb attraction.

electrostatic cathode-ray tube A cathode-ray tube in which electrostatic deflection is used on the electron beam.

electrostatic copying See electrostatography.

electrostatic deflection The deflection of an electron beam by means of an electrostatic field produced by electrodes on opposite sides of the beam; used chiefly in cathode-ray tubes for oscilloscopes.

electrostatic detection The detection and location of any type of solid body, such as a mineral deposit or a mine, by measuring the associated electrostatic field which arises spontaneously or is induced by the detection equipment.

electrostatic energy The potential energy which a collection of electric charges possesses by virtue of their positions relative to each other.

electrostatic error See antenna effect.

electrostatic field A time-independent electric field, such as that produced by stationary charges.

electrostatic focus Production of a focused electron beam in a cathode-ray tube by the application of an electric field.

electrostatic force Force on a charged particle due to an electrostatic field, equal to the electric field vector times the charge of the particle.

electrostatic generator Any machine which produces electric charges by friction or (more commonly) electrostatic induction.

electrostatic induction The process of charging an object electrically by bringing it near another charged object, then touching it to ground. Also known as induction.

electrostatic instrument A meter that depends for its operation on the forces of attraction and repulsion between electrically charged bodies.

electrostatic interactions See Coulomb interactions.

electrostatic lens An arrangement of electrostatic fields which acts upon beams of charged particles similar to the way a glass lens acts on light beams.

electrostatic memory See electrostatic storage.

electrostatic octupole lens A device for controlling beams of electrons or other charged particles, consisting of eight electrodes arranged in a circular pattern with alternating polarities; commonly used to correct aberrations of quadrupole lens systems.

electrostatic potential See electric potential.

electrostatic quadrupole lens A device for focusing beams of electrons or other charged particles, consisting of four electrodes arranged in a circular pattern with alternating polarities.

electrostatic repulsion See Coulomb repulsion.

electrostatics The study of electric charges at rest, their electric fields, and potentials.

electrostatic scanning Scanning that involves electrostatic deflection of an electron beam.

electrostatic shielding The placing of a grounded metal screen, sheet, or enclosure around a device or between two devices to prevent electric fields from interacting.

electrostatic storage A storage in which information is retained as the presence or absence of electrostatic charges at specific spot locations, generally on the screen of a special type of cathode-ray tube known as a storage tube. Also known as electrostatic memory.

144   electrostatic storage tube

electrostatic storage tube See storage tube.

electrostatic stress An electrostatic field acting on an insulator, which produces polarization in the insulator and causes electrical breakdown if raised beyond a certain intensity.

electrostatic units A centimeter-gram-second system of electric and magnetic units in which the unit of charge is that charge which exerts a force of 1 dyne on another unit charge when separated from it by a distance of 1 centimeter in vacuum; other units are derived from this definition by assigning unit coefficients in equations relating electric and magnetic quantities. Abbreviated esu.

electrothermal recording Type of electrochemical recording, used in facsimile equipment, wherein the chemical change is produced principally by signal-controlled thermal action.

element 1. A part of an electron tube, semiconductor device, or antenna array that contributes directly to the electrical performance. 2. A radiator, active or parasitic, that is a part of an antenna. 3. See component.

elemental area See picture element.

elevation angle The angle that a radio, radar, or other such beam makes with the horizontal.

elevation-angle error In radar, the error in the measurement of the elevation angle of a target resulting from the vertical bending or refraction of radio energy in traveling through the atmosphere. Also known as elevation error.

elevation error See elevation-angle error.

eliminator Device that takes the place of batteries, generally consisting of a rectifier operating from alternating current.

E lines Contour lines of constant electrostatic field strength referred to some reference base.

ellipsoidal floodlight A lighting unit used in theatrical lighting consisting of an ellipsoidal reflector with fixed spacing and a lamp; power requirements are 250–5000 watts and the reflector diameter is 10–24 inches (25–61 centimeters). Also known as scoop.

ellipsoidal spotlight A lighting unit consisting of a reflector, lamp, single or multiple lens system, and framing device; power requirements are 250–2000 watts.

elliptical polarization Polarization of an electromagnetic wave in which the electric field vector at any point in space describes an ellipse in a plane perpendicular to the propagation direction.

ellipticity See axial ratio.

emanation security The protection resulting from all measures designed to deny unauthorized persons information of value which might be derived from unintentional emissions from other than telecommunications systems.

emergency power supply A source of power that becomes available, usually automatically, when normal power line service fails.

emission Any radiation of energy by means of electromagnetic waves, as from a radio transmitter.

emission characteristics Relation, usually shown by a graph, between the emission and a factor controlling the emission, such as temperature, voltage, or current of the filament or heater.

emission electron microscope An electron microscope in which thermionic, photo, secondary, or field electrons emitted from a metal surface are projected on a fluorescent screen, with or without focusing.

emission security That component of communications security which results from all measures taken to protect any unintentional emissions of a telecommunications system from any form of exploitation other than cryptanalysis.

emitter A transistor region from which charge carriers that are minority carriers in the base are injected into the base, thus controlling the current flowing through the collector; corresponds to the cathode of an electron tube. Symbolized E. Also known as emitter region.

emitter barrier One of the regions in which rectification takes place in a transistor, lying between the emitter region and the base region.

emitter bias A bias voltage applied to the emitter electrode of a transistor.

emitter-coupled logic A form of current-mode logic in which the emitters of two transistors are connected to a single current-carrying resistor in such a way that only one transistor conducts at a time. Abbreviated ECL.

emitter follower A grounded-collector transistor amplifier which provides less than unity voltage gain but high input resistance and low output resistance, and which is similar to a cathode follower in its operations.

emitter junction A transistor junction normally biased in the low-resistance direction to inject minority carriers into a base.

emitter region See emitter.

emitter resistance The resistance in series with the emitter lead in an equivalent circuit representing a transistor.

E mode See transverse magnetic mode.

emphasizer See preemphasis network.

emu See electromagnetic system of units.

enabling pulse A pulse that prepares a circuit for some subsequent action.

enclosed arc lamp An arc lamp in which the arc produced by carbon electrodes is protected from the atmosphere by a translucent enclosure.

encoder 1. In an electronic computer: a network or system in which only one input is excited at a time and each input produces a combination of outputs. 2. See matrix.

end cell One of a group of cells in series with a storage battery, which can be switched in to maintain the output voltage of the battery when it is not being charged.

end-cell rectifier Small trickle charge rectifier used to maintain voltage of the storage battery end cells.

end effect The effect of capacitance at the ends of an antenna; it requires that the actual length of a half-wave antenna be about 5% less than a half wavelength.

end-fire array A linear array whose direction of maximum radiation is along the axis of the array; it may be either unidirectional or bidirectional; the elements of the array are parallel and in the same plane, as in a fishbone antenna. Also known as end-fire antenna.

end instrument A pickup used in telemetering to convert a physical quantity to an inductance, resistance, voltage, or other electrical quantity that can be transmitted over wires or by radio.

end loss The difference between the actual and the effective lengths of a radiating antenna element.

energized Electrically connected to a voltage source. Also known as alive; hot; live.

energy efficiency ratio A value that represents the relative electrical efficiency of air conditioners; it is the quotient obtained by dividing Btu-per-hour output by electrical-watts input during cooling.

energy gap A range of forbidden energies in the band theory of solids.

energy of a charge Charge energy measured in ergs according to the equation $E = QV$, where Q is the charge and V is the potential in electrostatic units.

energy product curve Curve obtained by plotting the product of the values of magnetic induction B and demagnetizing force H for each point on the demagnetization curve of a permanent magnet material; usually shown with the demagnetization curve.

engine starter The electric motor in the electric system of an automobile that cranks the engine for starting. Also known as starter; starting motor.

enhancement An increase in the density of charged carriers in a particular region of a semiconductor.

enhancement mode Operation of a field-effect transistor in which no current flows when zero gate voltage is applied, and increasing the gate voltage increases the current.

entrance cable Cable that brings power from an outside power line into a building.

envelope delay distortion *See* delay distortion.

envelope detector *See* detector.

epitaxial diffused-junction transistor A junction transistor produced by growing a thin, high-purity layer of semiconductor material on a heavily doped region of the same type.

epitaxial diffused-mesa transistor A diffused-mesa transistor in which a thin, high-resistivity epitaxial layer is deposited on the substrate to serve as the collector.

epitaxial layer A semiconductor layer having the same crystalline orientation as the substrate on which it is grown.

epitaxial transistor Transistor with one or more epitaxial layers.

E-plane antenna An antenna which lies in a plane parallel to the electric field vector of the radiation that it emits.

E-plane bend *See* E bend.

E-plane T junction Waveguide T junction in which the change in structure occurs in the plane of the electric field. Also known as series T junction.

epsilon structure The hexagonal close-packed structure of the ϵ-phase of an electron compound.

equalization The effect of all frequency-discriminating means employed in transmitting, recording, amplifying, or other signal-handling systems to obtain a desired overall frequency response. Also known as frequency-response equalization.

equalizer A network designed to compensate for an undesired amplitude-frequency or phase-frequency response of a system or component; usually a combination of coils, capacitors, and resistors. Also known as equalizing circuit.

equalizing circuit *See* equalizer.

equalizing current Current that circulates between two parallel-connected compound generators to equalize their output.

equalizing pulses In television, pulses at twice the line frequency, occurring just before and after the vertical synchronizing pulses, which minimize the effect of line frequency pulses on the interlace.

equiangular spiral antenna A frequency-independent broad-band antenna, cut from sheet metal, that radiates a very broad, circularly polarized beam on both sides of its surface; this bidirectional radiation pattern is its chief limitation.

equilibrium brightness Viewing screen brightness occurring when a display storage tube is in a fully written condition.

equipotential cathode *See* indirectly heated cathode.

equipotential surface A surface on which the electric potential is the same at every point.

equisignal surface Surface around an antenna formed by all points at which, for transmission, the field strength (usually measured in volts per meter) is constant.

equivalent circuit A circuit whose behavior is identical to that of a more complex circuit or device over a stated range of operating conditions.

equivalent noise conductance Spectral density of a noise current generator measured in conductance units at a specified frequency.

equivalent noise resistance Spectral density of a noise voltage generator measured in ohms at a specified frequency.

equivalent noise temperature Absolute temperature at which a perfect resistor, of equal resistance to the component, would generate the same noise as does the component at room temperature.

equivalent periodic line Of a uniform line, a periodic line having the same electrical behavior, at a given frequency, as the uniform line when measured at its terminals or at corresponding section junctions.

equivalent resistance Concentrated or lumped resistance that would cause the same power loss as the actual small resistance values distributed throughout a circuit.

erase 1. To remove recorded material from magnetic tape by passing the tape through a strong, constant magnetic field (dc erase) or through a high-frequency alternating magnetic field (ac erase). 2. To eliminate previously stored information in a charge-storage tube by charging or discharging all storage elements.

erase oscillator The oscillator used in a magnetic recorder to provide the high-frequency signal needed to erase a recording on magnetic tape; the bias oscillator usually serves also as the erase oscillator.

erasing head A magnetic head used to obliterate material previously recorded on magnetic tape.

erasing speed In charge-storage tubes, the rate of erasing successive storage elements.

ERP *See* effective radiated power .

error correction Correction of time errors in interconnected alternating-current power systems resulting from deviations from normal frequency, in order to make all areas synchronous.

error signal 1. A voltage that depends on the signal received from the target in a tracking system, having a polarity and magnitude dependent on the angle between the target and the center of the scanning beam. 2. *See* error voltage.

error voltage A voltage, usually obtained from a selsyn, that is proportional to the difference between the angular positions of the input and output shafts of a servosystem; this voltage acts on the system to produce a motion that tends to reduce the error in position. Also known as error signal.

Esaki tunnel diode *See* tunnel diode.

E scan *See* E scope.

E scope A cathode-ray scope on which signals appear as spots, with range as the horizontal coordinate and elevation angle or height as the vertical coordinate. Also known as E indicator; E scan.

148 Eshelby twist

Eshelby twist A torsional deformation of a crystal whisker resulting from a screw dislocation along the whisker axis.

Essen coefficient The torque exerted on the moving part of an electric rotating machine divided by the volume enclosed by the air gap.

esu See electrostatic units.

ether The medium postulated to carry electromagnetic waves, similar to the way a gas carries sound waves.

ether drag The hypothesis, advanced unsuccessfully to account for results of the Michelson-Morley experiment, that ether is dragged along with matter.

ether drift Hypothetical motion of the ether relative to the earth.

E transformer A transformer consisting of two coils wound around a laminated iron core in the shape of an E, with the primary and secondaries occupying the center and outside legs respectively.

E vector Vector representing the electric field of an electromagnetic wave.

Evjen method Method of calculating lattice sums in which groups of charges whose total charge is zero are taken together, so that the contribution of each group is small and the series rapidly converges.

EVR See electronic video recording.

Ewald-Kornfeld method An extension of the Ewald method to calculate Coulomb energies of dipole arrays.

Ewald method Method of calculating lattice sums in which certain mathematical techniques are employed to make series converge rapidly.

Ewald sphere A sphere superimposed on the reciprocal lattice of a crystal, used to determine the directions in which an x-ray or other beam will be reflected by a crystal lattice.

E wave See transverse magnetic wave.

Ewing theory of ferromagnetism Theory of ferromagnetic phenomena which assumes each atom is a permanent magnet which can turn freely about its center under the influence of applied fields and other magnets.

exalted-carrier receiver Receiver that counteracts selective fading by maintaining the carrier at a high level at all times; this minimizes the second harmonic distortion that would otherwise occur when the carrier drops out while leaving most of the sidebands at their normal amplitudes.

except gate A gate that produces an output pulse only for a pulse on one or more input lines and the absence of a pulse on one or more other lines.

excess conduction Electrical conduction by excess electrons in a semiconductor.

excess electron Electron introduced into a semiconductor by a donor impurity and available for conduction.

exchange anisotropy Phenomenon observed in certain mixtures of magnetic materials under certain conditions, in which magnetization is favored in some direction (rather than merely along some axis); thought to be caused by exchange coupling across the interface between compounds when one is ferromagnetic and one is antiferromagnetic.

exchange cable Lead covered, nonquadded, paper-insulated cable used within a given area to provide cable pairs between local subscribers and a central office.

exchange current The magnitude of the current which flows through a galvanic cell when it is operating in a reversible manner.

expandor 149

exchange line Line joining a subscriber or switchboard to a commercial exchange.

excitation 1. The application of voltage to field coils to produce a magnetic field, as required for the operation of an excited-field loudspeaker or a generator. 2. The signal voltage that is applied to the control electrode of an electron tube. Also known as drive. 3. Application of signal power to a transmitting antenna.

excitation anode An anode used to maintain a cathode spot on a pool cathode of a gas tube when output current is zero.

excitation loss *See* core loss.

excitation voltage Nominal voltage required for excitation of a circuit.

excited-state effect The motion of a crystal defect through a process in which the defect is first raised into an excited state and then decays, together with the surroundings, into a state in which motion of the defect readily occurs.

exciter 1. A small auxiliary generator that provides field current for an alternating-current generator. 2. A crystal oscillator or self-excited oscillator used to generate the carrier frequency of a transmitter. 3. The portion of a directional transmitting antenna system that is directly connected to the transmitter. 4. A loop or probe extending into a resonant cavity or waveguide. 5. *See* exciter lamp.

exciter lamp A bright incandescent lamp having a concentrated filament, used to excite a phototube or photocell in sound movie and facsimile systems. Also known as exciter.

exciter response In electrical rotating machinery, the rate of increase or decrease of the main exciter voltage when resistance is suddenly removed from or inserted in the main exciter field circuit.

exciting current *See* magnetizing current.

exciton An excited state of an insulator or semiconductor which allows energy to be transported without transport of electric charge; may be thought of as an electron and a hole in a bound state.

excitron A single-anode mercury-pool tube provided with means for maintaining a continuous cathode spot.

exhaustion region A layer in a semiconductor, adjacent to its contact with a metal, in which there is almost complete ionization of atoms in the lattice and few charge carriers, resulting in a space-charge density.

Exide ironclad battery A portable storage battery designed for propelling electric vehicles; a lead-antimony frame (positive plate) supports perforated hard-rubber tubes containing irregular lead-antimony cores packed with lead peroxide paste.

exogenous electrification The separation of electric charge in a conductor placed in a preexisting electric field, especially applied to the charge separation observed on metal-covered aircraft, resulting from induction effects, and by itself does not create any net total charge on the conductor.

expanded position indicator display Display of an expanded sector from a plan position indicator presentation.

expanded scope Magnified portion of a given type of cathode-ray tube presentation.

expanded sweep A cathode-ray sweep in which the movement of the electron beam across the screen is speeded up during a selected portion of the sweep time.

expander A transducer that, for a given input amplitude range, produces a larger output range.

expandor The part of a compandor that is used at the receiving end of a circuit to return the compressed signal to its original form; attenuates weak signals and amplifies strong signals.

expansion A process in which the effective gain of an amplifier is varied as a function of signal magnitude, the effective gain being greater for large signals than for small signals; the result is greater volume range in an audio amplifier and greater contrast range in facsimile.

expansion ellipsoid An ellipsoid whose axes have lengths which are proportional to the coefficient of linear expansion in the corresponding direction in a crystal.

exploring coil A small coil used to measure a magnetic field or to detect changes produced in a magnetic field by a hidden object; the coil is connected to an indicating instrument either directly or through an amplifier. Also known as magnetic test coil; search coil.

exponential amplifier An amplifier capable of supplying an output signal proportional to the exponential of the input signal.

exponential transmission line A two-conductor transmission line whose characteristic impedance varies exponentially with electrical length along the line.

exposure voltage The voltage at which the document-illuminating lamps are operated during exposure.

expulsion fuse *See* expulsion-fuse unit.

expulsion-fuse unit A vented fuse unit in which the arc is extinguished by the expulsion of gases generated by the arc and lining of the fuse holder, sometimes with the aid of a spring. Also known as expulsion fuse.

extended-interaction tube Microwave tube in which a moving electron stream interacts with a traveling electric field in a long resonator; bandwidth is between that of klystrons and traveling-wave tubes.

extender A male or female receptacle connected by a short cable to make a test point more conveniently accessible to a test probe.

extension cord A line cord having a plug at one end and an outlet at the other end.

external armature Armature for a machine of special design in which the armature is a ring which rotates around the magnetic poles.

external photoelectric effect *See* photoemission.

external Q The inverse of the difference between the loaded and unloaded Q values of a microwave tube.

extinction voltage The lowest anode voltage at which a discharge is sustained in a gas tube.

extra-high tension British term for the high direct-current voltage applied to the second anode in a cathode-ray tube, ranging from about 4000 to 50,000 volts in various sizes of tubes. Abbreviated eht.

extra-high voltage A voltage above 345 kilovolts used for power transmission. Abbreviated ehv.

extraneous emission Any emission of a transmitter or transponder, other than the output carrier fundamental, plus only those sidebands intentionally employed for the transmission of intelligence.

extraneous response Any undersired response of a receiver, recorder, or other susceptible device, due to the desired signals, undersired signals, or any combination or interaction among them.

extraterrestrial noise Cosmic and solar noise; radio disturbances from other sources other than those related to the earth.

extreme ultraviolet radiation *See* vacuum ultraviolet radiation.

extrinsic semiconductor 151

extrinsic photoconductivity Photoconductivity that occurs for photon energies smaller than the band gap and corresponds to optical excitation from an occupied imperfection level to the conduction band, or to an unoccupied imperfection level from the valence band, of a material.

extrinsic photoemission Photoemission by an alkali halide crystal in which electrons are ejected directly from negative ion vacancies, forming color centers. Also known as direct ionization.

extrinsic properties The properties of a semiconductor as modified by impurities or imperfections within the crystal.

extrinsic semiconductor A semiconductor whose electrical properties are dependent on impurities added to the semiconductor crystal, in contrast to an intrinsic semiconductor, whose properties are characteristic of an ideal pure crystal.

F

F *See* farad.

fA *See* femtoampere.

Faber flaw A deformation in a superconducting material that acts as a nucleation center for the growth of a superconducting region.

face *See* faceplate.

face-bonding Method of assembling hybrid microcircuits wherein semiconductor chips are provided with small mounting pads, turned facedown, and bonded directly to the ends of the thin-film conductors on the passive substrate.

faceplate The transparent or semitransparent glass front of a cathode-ray tube, through which the image is viewed or projected; the inner surface of the face is coated with fluorescent chemicals that emit light when hit by an electron beam. Also known as face.

facsimile receiver The receiver used to translate the facsimile signal from a wire or radio communication channel into a facsimile record of the subject copy.

facsimile recorder The section of a facsimile receiver that performs the final conversion of electric signals to an image of the subject copy on the record medium.

facsimile signal level Maximum facsimile signal power or voltage (root mean square or direct current) measured at any point in a facsimile system.

facsimile synchronizing Maintenance of predetermined speed relations between the scanning spot and the recording spot within each scanning line.

facsimile transmitter The apparatus used to translate the subject copy into facsimile signals suitable for delivery over a communication system.

fade chart Graph on which the null areas of an air-search radar antenna are plotted as an aid to estimating target altitude.

fader A multiple-unit level control used for gradual changeover from one microphone, audio channel, or television camera to another.

Fahnestock clip A spring-type terminal to which a temporary connection can readily be made.

fall time Measure of time required for a circuit to change its output from a high level to a low level.

false alarm In radar, an indication of a detected target even though one does not exist, due to noise or interference levels exceeding the set threshold of detection.

false pyroelectricity *See* tertiary pyroelectricity.

false target A nonexistent target which shows up on a radar scope as the result of time delay.

false target generator An electronic countermeasure device that generates a delayed return signal on an enemy radar frequency to give erroneous position information.

fan Volume of space periodically energized by a radar beam (or beams) repeatedly traversing an established pattern.

fan antenna An array of folded dipoles of different length forming a wide-band ultra-high-frequency or very-high-frequency antenna.

fan beam 1. A radio beam having an elliptically shaped cross section in which the ratio of the major to the minor axis usually exceeds 3 to 1; the beam is broad in the vertical plane and narrow in the horizontal plane. 2. A radar beam having the shape of a fan.

fan-in The number of inputs that can be connected to a logic circuit.

fanned-beam antenna Unidirectional antenna so designed that transverse cross sections of the major lobe are approximately elliptical.

fanning beam Narrow antenna beam which is repeatedly scanned over a limited arc.

fanning strip Insulated board, often of wood, which serves to spread out the wires of a cable for distribution to a terminal board.

fan-out The number of parallel loads that can be driven from one output mode of a logic circuit.

farad The unit of capacitance in the meter-kilogram-second system, equal to the capacitance of a capacitor which has a potential difference of 1 volt between its plates when the charge on one of its plates is 1 coulomb, there being an equal and opposite charge on the other plate. Symbolized F.

Faraday cage *See* Faraday shield.

Faraday cylinder 1. A closed, or nearly closed, hollow conductor, usually grounded, within which apparatus is placed to shield it from electrical fields. 2. A nearly closed, insulated, hollow conductor, usually shielded by a second grounded cylinder, used to collect and detect a beam of charged particles.

Faraday dark space The relatively nonluminous region that separates the negative glow from the positive column in a cold-cathode glow-discharge tube.

Faraday disk machine A device for demonstrating electromagnetic induction, consisting of a copper disk in which a radial electromotive force is induced when the disk is rotated between the poles of a magnet. Also known as Faraday generator.

Faraday generator *See* Faraday disk machine.

Faraday ice bucket experiment Experiment in which one lowers a charged metal body into a pail and observes the effect on an electroscope attached to the pail, with and without contact between body and pail; the experiment shows that charge resides on a conductor's outside surface.

Faraday's law of electromagnetic induction The law that the electromotive force induced in a circuit by a changing magnetic field is equal to the negative of the rate of change of the magnetic flux linking the circuit. Also known as law of electromagnetic induction.

Faraday rotation isolator *See* ferrite isolator.

Faraday screen *See* Faraday shield.

Faraday shield Electrostatic shield composed of wire mesh or a series of parallel wires, usually connected at one end to another conductor which is grounded. Also known as Faraday cage; Faraday screen.

faradic current An intermittent and nonsymmetrical alternating current like that obtained from the secondary winding of an induction coil; used in electrobiology.

far field *See* Fraunhofer region.

far-infrared radiation Infrared radiation the wavelengths of which are the longest of those in the infrared region, about 50–1000 micrometers; requires diffraction gratings for spectroscopic analysis.

Farnsworth image dissector tube *See* image dissector tube.

far region *See* Fraunhofer region.

far-ultraviolet radiation Ultraviolet radiation in the wavelength range of 200–300 nanometers; germicidal effects are greatest in this range.

far zone *See* Fraunhofer region.

fast automatic gain control Radar automatic gain control method characterized by a response time that is long with respect to a pulse width, and short with respect to the time on target.

fast time constant 1. An electric circuit which combines resistance and capacitance to give a short time constant for capacitor discharge through the resistor. 2. Circuit with short time constant used to emphasize signals of short duration to produce discrimination against low-frequency components of clutter in radar.

fatigue The decrease of efficiency of a luminescent or light-sensitive material as a result of excitation.

fault 1. A defect, such as an open circuit, short circuit, or ground, in a circuit, component, or line. Also known as electrical fault; faulting. 2. Any physical condition that causes a component of a data-processing system to fail in performance.

fault current *See* fault electrode current.

fault electrode current The current to an electrode under fault conditions, such as during arc-backs and load short circuits. Also known as fault current; surge electrode current.

faulting *See* fault.

Faure storage battery A storage battery in which the plates consist of lead-antimony supporting grids covered with a lead oxide paste, immersed in weak sulfuric acid. Also known as pasted-plate storage battery.

F band The optical absorption band arising from F centers.

F center A color center consisting of an electron trapped by a negative ion vacancy in an ionic crystal, such as an alkali halide or an alkaline-earth fluoride or oxide.

F display A rectangular display in which a target appears as a centralized blip when the radar antenna is aimed at it; horizontal and vertical aiming errors are respectively indicated by the horizontal and vertical displacement of the blip.

feed 1. To supply a signal to the input of a circuit, transmission line, or antenna. 2. The part of a radar antenna that is connected to or mounted on the end of the transmission line and serves to radiate radio-frequency electromagnetic energy to the reflector or receive energy therefrom.

feedback The return of a portion of the output of a circuit or device to its input.

feedback admittance Short-circuit transadmittance from the output electrode to the input electrode of an electron tube.

feedback amplifier An amplifier in which a passive network is used to return a portion of the output signal to its input so as to change the performance characteristics of the amplifier.

feedback circuit A circuit that returns a portion of the output signal of an electronic circuit or control system to the input of the circuit or system.

feedback factor The fraction of the output voltage of an oscillator which is applied to the feedback network.

feedback oscillator An oscillating circuit, including an amplifier, in which the output is fed back in phase with the input; oscillation is maintained at a frequency determined by the values of the components in the amplifier and the feedback circuits.

feedback winding A winding to which feedback connections are made in a magnetic amplifier.

feeder 1. A transmission line used between a transmitter and an antenna. 2. A conductor, or several conductors, connecting generating stations, substations, or feeding points in an electric power distribution system. 3. A group of conductors in an interior wiring system which link a main distribution center with secondary or branch-circuit distribution centers.

feeder panel The part of a switchboard in an electric power distribution system where feeder connections are made.

feeder reactor A small inductor connected in series with a feeder in order to limit and localize the disturbances due to faults on the feeder.

feedthrough A conductor that connects patterns on opposite sides of a printed circuit board. Also known as interface connection.

feedthrough capacitor A feedthrough terminal that provides a desired value of capacitance between the feedthrough conductor and the metal chassis or panel through which the conductor is passing; used chiefly for bypass purposes in ultra-high-frequency circuits.

feedthrough insulator *See* feedthrough terminal.

feedthrough terminal An insulator designed for mounting in a hole in a panel, wall, or bulkhead, with a conductor in the center on the insulator to permit feeding electricity through the partition. Also known as feedthrough insulator.

female connector A connector having one or more contacts set into recessed openings; jacks, sockets, and wall outlets are examples.

femitrons Class of field-emission microwave devices.

femtoampere A unit of current equal to 10^{-15} ampere. Abbreviated fA.

femtovolt A unit of voltage equal to 10^{-15} volt. Abbreviated fV.

Fermi distribution Distribution of energies of electrons in a semiconductor or metal as given by the Fermi-Dirac distribution function; nearly all energy levels below the Fermi level are filled, and nearly all above this level are empty.

Fermi hole A region surrounding an electron in a solid in which the energy band theory predicts that the probability of finding other electrons is less than the average over the volume of the solid.

Fermi surface A constant-energy surface in the space containing the wave vectors of states of members of an assembly of independent fermions, such as electrons in a semiconductor or metal, whose energy is that of the Fermi level.

Ferranti effect A rise in voltage occurring at the end of a long transmission line when its load is disconnected.

ferreed switch A switch whose contacts are mounted on magnetic blades or reeds sealed into an evacuated tubular glass housing, the contacts being operated by external electromagnets or permanent magnets.

ferrimagnet *See* ferrimagnetic material.

ferrimagnetic amplifier A microwave amplifier using ferrites.

ferrite switch 157

ferrimagnetic limiter Power limiter used in microwave systems to replace transmit-receive tubes; uses ferrimagnetic material (such as a piece of ferrite or garnet) that exhibits nonlinear properties.

ferrimagnetic material A material displaying ferrimagnetism; the ferrites are the principal example. Also known as ferrimagnet.

ferrimagnetism A type of magnetism in which the magnetic moments of neighboring ions tend to align nonparallel, usually antiparallel, to each other, but the moments are of different magnitudes, so there is an appreciable resultant magnetization.

ferristor A miniature, two-winding, saturable reactor that operates at a high carrier frequency and may be connected as a coincidence gate, current discriminator, free-running multivibrator, oscillator, or ring counter.

ferrite Any ferrimagnetic material having high electrical resistivity which has a spinel crystal structure and the chemical formula XFe_2O_4, where X represents any divalent metal ion whose size is such that it will fit into the crystal structure.

ferrite attenuator *See* ferrite limiter.

ferrite bead Magnetic information storage device consisting of ferrite powder mixtures in the form of a bead fired on the current-carrying wires of a memory matrix.

ferrite circulator A combination of two dual-mode transducers and a 45° ferrite rotator, used with rectangular waveguides to control and switch microwave energy. Also known as ferrite phase-differential circulator.

ferrite core A magnetic core made of ferrite material. Also known as dust core; powdered-iron core.

ferrite-core memory A magnetic memory consisting of a matrix of tiny toroidal cores molded from a square-loop ferrite, through which are threaded the pulse-carrying wires and the sense wire.

ferrite device An electrical device whose principle of operation is based upon the use of ferrites in powdered, compressed, sintered form, making use of their ferrimagnetism and their high electrical resistivity, which makes eddy-current losses extremely low at high frequencies.

ferrite isolator A device consisting of a ferrite rod, centered on the axis of a short length of circular waveguide, located between rectangular-waveguide sections displaced 45° with respect to each other, which passes energy traveling through the waveguide in one direction while absorbing energy from the opposite direction. Also known as Faraday rotation isolator.

ferrite limiter A passive, low-power microwave limiter having an insertion loss of less than 1 decibel when operating in its linear range, with minimum phase distortion; the input signal is coupled to a single-crystal sample of either yttrium iron garnet or lithium ferrite, which is biased to resonance by a magnetic field. Also known as ferrite attenuator.

ferrite phase-differential circulator *See* ferrite circulator.

ferrite-rod antenna An antenna consisting of a coil wound on a rod of ferrite; used in place of a loop antenna in radio receivers. Also known as ferrod; loopstick antenna.

ferrite rotator A gyrator consisting of a ferrite cylinder surrounded by a ring-type permanent magnet, inserted in a waveguide to rotate the plane of polarization of the electromagnetic wave passing through the waveguide.

ferrite switch A ferrite device that blocks the flow of energy through a waveguide by rotating the electric field vector 90°; the switch is energized by sending direct current through its magnetizing coil; the rotated electromagnetic wave is then reflected from a reactive mismatch or absorbed in a resistive card.

ferrite-tuned oscillator An oscillator in which the resonant characteristic of a ferrite-loaded cavity is changed by varying the ambient magnetic field, to give electronic tuning.

ferroacoustic storage A delay-line type of storage consisting of a thin tube of magnetostrictive material, a central conductor passing through the tube, and an ultrasonic driving transducer at one end of the tube.

ferrod See ferrite-rod antenna.

ferroelectric A crystalline substance displaying ferroelectricity, such as barium titanate, potassium dihydrogen phosphate, and Rochelle salt; used in ceramic capacitors, acoustic transducers, and dielectric amplifiers. Also known as seignette-electric.

ferroelectric converter A converter that transforms thermal energy into electric energy by utilizing the change in the dielectric constant of a ferroelectric material when heated beyond its Curie temperature.

ferroelectric crystal A crystal of a ferroelectric material.

ferroelectric domain A region of a ferroelectric material within which the spontaneous polarization is constant.

ferroelectric hysteresis The dependence of the polarization of ferroelectric materials not only on the applied electric field but also on their previous history; analogous to magnetic hysteresis in ferromagnetic materials. Also known as dielectric hysteresis; electric hysteresis.

ferroelectric hysteresis loop Graph of polarization or electric displacement versus applied electric field of a material displaying ferroelectric hysteresis.

ferroelectricity Spontaneous electric polarization in a crystal; analogous to ferromagnetism.

ferromagnetic amplifier A parametric amplifier based on the nonlinear behavior of ferromagnetic resonance at high radio-frequency power levels; incorrectly known as garnet maser.

ferromagnetic ceramic See ceramic magnet.

ferromagnetic crystal A crystal of a ferromagnetic material. Also known as polar crystal.

ferromagnetic domain A region of a ferromagnetic material within which atomic or molecular magnetic moments are aligned parallel. Also known as magnetic domain.

ferromagnetic film See magnetic thin film.

ferromagnetic material A material displaying ferromagnetism, such as the various forms of iron, steel, cobalt, nickel, and their alloys.

ferromagnetic resonance Magnetic resonance of a ferromagnetic material.

ferromagnetics The science that deals with the storage of binary information and the logical control of pulse sequences through the utilization of the magnetic polarization properties of materials.

ferromagnetic tape A tape made of magnetic material for use in winding closed magnetic cores of toroids and transformers.

ferromagnetism A property, exhibited by certain metals, alloys, and compounds of the transition (iron group) rare-earth and actinide elements, in which the internal magnetic moments spontaneously organize in a common direction; gives rise to a permeability considerably greater than that of vacuum, and to magnetic hysteresis.

ferroresonant circuit A resonant circuit in which a saturable reactor provides nonlinear characteristics, with tuning being accomplished by varying circuit voltage or current.

field emission   159

ferroresonant static inverter A static inverter consisting of a simple square-wave inverter system and a tuned output transformer that performs filtering, voltage regulation, and current limiting.

FET See field-effect transistor.

field 1. That part of an electric motor or generator which produces the magnetic flux which reacts with the armature, producing the desired machine action. 2. One of the equal parts into which a frame is divided in interlaced scanning for television; includes one complete scanning operation from top to bottom of the picture and back again.

field coil A coil used to produce a constant-strength magnetic field in an electric motor, generator, or excited-field loudspeaker; depending on the type of motor or generator, the field core may be on the stator or the rotor. Also known as field winding.

field desorption A technique which tears atoms from a surface by an electric field applied at a sharp dip to produce very well-ordered, clean, plane surfaces of many crystallographic orientations.

field-desorption microscope A type of field-ion microscope in which the tip specimen is imaged by ions that are field-desorbed or field-evaporated directly from the surface rather than by ions obtained from an externally supplied gas.

field discharge A spark discharge due to high potential across a gap.

field discharge switch A special type of switch that is connected in series with the field winding of an electrical machine, and that is operated to connect a resistor in parallel with the field winding before the main supply contacts are opened, in order to prevent the self-induced electromotive force in the field winding from reaching dangerous levels.

field distortion Any alteration in the direction of an electric or magnetic field; in particular, distortion of the magnetic fields between the north and south poles of a generator due to the counter electromotive force in the armature winding.

field-effect capacitor A capacitor in which the effective dielectric is a region of semiconductor material that has been depleted or inverted by the field effect.

field-effect device A semiconductor device whose properties are determined largely by the effect of an electric field on a region within the semiconductor.

field-effect diode A semiconductor diode in which the charge carriers are of only one polarity.

field-effect phototransistor A field-effect transistor that responds to modulated light as the input signal.

field-effect tetrode Four-terminal device consisting of two independently terminated semiconducting channels so displaced that the conductance of each is modulated along its length by the voltage conditions in the other.

field-effect transistor A transistor in which the resistance of the current path from source to drain is modulated by applying a transverse electric field between grid or gate electrodes; the electric field varies the thickness of the depletion layer between the gates, thereby reducing the conductance. Abbreviated FET.

field-effect-transistor resistor A field-effect transistor in which the gate is generally tied to the drain; the resultant structure is used as a resistance load for another transistor.

field-effect varistor A passive, two-terminal, nonlinear semiconductor device that maintains constant current over a wide voltage range.

field emission The emission of electrons from the surface of a metallic conductor into a vacuum (or into an insulator) under influence of a strong electric field; electrons

field-emission microscope

penetrate through the surface potential barrier by virtue of the quantum-mechanical tunnel effect. Also known as cold emission.

field-emission microscope A device that uses field emission of electrons or of positive ions (field-ion microscope) to produce a magnified image of the emitter surface on a fluorescent screen.

field-emission tube A vacuum tube within which field emission is obtained from a sharp metal point; must be more highly evacuated than an ordinary vacuum tube to prevent contamination of the point.

field-enhanced emission An increase in electron emission resulting from an electric field near the surface of the emitter.

field-free emission current Electron current emitted by a cathode when the electric field at the surface of the cathode is zero. Also known as zero-field emission.

field frequency The number of fields transmitted per second in television; equal to the frame frequency multiplied by the number of fields that make up one frame. Also known as field repetition rate.

field ionization The ionization of gaseous atoms and molecules by an intense electric field, often at the surface of a solid.

field-ion microscope A microscope in which atoms are ionized by an electric field near a sharp tip; the field then forces the ions to a fluorescent screen, which shows an enlarged image of the tip, and individual atoms are made visible; this is the most powerful microscope yet produced. Also known as ion microscope.

field magnet The magnet which creates a magnetic field in an electric machine or device.

field of search The space that a radar set or installation can cover effectively.

field pattern *See* radiation pattern.

field pole A structure of magnetic material on which a field coil of a loudspeaker, motor, generator, or other electromagnetic device may be mounted.

field-programmable logic array A programmed logic array in which the internal connections of the logic gates can be programmed once in the field by passing high current through fusible links, by using avalanche-induced migration to short base-emitter junctions at desired interconnections, or by other means. Abbreviated FPLA. Also known as programmable logic array.

field quenching Decrease in the emission of light of a phosphor excited by ultraviolet radiation, x-rays, alpha particles, or cathode rays when an electric field is simultaneously applied.

field repetition rate *See* field frequency.

field rheostat A rheostat used to adjust the current in the field winding of an electric machine.

field scan Television term denoting the vertical excursion of an electron beam downward across a cathode-ray tube face, the excursion being made in order to scan alternate lines.

field waveguide A single wire, threaded or coated with dielectric, which guides an electromagnetic field. Also known as G string.

field winding *See* field coil.

field wire An insulated flexible wire or cable used in field telephone and telegraph systems.

figure of merit A performance rating that governs the choice of a device for a particular application; for example, the figure of merit of a magnetic amplifier is the ratio of usable power gain to the control time constant.

filter discrimination 161

filament 1. Metallic wire or ribbon which is heated in an incandescent lamp to produce light, by passing an electric current through the filament. 2. A cathode made of resistance wire or ribbon, through which an electric current is sent to produce the high temperature required for emission of electrons in a thermionic tube. Also known as directly heated cathode; filamentary cathode; filament-type cathode.

filamentary cathode See filament.

filament current The current supplied to the filament of an electron tube for heating purposes.

filament emission Liberation of electrons from a heated filament wire in an electron tube.

filament lamp See incandescent lamp.

filament saturation See temperature saturation.

filament transformer A small transformer used exclusively to supply filament or heater current for one or more electron tubes.

filament-type cathode See filament.

filament winding The secondary winding of a power transformer that furnishes alternating-current heater or filament voltage for one or more electron tubes.

filled band An energy band, each of whose energy levels is occupied by an electron.

film The layer adjacent to the valve metal in an electrochemical valve, in which is located the high voltage drop when current flows in the direction of high impedance.

film integrated circuit An integrated circuit whose elements are films formed in place on an insulating substrate.

film reader A device for converting a pattern of transparent or opaque spots on a photographic film into a series of electric pulses.

film recorder A device which places data, usually in the form of transparent and opaque spots or light and dark spots, on photographic film.

film resistor A fixed resistor in which the resistance element is a thin layer of conductive material on an insulated form; the conductive material does not contain binders or insulating material.

film scanning The process of converting motion picture film into corresponding electric signals that can be transmitted by a television system.

filter Any transmission network used in electrical systems for the selective enhancement of a given class of input signals. Also known as electric filter; electric-wave filter.

filter capacitor A capacitor used in a power-supply filter system to provide a low-reactance path for alternating currents and thereby suppress ripple currents, without affecting direct currents.

filter choke An iron-core coil used in a power-supply filter system to pass direct current while offering high impedance to pulsating or alternating current.

filter crystal Quartz crystal which is used in an electrical circuit designed to pass energy of certain frequencies.

filter design The design of electrical networks in which the principle of electrical resonance is used to make the network accept wanted frequencies while rejecting unwanted ones.

filter discrimination Difference between the minimum insertion loss at any frequency in a filter attenuation band and the maximum insertion loss at any frequency in the operating range of a filter transmission band.

filtered radar data Radar data from which unwanted returns have been removed by mapping.

filter impedance compensator Impedance compensator which is connected across the common terminals of electric wave filters when the latter are used in parallel to compensate for the effects of the filters on each other.

filter pass band See filter transmission band.

filter reactor A reactor used for reducing the harmonic components of voltage in an alternating-current or direct-current circuit.

filter section A simple RC, RL, or LC network used as a broad-band filter in a power supply, grid-bias feed, or similar device.

filter slot Choke in the form of a slot designed to suppress unwanted modes in a waveguide.

filter transmission band Frequency band of free transmission; that is, frequency band in which, if dissipation is neglected, the attenuation constant is zero. Also known as filter pass band.

final amplifier The transmitter stage that feeds the antenna.

F indicator See F scope.

finding circuit See lockout circuit.

finite clipping Clipping in which the threshold level is large but is below the peak input signal amplitude.

Finsen lamp A high-temperature carbon arc or mercury arc lamp that produces a mixture of blue, violet, and near-ultraviolet light; used to treat certain skin disorders and to test paints and other protective coatings.

finsen unit A unit of intensity of ultraviolet radiation, equal to the intensity of ultraviolet radiation at a specified wavelength whose energy flux is 100,000 watts per square meter; the wavelength usually specified is 296.7 nanometers. Abbreviated FU.

fin waveguide Waveguide containing a thin longitudinal metal fin that serves to increase the wavelength range over which the waveguide will transmit signals efficiently; usually used with circular waveguides.

fire-control circuit An electric circuit in a fire-control system.

fired state The "on" state of a silicon controlled rectifier or other semiconductor switching device, occurring when a suitable triggering pulse is applied to the gate.

firing 1. The gas ionization that initiates current flow in a gas-discharge tube. 2. Excitation of a magnetron or transmit-receive tube by a pulse. 3. The transition from the unsaturated to the saturated state of a saturable reactor.

firing box A boxlike item in which are mounted switches, cables, fuses, plugs, indicator lights, batteries, and the like, specifically designed for firing a rocket or guided missile from a remote position.

firing button A button or switch for firing guns or rockets.

firing circuit 1. Circuit used with an ignitron to deliver a pulse of current of 5–50 amperes in the forward direction, from the igniter to the mercury, to start a cathode spot and to control the time of firing. 2. By analogy, a similar control circuit of silicon-controlled rectifiers and like devices.

firing point See critical grid voltage.

firing potential Controlled potential at which conduction through a gas-filled tube begins.

first detector See mixer.

first Fresnel zone Circular portion of a wavefront transverse to the line between an emitter and a more distant point, where the resultant disturbance is being observed, whose center is the intersection of the front with the direct ray, and whose radius is such that the shortest path from the emitter through the periphery to the receiving point is one-half wavelength longer than the direct ray.

first selector Selector which immediately follows a line finder in a switch train and which responds to dial pulses of the first digit of the called telephone number.

Fischer-Hinnen method Method of analysis of a complex waveform which has like loops above and below the time axis, in which the amplitude and phase of the n-th harmonic is determined from the ordinates of the resultant wave at a series of times which divide the half wave into $2n$ equal time intervals.

fish-bone antenna 1. Antenna consisting of a series of coplanar elements arranged in collinear pairs, loosely coupled to a balanced transmission line. 2. Directional antenna in the form of a plane array of doublets arranged transversely along both sides of a transmission line.

fishpole antenna *See* whip antenna.

five-wire line A transmission line which has four conductors, all in phase, at the corners of a square and a fifth conductor at the center of the square which is out of phase with the others.

fixed attenuator *See* pad.

fixed bias A constant value of bias voltage, independent of signal strength.

fixed-bias transistor circuit A transistor circuit in which a current flowing through a resistor is independent of the quiescent collector current.

fixed capacitor A capacitor having a definite capacitance value that cannot be adjusted.

fixed contact A relatively immovable contact that is engaged and disengaged by a moving contact to make and break a circuit, as in a switch or relay.

fixed echo An echo indication that remains stationary on a radar plan-position indicator display, indicating the presence of a fixed target.

fixed inductor An inductor whose coils are wound in such a manner that the turns remain fixed in position with respect to each other, and which either has no magnetic core or has a core whose air gap and position within the coil are fixed.

fixed resistor A resistor that has no provision for varying its resistance value.

fixed transmitter Transmitter that is operated in a fixed or permanent location.

flag A small metal tab that holds the getter during assembly of an electron tube.

flame arc lamp An arc lamp in which carbon electrodes are impregnated with chemicals, such as calcium, barium, or titanium, which are more volatile than the carbon and radiate light when driven into the arc.

flap attenuator A waveguide attenuator in which a contoured sheet of dissipative material is moved into the guide through a nonradiating slot to provide a desired amount of power absorption. Also known as vane attenuator.

flare 1. A radar screen target indication having an enlarged and distorted shape due to excessive brightness. 2. *See* horn antenna.

flash arc A sudden increase in the emission of large thermionic vacuum tubes, probably due to irregularities in the cathode surface.

flashback voltage Inverse peak voltage at which ionization takes place in a gas tube.

flash barrier A fireproof structure between conductors of an electric machine, designed to minimize flashover or the damage caused by flashover.

flasher A switch, generally either motor-driven or using a combination heater element and bimetallic strip, that turns lamps on and off rapidly.

flashing over Accidental formation of an arc over the surface of a rotating commutator from brush-to-brush; usually caused by faulty insulation between commutator segments.

flash lamp A gaseous-discharge lamp used in a photoflash unit to produce flashes of light of short duration and high intensity for stroboscopic photography. Also known as stroboscopic lamp.

flash magnetization Magnetization of a ferromagnetic object by a current impulse of short duration.

flashover An electric discharge around or over the surface of an insulator.

flashover voltage The voltage at which an electric discharge occurs between two electrodes that are separated by an insulator; the value depends on whether the insulator surface is dry or wet. Also known as sparkover voltage.

flash test A method of testing insulation by applying momentarily a voltage much higher than the rated working voltage.

flat cable A cable made of round or rectangular, parallel copper wires arranged in a plane and laminated or molded into a ribbon of flexible insulating plastic.

flat-conductor cable A cable made of wide, flat conductors arranged side by side in a plane and protected by ribbons of insulating plastic.

flat line A radio-frequency transmission line, or part thereof, having essentially 1-to-1 standing wave ratio.

flatpack Semiconductor network encapsulated in a thin, rectangular package, with the necessary connecting leads projecting from the edges of the unit.

flat-top antenna An antenna having two or more lengths of wire parallel to each other and in a plane parallel to the ground, each fed at or near its midpoint.

flat top response *See* band-pass response.

flat tuning Tuning of a radio receiver in which a change in frequency of the received waves produces only a small change in the current in the tuning apparatus.

Fleming's rule *See* left-hand rule; right-hand rule.

Fleming tube The original diode, consisting of a heated filament and a cold metallic electrode in an evacuated glass envelope; negative current flows from the filament to the cold electrode, but not in the reverse direction.

flexible circuit A printed circuit made on a flexible plastic sheet that is usually die-cut to fit between large components.

flexible coupling A coupling designed to allow a limited angular movement between the axes of two waveguides.

flexible resistor A wire-wound resistor having the appearance of a flexible lead; made by winding the Nichrome resistance wire around a length of asbestos or other heat-resistant cord, then covering the winding with asbestos and braided insulating covering.

flexible waveguide A waveguide that can be bent or twisted without appreciably changing its electrical properties.

flicker effect Random variations in the output current of an electron tube having an oxide-coated cathode, due to random changes in cathode emission.

flip chip A tiny semiconductor die having terminations all on one side in the form of solder pads or bump contacts; after the surface of the chip has been passivated or otherwise treated, it is flipped over for attaching to a matching substrate.

flip coil A small coil used to measure the strength of a magnetic field; it is placed in the field, connected to a ballistic galvanometer or other instrument, and suddenly flipped over 180°; alternatively, the coil may be held stationary and the magnetic field reversed.

flip-flop circuit *See* bistable multivibrator.

flip-open cutout fuse *See* dropout fuse.

flip-over process *See* Umklapp process.

floating The condition wherein a device or circuit is not grounded and not tied to an established voltage supply.

floating battery A storage battery connected permanently in parallel with another power source; the battery normally handles only small charging or discharging currents, but takes over the entire load upon failure of the main supply.

floating charge Application of a constant voltage to a storage battery, sufficient to maintain an approximately constant state of charge while the battery is idle or on light duty.

floating grid Vacuum-tube grid that is not connected to any circuit; it assumes a negative potential with respect to the cathode. Also known as free grid.

floating input Isolated input circuit not connected to ground at any point.

floating neutral Neutral conductor whose voltage to ground is free to vary when circuit conditions change.

flood To direct a large-area flow of electrons toward a storage assembly in a charge storage tube.

floodlight A light projector used for outdoor lighting of buildings, parking lots, sports fields, and the like, usually having a filament lamp or mercury-vapor lamp and a parabolic reflector.

floor outlet An electrical outlet whose face is level with or recessed into a floor. Also known as floor plug.

floor plug *See* floor outlet.

flopover A defect in television reception in which a series of frames move vertically up or down the screen, caused by lack of synchronization between the vertical and horizontal sweep frequencies.

fluctuating current Direct current that changes in value but not at a steady rate.

fluorescent lamp A tubular discharge lamp in which ionization of mercury vapor produces radiation that activates the fluorescent coating on the inner surface of the glass.

fluoroscopic image intensifier An electron-beam tube that converts a relatively feeble fluoroscopic image on the fluorescent input phosphor into a much brighter image on the output phosphor.

flute storage Ferrite storage consisting of a number of parallel lengths of fine prism-shaped tubing, each surrounding an insulated axial conductor that acts as a word line; the lengths of tubing are intersected at right angles by parallel sets of insulated wire bit lines that are displaced slightly from the word lines; each intersection stores one bit.

flutter A fast-changing variation in received signal strength, such as may be caused by antenna movements in a high wind or interaction with a signal or another frequency.

flutter echo A radar echo consisting of a rapid succession of reflected pulses resulting from a single transmitted pulse.

flux The electric or magnetic lines of force in a region.

flux-gate magnetometer A magnetometer in which the degree of saturation of the core by an external magnetic field is used as a measure of the strength of the earth's magnetic field; the essential element is the flux gate.

flux jumping *See* Meissner effect.

flux leakage Magnetic flux that does not pass through an air gap or other part of a magnetic circuit where it is required.

flux linkage The product of the number of turns in a coil and the magnetic flux passing through the coil. Also known as linkage.

flux path A path which is followed by magnetic lines of force and in which the magnetic flux density is significant.

flux refraction The abrupt change in direction of magnetic flux lines at the boundary between two media having different permeabilities, or of the electric flux lines at the boundary between two media having different dielectric constants, when these lines are oblique to the boundary.

flyback The time interval in which the electron beam of a cathode-ray tube returns to its starting point after scanning one line or one field of a television picture or after completing one trace in an oscilloscope. Also known as retrace; return trace.

flyback power supply A high-voltage power supply used to produce the direct-current voltage of about 10,000–25,000 volts required for the second anode of a cathode-ray tube in a television receiver or oscilloscope.

flyback transformer *See* horizontal output transformer.

flying-aperture scanner An optical scanner, used in character recognition, in which a document is flooded with light, and light is collected sequentially spot by spot from the illuminated image.

flying head A read/write head used on magnetic disks and drums, so designed that it flies a microscopic distance off the moving magnetic surface and is supported by a film of air.

flying spot A small point of light, controlled mechanically or electrically, which moves rapidly in a rectangular scanning pattern in a flying-spot scanner.

flying-spot scanner A scanner used for television film and slide transmission, electronic writing, and character recognition, in which a moving spot of light, controlled mechanically or electrically, scans the image field, and the light reflected from or transmitted by the image field is picked up by a phototube to generate electric signals. Also known as optical scanner.

flywheel synchronization Automatic frequency control of a scanning system by using the average timing of the incoming sync signals, rather than by making each pulse trigger the scanning circuit; used in high-sensitivity television receivers designed for fringe-area reception, when noise pulses might otherwise trigger the sweep circuit prematurely.

FM/AM multiplier Multiplier in which the frequency deviation from the central frequency of a carrier is proportional to one variable, and its amplitude is proportional to the other variable; the frequency-amplitude-modulated carrier is then consecutively demodulated for frequency modulation (FM) and for amplitude modulation (AM); the final output is proportional to the product of the two variables.

focus To control convergence or divergence of the electron paths within one or more beams, usually by adjusting a voltage or current in a circuit that controls the electric or magnetic fields through which the beams pass, in order to obtain a desired image or a desired current density within the beam.

focus control A control that adjusts spot size at the screen of a cathode-ray tube to give the sharpest possible image; it may vary the current through a focusing coil or change the position of a permanent magnet.

focusing anode An anode used in a cathode-ray tube to change the size of the electron beam at the screen; varying the voltage on this anode alters the paths of electrons in the beam and thus changes the position at which they cross or focus.

focusing coil A coil that produces a magnetic field parallel to an electron beam for the purpose of focusing the beam.

focusing electrode An electrode to which a potential is applied to control the cross-sectional area of the electron beam in a cathode-ray tube.

focusing magnet A permanent magnet used to produce a magnetic field for focusing an electron beam.

focus lamp 1. A lamp whose filament has a spiral or zigzag form in order to reduce its size, so that it can be brought into the focus of a lens or mirror. 2. An arc lamp whose feeding mechanism is designed to hold the arc in a constant position with respect to an optical system that is used to focus its rays.

focus projection and scanning Method of magnetic focusing and electrostatic deflection of the electron beam of a hybrid vidicon; a transverse electrostatic field is used for beam deflection; this field is immersed with an axial magnetic field that focuses the electron beam.

foil electret A thin film of strongly insulating material capable of trapping charge carriers, such as polyfluoroethylenepropylene, that is electrically charged to produce an external electric field; in the conventional design, charge carriers of one sign are injected into one surface, and a compensation charge of opposite sign forms on the opposite surface or an adjacent electrode.

folded cavity Arrangement used in a klystron repeater to make the incoming wave act on the electron stream from the cathode at several places and produce a cumulative effect.

folded dipole See folded-dipole antenna.

folded-dipole antenna A dipole antenna whose outer ends are folded back and joined together at the center; the impedance is about 300 ohms, as compared to 70 ohms for a single-wire dipole; widely used with television and frequency-modulation receivers. Also known as folded dipole.

foldover Picture distortion seen as a white line on the side, top, or bottom of a television picture; generally caused by nonlinear operation in either the horizontal or vertical deflection circuits of a receiver.

follow current The current at power frequency that passes through a surge diverter or other discharge path after a high-voltage surge has started the discharge.

follow spot A high-intensity spotlight used to follow action in arenas and stadiums and on large stages; it is equipped with adjustable iris and shutter controls, and its light source is either a carbon arc or an incandescent bulb.

forbidden band A range of unallowed energy levels for an electron in a solid.

fork oscillator An oscillator that uses a tuning fork as the frequency-determining element.

form factor 1. The ratio of the effective value of a periodic function, such as an alternating current, to its average absolute value. 2. A factor that takes the shape of a coil into account when computing its inductance. Also known as shape factor.

forming Application of voltage to an electrolytic capacitor, electrolytic rectifier, or semiconductor device to produce a desired permanent change in electrical characteristics as a part of the manufacturing process.

168 form-wound coil

form-wound coil Armature coil that is formed or shaped over a fixture before being placed on the armature of a motor or generator.

forty-four-type repeater Type of telephone repeater employing two amplifiers and no hybrid arrangements; used in a four-wire system.

forward-acting regulator Transmission regulator in which the adjustment made by the regulator does not affect the quantity which caused the adjustment.

forward bias A bias voltage that is applied to a pn-junction in the direction that causes a large current flow; used in some semiconductor diode circuits.

forward coupler Directional coupler used to sample incident power.

forward current Current which flows upon application of forward voltage.

forward direction Of a semiconductor diode, the direction of lower resistance to the flow of steady direct current.

forward drop The voltage drop in the forward direction across a rectifier.

forward recovery time Of a semiconductor diode, the time required for the forward current or voltage to reach a specified value after instantaneous application of a forward bias in a given circuit.

forward-scatter propagation *See* scatter propagation.

forward voltage drop *See* diode forward voltage.

forward wave Wave whose group velocity is the same direction as the electron stream motion.

Foster-Seely discriminator *See* phase-shift discriminator.

Foucault current *See* eddy current.

four-layer device A $pnpn$ semiconductor device, such as a silicon controlled rectifier, that has four layers of alternating p- and n-type material to give three pn junctions.

four-layer diode A semiconductor diode having three junctions, terminal connections being made to the two outer layers that form the junctions; a Shockley diode is an example.

four-layer transistor A junction transistor having four conductivity regions but only three terminals; a thyristor is an example.

four-pole double-throw A 12-terminal switch or relay contact arrangement that simultaneously connects two pairs of terminals to either of two other pairs of terminals. Abbreviated 4PDT.

four-vector potential A four-vector whose space components are the magnetic vector potential and whose time component is the electric scalar potential.

four-way switch An electric switch employed in house wiring, that makes it possible to turn a light on or off at three or more places.

four-wire line A transmission line in which four conductors lie at the corners of a rectangle, and each conductor is in phase with the conductor at the opposite corner and out of phase with the conductors at adjacent corners.

four-wire repeater Telephone repeater for use in a four-wire circuit and in which there are two amplifiers, one serving to amplify the telephone currents in one side of the four-wire circuit, and the other serving to amplify the telephone currents in the other side of the four-wire circuit.

four-wire terminating set Hybrid arrangement by which four-wire circuits are terminated on a two-wire basis for interconnection with two-wire circuits.

Fowler-DuBridge theory Theory of photoelectric emission from a metal based on the Sommerfeld model, which takes into account the thermal agitation of electrons in the metal and predicts the photoelectric yield and the energy spectrum of photoelectrons as functions of temperature and the frequency of incident radiation.

Fowler function A mathematical function used in the Fowler-DuBridge theory to calculate the photoelectric yield.

FPLA *See* field-programmable logic array.

Fr *See* statcoulomb.

fractional horsepower motor Any motor built into a frame smaller than that for a motor having an open construction and a continuous rating of 1 horsepower (745.7 watts) at 1800 revolutions per minute.

frame 1. One complete coverage of a television picture. 2. A rectangular area representing the size of copy handled by a facsimile system.

frame frequency The number of times per second that the frame is completely scanned in television. Also known as picture frequency.

frame period A time interval equal to the reciprocal of the frame frequency.

framer Device for adjusting facsimile equipment so the start and end of a recorded line are the same as on the corresponding line of the subject copy.

framing 1. Adjusting a television picture to a desired position on the screen of the picture tube. 2. Adjusting a facsimile picture to a desired position in the direction of line progression. Also known as phasing.

framing control 1. A control that adjusts the centering, width, or height of the image on a television receiver screen. 2. A control that shifts a received facsimile picture horizontally.

Franck-Hertz experiment Experiment for measuring the kinetic energy lost by electrons in inelastic collisions with atoms; it established the existence of discrete energy levels in atoms, and can be used to determine excitation and ionization potentials.

Franklin centimeter A unit of electric dipole moment, equal to the dipole moment of a charge distribution consisting of positive and negative charges of 1 statcoulomb separated by a distance of 1 centimeter.

Fraunhofer region The region far from an antenna compared to the dimensions of the antenna and the wavelength of the radiation. Also known as far field; far region; far zone; radiation zone.

free admittance The reciprocal of the blocked impedance of a transducer.

free charge Electric charge which is not bound to a definite site in a solid, in contrast to the polarization charge.

free electromagnetic field An electromagnetic field in empty space that does not interact with matter.

free-electron theory of metals A model of a metal in which the free electrons, that is, those giving rise to the conductivity, are regarded as moving in a potential (due to the metal ions in the lattice and to all the remaining free electrons) which is approximated as constant everywhere inside the metal. Also known as Sommerfeld model; Sommerfeld theory.

free grid *See* floating grid.

free hole Any hole which is not bound to an impurity or to an exciton.

free impedance Impedance at the input of the transducer when the impedance of its load is made zero. Also known as normal impedance.

free motional impedance Of a transducer, the complex remainder after the blocked impedance has been subtracted from the free impedance.

free-running frequency Frequency at which a normally driven oscillator operates in the absence of a driving signal.

free-running multivibrator See astable multivibrator.

free-running sweep Sweep triggered continuously by an internal trigger generator.

free-space field intensity Radio field intensity that would exist at a point in a uniform medium in the absence of waves reflected from the earth or other objects.

free space loss The theoretical radiation loss, depending only on frequency and distance, that would occur if all variable factors were disregarded when transmitting energy between two antennas.

free-space propagation Propagation of electromagnetic radiation over a straight-line path in a vacuum or ideal atmosphere, sufficiently removed from all objects that affect the wave in any way.

free-space radar equation Equation that governs a radar signal characteristic when it is propagated between a radar set and a reflecting object or target in otherwise empty space.

free-space radiation pattern Radiation pattern that an antenna would have if it were in free space where there is nothing to reflect, refract, or absorb the radiated waves.

free space wave An electromagnetic wave propagating in a vacuum, free from boundary effects.

Frenkel defect A crystal defect consisting of a vacancy and an interstitial which arise when an atom is plucked out of a normal lattice site and forced into an interstitial position. Also known as Frenkel pair.

Frenkel exciton A tightly bound exciton in which the electron and the hole are usually on the same atom, although the pair can travel anywhere in the crystal.

Frenkel pair See Frenkel defect.

frequency analyzer A device which measures the intensity of many different frequency components in some oscillation, as in a radio band; used to identify transmitting sources.

frequency-azimuth intensity Type of radar display in which frequency, azimuth, and strobe intensity are correlated.

frequency bridge A bridge in which the balance varies with frequency in a known manner, such as the Wien bridge; used to measure frequency.

frequency changer See frequency converter.

frequency-changer station An installation at which power is transmitted between two alternating-current electric power systems operating at different frequencies by a direct-current link.

frequency compensation See compensation.

frequency conversion Converting the carrier frequency of a received signal from its original value to the intermediate frequency value in a superheterodyne receiver.

frequency converter A circuit, device, or machine that changes an alternating current from one frequency to another, with or without a change in voltage or number of phases. Also known as frequency changer; frequency translator.

frequency counter An electronic counter used to measure frequency by counting the number of cycles in an electric signal during a preselected time interval.

frequency cutoff The frequency at which the current gain of a transistor drops 3 decibels below the low-frequency gain value.

frequency relay   171

frequency discriminator A discriminator circuit that delivers an output voltage which is proportional to the deviations of a signal from a predetermined frequency value.

frequency distortion Distortion in which the relative magnitudes of the different frequency components of a wave are changed during transmission or amplification. Also known as amplitude distortion; amplitude-frequency distortion; waveform-amplitude distortion.

frequency divider A harmonic conversion transducer in which the frequency of the output signal is an integral submultiple of the input frequency. Also known as counting-down circuit.

frequency-domain reflectometer A tuned reflectometer used for measuring reflection coefficients and impedance of waveguides over a wide frequency range, by sweeping a band of frequencies and analyzing the reflected returns.

frequency doubler An amplifier stage whose resonant anode circuit is tuned to the second harmonic of the input frequency; the output frequency is then twice the input frequency. Also known as doubler.

frequency drift A gradual change in the frequency of an oscillator or transmitter due to temperature or other changes in the circuit components that determine frequency.

frequency-modulated jamming Jamming technique consisting of a constant amplitude radio-frequency signal that is varied in frequency about a center frequency to produce a signal over a band of frequencies.

frequency-modulation detector A device, such as a Foster-Seely discriminator, for the detection or demodulation of a frequency-modulated wave.

frequency-modulation receiver A radio receiver that receives frequency-modulated waves and delivers corresponding sound waves.

frequency-modulation receiver deviation sensitivity Least frequency deviation that produces a specified output power.

frequency-modulation transmitter A radio transmitter that transmits a frequency-modulated wave.

frequency-modulation tuner A tuner containing a radio-frequency amplifier, converter, intermediate-frequency amplifier, and demodulator for frequency-modulated signals, used to feed a low-level audio-frequency signal to a separate af amplifier and loudspeaker.

frequency modulator A circuit or device for producing frequency modulation.

frequency monitor An instrument for indicating the amount of deviation of the carrier frequency of a transmitter from its assigned value.

frequency multiplier A harmonic conversion transducer in which the frequency of the output signal is an exact integral multiple of the input frequency. Also known as multiplier.

frequency-offset transponder Transponder that changes the signal frequency by a fixed amount before retransmission.

frequency pulling A change in the frequency of an oscillator due to a change in load impedance.

frequency recorder An instrument which uses a frequency bridge to sense the frequency of an alternating current, and which makes a graphical record of this frequency as a function of time.

frequency regulator A device that maintains the frequency of an alternating-current generator at a predetermined value.

frequency relay Relay which functions at a predetermined value of frequency; may be an over-frequency relay, an under-frequency relay, or a combination of both.

172 frequency-response equalization

frequency-response equalization *See* equalization.

frequency run A series of tests made to determine the amplitude-frequency response characteristic of a transmission line, circuit, or device.

frequency scan antenna A radar antenna similar to a phased array antenna in which one dimensional scanning is accomplished through frequency variation.

frequency scanning Type of system in which output frequency is made to vary at a mechanical rate over a desired frequency band.

frequency-selective device *See* electric filter.

frequency separation multiplier Multiplier in which each of the variables is split into a low-frequency part and a high-frequency part that are multiplied separately, and the results added to give the required product; this system makes it possible to get high accuracy and broad bandwidth.

frequency separator The circuit that separates the horizontal and vertical synchronizing pulses in a monochrome or color television receiver.

frequency shift A change in the frequency of a radio transmitter or oscillator. Also known as radio-frequency shift.

frequency-shift converter A device that converts a received frequency-shift signal to an amplitude-modulated signal or a direct-current signal.

frequency-shift keyer A lever to effect a frequency shift, that is, a change in the frequency of a radio transmitter, oscillator, or receiver.

frequency splitting One condition of operation of a magnetron which causes rapid alternating from one mode of operation to another; this results in a similar rapid change in oscillatory frequency and consequent loss in power at the desired frequency.

frequency stability The ability of an oscillator to maintain a desired frequency; usually expressed as percent deviation from the assigned frequency value.

frequency standard A stable oscillator, usually controlled by a crystal or tuning fork, that is used primarily for frequency calibration.

frequency synthesizer A device that provides a choice of a large number of different frequencies by combining frequencies selected from groups of independent crystals, frequency dividers, and frequency multipliers.

frequency-time-intensity Type of radar display in which the frequency, time, and strobe intensity are correlated.

frequency tolerance Of a radio transmitter, extent to which the carrier frequency of the transmitter may be permitted to depart from the frequency assigned.

frequency-to-voltage converter A converter that provides an analog output voltage which is proportional to the frequency or repetition rate of the input signal derived from a flowmeter, tachometer, or other alternating-current generating device. Abbreviated F/V converter.

frequency translator *See* frequency converter.

frequency-type telemeter Telemeter that employs frequency of an alternating current or voltage as the translating means.

frequency variation The change over time of the deviation from assigned frequency of a radio-frequency carrier (or power supply system); usually tightly controlled because of national or industry standards.

Fresnel region The region between the near field of an antenna (close to the antenna compared to a wavelength) and the Fraunhofer region.

Fresnel spotlight A lighting instrument that is composed of a lamp and a Fresnel (stepped planoconvex) lens; the unit can be made with or without reflectors and has a system to adjust the spacing between the lamp and the lens so as to control the light beam; models range from 100 to 5000 watts.

Fresnel zones Circular portions of a wavefront transverse to a line between an emitter and a point where the disturbance is being observed; the nth zone includes all paths whose lengths are between $n-1$ and n half-wavelengths longer than the line-of-sight path. Also known as half-period zones.

frictional electricity The electric charges produced on two different objects, such as silk and glass or catskin and ebonite, by rubbing them together. Also known as triboelectricity.

fringe magnetic field The part of the magnetic field of a horseshoe magnet that extends outside the space between its poles.

frogging repeater Carrier repeater having provisions for frequency frogging to permit use of a single multipair voice cable without having excessive crosstalk.

front-to-back ratio 1. Ratio of the effectiveness of a directional antenna, loudspeaker, or microphone toward the front and toward the rear. 2. Ratio of resistance of a crystal to current flowing in the normal direction to current flowing in the opposite direction.

frying noise Noise in telephone transmission even when no conversation is taking place; caused by signal current flowing across a resistance element having multiple intermittent paths. Also known as transmitter noise.

F scan *See* F scope.

F scope A cathode-ray scope on which a single signal appears as a spot with bearing error as the horizontal coordinate and elevation angle error as the vertical coordinate, with cross hairs on the scope face to assist in bringing the system to bear on the target. Also known as F indicator; F scan.

FU *See* finsen unit.

fuel cell A cell that converts chemical energy directly into electric energy, with electric power being produced as a part of a chemical reaction between the electrolyte and a fuel such as kerosine or industrial fuel gas.

full adder A logic element which operates on two binary digits and a carry digit from a preceding stage, producing as output a sum digit and a new carry digit.

full load The greatest load that a circuit or piece of equipment is designed to carry under specified conditions.

full-load current The greatest current that a circuit or piece of equipment is designed to carry under specified conditions.

full-pitch winding An armature winding in which the distance between two active conductors of a coil equals the pole pitch.

full section filter A filter network whose graphical representation has the shape of the Greek letter pi, connoting capacitance in the upright legs and inductance or reactance in the horizontal member.

full subtracter A logic element which operates on three binary input signals representing a minuend, subtrahend, and borrow digit, producing as output a different digit and a new borrow digit. Also known as three-input subtracter.

full-wave amplifier An amplifier without any clipping.

full-wave bridge A circuit having a bridge with four diodes, which provides full-wave rectification and gives twice as much direct-current output voltage for a given alternating-current input voltage as a conventional full-wave rectifier.

174   full-wave control

full-wave control Phase control that acts on both halves of each alternating-current cycle, for varying load power over the full range from 0 to the full-wave maximum value.

full-wave rectification Rectification in which output current flows in the same direction during both half cycles of the alternating input voltage.

full-wave rectifier A double-element rectifier that provides full-wave rectification; one element functions during positive half cycles and the other during negative half cycles.

full-wave vibrator A vibrator having an armature that moves back and forth between two fixed contacts so as to change the direction of direct-current flow through a transformer at regular intervals and thereby permit voltage stepup by the transformer; used in battery-operated power supplies for mobile and marine radio equipment.

functional generator *See* function generator.

functional multiplier *See* function multiplier.

functional switching circuit One of a relatively small number of types of circuits which implements a Boolean function and constitutes a basic building block of a switching system; examples are the AND, OR, NOT, NAND, and NOR circuits.

function generator Also known as functional generator. 1. An analog computer device that indicates the value of a given function as the independent variable is increased. 2. A signal generator that delivers a choice of a number of different waveforms, with provisions for varying the frequency over a wide range.

function multiplier An analog computer device that takes in the changing values of two functions and puts out the changing value of their product as the independent variable is changed. Also known as functional multiplier.

function switch A network having a number of inputs and outputs so connected that input signals expressed in a certain code will produce output signals that are a function of the input information but in a different code.

fundamental mode The waveguide mode having the lowest critical frequency. Also known as dominant mode; principal mode.

fuse An expandable device for opening an electric circuit when the current therein becomes excessive, containing a section of conductor which melts when the current through it exceeds a rated value for a definite period of time. Also known as electric fuse.

fuse alarm Circuit that produces a visual or audible signal to indicate a blown fuse.

fuse block An insulating base on which are mounted fuse clips or other contacts for fuses. Also known as fuseboard.

fuseboard *See* fuse block.

fuse clip A spring contact used to hold and make connection to a cartridge-type fuse.

fuse cutout Assembly of a fuse support and a fuse holder which may or may not include the fuse link.

fuse diode A diode that opens under specified current surge conditions.

fuse disconnecting switch Disconnecting switch in which a fuse unit forms a part of the blade.

fused junction *See* alloy junction.

fused-junction diode *See* alloy junction diode.

fused-junction transistor *See* alloy-junction transistor.

F/V converter   175

fused semiconductor Junction formed by recrystallization on a base crystal from a liquid phase of one or more components and the semiconductor.

fuse link Part of a fuse that carries the current of the circuit and all or part of which melts when the current exceeds a predetermined value.

fuse wire Wire made from an alloy that melts at a relatively low temperature and overheats to this temperature when carrying a particular value of overload current.

fusible resistor A resistor designed to protect a circuit against overload; its resistance limits current flow and thereby protects against surges when power is first applied to a circuit; its fuse characteristic opens the circuit when current drain exceeds design limits.

fV *See* femtovolt.

F/V converter *See* frequency-to-voltage converter.

G

G *See* conductance.

G$_m$ *See* transconductance.

g$_m$ *See* transconductance.

gain 1. The increase in signal power that is produced by an amplifier; usually given as the ratio of output to input voltage, current, or power, expressed in decibels. Also known as transmission gain. 2. *See* antenna gain.

gain-bandwidth product The midband gain of an amplifier stage multiplied by the bandwidth in megacycles.

gain control A device for adjusting the gain of a system or component.

gain reduction Diminution of the output of an amplifier, usually achieved by reducing the drive from feed lines by use of equalizer pads or reducing amplification by a volume control.

gain sensitivity control *See* differential gain control.

gain turndown A receiver gain control incorporated in a transponder to protect the transmitter from overload.

galactic radio waves Radio waves emanating from the Milky Way Galaxy.

gallium arsenide semiconductor A semiconductor having a forbidden-band gap of 1.4 electron volts and a maximum operating temperature of 400°C when used in a transistor.

gallium phosphide semiconductor A semiconductor having a forbidden-band gap of 2.4 electronvolts and a maximum operating temperature of 870°C when used in a transistor.

galvanic Pertaining to electricity flowing as a result of chemical action.

galvanic battery A galvanic cell, or two or more such cells electrically connected to produce energy.

galvanic cell An electrolytic cell that is capable of producing electric energy by electrochemical action.

galvanic couple A pair of unlike substances, such as metals, which generate a voltage when brought in contact with an electrolyte.

galvanic current A steady direct current.

galvanomagnetic effect One of the electrical or thermal phenomena occurring when a current-carrying conductor or semiconductor is placed in a magnetic field; examples are the Hall effect, Ettingshausen effect, transverse magnetoresistance, and Nernst effect. Also known as magnetogalvanic effect.

galvanometer constant Number by which a certain function of the reading of a galvanometer must be multiplied to obtain the current value in ordinary units.

galvanometer shunt Resistor connected in parallel with a galvanometer to increase its range under certain conditions; it allows only a known fraction of the current to pass through the galvanometer.

galvanostat A device to deliver constant current from a high-voltage battery.

gamma A unit of magnetic field strength, equal to 10 microersteds, or 0.00001 oersted.

gamma structure A Hume-Rothery designation for structurally analogous phases or intermetallic phases having 21 valence electrons to 13 atoms, analogous to the γ-brass structure.

gang capacitor A combination of two or more variable capacitors mounted on a common shaft to permit adjustment by a single control.

ganged control Controls of two or more circuits mounted on a common shaft to permit simultaneous control of the circuits by turning a single knob.

gang switch A combination of two or more switches mounted on a common shaft to permit operation by a single control. Also known as deck switch.

gap 1. The spacing between two electric contacts. 2. A break in a closed magnetic circuit, containing only air or filled with a nonmagnetic material.

gap factor Ratio of the maximum energy gained in volts to the maximum gap voltage in a tube employing electron accelerating gaps, that is, a traveling-wave tube.

gap filling Electrical or mechanical rearrangement of an antenna array, or the use of a supplementary array, to produce lobes where gaps previously occurred

garnet maser A name incorrectly applied to a ferromagnetic amplifier.

gas-activated battery A reserve battery which is activated by introducing a gas which reacts with a material between the electrodes of the battery to form an electrolyte.

gas-bubble protective device *See* Buchholz protective device.

gas capacitor A capacitor consisting of two or more electrodes separated by a gas, other than air, that serves as a dielectric.

gas cell Cell in which the action depends on the absorption of gases by the electrodes.

gas current A positive-ion current produced by collisions between electrons and residual gas molecules in an electron tube. Also known as ionization current.

gas discharge Conduction of electricity in a gas, due to movements of ions produced by collisions between electrons and gas molecules.

gas-discharge display A display in which seven or more cathode elements form the segments of numerical or alphameric characters when energized by about 160 volts direct current; the segments are vacuum-sealed in a neon-mercury gas mixture.

gas-discharge lamp *See* discharge lamp.

gas doping The introduction of impurity atoms into a semiconductor material by epitaxial growth, by using streams of gas that are mixed before being fed into the reactor vessel.

gas-filled cable A coaxial or other cable containing gas under pressure to serve as insulation and keep out moisture.

gas-filled diode A gas tube which is a diode, such as a cold-cathode rectifier or phanotron.

gas-filled rectifier *See* cold-cathode rectifier.

gas-filled triode A gas tube which has a grid or other control element, such as a thyratron or ignitron.

gate-controlled switch 179

gas focusing A method of concentrating an electron beam by utilizing the residual gas in a tube; beam electrons ionize the gas molecules, forming a core of positive ions along the path of the beam which attracts beam electrons and thereby makes the beam more compact. Also known as ionic focusing.

gas-insulated substation An electric power substation in which all live equipment and busbars are housed in grounded metal enclosures sealed and filled with sulfur hexafluoride gas.

gas ionization Removal of the planetary electrons from the atoms of gas filling an electron tube, so that the resulting ions participate in current flow through the tube.

gas magnification Increase in current through a phototube due to ionization of the gas in the tube.

gas phototube A phototube into which a quantity of gas has been introduced after evacuation, usually to increase its sensitivity.

gas scattering The scattering of electrons or other particles in a beam by residual gas in the vacuum system.

gassiness Presence of unwanted gas in a vacuum tube, usually in relatively small amounts, caused by the leakage from outside or evolution from the inside walls or elements of the tube.

gassing The evolution of gas in the form of small bubbles in a storage battery when charging continues after the battery has been completely charged.

gassy tube A vacuum tube that has not been fully evacuated or has lost part of its vacuum due to release of gas by the electrode structure during use, so that enough gas is present to impair operating characteristics appreciably. Also known as soft tube.

gas tetrode *See* tetrode thyratron.

gas thermostatic switch A thermostatic switch in which heat causes the pressure of gas in a sealed metal bellows to increase, thereby moving the bellows and closing the contacts of a switch.

gas tube An electron tube into which a small amount of gas or vapor is admitted after the tube has been evacuated; ionization of gas molecules during operation greatly increases current flow.

gas vacuum breakdown Ionization of residual gas in a vacuum, causing reverse conduction in an electron tube.

gate 1. A circuit having an output and a multiplicity of inputs and so designed that the output is energized only when a certain combination of pulses is present at the inputs. 2. A circuit in which one signal, generally a square wave, serves to switch another signal on and off. 3. One of the electrodes in a field-effect transistor. 4. An output element of a cryotron. 5. To control the passage of a pulse or signal. 6. In radar, an electric waveform which is applied to the control point of a circuit to alter the mode of operation of the circuit at the time when the waveform is applied. Also known as gating waveform.

gate-array device An integrated logic circuit that is manufactured by first fabricating a two-dimensional array of logic cells, each of which is equivalent to one or a few logic gates, and then adding final layers of metallization that determine the exact function of each cell and interconnect the cells to form a specific network when the customer orders the device.

gate-controlled rectifier A three-terminal semiconductor device, such as a silicon controlled rectifier, in which the unidirectional current flow between the rectifier terminals is controlled by a signal applied to a third terminal called the gate.

gate-controlled switch A semiconductor device that can be switched from its nonconducting or "off" state to its conducting or "on" state by applying a negative pulse to

gated-beam tube

its gate terminal and that can be turned off at any time by applying reverse drive to the gate. Abbreviated GCS.

gated-beam tube A pentode electron tube having special electrodes that form a sheet-shaped beam of electrons; this beam may be deflected away from the anode by a relatively small voltage applied to a control electrode, thus giving extremely sharp cutoff of anode current.

gated sweep Sweep in which the duration as well as the starting time is controlled to exclude undesired echoes from the indicator screen.

gate equivalent circuit A unit of measure for specifying relative complexity of digital circuits, equal to the number of individual logic gates that would have to be interconnected to perform the same function as the digital circuit under evaluation.

gate generator A circuit used to generate gate pulses; in one form it consists of a multivibrator having one stable and one unstable position.

gate multivibrator Rectangular-wave generator designed to produce a single positive or negative gate voltage when triggered and then to become inactive until the next trigger pulse.

gate pulse A pulse that triggers a gate circuit so it will pass a signal.

gate turnoff A *pnpn* switching device comparable to a silicon-controlled rectifier, but having a more complex gate structure that permits easy and fast turnoff as well as turn-on from its gate input terminal, at frequencies up to 100 kilohertz.

gate-turnoff silicon-controlled rectifier A silicon-controlled rectifier that can be turned off by applying a current to its gate; used largely for direct-current switching, because turnoff can be achieved in a fraction of a microsecond.

gate winding A winding used in a magnetic amplifier to produce on-off action of load current.

gating The process of selecting those portions of a wave that exist during one or more selected time intervals or that have magnitudes between selected limits.

gating waveform *See* gate.

gauge One of the family of possible choices for the electric scalar potential and magnetic vector potential, given the electric and magnetic fields.

gauge invariance The invariance of electric and magnetic fields and electrodynamic interactions under gauge transformations.

gauge transformation The addition of the gradient of some function of space and time to the magnetic vector potential, and the addition of the negative of the partial derivative of the same function with respect to time, divided by the speed of light, to the electric scalar potential; this procedure gives different potentials but leaves the electric and magnetic fields unchanged.

gauss Unit of magnetic induction in the electromagnetic and Gaussian systems of units, equal to 1 maxwell per square centimeter, or 10^{-4} weber per square meter. Also known as abtesla (abt).

Gaussian noise generator A signal generator that produces a random noise signal whose frequency components have a Gaussian distribution centered on a predetermined frequency value.

Gaussian system A combination of the electrostatic and electromagnetic systems of units (esu and emu), in which electrostatic quantities are expressed in esu and magnetic and electromagnetic quantities in emu, with appropriate use of the conversion constant c (the speed of light) between the two systems. Also known as Gaussian units.

Gaussian units *See* Gaussian system.

ghost mode   181

Gauss positions The Gauss A and B positions; that is, a point on the axis of a bar magnet is in Gauss A position, and a point on the magnetic equator of the magnet is in Gauss B position, with respect to the magnet.

Gauss' law of flux The law that the total electric flux which passes out from a closed surface equals (in rationalized units) the total charge within the surface.

GCS *See* gate-controlled switch.

G display A rectangular radar display in which horizontal and vertical aiming errors are indicated by horizontal and vertical displacement, respectively, and range is indicated by the length of wings appearing on the blip, with length increasing as range decreases.

Geissler tube An experimental discharge tube with two electrodes at opposite ends, used to demonstrate and study the luminous effects of electric discharges through various gases at low pressures.

gelled cell A lead-acid cell with a nonspillable gelled electrolyte for portable use.

gemmho A unit of conductance, equal to the conductance of a substance which has a resistance of 10^6 ohms, or to 10^{-6} mho.

general-purpose automatic test system Modular, computer-type, automatic electronic checkout system capable of finding faults in electronic equipment at the system, subsystem, line replaceable unit, module, and piece part levels.

generation rate In a semiconductor, the time rate of creation of electron-hole pairs.

generator 1. A machine that converts mechanical energy into electrical energy; in its commonest form, a large number of conductors are mounted on an armature that is rotated in a magnetic field produced by field coils. Also known as dynamo; electric generator. 2. A vacuum-tube oscillator or any other nonrotating device that generates an alternating voltage at a desired frequency when energized with direct-current power or low-frequency alternating-current power. 3. A circuit that generates a desired repetitive or nonrepetitive waveform, such as a pulse generator.

generator field control Method of regulating the output voltage of a generator by controlling the voltage which excites the field of the generator.

generator reactor A small inductor connected between power-plant generators and the rest of an electric power system in order to limit and localize the effects of voltage transients.

generator resistance The resistance of the current source in a network; usually much smaller than the load but taken into account in some network calculations.

geophone A transducer, used in seismic work, that responds to motion of the ground at a location on or below the surface of the earth.

germanium diode A semiconductor diode that uses a germanium crystal pellet as the rectifying element. Also known as germanium rectifier.

germanium rectifier *See* germanium diode.

germanium transistor A transistor in which the semiconductor material is germanium, to which electric contacts are made.

getter sputtering The deposition of high-purity thin films at ordinary vacuum levels by using a getter to remove contaminants remaining in the vacuum.

ghost image 1. An undesired duplicate image at the right of the desired image on a television receiver; due to multipath effect, wherein a reflected signal traveling over a longer path arrives slightly later than the desired signal. 2. *See* ghost pulse.

ghost mode Waveguide mode having a trapped field associated with an imperfection in the wall of the waveguide; a ghost mode can cause trouble in a waveguide operating close to the cutoff frequency of a propagation mode.

ghost pulse An unwanted signal appearing on the screen of a radar indicator and caused by echoes which have a basic repetition frequency differing from that of the desired signals. Also known as ghost image; ghost signal.

ghost signal 1. The reflection-path signal that produces a ghost image on a television receiver. Also known as echo. 2. *See* ghost pulse.

gigawatt One billion watts, or 10^9 watts. Abbreviated GW.

gigohm One billion ohms, or 10^9 ohms.

gilbert The unit of magnetomotive force in the electromagnetic system, equal to the magnetomotive force of a closed loop of one turn in which there is a current of $1/4\pi$ abamp.

Gilbert circuit A circuit that compensates for nonlinearities and instabilities in a monolithic variable-transconductance circuit by using the logarithmic properties of diodes and transistors.

gimmick Length of twisted two-conductor cable, used as a variable capacitive load, in which the capacitance is varied by untwisting and separating the individual conductors.

G indicator *See* G scope.

Ginzburg-London superconductivity theory A modification of the London superconductivity theory to take into account the boundary energy.

glass capacitor A capacitor whose dielectric material is glass.

glassivation Method of transistor passivation by a pyrolytic glass-deposition technique, whereby silicon semiconductor devices, complete with metal contact systems, are fully encapsulated in glass.

glass-plate capacitor High-voltage capacitor in which the metal plates are separated by sheets of glass serving as the dielectric, with the complete assembly generally immersed in oil.

glass resistor A glass tube with a helical carbon resistance element painted on it.

glass switch An amorphous solid-state device used to control the flow of electric current. Also known as ovonic device.

glass-to-metal seal An airtight seal between glass and metal parts of an electron tube, made by fusing together a special glass and special metal alloy having nearly the same temperature coefficients of expansion.

glassy state *See* vitreous state.

G line A single dielectric-coated, round wire used for transmitting microwave energy.

glint 1. Pulse-to-pulse variation in amplitude of a reflected radar signal, owing to the reflection of the radar from a body that is rapidly changing its reflecting surface, for example, a spinning airplane propeller. 2. The use of this effect to degrade tracking or seeking functions of an enemy weapons system.

glissile dislocation *See* Shockley partial dislocation.

glitch 1. An undesired transient voltage spike occurring on a signal being processed. 2. A minor technical problem arising in electronic equipment.

Globar lamp A lamp whose illuminating element is a silicon carbide rod which gives off blackbody radiation when heated.

glow discharge A discharge of electricity through gas at relatively low pressure in an electron tube, characterized by several regions of diffuse, luminous glow and a voltage drop in the vicinity of the cathode that is much higher than the ionization voltage of the gas. Also known as cold-cathode discharge.

graphical design 183

glow-discharge cold-cathode tube See glow-discharge tube.

glow-discharge tube A gas tube that depends for its operation on the properties of a glow discharge. Also known as glow-discharge cold-cathode tube; glow tube.

glow-discharge voltage regulator Gas tube that varies in resistance, depending on the value of the applied voltage; used for voltage regulation.

glow lamp A two-electrode electron tube containing a small quantity of an inert gas, in which light is produced by a negative glow close to the negative electrode when voltage is applied between the electrodes.

glow potential The potential across a glow discharge, which is greater than the ionization potential and less than the sparking potential, and is relatively constant as the current is varied across an appreciable range.

glow tube See glow-discharge tube.

glow-tube oscillator A circuit using a glow-discharge tube which functions as a simple relaxation oscillator, generating a fixed-amplitude periodic sawtooth waveform.

gnd See ground.

gold doping A technique for controlling the lifetime of minority carriers in a transistor; gold is diffused into the base and collector regions to reduce storage time in transistor circuits.

gold-leaf electroscope An electroscope in which two narrow strips of gold foil or leaf suspended in a glass jar spread apart when charged; the angle between the strips is related to the charge.

Goldschmidt's law The law that crystal structure is determined by the ratios of the numbers of the constituents, the ratios of their sizes, and their polarization properties.

goniometer An instrument for determining the direction of maximum response to a received radio signal, or selecting the direction of maximum radiation of a transmitted radio signal; consists of two fixed perpendicular coils, each attached to one of a pair of loop antennas which are also perpendicular, and a rotatable coil which bears the same space relationship to the coils as the direction of the signal to the antennas.

goniometric locator In radio detection finding, a rotating device that samples signals from orthogonal fixed antennas.

Goto pair Two tunnel diodes connected in series in such a way that when one is in the forward conduction region, the other is in the reverse tunneling region; used in high-speed gate circuits.

g parameter One of a set of four transistor equivalent-circuit parameters; they are the inverse of the h parameters.

graded-junction transistor See rate-grown transistor.

graded periodicity technique A technique for modifying the response of a surface acoustic wave filter by varying the spacing between successive electrodes of the interdigital transducer.

graphechon A storage tube having two electron guns, one for writing and the other for reading and simultaneous erasing, on opposite sides of the storage medium, which consists of an insulator or semiconductor deposited on a thin substratum of metal supported by a fine mesh.

graphical design Methods of obtaining operating data for an electron tube or semiconductor circuit by using graphs which plot the relationship between two variables, such as plate voltage and grid voltage, while another variable, such as plate current, is held constant.

184 graphical symbol

graphical symbol A true symbol, rather than a coarse picture, representing an element in an electrical diagram.

graphic display The display of data in graphical form on the screen of a cathode-ray tube.

graphic terminal A cathode-ray-tube or other type of computer terminal capable of producing some form of line drawing based on data being processed by or stored in a computer.

graphite anode 1. The rod of graphite which is inserted into the mercury-pool cathode of an ignitron to start current flow. 2. The collector of electrons in a beam power tube or other high-current tube.

graphite resistor A resistor made of carbon for resistance heating and suitable for any temperature that can be used.

grass Clutter due to circuit noise in a radar receiver, seen on an A scope as a pattern resembling a cross section of turf. Also known as hash.

grasshopper fuse Small fuse incorporating a spring which, upon release by the fusing wire, connects an auxiliary circuit to operate an alarm.

grating 1. An arrangement of fine, parallel wires used in waveguides to pass only a certain type of wave. 2. An arrangement of crossed metal ribs or wires that acts as a reflector for a microwave antenna and offers minimum wind resistance.

Gratz rectifier Three-phase, full-wave rectifying circuit using six rectifiers connected in a bridge circuit.

grid 1. A metal plate with holes or ridges, used in a storage cell or battery as a conductor and a support for the active material. 2. Any systematic network, such as of telephone lines or power lines. 3. An electrode located between the cathode and anode of an electron tube, which has one or more openings through which electrons or ions can pass, and serves to control the flow of electrons from cathode to anode.

grid-anode transconductance See transconductance.

grid battery See C battery.

grid bias The direct-current voltage applied between the control grid and cathode of an electron tube to establish the desired operating point. Also known as bias; C bias; direct grid bias.

grid-bias cell See bias cell.

grid blocking 1. Method of keying a circuit by applying negative grid bias several times cutoff value to the grid of a tube during key-up conditions; when the key is down, the blocking bias is removed and normal current flows through the keyed circuit. 2. Blocking of capacitance-coupled stages in an amplifier caused by the accumulation of charge on the coupling capacitors due to grid current passed during the reception of excessive signals.

grid blocking capacitor See grid capacitor.

grid cap A top-cap terminal for the control grid of an electron tube.

grid capacitor A small capacitor used in the grid circuit of an electron tube to pass signal current while blocking the direct-current anode voltage of the preceding stage. Also known as grid blocking capacitor; grid condenser.

grid cathode capacitance Capacitance between the grid and the cathode in a vacuum tube.

grid characteristic Relationship of grid current to grid voltage of a vacuum tube.

grid circuit The circuit connected between the grid and cathode of an electron tube.

grid condenser *See* grid capacitor.

grid conductance *See* electrode conductance.

grid control Control of anode current of an electron tube by variation (control) of the control grid potential with respect to the cathode of the tube.

grid-controlled mercury-arc rectifier A mercury-arc rectifier in which one or more electrodes are employed exclusively to control the starting of the discharge. Also known as grid-controlled rectifier.

grid-controlled rectifier *See* grid-controlled mercury-arc rectifier.

grid control tube Mercury-vapor-filled thermionic vacuum tube with an external grid control.

grid current Electron flow to a positive grid in an electron tube.

grid-dip meter A multiple-range electron-tube oscillator incorporating a meter in the grid circuit to indicate grid current; the meter reading dips (reads lower grid current) when an external resonant circuit is tuned to the oscillator frequency. Also known as grid-dip oscillator.

grid-dip oscillator *See* grid-dip meter.

grid drive A signal applied to the grid of a transmitting tube.

grid driving power Average product of the instantaneous value of the grid current and of the alternating component of the grid voltage over a complete cycle; this comprises the power supplied to the biasing device and to the grid.

grid element A sinuous resistor used to heat a furnace, made of heavy wire, strap, or casting and suspended from refractory or stainless supports built into the furnace walls, floor, and roof.

grid-glow tube A glow-discharge tube in which one or more control electrodes initiate but do not limit the anode current except under certain operating conditions.

gridistor Field-effect transistor which uses the principle of centripetal striction and has a multichannel structure, combining advantages of both field effect transistors and minority carrier injection transistors.

grid leak A resistor used in the grid circuit of an electron tube to provide a discharge path for the grid capacitor and for charges built up on the control grid.

grid-leak detector A detector in which the desired audio-frequency voltage is developed across a grid leak and grid capacitor by the flow of modulated radio-frequency current; the circuit provides square-law detection on weak signals and linear detection on strong signals, along with amplification of the audio-frequency signal.

grid limiter Limiter circuit which operates by limiting positive grid voltages by means of a large ohmic value resistor; as the exciting signal moves in a positive direction with respect to the cathode, current through the resistor causes an IR drop which holds the grid voltage essentially at cathode potential; during negative excursions no current flows in the grid circuit, so no voltage drop occurs across the resistor.

grid locking Defect of tube operation in which the grid potential becomes continuously positive due to excessive grid emission.

grid modulation Modulation produced by feeding the modulating signal to the control-grid circuit of any electron tube in which the carrier is present.

grid neutralization Method of amplifier neutralization in which a portion of the grid-cathode alternating-current voltage is shifted 180° and applied to the plate-cathode circuit through a neutralizing capacitor.

grid-plate capacitance Direct capacitance between the grid and the plate in a vacuum tube.

186 grid-plate transconductance

grid-plate transconductance See transconductance.

grid-pool tube An electron tube having a mercury-pool cathode, one or more anodes, and a control electrode or grid that controls the start of current flow in each cycle; the excitron and ignitron are examples.

grid pulse modulation Modulation produced in an amplifier or oscillator by applying one or more pulses to a grid circuit.

grid pulsing Circuit arrangement of a radio-frequency oscillator in which the grid of the oscillator is biased so negatively that no oscillation takes place even when full plate voltage is applied; pulsing is accomplished by removing this negative bias through the application of a positive pulse on the grid.

grid resistor A general term used to denote any resistor in the grid circuit.

grid return External conducting path for the return grid current to the cathode.

grid suppressor Resistor of low ohmic value inserted in the grid circuit of a radio-frequency amplifier to prevent low-frequency parasitic oscillations.

grid swing Total variation in grid-cathode voltage from the positive peak to the negative peak of the applied signal voltage.

grid transformer Transformer to supply an alternating voltage to a grid circuit or circuits.

grid-type level detector A detector using a vacuum tube with input applied to a grid.

grid voltage The voltage between a grid and the cathode of an electron tube.

Griebe-Schiebe method A method of observing the piezoelectric behavior of small crystals, in which the crystals are placed between two electrodes connected to the resonant circuit of an oscillator, and tuning of the resonant circuit results in jumps in the oscillator frequency which produce clicks in headphones or a loudspeaker attached to the plate circuit of the oscillator.

Griebhard's rings A method of producing lines of constant color on a copper sheet, coinciding with the equipotential lines of an electric field.

grinding 1. A mechanical operation performed on silicon substrates of semiconductors to provide a smooth surface for epitaxial deposition or diffusion of impurities. 2. A mechanical operation performed on quartz crystals to alter their physical size and hence their resonant frequencies.

ground 1. A conducting path, intentional or accidental, between an electric circuit or equipment and the earth, or some conducting body serving in place of the earth. Abbreviated gnd. Also known as earth (British usage); earth connection. 2. To connect electrical equipment to the earth or to some conducting body which serves in place of the earth.

ground absorption Loss of energy in transmission of radio waves, due to dissipation in the ground.

ground cable A heavy cable connected to earth for the purpose of grounding electric equipment.

ground circuit A telephone or telegraph circuit part of which passes through the ground.

ground clutter Clutter on a ground or airborne radar due to reflection of signals from the ground or objects on the ground. Also known as ground flutter; ground return; land return; terrain echoes.

ground conductivity The effective conductivity of the ground, used in calculating the attenuation of radio waves.

ground current See earth current.

ground indicator

ground detector An instrument or equipment used for indicating the presence of a ground on an ungrounded system. Also known as ground indicator.

ground dielectric constant Dielectric constant of the earth at a given location.

grounded-anode amplifier See cathode follower.

grounded-base amplifier An amplifier that uses a transistor in a grounded-base connection.

grounded-base connection A transistor circuit in which the base electrode is common to both the input and output circuits; the base need not be directly connected to circuit ground. Also known as common-base connection.

grounded-cathode amplifier Electron-tube amplifier with a cathode at ground potential at the operating frequency, with input applied between control grid and ground, and with the output load connected between plate and ground.

grounded-collector connection A transistor circuit in which the collector electrode is common to both the input and output circuits; the collector need not be directly connected to circuit ground. Also known as common-collector connection.

grounded-emitter amplifier An amplifier that uses a transistor in a grounded-emitter connection.

grounded-emitter connection A transistor circuit in which the emitter electrode is common to both the input and output circuits; the emitter need not be directly connected to circuit ground. Also known as common-emitter connection.

grounded-gate amplifier Amplifier that uses thin-film transistors in which the gate electrode is connected to ground; the input signal is fed to the source electrode and the output is obtained from the drain electrode.

grounded-grid amplifier An electron-tube amplifier circuit in which the control grid is at ground potential at the operating frequency; the input signal is applied between cathode and ground, and the output load is connected between anode and ground.

grounded-grid-triode circuit Circuit in which the input signal is applied to the cathode and the output is taken from the plate; the grid is at radio-frequency ground and serves as a screen between the input and output circuits.

grounded-grid-triode mixer Triode in which the grid forms part of a grounded electrostatic screen between the anode and cathode, and is used as a mixer for centimeter wavelengths.

grounded-plate amplifier See cathode follower.

grounded system Any conducting apparatus connected to ground. Also known as earthed system.

ground electrode A conductor buried in the ground, used to maintain conductors connected to it at ground potential and dissipate current conducted to it into the earth, or to provide a return path for electric current in a direct-current power transmission system.

ground equalizer inductors Coils, having relatively low inductance, inserted in the circuit to one or more of the grounding points of an antenna to distribute the current to the various points in any desired manner.

ground fault Accidental grounding of a conductor.

ground fault interrupter A fast-acting circuit breaker that also senses very small ground fault currents such as might flow through the body of a person standing on damp ground while touching a hot alternating-current line wire.

ground flutter See ground clutter.

ground indicator See ground detector.

grounding Intentional electrical connection to a reference conducting plane, which may be earth, but which more generally consists of a specific array of interconnected electrical conductors referred to as the grounding conductor.

grounding conductor An array of interconnected electric conductors at a uniform potential, to which electrical connections are made for the purpose of grounding.

grounding electrode Conductor embedded in the earth, used for maintaining ground potential on conductors connected to it, and for dissipating into the earth, current conducted to it.

grounding plate An electrically grounded metal plate on which a person stands to discharge static electricity picked up by his body, or a similar plate buried in the ground to act as a ground rod.

grounding reactor A reactor sometimes used in a grounded alternating-current system which joins a conductor or neutral point to ground and serves to limit ground current in case of a fault. Also known as earthing reactor (British usage).

grounding receptacle A receptacle which has an extra contact that accepts the third round or U-shaped prong of a grounding attachment plug and is connected internally to a supporting strap, providing a ground both through the outlet box and the grounding conductor, armor, or raceway of the wiring system.

grounding transformer Transformer intended primarily for the purpose of providing a neutral point for grounding purposes.

ground junction *See* grown junction.

ground loop Potentially detrimental loop formed when two or more points in an electric system that are nominally at ground potential are connected by a conducting path.

ground outlet Outlet equipped with a receptacle of the polarity type having, in addition to the current-carrying contacts, one grounded contact which can be used for the connection of an equipment-grounding conductor.

ground plane A grounding plate, aboveground counterpoise, or arrangement of buried radial wires required with a ground-mounted antenna that depends on the earth as the return path for radiated radio-frequency energy.

ground plane antenna Vertical antenna combined with a grounded horizontal disk, turnstile element, or similar ground plane simulation; such antennas may be mounted several wavelengths above the ground, and provide a low radiation angle.

ground plate A plate of conducting material embedded in the ground to act as a ground electrode.

ground potential Zero potential with respect to the ground or earth.

ground protection Protection provided a circuit by a device which opens the circuit when a fault to ground occurs.

ground recharge The flow of electrons from the ground, in reference to lightning effects.

ground-reflected wave Component of the ground wave that is reflected from the ground.

ground resistance Opposition of the earth to the flow of current through it; its value depends on the nature and moisture content of the soil, on the material, composition, and nature of connections to the earth, and on the electrolytic action present.

ground return 1. Use of the earth as the return path for a transmission line. 2. An echo received from the ground by an airborne radar set. 3. *See* ground clutter.

ground-return circuit Circuit which has a conductor (or two or more in parallel) between two points and which is completed through the ground or earth.

ground rod A rod that is driven into the earth to provide a good ground connection.

ground system The portion of an antenna that is closely associated with an extensive conducting surface, which may be the earth itself.

ground wire A conductor used to connect electric equipment to a ground rod or other grounded object.

group A kits Normally those items of electronic equipment which may be permanently or semipermanently installed in an aircraft for supporting, securing, or interconnecting the components and controls of the equipment, and which will not in any manner compromise the security classification of the equipment.

group B kits Normally, the operating or operable component of the electronic equipment in an aircraft which, when installed on or in connection with group A parts, constitute the complete operable equipment.

group bus A scheme of electrical connections for a generating station in which more than two feeder lines are supplied by two bus-selector circuit breakers which lead to a main bus and an auxiliary bus.

group frequency Frequency corresponding to group velocity of propagated waves in a transmission line or waveguide.

grove cell Primary cell, having a platinum electrode in an electrolyte of nitric acid within a porous cup, outside of which is a zinc electrode in an electrolyte of sulfuric acid; it normally operates on a closed circuit.

growler An electromagnetic device consisting essentially of two field poles arranged as in a motor, used for locating short-circuited coils in the armature of a generator or motor and for magnetizing or demagnetizing objects; a growling noise indicates a short-circuited coil.

grown-diffused transistor A junction transistor in which the final junctions are formed by diffusion of impurities near a grown junction.

grown junction A junction produced by changing the types and amounts of donor and acceptor impurities that are added during the growth of a semiconductor crystal from a melt. Also known as ground junction.

grown-junction photocell A photodiode consisting of a bar of semiconductor material having a pn junction at right angles to its length and an ohmic contact at each end of the bar.

grown-junction transistor A junction transistor in which different impurities are placed in the melt in sequence as the silicon or germanium seed crystal is slowly withdrawn, to produce the alternate pn and np junctions.

Grüneisen constant Three times the bulk modulus of a solid times its linear expansion coefficient, divided by its specific heat per unit volume; it is reasonably constant for most cubic crystals. Also known as Grüneisen gamma.

Grüneisen gamma *See* Grüneisen constant.

Grüneisen relation The relation stating that the electrical resistivity of a very pure metal is proportional to a mathematical function which depends on the ratio of the temperature to a characteristic temperature.

G scan *See* G scope.

G scope A cathode-ray scope on which a single signal appears as a spot on which wings grow as the distance to the target is decreased, with bearing error as the horizontal coordinate and elevation angle error as the vertical coordinate. Also known as G indicator; G scan.

G string *See* field waveguide.

guard arm 1. Crossarm placed across and in line with a cable to prevent damage to the cable. 2. Crossarm located over wires to prevent foreign wires from falling into them.

guard band A narrow frequency band provided between adjacent channels in certain portions of the radio spectrum to prevent interference between stations.

guarding A method of eliminating surface-leakage effects from measurements of electrical resistance which employs a low-resistance conductor in the vicinity of one of the terminals or a portion of the measuring circuit.

guard relay Used in the linefinder circuit to make sure that only one linefinder can be connected to any line circuit when two or more line relays are operated simultaneously.

guard ring 1. A ring-shaped auxiliary electrode surrounding one of the plates of a parallel-plate capacitor to reduce edge effects. 2. A ring-shaped auxiliary electrode used in an electron tube or other device to modify the electric field or reduce insulator leakage; in a counter tube or ionization chamber a guard ring may also serve to define the sensitive volume.

guard shield Internal floating shield that surrounds the entire input section of an amplifier; effective shielding is achieved only when the absolute potential of the guard is stabilized with respect to the incoming signal.

guard wire A grounded conductor placed beneath an overhead transmission line in order to ground the line, in case it breaks, before reaching the ground.

Gudden-Pohl effect The momentary illumination produced when an electric field is applied to a phosphor previously excited by ultraviolet radiation.

guided wave A wave whose energy is concentrated near a boundary or between substantially parallel boundaries separating materials of different properties and whose direction of propagation is effectively parallel to these boundaries; waveguides transmit guided waves.

guide wavelength Wavelength of electromagnetic energy conducted in a waveguide; guide wavelength for all air-filled guides is always longer than the corresponding free-space wavelength.

guiding center A slowly moving point about which a charged particle rapidly revolves; this is used in an approximation for the motion of a charged particle in slowly varying electric and magnetic fields.

Guillemin effect The tendency of a bent magnetorestrictive rod to straighten in a magnetic field parallel to its length.

Guillemin line A network or artificial line used in high-level pulse modulation to generate a nearly square pulse, with steep rise and fall; used in radar sets to control pulse width.

Gunn amplifier A microwave amplifier in which a Gunn oscillator functions as a negative-resistance amplifier when placed across the terminals of a microwave source.

Gunn diode *See* Gunn oscillator.

Gunn effect Development of a rapidly fluctuating current in a small block of a semiconductor (perhaps n-type gallium arsenide) when a constant voltage above a critical value is applied to contacts on opposite faces.

Gunn oscillator A microwave oscillator utilizing the Gunn effect. Also known as Gunn diode.

Gurevich effect An effect observed in electric conductors in which phonon-electron collisions are important, in the presence of a temperature gradient, in which phonons carrying a thermal current tend to drag the electrons with them from hot to cold.

GW *See* gigawatt.

gyration tensor A tensor characteristic of an optically active crystal, whose product with a unit vector in the direction of propagation of a light ray gives the gyration vector.

gyrator A waveguide component that uses a ferrite section to give zero phase shift for one direction of propagation and 180° phase shift for the other direction; in other words, it causes a reversal of signal polarity for one direction of propagation but not for the other direction. Also known as microwave gyrator.

gyrator filter A highly selective active filter that uses a gyrator which is terminated in a capacitor so as to have an inductive input impedance.

gyrofrequency *See* cyclotron frequency.

gyromagnetic coupler A coupler in which a single-crystal yig (yttrium iron garnet) resonator provides coupling at the required low signal levels between two crossed strip-line resonant circuits.

gyromagnetic effect The rotation induced in a body by a change in its magnetization, or the magnetization resulting from a rotation.

gyromagnetic radius *See* Larmor radius.

gyromagnetics The study of the relation between the angular momentum and the magnetization of a substance as exhibited in the gyromagnetic effect.

gyrotron 1. A device that detects motion of a system by measuring the phase distortion that occurs when a vibrating tuning fork is moved. 2. A type of microwave tube in which microwave amplification or generation results from cyclotron resonance coupling between microwave fields and an electron beam in vacuum. Also known as cyclotron resonance maser.

H

H *See* henry.

Hahn technique A method of studying changes in solids under various treatments that involves incorporating small amounts of radium into the solid and measuring the emanating power.

halation An area of glow surrounding a bright spot on a fluorescent screen, due to scattering by the phosphor or to multiple reflections at front and back surfaces of the glass faceplate.

half-adder A logic element which operates on two binary digits (but no carry digits) from a preceding stage, producing as output a sum digit and a carry digit.

half-bridge A bridge having two power supplies, located in two of the bridge arms, to replace the single power supply of a conventional bridge.

half-duplex repeater Duplex telegraph repeater provided with interlocking arrangements which restrict the transmission of signals to one direction at a time.

half-period zones *See* Fresnel zones.

half-power frequency One of the two values of frequency, on the sides of an amplifier response curve, at which the voltage is $1/\sqrt{2}$ (70.7%) of a midband or other reference value. Also known as half-power point.

half-power point 1. A point on the graph of some quantity in an antenna, network, or control system, versus frequency, distance, or some other variable at which the power is half that of a nearby point at which power is a maximum. 2. *See* half-power frequency.

half-pulse-repetition-rate delay In the loran navigation system, an interval of time equal to half the pulse repetition rate of a pair of loran transmitting stations, introduced as a delay between transmission of the master and slave signals, to place the slave station signal on the B trace when the master station signal is mounted on the A trace pedestal.

half-shift register Logic circuit consisting of a gated input storage element, with or without an inverter.

half-subtracter A logic element which operates on two digits from a preceding stage, producing as output a difference digit and a borrow digit. Also known as one-digit subtracter; two-input subtracter.

half tap Bridge placed across conductors without disturbing their continuity.

half-wave 1. Pertaining to half of one cycle of a wave. 2. Having an electrical length of a half wavelength.

half-wave amplifier A magnetic amplifier whose total induced voltage has a frequency equal to the power supply frequency.

half-wave antenna An antenna whose electrical length is half the wavelength being transmitted or received.

half-wave dipole See dipole antenna.

half-wavelength The distance corresponding to an electrical length of half a wavelength at the operating frequency of a transmission line, antenna element, or other device.

half-wave rectification Rectification in which current flows only during alternate half cycles.

half-wave rectifier A rectifier that provides half-wave rectification.

half-wave transmission line Transmission line which has an electrical length equal to one-half the wavelength of the signal being transmitted or received.

half-wave vibrator A vibrator having only one pair of contacts; interrupts the flow of direct current through the primary of a power transformer, but does not reverse the current.

Hall angle The electric field, resulting from the Hall effect, perpendicular to a current, divided by the electric field generating the current.

Hall coefficient A measure of the Hall effect, equal to the transverse electric field (Hall field) divided by the product of the current density and the magnetic induction. Also known as Hall constant.

Hall constant See Hall coefficient.

Hall effect The development of a transverse electric field in a current-carrying conductor placed in a magnetic field; ordinarily the conductor is positioned so that the magnetic field is perpendicular to the direction of current flow and the electric field is perpendicular to both.

Hall-effect isolator An isolator that makes use of the Hall effect in a semiconductor plate mounted in a magnetic field, to provide greater loss in one direction of signal travel through a waveguide than in the other direction.

Hall-effect modulator A Hall-effect multiplier used as a modulator to give an output voltage that is proportional to the product of two input voltages or currents.

Hall-effect multiplier A multiplier based on the Hall effect, used in analog computers to solve such problems as finding the square root of the sum of the squares of three independent variables.

Hall-effect switch A magnetically activated switch that uses a Hall generator, trigger circuit, and transistor amplifier on a silicon chip.

Hall generator A generator using the Hall effect to give an output voltage proportional to magnetic field strength.

Hall mobility The product of conductivity and the Hall constant for a conductor or semiconductor; a measure of the mobility of the electrons or holes in a semiconductor.

Hall voltage The no-load voltage developed across a semiconductor plate due to the Hall effect, when a specified value of control current flows in the presence of a specified magnetic field.

halo An undesirable bright or dark ring surrounding an image on the fluorescent screen of a television cathode-ray tube; generally due to overloading or maladjustment of the camera tube.

ham radio See amateur radio.

hand generator A manually cranked dynamo or alternator, usually used as the prime mover for emergency radio transmitters.

harmonic producer 195

handie-talkie Two-way radio communications unit small enough to be carried in the hand.

hand-reset Pertaining to a relay in which the contacts must be reset manually to their original positions when normal conditions are resumed.

hand rule The rule that, when grasping the conductor in the right hand with the thumb pointing in the direction of the current, the fingers will then point in the direction of the lines of flux.

hardened circuit A circuit that uses components whose tolerance to radiation released by a nuclear explosion has been increased by various radiation-hardening procedures.

hardness That quality which determines the penetrating ability of x-rays; the shorter the wavelength, the harder and more penetrating the rays.

hard-wire To connect electric components with solid, metallic wires as opposed to radio links, and the like.

hard x-ray An x-ray having high penetrating power.

Harker-Kasper inequalities Inequalities used in the analysis of crystal structure by x-ray diffraction which relate the structure factors and help to determine their phase factors.

harmonica bug A surreptitious interception technique applied to telephone lines; the target instrument is modified so that a tuned relay bypasses the switch hook and ringing circuit when a 500-hertz tone is received; this tone was originally generated by use of a harmonica.

harmonic analyzer An instrument that measures the strength of each harmonic in a complex wave. Also known as harmonic wave analyzer.

harmonic antenna An antenna whose electrical length is an integral multiple of a half-wavelength at the operating frequency of the transmitter or receiver.

harmonic attenuation Attenuation of an undesired harmonic component in the output of a transmitter.

harmonic conversion transducer A conversion transducer of which the useful output frequency is a multiple or a submultiple of the input frequency.

harmonic detector Voltmeter circuit so arranged as to measure only a particular harmonic of the fundamental frequency.

harmonic distortion Nonlinear distortion in which undesired harmonics of a sinusoidal input signal are generated because of circuit nonlinearity.

harmonic fields The sinusoidal Fourier components of a magnetic or other field confined to a finite region of space; their half-wavelengths are integral divisors of the length of the space in which the field is confined.

harmonic filter A filter that is tuned to suppress an undesired harmonic in a circuit.

harmonic generator A generator operated under conditions such that it generates strong harmonics along with the fundamental frequency.

harmonic loss Energy loss in a generator due to space harmonics of the magnetomotive force produced by armature current, especially losses resulting from the fifth and seventh harmonics.

harmonic oscillator *See* sinusoidal oscillator.

harmonic producer Tuning-fork controlled oscillator device capable of producing odd and even harmonics of the fundamental tuning-fork frequency; used to provide carrier frequencies for broad-band carrier systems.

harmonic telephone ringer

harmonic telephone ringer Telephone ringer which responds only to alternating current within a very narrow frequency band.

harmonic wave analyzer See harmonic analyzer.

harness Wire and cables so arranged and tied together that they may be inserted and connected, or may be removed after disconnection, as a unit.

Harris flow Electron flow in a cylindrical beam in which a radial electric field is used to overcome space charge divergence.

Hartley oscillator A vacuum-tube oscillator in which the parallel-tuned tank circuit is connected between grid and anode; the tank coil has an intermediate tap at cathode potential, so the grid-cathode portion of the coil provides the necessary feedback voltage.

Hartree equation An equation which gives the lowest anode voltage at which it is theoretically possible to maintain oscillation in the different modes of a magnetron.

hash 1. Electric noise produced by the contacts of a vibrator or by the brushes of a generator or motor. 2. See grass.

H attenuator See H network.

Hay bridge A four-arm alternating-current bridge used to measure inductance in terms of capacitance, resistance, and frequency; bridge balance depends on frequency.

H bend See H-plane bend.

head The photoelectric unit that converts the sound track on motion picture film into corresponding audio signals in a motion picture projector.

header A mounting plate through which the insulated terminals or leads are brought out from a hermetically sealed relay, transformer, transistor, tube, or other device.

heading-upward plan position indicator A plan position indicator in which the heading of the craft appears at the top of the indicator at all times.

heads-up display An electronic display that presents critical aircraft performance, such as speed and altitude, on a combining glass at the wind screen for pilot monitoring, while permitting the pilot to look out the window for other aircraft on the runway.

heat coil Protective device which uses a mechanical element which is allowed to move when the fusible substance that holds it in place is heated above a predetermined temperature by current in the circuit.

heater An electric heating element for supplying heat to an indirectly heated cathode in an electron tube. Also known as electron-tube heater.

heater-type cathode See indirectly heated cathode.

heating element The part of a heating appliance in which electrical energy is transformed into heat.

heat lamp An infrared lamp used for brooders in farming, for drying paint or ink, for keeping food warm, and for therapeutic and other applications requiring heat with or without some visible light.

heat of emission Additional heat energy that must be supplied to an electron-emitting surface to maintain it at a constant temperature.

heat run A series of temperature measurements made on an electric device during operating tests under various conditions.

heatsink A mass of metal that is added to a device for the purpose of absorbing and dissipating heat; used with power transistors and many types of metallic rectifiers. Also known as dissipator.

heat wave Infrared radiation, much higher in frequency than radio waves.

Heaviside-Lorentz system A system of electrical units which is the same as the Gaussian system except that the units of charge and current are smaller by a factor of $1/\sqrt{4\pi}$, and those of electric and magnetic field are larger by a factor by $\sqrt{4\pi}$. Also known as Lorentz-Heaviside system.

heavy-ion source Any source of ionized molecules or atoms of elements heavier than helium.

Hedvall effect I A discontinuous change in the temperature dependence of the chemical reaction rate of certain substances at the Curie temperture.

Hedvall effect II A discontinuous change in the activation energy of certain substances at the Curie temperature.

Heidelberg capsule A radio pill for telemetering pH values of gastric acidity.

height control The television receiver control that adjusts picture height.

height gain A radio-wave interference phenomenon which results in a more or less periodic signal strength variation with height; this specifically refers to interference between direct and surface-reflected waves; maxima or minima in these height-gain curves occur at those elevations at which the direct and reflected waves are exactly in phase or out of phase respectively.

height input Radar height information on target received by a computer from height finders and relayed via ground-to-ground data link or telephone.

height overlap coverage Height-finder coverage within which there is an area of overlapping coverage from adjacent height finders or other radar stations.

height-position indicator Radar display which shows simultaneously angular elevation, slant range, and height of objects detected in the vertical sight plane.

height-range indicator 1. Radar display which shows an echo as a bright spot on a rectangular field, slant range being indicated along the X axis, height above the horizontal plane being indicated (on a magnified scale) along the Y axis, and height above the earth being shown by a cursor. 2. Cathode-ray tube from which altitude and range measurements of flightborne objects may be viewed.

Heisenberg exchange coupling The exchange forces between electrons in neighboring atoms which give rise to ferromagnetism in the Heisenberg theory.

Heisenberg theory of ferromagnetism A theory in which exchange forces between electrons in neighboring atoms are shown to depend on relative orientations of electron spins, and ferromagnetism is explained by the assumption that parallel spins are favored so that all the spins in a lattice have a tendency to point in the same direction.

helical antenna An antenna having the form of a helix. Also known as helix antenna.

helical line A transmission line with a helical inner conductor.

helical potentiometer A multiturn precision potentiometer in which a number of complete turns of the control knob are required to move the contact arm from one end of the helically wound resistance element to the other end.

helical resonator A cavity resonator with a helical inner conductor.

helical traveling-wave tube *See* helix tube.

helimagnet A metal, alloy, or salt that possesses helimagnetism.

helimagnetism A property possessed by some metals, alloys, and salts of transition elements or rare earths, in which the atomic magnetic moments, at sufficiently low temperatures, are arranged in ferromagnetic planes, the direction of the magnetism varying in a uniform way from plane to plane.

helitron An electrostatically focused, low-noise backward-wave oscillator; the microwave output signal frequency can be swept rapidly over a wide range by varying the voltage applied between the cathode and the associated radio-frequency circuit.

helix A spread-out, single-layer coil of wire, either wound around a supporting cylinder or made of stiff enough wire to be self-supporting.

helix tube A traveling-wave tube in which the electromagnetic wave travels along a wire wound in a spiral about the path of the beam, so that the wave travels along the tube at a velocity approximately equal to the beam velocity. Also known as helical traveling-wave tube.

helmet-mounted display An electronic display that presents, on a combining glass within the visor of the helmet of a helicopter gunner, primary information for directing firepower; the angular direction of the helmet is sensed and used to control weapons to point in the same direction as the gunner is looking. Also known as visually coupled display.

helmholtz A unit of dipole moment per unit area, equal to 1 Debye unit per square angstrom, or approximately 3.335×10^{-10} coulomb per meter.

Helmholtz coils A pair of flat, circular coils having equal numbers of turns and equal diameters, arranged with a common axis, and connected in series; used to obtain a magnetic field more nearly uniform than that of a single coil.

Helmholtz's theorem *See* Thévenin's theorem.

henry The mks unit of self and mutual inductance, equal to the self-inductance of a circuit or the mutual inductance between two circuits if there is an induced electromotive force of 1 volt when the current is changing at the rate of 1 ampere per second. Symbolized H.

heptode A seven-electrode electron tube containing an anode, a cathode, a control electrode, and four additional electrodes that are ordinarily grids. Also known as pentagrid.

hermaphroditic connector A connector in which both mating parts are exactly alike at their mating surfaces. Also known as sexless connector.

herringbone pattern An interference pattern sometimes seen on television receiver screens, consisting of a horizontal band of closely spaced V- or S-shaped lines.

Hertz antenna An ungrounded half-wave antenna.

Hertz effect Increase in the length of a spark induced across a spark gap when the gap is irradiated with ultraviolet light.

Hertzian oscillator 1. A generator of electric dipole radiation; consists of two capacitors joined by a conducting rod having a small spark gap; an oscillatory discharge occurs when the two halves of the oscillator are raised to a sufficiently high potential difference. 2. A dumbbell-shaped conductor in which electrons oscillate from one end to the other, producing electric dipole radiation.

Hertzian wave *See* electric wave.

Hertz vector *See* polarization potential.

heterodyne To mix two alternating-current signals of different frequencies in a nonlinear device for the purpose of producing two new frequencies, the sum of and difference between the two original frequencies.

heterodyne conversion transducer *See* converter.

heterodyne detector A detector in which an unmodulated carrier frequency is combined with the signal of a local oscillator having a slightly different frequency, to provide an audio-frequency beat signal that can be heard with a loudspeaker or headphones; used chiefly for code reception.

high-frequency voltmeter 199

heterodyne frequency meter A frequency meter in which a known frequency, which may be adjustable or fixed, is heterodyned with an unknown frequency to produce a zero beat or an audio-frequency signal whose value is measured by other means. Also known as heterodyne wavemeter.

heterodyne measurement A measurement carried out by a type of harmonic analyzer which employs a highly selective filter, at a frequency well above the highest frequency to be measured, and a heterodyning oscillator.

heterodyne modulator See mixer.

heterodyne oscillator 1. A separate variable-frequency oscillator used to produce the second frequency required in a heterodyne detector for code reception. 2. See beat-frequency oscillator.

heterodyne reception Radio reception in which the incoming radio-frequency signal is combined with a locally generated rf signal of different frequency, followed by detection. Also known as beat reception.

heterodyne repeater A radio repeater in which the received radio signals are converted to an intermediate frequency, amplified, and reconverted to a new frequency band for transmission over the next repeater section.

heterodyne wavemeter See heterodyne frequency meter.

heterojunction The boundary between two different semiconductor materials, usually with a negligible discontinuity in the crystal structure.

heteropolar generator A generator whose active conductors successively pass through magnetic fields of opposite direction.

heterostatic Pertaining to the measurement of one electrostatic potential by means of a different potential.

hexode A six-electrode electron tube containing an anode, a cathode, a control electrode, and three additional electrodes that are ordinarily grids.

hickey A threaded coupling for attaching an electrical fixture to an outlet box, used when wires from the fixture come out of the end of a stem on the fixture, rather than through an opening in the side of the stem.

high-current rectifier A solid-state device, gas tube, or vacuum tube used to convert alternating to direct current for powering low-impedance loads.

high-current switch A switch used to redirect heavy current flow; usually has a make-before-break feature to prevent excessive arcing.

higher mode A waveguide mode whose frequency is higher than the lowest one.

high-frequency compensation Increasing the amplification at high frequencies with respect to that at low and middle frequencies in a given band, such as in a video band or an audio band. Also known as high boost.

high-frequency resistance The total resistance offered by a device in an alternating-current circuit, including the direct-current resistance and the resistance due to eddy current, hysteresis, dielectric, and corona losses. Also known as alternating-current resistance; effective resistance; radio-frequency resistance.

high-frequency transformer A transformer which matches impedances and transmits a frequency band in the carrier (or higher) frequency ranges.

high-frequency triode A triode designed for operation at high frequency, having small spacings between the grid and the cathode and anode, large emission and power densities, and low active and inactive capacitances.

high-frequency voltmeter A voltmeter designed to measure currents alternating at high frequencies.

high-impedance voltmeter A voltage-measuring device with a high-impedance input to reduce load on the unit under test; a vacuum-tube voltmeter is one type.

high-K capacitor A capacitor whose dielectric material is a ferroelectric having a high dielectric constant, up to about 6000.

high level The more positive of the two logic levels or states in a binary digital logic system.

highlights Bright areas occurring in a television image.

high-low bias test A routine maintenance procedure that tests equipment over and under normal operating conditions in order to detect defective units.

high-mu tube A tube having a very high amplification factor.

high-pass filter A filter that transmits all frequencies above a given cutoff frequency and substantially attenuates all others.

high-potting Testing with a high voltage, generally on a production line.

high-pressure mercury-vapor lamp A discharge tube containing an inert gas and a small quantity of liquid mercury; the initial glow discharge through the gas heats and vaporizes the mercury, after which the discharge through mercury vapor produces an intensely brilliant light.

high Q A characteristic wherein a component has a high ratio of reactance to effective resistance, so that its Q factor is high.

high-Q cavity A cavity resonator which has a large Q factor, and thus has a small energy loss. Also known as high-Q resonator.

high-Q resonator *See* high-Q cavity.

high-resistance voltmeter A voltmeter having a resistance considerably higher than 1000 ohms per volt, so that it draws little current from the circuit in which a measurement is made.

high-side capacitance coupling Taking the output of an oscillator or amplifier from a point of high potential, using a capacitor to block direct current flow.

high-speed excitation system Excitation system capable of changing its voltage rapidly in response to a change in the excited generator field circuit.

high-speed oscilloscope An oscilloscope with a very fast sweep, capable of observing signals with rise times or periods on the order of nanoseconds.

high-speed relay A relay specifically designed for short operate time, short release time, or both.

high-temperature fuel cell A fuel cell which operates at temperatures above about 550°C, can use inexpensive hydrocarbon fuels, and usually uses a molten salt as an electrolyte.

high tension *See* high voltage.

high-vacuum rectifier Vacuum-tube rectifier in which conduction is entirely by electrons emitted from the cathode.

high-vacuum switching tube A microwave transmit-receive (TR) tube of the high-vacuum variety, as contrasted with gas-tube or semiconductor devices.

high-vacuum tube Electron tube evacuated to such a degree that its electrical characteristics are essentially unaffected by gaseous ionization. Also known as hard tube.

high voltage A voltage on the order of thousands of volts. Also known as high tension.

high-voltage direct current A long-distance direct-current power transmission system that uses direct-current voltages up to about 1 megavolt to keep transmission losses down. Abbreviated HVDC.

hole conduction   201

high-voltage electron microscope An electron microscope whose accelerating voltage is on the order of 10^6 volts, as compared with 40–100 kilovolts for an ordinary electron microscope; it has the advantages of increased specimen penetration, reduced specimen damage, better theoretical resolution, and more efficient dark-field operation.

high-voltage insulation Electrical insulation designed to prevent breakdown in a circuit operating at high voltages.

hill bandwidth The difference between the upper and lower frequencies at which the gain of an amplifier is 3 decibels less than its maximum value.

hi pot High potential voltage applied across a conductor to test the insulation or applied to an etched circuit to burn out tenuous conducting paths that might later fail in service.

hit A momentary electrical disturbance on a transmission line.

Hittorf dark space *See* cathode dark space.

Hittorf principle The principle that a discharge between electrodes in a gas at a given pressure does not necessarily occur between the closest points of the electrodes if the distance between these points lies to the left of the minimum on a graph of spark potential versus distance. Also known as short-path principle.

H mode *See* transverse electric mode.

H network An attenuation network composed of five branches and having the form of the letter H. Also known as H attenuator; H pad.

hoghorn antenna *See* horn antenna.

hold To maintain storage elements at equilibrium voltages in a charge storage tube by electron bombardment.

hold circuit A circuit in a sampled-data control system that converts the series of impulses, generated by the sampler, into a rectangular function, in order to smooth the signal to the motor or plant.

hold control A manual control that changes the frequency of the horizontal or vertical sweep oscillator in a television receiver, so that the frequency more nearly corresponds to that of the incoming synchronizing pulses.

holder A device that mechanically and electrically accommodates one or more crystals, fuses, or other components in such a way that they can readily be inserted or removed.

holding anode A small auxiliary anode used in a mercury-pool rectifier to keep a cathode spot energized during the intervals when the main-anode current is zero.

holding beam A diffused beam of electrons used to regenerate the charges stored on the dielectric surface of a cathode-ray storage tube.

holding coil A separate relay coil that is energized by contacts which close when a relay pulls in, to hold the relay in its energized position after the orginal operating circuit is opened.

holding current The minimum current required to maintain a switching device in a closed or conducting state after it is energized or triggered.

hold lamp Indicating lamp which remains lighted while a telephone connection is being held.

hole A vacant electron energy state near the top of an energy band in a solid; behaves as though it were a positively charged particle. Also known as electron hole.

hole conduction Conduction occurring in a semiconductor when electrons move into holes under the influence of an applied voltage and thereby create new holes.

202 hole injection

hole injection The production of holes in an n-type semiconductor when voltage is applied to a sharp metal point in contact with the surface of the material.

hole mobility A measure of the ability of a hole to travel readily through a semiconductor, equal to the average drift velocity of holes divided by the electric field.

hole trap A semiconductor impurity capable of releasing electrons to the conduction or valence bands, equivalent to trapping a hole.

hollow cathode A cathode which is hollow and closed at one end in a discharge tube filled with inert gas, designed so that radiation is emitted from the cathode glow inside the cathode.

hollow-pipe waveguide A waveguide consisting of a hollow metal pipe; electromagnetic waves are transmitted through the interior and electric currents flow on the inner surfaces.

Holtz machine *See* Toepler-Holtz machine

home To return to the starting position, as in a stepping relay or turning motor.

home-on-jam A feature that permits radar to track a jamming source in angle.

homing antenna A directional antenna array used in flying directly to a target that is emitting or reflecting radio or radar waves.

homing device A control device that automatically starts in the correct direction of motion or rotation to achieve a desired change, as in a remote-control tuning motor for a television receiver.

homing relay A stepping relay that returns to a specified starting position before each operating cycle.

homodyne reception A system of radio reception for suppressed-carrier systems of radiotelephony, in which the receiver generates a voltage having the original carrier frequency and combines it with the incoming signal. Also known as zero-beat reception.

homopolar 1. Electrically symmetrical. 2. Having equal distribution of charge.

homopolar crystal A crystal in which the bonds are all covalent.

homopolar generator A direct-current generator in which the poles presented to the armature are all of the same polarity, so that the voltage generated in active conductors has the same polarity at all times; a pure direct current is thus produced, without commutation. Also known as acyclic machine; homopolar machine; unipolar machine.

homopolar machine *See* homopolar generator.

homotaxial-base transistor Transistor manufactured by a single-diffusion process to form both emitter and collector junctions in a uniformly doped silicon slice; the resulting homogeneously doped base region is free from accelerating fields in the axial (collector-to-emitter) direction, which could cause undesirable high current flow and destroy the transistor.

honeycomb coil A coil wound in a crisscross manner to reduce distributed capacitance. Also known as duolateral coil; lattice-wound coil.

hook A circuit phenomenon occurring in four-zone transistors, wherein hole or electron conduction can occur in opposite directions to produce voltage drops that encourage other types of conduction.

hook collector transistor A transistor in which there are four layers of alternating n- and p-type semiconductor material and the two interior layers are thin compared to the diffusion length. Also known as hook transistor; pn hook transistor.

hook transistor *See* hook collector transistor.

horizontal output transformer 203

hookup An arrangement of circuits and apparatus for a particular purpose.

hookup wire Tinned and insulated solid or stranded soft-drawn copper wire used in making circuit connections.

Hopkinson's coefficient The average magnetic flux per turn of an induction coil divided by the average flux per turn of another coil linked with it.

horizontal blanking Blanking of a television picture tube during the horizontal retrace.

horizontal blanking pulse The rectangular pulse that forms the pedestal of the composite television signal between active horizontal lines and causes the beam current of the picture tube to be cut off during retrace. Also known as line-frequency blanking pulse.

horizontal centering control The centering control provided in a television receiver or cathode-ray oscilloscope to shift the position of the entire image horizontally in either direction on the screen.

horizontal convergence control The control that adjusts the amplitude of the horizontal dynamic convergence voltage in a color television receiver.

horizontal definition *See* horizontal resolution.

horizontal deflection electrode One of a pair of electrodes that move the electron beam horizontally from side to side on the fluorescent screen of a cathode-ray tube employing electrostatic deflection.

horizontal deflection oscillator The oscillator that produces, under control of the horizontal synchronizing signals, the sawtooth voltage waveform that is amplified to feed the horizontal deflection coils on the picture tube of a television receiver. Also known as horizontal oscillator.

horizontal drive control The control in a television receiver, usually at the rear, that adjusts the output of the horizontal oscillator. Also known as drive control.

horizontal field-strength diagram Representation of the field strength at a constant distance from an antenna and in a horizontal plane; unless otherwise specified, this plane is that passing through the antenna.

horizontal flyback Flyback in which the electron beam of a television picture tube returns from the end of one scanning line to the beginning of the next line. Also known as horizontal retrace.

horizontal frequency *See* line frequency.

horizontal hold control The hold control that changes the free-running period of the horizontal deflection oscillator in a television receiver, so that the picture remains steady in the horizontal direction.

horizontal linearity control A linearity control that permits narrowing or expanding of the width of the left half of a television receiver image, to give linearity in the horizontal direction so that circular objects appear as true circles.

horizontal line frequency *See* line frequency.

horizontal oscillator *See* horizontal deflection oscillator.

horizontal output stage The television receiver stage that feeds the horizontal deflection coils of the picture tube through the horizontal output transformer; may also include a part of the second-anode power supply for the picture tube.

horizontal output transformer A transformer used in a television receiver to provide the horizontal deflection voltage, the high voltage for the second-anode power supply of the picture tube, and the filament voltage for the high-voltage rectifier tube. Also known as flyback transformer; horizontal sweep transformer.

204 horizontal resolution

horizontal resolution The number of individual picture elements or dots that can be distinguished in a horizontal scanning line of a television or facsimile image. Also known as horizontal definition.

horizontal retrace *See* horizontal flyback.

horizontal scanning frequency The number of horizontal lines scanned by the electron beam in a television receiver in 1 second.

horizontal sweep The sweep of the electron beam from left to right across the screen of a cathode-ray tube.

horizontal sweep transformer *See* horizontal output transformer.

horizontal synchronizing pulse The rectangular pulse transmitted at the end of each line in a television system, to keep the receiver in line-by-line synchronism with the transmitter. Also known as line synchronizing pulse.

horizontal vee An antenna consisting of two linear radiators in the form of the letter V, lying in a horizontal plane.

horn *See* horn antenna.

horn antenna A microwave antenna produced by flaring out the end of a circular or rectangular waveguide into the shape of a horn, for radiating radio waves directly into space. Also known as electromagnetic horn; flare (British usage); hoghorn antenna (British usage); horn; horn radiator.

horn arrester A lightning arrester in which the spark gap has thick wire horns that spread outward and upward; the arc forms at the narrowest bottom part of the gap, travels upward, and extinguishes itself when it reaches the widest part of the gap. Also known as horn lightning arrester.

horn gap Type of spark gap which is provided with divergent electrodes.

horn-gap switch Form of air switch provided with arcing horns.

horn lightning arrester *See* horn arrester.

horn radiator *See* horn antenna.

horseshoe magnet A permanent magnet or electromagnet in which the core is horseshoe-shaped or U-shaped, to bring the two poles near each other.

hot *See* energized.

hot carrier A carrier, which may be either an electron or a hole, that has relatively high energy with respect to the carriers normally found in majority-carrier devices such as thin-film transistors.

hot-carrier diode *See* Schottky barrier diode.

hot cathode A cathode in which electron or ion emission is produced by heat. Also known as thermionic cathode.

hot-cathode gas-filled tube *See* thyratron.

hot-cathode tube *See* thermionic tube.

hot electron An electron that is in excess of the thermal equilibrium number and, for metals, has an energy greater than the Fermi level; for semiconductors, the energy must be a definite amount above that of the edge of the conduction band.

hot-electron triode Solid-state, evaporated thin-film structure directly equivalent to a vacuum triode.

hot-filament ionization gage An ionization gage in which electrons emitted by an incandescent filament, and attracted toward a positively charged grid electrode, collide with gas molecules to produce ions which are then attracted to a negatively charged electrode; the ion current is a measure of the number of gas molecules.

hot hole A hole that can move at much greater velocity than normal holes in a semiconductor.

hot junction The heated junction of a thermocouple.

howler An audio device used to warn a radar operator that signals are appearing on a radar screen.

H pad *See* H network.

h parameter One of a set of four transistor equivalent-circuit parameters that conveniently specify transistor performance for small voltages and currents in a particular circuit. Also known as hybrid parameter.

H plane The plane of an antenna in which lies the magnetic field vector of linearly polarized radiation.

H-plane bend A rectangular waveguide bend in which the longitudinal axis of the waveguide remains in a plane parallel to the plane of the magnetic field vector throughout the bend. Also known as H bend.

H-plane T junction Waveguide T junction in which the change in structure occurs in the plane of the magnetic field. Also known as shunt T junction.

hue control A control that varies the phase of the chrominance signals with respect to that of the burst signal in a color television receiver, in order to change the hues in the image. Also known as phase control.

hum 1. A sound produced by an iron core of a transformer due to loose laminations or magnetostrictive effects; the frequency of the sound is twice the power line frequency. 2. An electrical disturbance occurring at the power supply frequency or its harmonics, usually 60 or 120 hertz in the United States.

hum bar A dark horizontal band extending across a television picture due to excessive hum in the video signal applied to the input of the picture tube.

Hume-Rothery rule The rule that the ratio of the number of valence electrons to the number of atoms in a given phase of an electron compound depends only on the phase, and not on the elements making up the compounds.

humidity capacitor A device for measuring ambient relative humidity by sensing a change in capacitance.

hum modulation Modulation of a radio-frequency signal or detected audio-frequency signal by hum; heard in a radio receiver only when a station is tuned in.

hunting Operation of a selector in moving from terminal to terminal until one is found which is idle.

hunting circuit *See* lockout circuit.

HVDC *See* high-voltage direct current.

H vector A vector that is the magnetic field. For a plane wave in free space, it is perpendicular to the E vector and to the direction of propagation.

H wave *See* transverse electric wave.

hybrid balance Loss between two conjugate sides of a hybrid set less the same loss when one of the other sides is open or shorted.

hybrid circuit A circuit in which two or more basically different types of components, such as tubes and transistors, performing similar functions are used together.

hybrid coil *See* hybrid transformer.

hybrid electromagnetic wave Wave which has both transverse and longitudinal components of displacement.

hybrid integrated circuit A circuit in which one or more discrete components are used in combination with integrated-circuit construction.

hybrid junction A transformer, resistor, or waveguide circuit or device that has four pairs of terminals so arranged that a signal entering at one terminal pair divides and emerges from the two adjacent terminal pairs, but is unable to reach the opposite terminal pair. Also known as bridge hybrid.

hybrid microcircuit Microcircuit in which thin-film, thick-film, or diffusion techniques are combined with separately attached semiconductor chips to form the circuit.

hybrid parameter See h parameter.

hybrid relay A relay in which solid-state elements are combined with moving contacts.

hybrid repeater See hybrid transformer.

hybrid set Two or more transformers interconnected to form a hybrid junction. Also known as transformer hybrid.

hybrid tee A microwave hybrid junction composed of an E-H tee with internal matching elements; it is reflectionless for a wave propagating into the junction from any arm when the other three arms are match-terminated. Also known as magic tee.

hybrid thin-film circuit Microcircuit formed by attaching discrete components and semiconductor devices to networks of passive components and conductors that have been vacuum-deposited on glazed ceramic, sapphire, or glass substrates.

hybrid transformer A single transformer that performs the essential functions of a hybrid set. Also known as bridge transformer; hybrid coil; hybrid repeater.

hydroelectricity Electric power produced by hydroelectric generators.

hydrogen-discharge lamp A discharge lamp containing hydrogen and used as a source of ultraviolet radiation.

hydrogen thyratron A thyratron containing hydrogen instead of mercury vapor to give freedom from effects of changes in ambient temperature; used in radar pulse circuits and stroboscopic photography.

hydrophone noise Any unwanted disturbance in the electric waves delivered by a hydrophone.

hydrophone response The electric waves delivered by a hydrophone in response to waterborne sound waves.

hygristor A resistor whose resistance varies with humidity; used in some types of recording hygrometers.

hyperbolic antenna A radiator whose reflector in cross section describes a half hyperbola.

hyperbolic distance A function of pairs of points within a unit circle, where the interior of this circle is a conformal or projective representation of a hyperbolic space used in transmission line theory and waveguide analysis.

hyperbolic sweep generator A sweep generator that generates a waveform resembling a hyperbola.

hyperbolic waveform A waveform which is an approximate hyperbola.

hyperfrequency waves Microwaves having wavelengths in the range from 1 centimeter to 1 meter.

hyperpure germanium detector A variant of the lithium-drifted germanium crystal which uses high-purity germanium, making it possible to store the detector at room temperature rather than liquid nitrogen temperature.

hysteresis motor 207

hypersensor Single-component, resettable circuit breaker which operates as a majority-carrier tunneling device, and is used for overcurrent or overvoltage protection of integrated circuits.

hysteresis 1. An oscillator effect wherein a given value of an operating parameter may result in multiple values of output power or frequency. 2. *See* magnetic hysteresis.

hysteresis motor A synchronous motor without salient poles and without direct-current excitation which utilizes the hysteresis and eddy-current losses induced in its hardened-steel rotor to produce rotor torque.

I

IC *See* integrated circuit.

iconocenter The image of the reflection coefficient of a matched load as plotted on an Argand diagram.

iconoscope A television camera tube in which a beam of high-velocity electrons scans a photoemissive mosaic that is capable of storing an electric charge pattern corresponding to an optical image focused on the mosaic. Also known as storage camera; storage-type camera tube.

ICS system *See* intercarrier sound system.

ideal bunching Theoretical condition in which the bunching of electrons in a velocity-modulated tube would give a single infinitely large current peak during each cycle.

ideal dielectric Dielectric in which all the energy required to establish an electric field in the dielectric is returned to the source when the field is removed. Also known as perfect dielectric.

ideal network An interconnection of lumped, constant electrical quantities analyzed without consideration of noise and distributed parameters that would exist in actual settings.

ideal transducer Hypothetical passive transducer which transfers the maximum possible power from the source to the load.

ideal transformer A hypothetical transformer that neither stores nor dissipates energy, has unity coefficient of coupling, and has pure inductances of infinitely great value.

I demodulator Stage of a color television receiver which combines the chrominance signal with the color oscillator output to restore the I signal.

I display A radarscope display in which a target appears as a complete circle when the radar antenna is correctly pointed at it, the radius of the circle being proportional to target distance; when the antenna is not aimed at the target, the circle reduces to a circle segment.

idle component *See* reactive component.

idle current *See* reactive current.

idler frequency Of a parametric device, a sum or difference frequency generated within the parametric device other than the input, output, or pump frequencies which require specific circuit consideration to achieve the desired device performance; it is called an idler frequency since, in conventional parametric amplifiers, it is more or less a useless by-product of the parametric process.

idle trunk lamp Signal lamp associated with an outgoing trunk to indicate that the trunk is not busy.

i-f *See* intermediate frequency.

i-f amplifier See intermediate-frequency amplifier.

IF canceler In radar, a moving-target indicator canceler that operates at intermediate frequencies.

i-f transformer See intermediate-frequency transformer.

ignition coil A coil in an ignition system which stores energy in a magnetic field relatively slowly and releases it suddenly to ignite a fuel mixture.

ignitor 1. An electrode used to initiate and sustain the discharge in a switching tube. Also known as keep-alive electrode (deprecated). 2. A pencil-shaped electrode, made of carborundum or some other conducting material that is not wetted by mercury, partly immersed in the mercury-pool cathode of an ignitron and used to initiate conduction at the desired point in each alternating-current cycle.

ignitron A single-anode pool tube in which an ignitor electrode is employed to initiate the cathode spot on the surface of the mercury pool before each conducting period.

ignitron contactor A circuit containing an ignitron and control contacts that serves as a heavy-duty switch in the primary of a resistance-welding transformer.

I indicator See I scope.

I²L See integrated injection logic.

illumination 1. The geometric distribution of power reaching various parts of a dish reflector in an antenna system. 2. The power distribution to elements of an antenna array.

illumination control A photoelectric control that turns on lights when outdoor illumination decreases below a predetermined level.

image 1. The input reflection coefficient corresponding to the reflection coefficient of a specified load when the load is placed on one side of a waveguide junction and a slotted line is placed on the other. 2. See electric image.

image antenna A fictitious electrical counterpart of an actual antenna, acting mathematically as if it existed in the ground directly under the real antenna and served as the direct source of the wave that is reflected from the ground by the actual antenna.

image attenuation constant The real part of the image transfer constant.

image converter See image tube.

image converter camera A camera consisting of an image tube and an optical system which focuses the image produced on the phosphorescent screen of the tube onto photographic film.

image dissection photography A method of high-speed photography in which an image is split in any one of various ways into interlaced space and time elements which can be unscrambled or played back through the system either to be viewed or to give a master negative.

image dissector tube A television camera tube in which an electron image produced by a photoemitting surface is focused in the plane of the defining aperture and is scanned past that aperture. Also known as Farnsworth image dissector tube.

image effect Effect produced on the field of an antenna due to the presence of the earth; electromagnetic waves are reflected from the earth's surface, and these reflections often are accounted for by an image antenna at an equal distance below the earth's surface.

image force The electrostatic force on a charge in the neighborhood of a conductor, which may be thought of as the attraction to the charge's electric image.

immersion electron microscope 211

image frequency An undesired carrier frequency that differs from the frequency to which a superheterodyne receiver is tuned by twice the intermediate frequency.

image iconoscope A camera tube in which an optical image is projected on a semitransparent photocathode, and the resulting electron image emitted from the other side of the photocathode is focused on a separate storage target; the target is scanned on the same side by a high-velocity electron beam, neutralizing the elemental charges in sequence to produce the camera output signal at the target. Also known as superemitron camera (British usage).

image impedance One of the impedances that, when connected to the input and output of a transducer, will make the impedances in both directions equal at the input terminals and at the output terminals.

image intensifier *See* light amplifier.

image isocon A television camera tube which is similar to the image orthicon but whose return beam consists of scanning beam electrons that are scattered by positive stored charges on the target.

image load Load parameters reflected back to the source by line discontinuities.

image orthicon A television camera tube in which an electron image is produced by a photoemitting surface and focused on one side of a separate storage tube that is scanned on its opposite side by a beam of low-velocity electrons; electrons that are reflected from the storage tube, after positive stored charges are neutralized by the scanning beam, form a return beam which is amplified by an electron multiplier.

image parameter design A method of filter design using image impedance and image transfer functions as the fundamental network functions.

image parameter filter A filter constructed by image parameter design.

image phase constant The imaginary part of the image transfer constant.

image ratio In a heterodyne receiver, the ratio of the image frequency signal input at the antenna to the desired signal input for identical outputs.

image reject mixer Combination of two balanced mixers and associated hybrid circuits designed to separate the image channel from the signal channels normally present in a conventional mixer; the arrangement gives image rejection up to 30 decibels without the use of filters.

image response The response of a superheterodyne receiver to an undesired signal at its image frequency.

image-storage array A solid-state panel or chip in which the image-sensing elements may be a metal oxide semiconductor or a charge-coupled or other light-sensitive device that can be manufactured in a high-density configuration.

image transfer constant One-half the natural logarithm of the complex ratio of the steady-state apparent power entering and leaving a network terminated in its image impedance.

image tube An electron tube that reproduces on its fluorescent screen an image of the optical image or other irradiation pattern incident on its photosensitive surface. Also known as electron image tube; image converter.

imitative deception Introduction of electromagnetic radiations into enemy channels which imitate their own emissions, in order to mislead them.

immersion electron microscope An emission electron microscope in which the specimen is a flat conducting surface which may be heated, illuminated, or bombarded by high-velocity electrons or ions so as to emit low-velocity thermionic, photo-, or secondary electrons; these are accelerated to a high velocity in an immersion objective or cathode lens and imaged as in a transmission electron microscope.

immersion electrostatic lens *See* bipotential electrostatic lens.

immersion heater An electric device for heating a liquid by direct immersion in the liquid.

immittance A term used to denote both impedance and admittance, as commonly applied to transmission lines, networks, and certain types of measuring instruments.

immittance bridge A modification of an admittance bridge which compares the output current of a four-terminal device with admittance standards in a T configuration in order to measure transfer admittance by a null method.

impact avalanche and transit time diode *See* IMPATT diode.

impact excitation Starting of damped oscillations in a radio circuit by a sudden surge, such as that produced by a spark discharge.

impact ionization Ionization produced by the impact of a high-energy charge carrier on an atom of semiconductor material; the effect is an increase in the number of charge carriers.

IMPATT amplifier A diode amplifier that uses an IMPATT diode; operating frequency range is from about 5 to 100 gigahertz, primarily in the C and X bands, with power output up to about 20 watts continuous-wave or 100 watts pulsed.

IMPATT diode A *pn* junction diode that has a depletion region adjacent to the junction, through which electrons and holes can drift, and is biased beyond the avalanche breakdown voltage. Derived from impact avalanche and transit time diode.

impedance *See* electrical impedance.

impedance-admittance matrix A four-element matrix used to describe analytically a transistor in terms of impedances or admittances.

impedance bridge A device similar to a Wheatstone bridge, used to compare impedances which may contain inductance, capacitance, and resistance.

impedance coil A coil of wire designed to provide impedance in an electric circuit.

impedance compensator Electric network designed to be associated with another network or a line with the purpose of giving the impedance of the combination a desired characteristic with frequency over a desired frequency range.

impedance component 1. Resistance or reactance. 2. A device such as a resistor, inductor, or capacitor designed to provide impedance in an electric circuit.

impedance coupling Coupling of two signal circuits with an impedance.

impedance drop The total voltage drop across a component or conductor of an alternating-current circuit, equal to the phasor sum of the resistance drop and the reactance drop.

impedance irregularities Discontinuities or abrupt changes which result from junctions between unlike sections of a transmission line or irregularities on a line.

impedance match The condition in which the external impedance of a connected load is equal to the internal impedance of the source or to the surge impedance of a transmission line, thereby giving maximum transfer of energy from source to load, minimum reflection, and minimum distortion.

impedance-matching network A network of two or more resistors, coils, or capacitors used to couple two circuits in such a manner that the impedance of each circuit will be equal to the impedance into which it locks. Also known as line-building-out network.

impedance matrix A matrix Z whose elements are the mutual impedances between the various meshes of an electrical network; satisfies the matrix equation $V = ZI$, where

incoming selector 213

V and I are column vectors whose elements are the voltages and currents in the meshes.

impedance meter See electrical impedance meter.

impedometer An instrument used to measure impedances in waveguides.

implanted atom An atom into semiconductor material by ion implantation.

implanted device A resistor or other device that is fabricated within a silicon or other semiconducting substrate by ion implantation.

impressed voltage Voltage applied to a circuit or device.

impulse excitation See shock excitation.

impulse generator An apparatus which produces very short surges of high-voltage or high-current power by discharging capacitors in parallel or in series. Also known as pulse generator.

impulse noise Noise characterized by transient short-duration disturbances distributed essentially uniformly over the useful passband of a transmission system.

impulse relay A relay that stores the energy of a short pulse, to operate the relay after the pulse ends.

impulse separator In a television receiver, the circuit that separates the horizontal synchronizing impulses in the received signal from the vertical synchronizing impulses.

impulse solenoid A solenoid that operates on pulse power, at speeds up to several hundred strokes per second.

impulse strength Voltage breakdown of insulation under voltage surges on the order of microseconds in duration.

impulse voltage A unidirectional voltage that rapidly rises to a peak value and then drops to zero more or less rapidly. Also known as pulse voltage.

impurity A substance that, when diffused into semiconductor metal in small amounts, either provides free electrons to the metal or accepts electrons from it.

impurity scattering Scattering of electrons by holes or phonons in the crystal.

impurity semiconductor A semiconductor whose properties are due to impurity levels produced by foreign atoms.

incandescent lamp An electric lamp that produces light when a metallic filament is heated white-hot in a vacuum by passing an electric current through the filament. Also known as filament lamp; light bulb.

incandescent readout A readout in which each character is formed by energizing an appropriate combination of seven bar-shaped incandescent lamps.

inching See jogging.

incident field intensity Field strength of a sky wave without including the effects of earth reflections at the receiving location.

incident power Product of the outgoing current and voltage, from a transmitter, traveling down a transmission line to the antenna.

incident wave A current or voltage wave that is traveling through a transmission line in the direction from source to load.

incoming first selector Connects incoming calls from outlying dial offices to local second selectors.

incoming selector Selector associated with trunk circuits from another central office.

214 incremental hysteresis loss

incremental hysteresis loss Hysteresis loss when a magnetic material is subjected to a pulsating magnetizing force.

incremental induction The quantity lying between the highest and lowest value of a magnetic induction at a point in a polarized material, when subjected to a small cycle of magnetization.

incremental permeability The ratio of a small cyclic change in magnetic induction to the corresponding cyclic change in magnetizing force when the average magnetic induction is greater than zero.

independent-sideband receiver A radio receiver designed for the reception of independent-sideband modulation, having provisions for restoring the carrier.

independent-sideband transmitter A transmitter which produces independent-sideband modulated signals.

index ratio The ratio of the radius of a conductor used in induction heating to its skin depth at the frequency used.

indicator A cathode-ray tube or other device that presents information transmitted or relayed from some other source, as from a radar receiver.

indicator element A component whose variability under conditions of manufacture or use is likely to cause the greatest variation in some measurable parameter.

indicator gate Rectangular voltage waveform which is applied to the grid or cathode circuit of an indicator cathode-ray tube to sensitize or desensitize it during a desired portion of the operating cycle.

indicator lamp A neon lamp whose on-off condition is used to convey information.

indicator tube An electron-beam tube in which useful information is conveyed by the variation in cross section of the beam at a luminescent target.

indirectly heated cathode A cathode to which heat is supplied by an independent heater element in a thermionic tube; this cathode has the same potential on its entire surface, whereas the potential along a directly heated filament varies from one end to the other. Also known as equipotential cathode; heater-type cathode; unipotential cathode.

indirect stroke A lightning stroke that induces a voltage in a power or communications system without actually striking it.

induced anisotropy A type of uniaxial anisotropy in a magnetic material produced by annealing the magnetic material in a magnetic field.

induced capacity *See* absolute permeability.

induced current A current produced in a conductor by a time-varying magnetic field, as in induction heating.

induced dipole An electric dipole produced by application of an electric field.

induced electromotive force An electromotive force resulting from the motion of a conductor through a magnetic field, or from a change in the magnetic flux that threads a conductor.

induced magnetism The magnetism acquired by magnetic material while it is in a magnetic field.

induced moment The average electric dipole moment per molecule which is produced by the action of an electric field on a dielectric substance.

induced potential *See* induced voltage.

induced voltage A voltage produced by electromagnetic or electrostatic induction. Also known as induced potential.

induction voltage regulator 215

inductance 1. That property of an electric circuit or of two neighboring circuits whereby an electromotive force is generated (by the process of electromagnetic induction) in one circuit by a change of current in itself or in the other. 2. Quantitatively, the ratio of the emf (electromotive force) to the rate of change of the current. 3. See coil.

inductance bridge 1. A device, similar to a Wheatstone bridge, for comparing inductances. 2. A four-coil alternating-current bridge circuit used for transmitting a mechanical movement to a remote location over a three-wire circuit; half of the bridge is at each location.

inductance coil See coil.

inductance measurement The determination of the self-inductance of a circuit or the mutual inductance of two circuits.

inductance meter A device which measures the self-inductance of a circuit or the mutual inductance of two circuits.

inductance standards Two equal, multilayer coils, wound on toroidal cores of non-magnetic materials, connected in series and located so that their interactions with external fields tend to cancel one another.

induction See electromagnetic induction; electrostatic induction.

induction charging Production of electric charge on a body by means of electrostatic induction.

induction coil A device for producing high-voltage alternating current or high-voltage pulses from low-voltage direct current, in which interruption of direct current in a primary coil, containing relatively few turns of wire, induces a high voltage in a secondary coil, containing many turns of wire wound over the primary.

induction disk relay A unit widely used in regulating and protective relays, in which alternating current applied to a coil produces torque to rotate a disk.

induction field A component of an electromagnetic field associated with an alternating current in a loop, coil, or antenna which carries energy alternately away from and back into the source, with no net loss, and which is responsible for self-inductance in a coil or mutual inductance with neighboring coils.

induction frequency converter Slip-ring induction machine which is driven by an external source of mechanical power and whose primary circuits are connected to a source of electric energy having a fixed frequency; the secondary circuits deliver energy at a frequency proportional to the relative speed of the primary magnetic field and the secondary member.

induction generator A nonsynchronous alternating-current generator whose construction is identical to that of an ac motor, and which is driven above synchronous speed by external sources of mechanical power.

induction machine An asynchronous alternating-current machine, such as an induction motor or induction generator, in which the windings of two electric circuits rotate with respect to each other and power is transferred from one circuit to the other by electromagnetic induction.

induction motor An alternating-current motor in which a primary winding on one member (usually the stator) is connected to the power source, and a secondary winding on the other member (usually the rotor) carries only current induced by the magnetic field of the primary.

induction problems The effects of potentials and currents induced in conductors of a telephone system by paralleling power facilities or power lines.

induction voltage regulator A type of transformer having a primary winding connected in parallel with a circuit and a secondary winding in series with the circuit; the relative positions of the primary and secondary windings are changed to vary the voltage or phase relations in the circuit.

induction watthour meter A watthour meter used with alternating current; the energy taken by a circuit over a period of time is proportional to the rotation in that period of a light aluminum disk, in which a driving torque is developed by the joint action of the alternating magnetic flux produced by the potential circuit and by the load current.

inductive charge The charge that exists on an object as a result of its being near another charged object.

inductive circuit A circuit containing a higher value of inductive reactance than capacitive reactance.

inductive coordination Measures to reduce induction problems.

inductive coupler A mutual inductance that provides electrical coupling between two circuits; used in radio equipment.

inductive coupling Coupling of two circuits by means of the mutual inductance provided by a transformer. Also known as transformer coupling.

inductive divider A device for incorporating a desired fraction of an inductance into a circuit, usually consisting of an autotransformer with an intermediate tap.

inductive feedback 1. Transfer of energy from the plate circuit to the grid circuit of a vacuum tube by means of induction. 2. Transfer of energy from the output circuit to the input circuit of an amplifying device through an inductor, or by means of inductive coupling.

inductive filter A low-pass filter used for smoothing the direct-current output voltage of a rectifier; consists of one or more sections in series, each section consisting of an inductor on one of the pair of conductors in series with a capacitor between the conductors. Also known as LC filter.

inductive grounding Use of grounding connections containing an inductance in order to reduce the magnitude of short-circuit currents created by line-to-ground faults.

inductive load A load that is predominantly inductive, so that the alternating load current lags behind the alternating voltage of the load. Also known as lagging load.

inductive neutralization Neutralizing an amplifier whereby the feedback susceptance due to an interelement capacitance is canceled by the equal and opposite susceptance of an inductor. Also known as coil neutralization; shunt neutralization.

inductive post Metal post or screw extending across a waveguide parallel to the E field, to add inductive susceptance in parallel with the waveguide for tuning or matching purposes.

inductive pressure transducer A type of pressure transducer in which changes in pressure cause a bourdon tube or other pressure-sensing element to move a magnetic core, and this results in a change in inductance of one or more windings of a coil surrounding the core.

inductive reactance Reactance due to the inductance of a coil or circuit.

inductive relay A relay displaying inductance, as opposed to one wound to be noninductive.

inductive spacing Spacing of parallel transmission lines so that there is transfer of energy by mutual inductance.

inductive surge A surge in voltage caused by sudden interruption of current in an inductive circuit.

inductive susceptance In a circuit containing almost no resistance, the part of the susceptance due to inductance.

inductive tuning Tuning involving the use of a variable inductance.

inductive waveform A graph or trace of the effect of current buildup across an inductive network; proportional to the exponential of the product of a negative constant and the time.

inductive window Conducting diaphragm extending into a waveguide from one or both sidewalls of the waveguide, to give the effect of an inductive susceptance in parallel with the waveguide.

inductometer A coil of wire of known inductance; the inductance may be fixed as in the case of primary standards, adjustable by means of switches, or continuously variable by means of a movable-coil construction.

inductor *See* coil.

inductor alternator A synchronous generator in which the field winding is fixed in magnetic position relative to the armature conductors.

inductor generator An alternating-current generator in which all the windings are fixed, and the flux linkages are varied by rotating an appropriately toothed ferromagnetic rotor; sometimes used for generating high power at frequencies up to several thousand hertz for induction heating.

inertia switch A switch that is actuated by an abrupt change in the velocity of the item on which it is mounted.

infinity transmitter A device used to tap a telephone; the telephone instrument is so modified that an interception device can be actuated from a distant source without the caller's becoming aware.

infradyne receiver A superheterodyne receiver in which the intermediate frequency is higher than the signal frequency, so as to obtain high selectivity.

infrared Pertaining to infrared radiation.

infrared absorption The taking up of energy from infrared radiation by a medium through which the radiation is passing.

infrared bolometer A bolometer adapted to detecting infrared radiation, as opposed to microwave radiation.

infrared communications set Components required to operate a two-way electronic system using infrared radiation to carry intelligence.

infrared detector A device responding to infrared radiation, used in detecting fires, or overheating in machinery, planes, vehicles, and people, and in controlling temperature-sensitive industrial processes.

infrared-emitting diode A light-emitting diode that has maximum emission in the near-infrared region, typically at 0.9 micrometer for *pn* gallium arsenide.

infrared heterodyne detector A heterodyne detector in which both the incoming signal and the local oscillator signal frequencies are in the infrared range and are combined in a photodetector to give an intermediate frequency in the kilohertz or megahertz range for conventional amplification.

infrared image converter A device for converting an invisible infrared image into a visible image, consisting of an infrared-sensitive, semitransparent photocathode on one end of an evacuated envelope and a phosphor screen on the other, with an electrostatic lens system between the two. Also known as infrared image tube.

infrared image tube *See* infrared image converter.

infrared jamming An attempt to confuse heat-seeking missiles by emissions which overload their inputs or misdirect them.

infrared lamp An incandescent lamp which operates at reduced voltage with a filament temperature of 4000°F (2200°C) so that it radiates electromagnetic energy primarily in the infrared region.

infrared optical material A material which is transparent to infrared radiation.

infrared phosphor A phosphor which, when exposed to infrared radiation during or even after decay of luminescence resulting from its usual or dominant activator, emits light having the same spectrum as that of the dominant activator; sulfide and selenide phosphors are the most important examples.

infrared photoconductor A conductor whose conductivity increases when it is exposed to infrared radiation.

infrared radiation Electromagnetic radiation whose wavelengths lie in the range from 0.75 or 0.8 micrometer (the long-wavelength limit of visible red light) to 1000 micrometers (the shortest microwaves).

infrared receiver A device that intercepts or demodulates infrared radiation that may carry intelligence. Also known as nancy receiver.

infrared scanner An infrared detector mounted on a motor-driven platform which causes it to scan a field of view line by line, much as in television.

infrared spectrum 1. The range of wavelengths of infrared radiation. 2. A display or graph of the intensity of infrared radiation emitted or absorbed by a material as a function of wavelength or some related parameter.

infrared thermistor A thermistor to measure the power of infrared radiation.

infrared transmitter A transmitter that emits energy in the infrared spectrum; may be modulated with intelligence signals.

infrared vidicon A vidicon whose photoconductor surface is sensitive to infrared radiation.

inhibit-gate Gate circuit whose output is energized only when certain signals are present and other signals are not present at the inputs.

inhibiting input A gate input which, if in its prescribed state, prevents any output which might otherwise occur.

inhibiting signal A signal, which when entered into a specific circuit will prevent the circuit from exercising its normal function; for example, an inhibit signal fed into an AND gate will prevent the gate from yielding an output when all normal input signals are present.

inhibit pulse A drive pulse that tends to prevent flux reversal of a magnetic cell by certain specified drive pulses.

initial inverse voltage Of a rectifier tube, the peak inverse anode voltage immediately following the conducting period.

initial surge voltage A spike of voltage experienced when a noncompensated load is first connected to a generator.

injection 1. The method of applying a signal to an electronic circuit or device. 2. The process of introducing electrons or holes into a semiconductor so that their total number exceeds the number present at thermal equilibrium.

injection efficiency A measure of the efficiency of a semiconductor junction when a forward bias is applied, equal to the current of injected minority carriers divided by the total current across the junction.

injection electroluminescence Radiation resulting from recombination of minority charge carriers injected in a *pn* or *pin* junction that is biased in the forward direction. Also known as Lossev effect; recombination electroluminescence.

injection grid Grid introduced into a vacuum tube in such a way that it exercises control over the electron stream without causing interaction between the screen grid and control grid.

injection locking The capture or synchronization of a free-running oscillator by a weak injected signal at a frequency close to the natural oscillator frequency or to one of its subharmonics; used for frequency stabilization in IMPATT or magnetron microwave oscillators, gas-laser oscillators, and many other types of oscillators.

injection luminescent diode Gallium arsenide diode, operating in either the laser or the noncoherent mode, that can be used as a visible or near-infrared light source for triggering such devices as light-activated switches.

injector An electrode through which charge carriers (holes or electrons) are forced to enter the high-field region in a spacistor.

in-line guns An arrangement of three electron guns in a horizontal line; used in color picture tubes that have a slot mask in front of vertical color phosphor stripes.

in-line tuning Method of tuning the intermediate-frequency strip of a superheterodyne receiver in which all the intermediate-frequency amplifier stages are made resonant to the same frequency.

in-phase component The component of the phasor representing an alternating current which is parallel to the phasor representing voltage.

in-phase rejection *See* common-mode rejection.

in-phase signal *See* common-mode signal.

input 1. The power or signal fed into an electrical or electronic device. 2. The terminals to which the power or signal is applied.

input admittance The admittance measured across the input terminals of a four-terminal network with the output terminals short-circuited.

input capacitance The short-circuited transfer capacitance that exists between the input terminals and all other terminals of an electron tube (except the output terminal) connected together.

input gap An interaction gap used to initiate a variation in an electron stream; in a velocity-modulated tube it is in the buncher resonator.

input impedance The impedance across the input terminals of a four-terminal network when the output terminals are short-circuited.

input resistance *See* transistor input resistance.

insertion gain The ratio of the power delivered to a part of the system following insertion of an amplifier, to the power delivered to that same part before insertion of the amplifier; usually expressed in decibels.

insertion loss The loss in load power due to the insertion of a component or device at some point in a transmission system; generally expressed as the ratio in decibels of the power received at the load before insertion of the apparatus, to the power received at the load after insertion.

installed capacity The maximum runoff of a hydroelectric facility that can be constantly maintained and utilized by equipment.

instantaneous automatic gain control Portion of a radar system that automatically adjusts the gain of an amplifier for each pulse to obtain a substantially constant output-pulse peak amplitude with different input-pulse peak amplitudes; the circuit is fast enough to act during the time a pulse is passing through the amplifier.

instantaneous carrying current The maximum value of current which a switch, circuit breaker, or similar apparatus can carry instantaneously.

instantaneous companding Companding in which the effective gain variations are made in response to instantaneous values of the signal wave.

instantaneous frequency-indicating receiver A radio receiver with a digital, cathode-ray, or other display that shows the frequency of a signal at the instant it is picked up anywhere in the band covered by the receiver.

instantaneous power The product of the instantaneous voltage and the instantaneous current for a circuit or component.

instant-on switch A switch that applies a reduced filament voltage to all tubes in a television receiver continuously, so the picture appears almost instantaneously after the set is turned on.

instant replay *See* video replay.

instrumentation amplifier An amplifier that accepts a voltage signal as an input and produces a linearly scaled version of this signal at the output; it is a closed-loop fixed-gain amplifier, usually differential, and has high input impedance, low drift, and high common-mode rejection over a wide range of frequencies.

instrument multiplier A highly accurate resistor used in series with a voltmeter to extend its voltage range. Also known as voltage multiplier; voltage-range multiplier.

instrument resistor A high-accuracy, four-terminal resistor used to bypass the major portion of currents around the low-current elements of an instrument, such as a direct-current ammeter.

instrument shunt A resistor designed to be connected in parallel with an ammeter to extend its current range.

instrument transformer A transformer that transfers primary current, voltage, or phase values to the secondary circuit with sufficient accuracy to permit connecting an instrument to the secondary rather than the primary; used so only low currents or low voltages are brought to the instrument.

instrument-type relay A relay constructed like a meter, with one adjustable contact mounted on the scale and the other contact mounted on the pointer. Also known as contact-making meter.

insulated Separated from other conducting surfaces by a nonconducting material.

insulated conductor A conductor surrounded by insulation to prevent current leakage or short circuits. Also known as insulated wire.

insulated-gate field-effect transistor *See* metal oxide semiconductor field-effect transistor.

insulated-return power system A system for distributing electric power to trains or other vehicles, in which both the outgoing and return conductors are insulated, in contrast to a track-return system.

insulated-substrate monolithic circuit Integrated circuit which may be either an all-diffused device or a compatible structure so constructed that the components within the silicon substrate are insulated from one another by a layer of silicon dioxide, instead of reverse-biased *pn* junctions used for isolation in other techniques.

insulated wire *See* insulated conductor.

insulating strength Measure of the ability of an insulating material to withstand electric stress without breakdown; it is defined as the voltage per unit thickness necessary to initiate a disruptive discharge; usually measured in volts per centimeter.

insulation A material having high electrical resistivity and therefore suitable for separating adjacent conductors in an electric circuit or preventing possible future contact between conductors. Also known as electrical insulation.

insulation coordination Steps taken to ensure that electric equipment is not damaged by overvoltages and that flashovers are localized in regions where no damage results from them.

insulation protection Use of devices to protect insulators of power transmission lines from damage by heavy arcs.

insulation resistance The electrical resistance between two conductors separated by an insulating material.

insulator 1. A device having high electrical resistance and used for supporting or separating conductors to prevent undesired flow of current from them to other objects. Also known as electrical insulator. 2. A substance in which the normal energy band is full and is separated from the first excitation band by a forbidden band that can be penetrated only by an electron having an energy of several electron volts, sufficient to disrupt the substance.

insulator arc-over Discharge of power current in the form of an arc, following a surface discharge over an insulator.

insulator arrangement The placement of insulators on a transmission mast.

integral discriminator A circuit which accepts only pulses greater than a certain minimum height.

integrated circuit An interconnected array of active and passive elements integrated with a single semiconductor substrate or deposited on the substrate by a continuous series of compatible processes, and capable of performing at least one complete electronic circuit function. Abbreviated IC. Also known as integrated semiconductor.

integrated-circuit capacitor A capacitor that can be produced in a silicon substrate by conventional semiconductor production processes.

integrated-circuit resistor A resistor that can be produced in or on an integrated-circuit substrate as part of the manufacturing process.

integrated electronics A generic term for that portion of electronic art and technology in which the interdependence of material, device, circuit, and system-design consideration is especially significant; more specifically, that portion of the art dealing with integrated circuits.

integrated injection logic Integrated-circuit logic that uses a simple and compact bipolar transistor gate structure which makes possible large-scale integration on silicon for logic arrays, memories, watch circuits, and various other analog and digital applications. Abbreviated I^2L. Also known as merged-transistor logic.

integrating amplifier An operational amplifier with a shunt capacitor such that mathematically the waveform at the output is the integral (usually over time) of the input.

integrating detector A frequency-modulation detector in which a frequency-modulated wave is converted to an intermediate-frequency pulse-rate modulated wave, from which the original modulating signal can be recovered by use of an integrator.

integrating filter A filter in which successive pulses of applied voltage cause cumulative buildup of charge and voltage on an output capacitor.

integrating network A circuit or network whose output waveform is the time integral of its input waveform. Also known as integrator.

integrator 1. A computer device that approximates the mathematical process of integration. 2. *See* integrating network.

intensifier electrode An electrode used to increase the velocity of electrons in a beam near the end of their trajectory, after deflection of the beam. Also known as postaccelerating electrode; postdeflection accelerating electrode.

intensifier image orthicon An image orthicon combined with an image intensifier that amplifies the electron stream originating at the photocathode before it strikes the target.

intensity control *See* brightness control.

intensity modulation Modulation of electron beam intensity in a cathode-ray tube in accordance with the magnitude of the received signal.

intensity of magnetization *See* intrinsic induction.

interaction space A region of an electron tube in which electrons interact with an alternating electromagnetic field.

interbase current The current that flows from one base connection of a junction tetrode transistor to the other, through the base region.

intercarrier noise suppression Means of suppressing the noise resulting from increased gain when a high-gain receiver with automatic volume control is tuned between stations; the suppression circuit automatically blocks the audio-frequency input of the receiver when no signal exists at the second detector. Also known as interstation noise suppression.

intercarrier sound system A television receiver arrangement in which the television picture carrier and the associated sound carrier are amplified together by the video intermediate-frequency amplifier and passed through the second detector, to give the conventional video signal plus a frequency-modulated sound signal whose center frequency is the 4.5 megahertz difference between the two carrier frequencies. Abbreviated ICS system.

interchange The current flowing into or out of a power system which is interconnected with one or more other power systems.

interconnection A link between power systems enabling them to draw on one another's reserves in time of need and to take advantage of energy cost differentials resulting from such factors as load diversity, seasonal conditions, time-zone differences, and shared investment in larger generating units.

interdigital magnetron Magnetron having axial anode segments around the cathode, alternate segments being connected together at one end, remaining segments connected together at the opposite end.

interdigital structure A structure in which the length of the region between two electrodes is increased by an interlocking-finger design for metallization of the electrodes. Also known as interdigitated structure.

interdigital transducer Two interlocking comb-shaped metallic patterns applied to a piezoelectric substrate such as quartz or lithium niobate, used for converting microwave voltages to surface acoustic waves, or vice versa.

interdigitated structure *See* interdigital structure.

interelectrode capacitance The capacitance between one electrode of an electron tube and the next electrode on the anode side. Also known as direct interelectrode capacitance.

interelectrode transit time Time required for an electron to traverse the distance between the two electrodes.

interface connection *See* feedthrough.

interfacial polarization *See* space-charge polarization.

interference analyzer An instrument that discloses the frequency and amplitude of unwanted input.

interference blanker Device that permits simultaneous operation of two or more pieces of radio or radar equipment without confusion of intelligence, or that suppresses undesired signals when used with a single receiver.

intermediate-frequency signal 223

interference filter 1. A filter used to attenuate artificial interference signals entering a receiver through its power line. 2. A filter used to attenuate unwanted carrier-frequency signals in the tuned circuits of a receiver.

interference pattern Pattern produced on a radarscope by interference signals.

interference prediction Process of estimating the interference level of a particular equipment as a function of its future electromagnetic environment.

interference reduction Reduction of interference from such causes as power lines and equipment, radio transmitters, and lightning, usually through the use of electric filters. Also known as interference suppression.

interference rejection Use of a filter to reject (to bypass to ground) unwanted input.

interference source suppression Techniques applied at or near the source to reduce its emission of undesired signals.

interference spectrum Frequency distribution of the jamming interference in the propagation medium external to the receiver.

interference suppression *See* interference reduction.

interferometer systems Method of determining the position of a target in azimuth by using an interferometer to compare the phases of signals at the output terminals of a pair of antennas receiving a common signal from a distant source.

interior distribution Distribution of electric power within a building or plant.

interlaced scanning A scanning process in which the distance from center to center of successively scanned lines is two or more times the nominal line width, so that adjacent lines belong to different fields. Also known as line interlace.

interleaved windings An arrangement of winding coils around a transformer core in which the coils are wound in the form of a disk, with a group of disks for the low-voltage windings stacked alternately with a group of disks for the high-voltage windings.

interlock relay A relay composed of two or more coils, each with its own armature and associated contacts, so arranged that movement of one armature or the energizing of its coil is dependent on the position of the other armature.

interlock switch A switch designed for mounting on a door, drawer, or cover so that it opens automatically when the door or other part is opened.

intermediate distributing frame Frame in a local telephone central office, the primary purpose of which is to cross-connect the subscriber line multiple to the subscriber line circuit; in a private exchange, the intermediate distributing frame is for similar purposes.

intermediate frequency The frequency produced by combining the received signal with that of the local oscillator in a superheterodyne receiver. Abbreviated i-f.

intermediate-frequency amplifier The section of a superheterodyne receiver that amplifies signals after they have been converted to the fixed intermediate-frequency value by the frequency converter. Abbreviated i-f amplifier.

intermediate-frequency jamming Form of continuous wave jamming that is accomplished by transmitting two continuous wave signals separated by a frequency equal to the center frequency of the radar receiver intermediate-frequency amplifier.

intermediate-frequency response ratio In a superheterodyne receiver, the ratio of the intermediate-frequency signal input at the antenna to the desired signal input for identical outputs. Also known as intermediate-interference ratio.

intermediate-frequency signal A modulated or continuous-wave signal whose frequency is the intermediate-frequency value of a superheterodyne receiver and is produced by frequency conversion before demodulation.

intermediate-frequency stage One of the stages in the intermediate-frequency amplifier of a superheterodyne receiver.

intermediate-frequency strip A receiver subassembly consisting of the intermediate-frequency amplifier stages, installed or replaced as a unit.

intermediate-frequency transformer The transformer used at the input and output of each intermediate-frequency amplifier stage in a superheterodyne receiver for coupling purposes and to provide selectivity. Abbreviated i-f transformer.

intermediate horizon Screening object (hill, mountain, ridge, building, and so on) similar to the radar horizon, but not the most distant.

intermediate-infrared radiation Infrared radiation having a wavelength between about 2.5 micrometers and about 50 micrometers; this range includes most molecular vibrations.

intermediate-interference ratio *See* intermediate-frequency response ratio.

intermediate repeater Repeater for use in a trunk or line at a point other than an end.

intermediate trunk distributing frame A frame which mounts terminal blocks for connecting linefinders and first selectors.

intermittent current A unidirectional current that flows and ceases to flow at irregular or regular intervals.

intermittent scanning Scans of an antenna beam at irregular intervals to increase difficulty of detection by intercept receivers.

intermodulation Modulation of the components of a complex wave by each other, producing new waves whose frequencies are equal to the sums and differences of integral multiples of the component frequencies of the original complex wave.

intermodulation distortion Nonlinear distortion characterized by the appearance of output frequencies equal to the sums and differences of integral multiples of the input frequency components; harmonic components also present in the output are usually not included as part of the intermodulation distortion.

intermodulation interference Interference that occurs when the signals from two undesired stations differ by exactly the intermediate-frequency value of a superheterodyne receiver, and both signals are able to pass through the preselector due to poor selectivity.

internal photoelectric effect A process in which the absorption of a photon in a semiconductor results in the excitation of an electron from the valence band to the conduction band.

internal resistance The resistance within a voltage source, such as an electric cell or generator.

international ampere The current that, when flowing through a solution of silver nitrate in water, deposits silver at a rate of 0.001118 gram per second; it has been superseded by the ampere as a unit of current, and is equal to approximately 0.999850 ampere.

international henry A unit of electrical inductance which has been superseded by the henry, and is equal to 1.00049 henry. Also known as quadrant; secohm.

international ohm A unit of resistance, equal to that of a column of mercury of uniform cross section that has a length of 160.3 centimeters and a mass of 14.4521 grams at the temperature of melting ice; it has been superseded by the ohm, and is equal to 1.00049 ohms.

international system of electrical units System of electrical units based on agreed fundamental units for the ohm, ampere, centimeter, and second, in use between 1893 and 1947, inclusive; in 1948, the Giorgi, or meter-kilogram-second-absolute system, was adopted for international use.

intrinsic induction 225

international volt A unit of potential difference or electromotive force, equal to 1/1.01858 of the electromotive force of a Weston cell at 20°C; it has been superseded by the volt, and is equal to 1.00034 volts.

interphase reactor A type of current-equalizing reactor that is connected between two parallel silicon controlled rectifier converters and provides balanced system operation when both converters are conducting by acting as an inductive voltage divider.

interphase transformer Autotransformer or a set of mutually coupled reactors used in conjunction with three-phase rectifier transformers to modify current relations in the rectifier system to increase the number of anodes of different phase relations which carry current at any instant.

interpole *See* commutating pole.

interrogator 1. A radar transmitter which sends out a pulse that triggers a transponder; usually combined in a single unit with a responsor, which receives the reply from a transponder and produces an output suitable for actuating a display of some navigational parameter. Also known as challenger; interrogator-transmitter. 2. *See* interrogator-responsor.

interrogator-responsor A transmitter and receiver combined, used for sending out pulses to interrogate a radar beacon and for receiving and displaying the resulting replies. Also known as interrogator.

interrogator-transmitter *See* interrogator.

interrupted current A current produced by opening and closing at regular intervals a circuit that would otherwise carry a steady current or one that varied continuously with time.

interrupter An electric, electronic, or mechanical device that periodically interrupts the flow of a direct current so as to produce pulses.

interrupter vibrator A mechanical device used to change direct current to alternating current.

interrupting capacity Maximum power in the arc that a circuit breaker or fuse can successfully interrupt without restrike or violent failure; rated in volt-amperes for alternating-current circuits and watts for direct-current circuits.

interstage transformer A transformer used to provide coupling between two stages.

interstation noise suppression *See* intercarrier noise suppression.

interstice A space or volume between atoms of a lattice, or between groups of atoms or grains of a solid structure.

interstitial impurity An atom which is not normally found in a solid, and which is located at a position in the lattice structure where atoms or ions normally do not exist.

intrinsic-barrier diode A *pin* diode, in which a thin region of intrinsic material separates the *p*-type region and the *n*-type region.

intrinsic-barrier transistor A *pnip* or *npin* transistor, in which a thin region of intrinsic material separates the base and collector.

intrinsic conductivity The conductivity of a semiconductor or metal in which impurities and structural defects are absent or have a very low concentration.

intrinsic contact potential difference True potential difference between two perfectly clean metals in contact.

intrinsic flux density *See* intrinsic induction.

intrinsic induction The vector difference between the magnetic flux density at a given point and the magnetic flux density which would exist there, for the same magnetic field strength, if the point were in a vacuum. Symbol Bi. Also known as intensity of magnetization; intrinsic flux density; magnetic polarization

intrinsic layer A layer of semiconductor material whose properties are essentially those of the pure undoped material.

intrinsic mobility The mobility of the electrons in an intrinsic semiconductor.

intrinsic photoconductivity Photoconductivity associated with excitation of charge carriers across the band gap of a material.

intrinsic photoemission Photoemission which can occur in an ideally pure and perfect crystal, in contrast to other types of photoemission which are associated with crystal defects.

intrinsic semiconductor A semiconductor in which the concentration of charge carriers is characteristic of the material itself rather than of the content of impurities and structural defects of the crystal. Also known as i-type semiconductor.

intrinsic temperature range In a semiconductor, the temperature range in which its electrical properties are essentially not modified by impurities or imperfections within the crystal.

inverse current The current resulting from an inverse voltage in a contact rectifier.

inverse direction The direction in which the electron flow encounters greater resistance in a rectifier, going from the positive to the negative electrode; the opposite of the conducting direction. Also known as reverse direction.

inverse electrode current Current flowing through an electrode in the direction opposite to that for which the tube is designed.

inverse limiter A transducer, the output of which is constant for input of instantaneous values within a specified range and a linear or other prescribed function of the input for inputs above and below that range.

inverse network Two two-terminal networks are said to be inverse when the product of their impedances is independent of frequency within the range of interest.

inverse peak voltage 1. The peak value of the voltage that exists across a rectifier tube or x-ray tube during the half cycle in which current does not flow. 2. The maximum instantaneous voltage value that a rectifier tube or x-ray tube can withstand in the inverse direction (with anode negative) without breaking down and becoming conductive.

inverse piezoelectric effect The contraction or expansion of a piezoelectric crystal under the influence of an electric field, as in crystal headphones; also occurs at pn junctions in some semiconductor materials.

inverse voltage The voltage that exists across a rectifier tube or x-ray tube during the half cycle in which the anode is negative and current does not normally flow.

inversion 1. The solution of certain problems in electrostatics through the use of the transformation in Kelvin's inversion theorem. 2. The production of a layer at the surface of a semiconductor which is of opposite type from that of the bulk of the semiconductor, usually as the result of an applied electric field.

inverted amplifier A two-tube amplifier in which the control grids are grounded and the input signal is applied between the cathodes; the grid then serves as a shield between the input and output circuits.

inverted L antenna An antenna consisting of one or more horizontal wires to which a connection is made by means of a vertical wire at one end.

inverted vee 1. A directional antenna consisting of a conductor which has the form of an inverted V, and which is fed at one end and connected to ground through an

ionization time   227

appropriate termination at the other. 2. A center-fed horizontal dipole antenna whose arms have ends bent downward 45°.

inverter 1. A device for converting direct current into alternating current; it may be electromechanical, as in a vibrator or synchronous inverter, or electronic, as in a thyratron inverter circuit. Also known as dc-to-ac converter; dc-to-ac inverter. 2. *See* phase inverter.

inverter circuit *See* NOT circuit.

inverting amplifier Amplifier whose output polarity is reversed as compared to its input; such an amplifier obtains its negative feedback by a connection from output to input, and with high gain is widely used as an operational amplifier.

inverting function A logic device that inverts the input signal, so that the output is out of phase with the input.

inverting parametric device Parametric device whose operation depends essentially upon three frequencies, a harmonic of the pump frequency and two signal frequencies, of which the higher signal frequency is the difference between the pump harmonic and the lower signal frequency.

inverting terminal The negative input terminal of an operational amplifier; a positive-going voltage at the inverting terminal gives a negative-going output voltage.

ion backscattering Large-angle elastic scattering of monoenergetic ions in a beam directed at a metallized film on silicon or some other thin multilayer system.

ion-beam scanning The process of analyzing the mass spectrum of an ion beam in a mass spectrometer either by changing the electric or magnetic fields of the mass spectrometer or by moving a probe.

ion burn *See* ion spot.

ion-exchange electrolyte cell Fuel cell which operates on hydrogen and oxygen in the air, similar to the standard hydrogen-oxygen fuel cell with the exception that the liquid electrolyte is replaced by an ion-exchange membrane; operation is at atmospheric pressure and room temperature.

ion gage *See* ionization gage.

ion gun *See* ion source.

ionic conduction Electrical conduction of a solid due to the displacement of ions within the crystal lattice.

ionic focusing *See* gas focusing.

ionic-heated cathode Hot cathode heated primarily by ionic bombardment of the emitting surface.

ionic semiconductor A solid whose electrical conductivity is due primarily to the movement of ions rather than that of electrons and holes.

ionization arc-over 1. Arcing across terminals or contacts due to ionization of the adjacent air or gas. 2. Arcing across satellite antenna terminals as the satellite passes through the ionized regions of the ionosphere.

ionization current *See* gas current.

ionization density The density of ions in a gas.

ionization gage An instrument for measuring low gas densities by ionizing the gas and measuring the ion current. Also known as ion gage; ionization vacuum gage.

ionization source *See* ion source.

ionization time Of a gas tube, the time interval between the initiation of conditions for and the establishment of conduction at some stated value of tube voltage drop.

ionization vacuum gage

ionization vacuum gage See ionization gage.

ion microscope See field-ion microscope.

ion migration Movement of ions produced in an electrolyte, semiconductor, and so on, by the application of an electric potential between electrodes.

ionospheric recorder A radio device for determining the distribution of virtual height with frequency, and the critical frequencies of the various layers of the ionosphere.

ionospheric wave See sky wave.

ion pump A vacuum pump in which gas molecules are first ionized by electrons that have been generated by a high voltage and are spiraling in a high-intensity magnetic field, and the molecules are then attracted to a cathode, or propelled by electrodes into an auxiliary pump or an ion trap.

ion source A device in which gas ions are produced, focused, accelerated, and emitted as a narrow beam. Also known as ion gun; ionization source.

ion spot Of a cathode-ray tube screen, an area of localized deterioration of luminescence caused by bombardment with negative ions. Also known as ion burn.

ion trap 1. An arrangement whereby ions in the electron beam of a cathode-ray tube are prevented from bombarding the screen and producing an ion spot, usually employing a magnet to bend the electron beam so that it passes through the tiny aperture of the electron gun, while the heavier ions are less affected by the magnetic field and are trapped inside the gun. 2. A metal electrode, usually of titanium, into which ions in an ion pump are absorbed.

IR drop See resistance drop.

iris A conducting plate mounted across a waveguide to introduce impedance; when only a single mode can be supported, an iris acts substantially as a shunt admittance and may be used for matching the waveguide impedance to that of a load. Also known as diaphragm; waveguide window.

I^2R loss See copper loss.

iron core A core made of solid or laminated iron, or some other magnetic material which may contain very little iron.

iron-core choke See iron-core coil.

iron-core coil A coil in which solid or laminated iron or other magnetic material forms part or all of the magnetic circuit linking its winding. Also known as iron-core choke; magnet coil.

iron-core transformer A transformer in which laminations of iron or other magnetic material make up part or all of the path for magnetic lines of force that link the transformer windings.

iron-dust core A core made by mixing finely powdered magnetic material with an insulating binder and molding under pressure to form a rod-shaped core that can be moved into or out of a coil or transformer to vary the inductance or degree of coupling for tuning purposes.

iron loss See core loss.

irradiance See radiant flux density.

I scan See I scope.

I scope A cathode-ray scope on which a single signal appears as a circular segment whose radius is proportional to the range and whose circular length is inversely proportional to the error of aiming the antenna, true aim resulting in a complete circle; the position of the arc, relative to the center, indicates the position of the target relative to the beam axis. Also known as broken circle indicator; I indicator; I scan.

I signal The in-phase component of the chrominance signal in color television, having a bandwidth of 0 to 1.5 megahertz, and consisting of $+0.74(R - Y)$ and $-0.27(B - Y)$, where Y is the luminance signal, R is the red camera signal, and B is the blue camera signal.

Ising coupling A model of coupling between two atoms in a lattice, used to study ferromagnetism, in which the spin component of each atom along some axis is taken to be $+1$ or -1, and the energy of interaction is proportional to the negative of the product of the spin components along this axis.

Ising model A crude model of a ferromagnetic material or an analogous system, used to study phase transitions, in which atoms in a one-, two-, or three-dimensional lattice interact via Ising coupling between nearest neighbors, and the spin components of the atoms are coupled to a uniform magnetic field.

isochronous circuits Circuits having the same resonant frequency.

isoclinic line A line joining points in a plate at which the principal stresses have parallel directions.

isoelectric Pertaining to a constant electric potential.

isograph An electronic calculator that ascertains both real and imaginary roots for algebraic equations.

isolate To disconnect a circuit or piece of equipment from an electric supply system.

isolated camera 1. A television camera that views a particular portion of a scene of action and produces a tape which can then be used either immediately for instant replay or for video replay at a later time. 2. The technique of video replay involving such a camera.

isolating switch A switch intended for isolating an electric circuit from the source of power; it has no interrupting rating and is intended to be operated only after the circuit has been opened by some other means.

isolation amplifier An amplifier used to minimize the effects of a following circuit on the preceding circuit.

isolation diode A diode used in a circuit to allow signals to pass in only one direction.

isolation network A network inserted in a circuit or transmission line to prevent interaction between circuits on each side of the insertion point.

isolation transformer A transformer inserted in a system to separate one section of the system from undesired influences of other sections.

isolator A passive attenuator in which the loss in one direction is much greater than that in the opposite direction; a ferrite isolator for waveguides is an example.

isolith Integrated circuit of components formed on a single silicon slice, but with the various components interconnected by beam leads and with circuit parts isolated by removal of the silicon between them.

isotope effect Variation of the transition temperatures of the isotopes of a superconducting element in inverse proportion to the square root of the atomic mass.

isotropic antenna *See* unipole.

isotropic dielectric A dielectric whose polarization always has a direction that is parallel to the applied electric field, and a magnitude which does not depend on the direction of the electric field.

isotropic gain of an antenna *See* absolute gain of an antenna.

isotropic noise Random noise radiation which reaches a location from all directions with equal intensity.

isotropic radiation Radiation which is emitted by a source in all directions with equal intensity, or which reaches a location from all directions with equal intensity.

iterative filter Four-terminal filter that provides iterative impedance.

iterative impedance Impedance that, when connected to one pair of terminals of a four-terminal transducer, will cause the same impedance to appear between the other two terminals.

i-type semiconductor *See* intrinsic semiconductor.

J

Jablochkoff candle An early type of arc lamp in which carbons were placed side by side and separated by plaster of Paris.

jack A connecting device into which a plug can be inserted to make circuit connections; may also have contacts that open or close to perform switching functions when the plug is inserted or removed.

jammer A transmitter used in jamming of radio or radar transmissions. Also known as electronic jammer.

jammer finder Radar which attempts to obtain the range of the target by training a highly directional pencil beam on a jamming source. Also known as burn-through.

jamming Radiation or reradiation of electromagnetic waves so as to impair the usefulness of a specific segment of the radio spectrum that is being used by the enemy for communication or radar. Also known as active jamming; electronic jamming.

J antenna Antenna having a configuration resembling a J, consisting of a half-wave antenna end-fed by a parallel-wire quarter-wave section.

jar A unit of capacitance equal to 1000 statfarads, or approximately 1.11265×10^{-9} farad; it is approximately equal to the capacitance of a Leyden jar; this unit is now obsolete.

J display A modified radarscope A display in which the time base is a circle; the target signal appears as an outward radial deflection from the time base.

JFET *See* junction field-effect transistor.

J indicator *See* J scope.

jitter Small, rapid variations in a waveform due to mechanical vibrations, fluctuations in supply voltages, control-system instability, and other causes.

J-K flip-flop A storage stage consisting only of transistors and resistors connected as flip-flops between input and output gates, and working with charge-storage transistors; gives a definite output even when both inputs are 1.

jogging Quickly repeated opening and closing of a circuit to produce small movements of the driven machine. Also known as inching.

Johnson and Lark-Horowitz formula A formula according to which the resistivity of a metal or degenerate semiconductor resulting from impurities which scatter the electrons is proportional to the cube root of the density of impurities.

Johnson noise *See* thermal noise.

joint A juncture of two wires or other conductive paths for current.

joint pole Pole used in common by two or more utility companies.

Joule heat The heat which is evolved when current flows through a medium having electrical resistance, as given by Joule's law.

Joule's law The law that when electricity flows through a substance, the rate of evolution of heat in watts equals the resistance of the substance in ohms times the square of the current in amperes.

J scan *See* J scope.

J scope A modification of an A scope in which the trace appears as a circular range scale near the circumference of the cathode-ray tube face, the signal appearing as a radial deflection of the range scale; no bearing indication is given. Also known as J indicator; J scan.

jumper A short length of conductor used to make a connection between two points or terminals in a circuit or to provide a path around a break in a circuit.

junction 1. A region of transition between two different semiconducting regions in a semiconductor device, such as a *pn* junction, or between a metal and a semiconductor. 2. A fitting used to join a branch waveguide at an angle to a main waveguide, as in a tee junction. Also known as waveguide junction. 3. *See* major node.

junction capacitor An integrated-circuit capacitor that uses the capacitance of a reverse-biased *pn* junction.

junction diode A semiconductor diode in which the rectifying characteristics occur at an alloy, diffused, electrochemical, or grown junction between *n*-type and *p*-type semiconductor materials. Also known as junction rectifier.

junction field-effect transistor A field-effect transistor in which there is normally a channel of relatively low-conductivity semiconductor joining the source and drain, and this channel is reduced and eventually cut off by junction depletion regions, reducing the conductivity, when a voltage is applied between the gate electrodes. Abbreviated JFET. Also known as depletion-mode field-effect transistor.

junction isolation Electrical isolation of a component on an integrated circuit by surrounding it with a region of a conductivity type that forms a junction, and reverse-biasing the junction so it has extremely high resistance.

junction phenomena Phenomena which occur at the boundary between two semiconductor materials, or a semiconductor and a metal, such as the existence of an electrostatic potential in the absence of current flow, and large injection currents which may arise when external voltages are applied across the junction in one direction.

junction point *See* branch point.

junction pole Pole at the end of a transposition section of an open-wire line or the pole common to two adjacent transposition sections.

junction rectifier *See* junction diode.

junction station Microwave relay station that joins a microwave radio leg or legs to the main or through route.

junction transistor A transistor in which emitter and collector barriers are formed between semiconductor regions of opposite conductivity type.

junction transposition Transposition located at the junction pole between two transposition sections of an open-wire line.

junctor In crossbar systems, a circuit extending between frames of a switching unit and terminating in a switching device on each frame.

K

K *See* cathode.

kA *See* kiloampere.

kΩ *See* kilohm.

Karnaugh map A truth table that has been rearranged to show a geometrical pattern of functional relationships for gating configurations; with this map, essential gating requirements can be recognized in their simplest form.

Karp circuit A slow-wave circuit used at millimeter wavelengths for backward-wave oscillators.

K band An optical absorption band which appears together with an F-band and has a lower intensity and shorter wavelength than the latter.

K display A modified radarscope A display in which a target appears as a pair of vertical deflections instead of as a single deflection; when the radar antenna is correctly pointed at the target in azimuth, the deflections are of equal height; when the antenna is not correctly pointed, the difference in pulse heights is an indication of direction and magnitude of azimuth pointing error.

keep-alive circuit A circuit used with a transmit-receive (TR) tube or anti-TR tube to produce residual ionization for the purpose of reducing the initiation time of the main discharge.

keep-alive electrode *See* ignitor.

keeper A bar of iron or steel placed across the poles of a permanent magnet to complete the magnetic circuit when the magnet is not in use, to avoid the self-demagnetizing effect of leakage lines. Also known as magnet keeper.

kelvin A name formerly given to the kilowatt-hour. Also known as thermal volt.

Kelvin balance An ammeter in which the force between two coils in series that carry the current to be measured, one coil being attached to one arm of a balance, is balanced against a known weight at the other end of the balance arm.

Kelvin bridge A specialized version of the Wheatstone bridge network designed to eliminate, or greatly reduce, the effect of lead and contact resistance, and thus permit accurate measurement of low resistance. Also known as double bridge; Kelvin network; Thomson bridge.

Kelvin guard-ring capacitor A capacitor with parallel circular plates, one of which has a guard ring separated from the plate by a narrow gap; it is used as a standard, whose capacitance can be accurately calculated from its dimensions.

Kelvin network *See* Kelvin bridge.

Kelvin's formula *See* Thomson formula.

kenotron A high-vacuum diode deigned to serve as a rectifier in appliances requiring high voltage and low current.

key 1. A hand-operated switch used for transmitting code signals. Also known as signaling key. **2.** A special lever-type switch used for opening or closing a circuit only as long as the handle is depressed. Also known as switching key.

keyboard send/receive A manual teleprinter that can transmit or receive. Abbreviated KSR. Also known as keyboard teleprinter.

keyboard teleprinter *See* keyboard send/receive.

key cabinet A case, installed on a customer's premises, to permit different lines to the control office to be connected to various telephone stations; it has signals to indicate originating calls and busy lines.

keyed clamp Clamping circuit in which the time of clamping is determined by a control signal.

keyed clamp circuit A clamp circuit in which the time of clamping is controlled by separate voltage or current sources, rather than by the signal itself. Also known as synchronous clamp circuit.

keyer Device which changes the output of a transmitter from one condition to another according to the intelligence to be transmitted.

keyer adapter Device which detects a modulated signal and produces the modulating frequency as a direct-current signal of varying amplitude.

keying The forming of signals, such as for telegraph transmission, by modulating a direct-current or other carrier between discrete values of some characteristic.

keying wave *See* marking wave.

keylock switch A switch that can be operated only by inserting and turning a key such as that used in ordinary locks.

keystoning Producing a keystone-shaped (wider at the top than at the bottom, or vice versa) scanning pattern because the electron beam in the television camera tube is at an angle with the principal axis of the tube.

kidney joint Flexible joint, or air-gap coupling, used in the waveguide of certain radars and located near the transmitting-receiving position.

killer circuit Vaccum tube or tubes and associated circuits in which are generated the blanking pulses used to temporarily disable a radar set.

killer pulse Blanking pulse generated by a killer circuit.

killer stage *See* color killer circuit.

kiloampere A metric unit of current flow equal to 1000 amperes. Abbreviated kA.

kilohm A unit of electrical resistance equal to 1000 ohms. Abbreviated kΩ; kohm.

kilovar A unit equal to 1000 volt-amperes reactive. Abbreviated kvar.

kilovolt A unit of potential difference equal to 1000 volts. Abbreviated kV.

kilovolt-ampere A unit of apparent power in an alternating-current circuit, equal to 1000 volt-amperes. Abbreviated kVA.

kilovoltmeter A voltmeter which measures potential differences on the order of several kilovolts.

kilovolts peak The peak voltage applied to an x-ray tube, expressed in kilovolts. Abbreviated kVp.

kilowatt-hour A unit of energy or work equal to 1000 watt-hours. Abbreviated kWh; kW-hr. Also known as Board of Trade Unit.

K indicator *See* K scope.

kinescope See picture tube.

Kirchhoff's current law The law that at any given instant the sum of the instantaneous values of all the currents flowing toward a point is equal to the sum of instantaneous values of all the currents flowing away from the point. Also known as Kirchhoff's first law.

Kirchhoff's first law See Kirchhoff's current law.

Kirchhoff's law Either of the two fundamental laws dealing with the relation of currents at a junction and voltages around closed loops in an electric network; they are known as Kirchhoff's current law and Kirchhoff's voltage law.

Kirchhoff's second law See Kirchhoff's voltage law.

Kirchhoff's voltage law The law that at each instant of time the algebraic sum of the voltage rises around a closed loop in a network is equal to the algebraic sum of the voltage drops, both being taken in the same direction around the loop. Also known as Kirchhoff's second law.

klystron An evacuated electron-beam tube in which an initial velocity modulation imparted to electrons in the beam results subsequently in density modulation of the beam; used as an amplifier in the microwave region or as an oscillator.

klystron generator Klystron tube used as a generator, with its second cavity or catcher directly feeding waves into a waveguide.

klystron oscillator See velocity-modulated oscillator.

klystron repeater Klystron tube operated as an amplifier and inserted directly in a waveguide in such a way that incoming waves velocity-modulate the electron stream emitted from a heated cathode; a second cavity converts the energy of the electron clusters into waves of the original type but of greatly increased amplitude and feeds them into the outgoing guide.

knife-edge refraction Radio propagation effect in which the atmospheric attenuation of a signal is reduced when the signal passes over and is diffracted by a sharp obstacle such as a mountain ridge.

knife switch An electric switch consisting of a metal blade hinged at one end to a stationary jaw, so that the blade can be pushed over to make contact between spring clips.

knob-and-tube wiring An electric wiring method used for light and power circuits that uses open insulated wiring on solid insulators; now obsolete and illegal in most countries.

knock-on atom An atom which is knocked out of its equilibrium position in a crystal lattice by an energetic bombarding particle, and is displaced many atomic distances away into an interstitial position, leaving behind a vacant lattice site.

kohm See kilohm.

Kramer's theorem The theorem that the states of a system consisting of an odd number of electrons in an external electrostatic field are at least twofold degenerate.

Kronig-Penney model An idealized one-dimensional model of a crystal in which the potential energy of an electron is an infinite sequence of periodically spaced square wells.

krypton lamp An arc lamp filled with krypton; one type pierces fog for 1000 feet (300 meters) or more and is used to light airplane runways at night.

K scan See K scope.

K scope A modified form of A scope on which one signal appears as two pips, the relative amplitudes of which indicate the error of aiming the antenna. Also known as K indicator; K scan.

k-space *See* wave-vector space.
kV *See* kilovolt.
kVA *See* kilovolt-ampere.
kvar *See* kilovar.
kVp *See* kilovolts peak.
kWh *See* kilowatt-hour.
kW-hr *See* kilowatt-hour.

L

labile oscillator An oscillator whose frequency is controlled from a remote location by wire or radio.

lacing Tying insulated wires together to support each other and form a single neat cable, with separately laced branches.

ladder attenuator A type of ladder network designed to introduce a desired, adjustable loss when working between two resistive impedances, one of which has a shunt arm that may be connected to any of various switch points along the ladder.

ladder network A network composed of a sequence of H, L, T, or pi networks connected in tandem; chiefly used as an electric filter. Also known as series-shunt network.

laddic Multiaperture magnetic structure resembling a ladder, used to perform logic functions; operation is based on a flux change in the shortest available path when adjacent rungs of the ladder are initially magnetized with opposite polarity.

lag A persistence of the electric charge image in a camera tube for a small number of frames.

lagging current An alternating current that reaches its maximum value up to 90° behind the voltage that produces it.

lagging load *See* inductive load.

lag time The time between the application of current and rupture of the circuit within the detonator.

Lalande cell A type of wet cell that uses a zinc anode and cupric oxide cathode cast as flat plates or hollow cylinders, and an electrolyte of sodium hydroxide in aqueous solution (caustic soda).

Lamb wave Electromagnetic wave propagated over the surface of a solid whose thickness is comparable to the wavelength of the wave.

laminated contact Switch contact made up of a number of laminations, each making individual contact with the opposite conducting surface.

laminated core An iron core for a coil transformer, armature, or other electromagnetic device, built up from laminations stamped from sheet iron or steel and more or less insulated from each other by surface oxides and sometimes also by application of varnish.

laminography *See* sectional radiography.

lamp bank A number of incandescent lamps connected in parallel or series to serve as a resistance load for full-load tests of electric equipment.

lamp cord Two twisted or parallel insulated wires, usually no. 18 or no. 20, used chiefly for connecting electric equipment to wall outlets.

lampholder A device designed to connect an electric lamp to a circuit and to support it mechanically.

land *See* terminal area.

Landau levels Energy levels of conduction electrons which occur in a metal subjected to a magnetic field at very low temperatures and which are quantized because of the quantization of the electron motion perpendicular to the field.

land effect *See* coastal refraction.

landline A communications cable on or under the earth's surface, in contrast to a submarine cable.

land return *See* ground clutter.

Langevin function A mathematical function, $L(x)$, which occurs in the expressions for the paramagnetic susceptibility of a classical (non-quantum-mechanical) collection of magnetic dipoles, and for the polarizability of molecules having a permanent electric dipole moment; given by $L(x) = \coth x - 1/x$.

Langevin ion-mobility theories Two theories developed to calculate the mobility of ions in gases; the first assumes that atoms and ions interact through a hard-sphere collision and have a constant mean free path, while the second assumes that there is an attraction between atoms and ions arising from the polarization of the atom in the ion's field, in addition to hard-sphere repulsion for close distances of approach.

Langevin ion-recombination theory A theory predicting the rate of recombination of negative with positive ions in an ionized gas on the assumption that ions of opposite sign approach one another under the influence of mutual attraction, and that their relative velocities are determined by ion mobilities; applicable at high pressures, above 1 or 2 atmospheres.

Langevin theory of diamagnetism A theory based on the idea that diamagnetism results from electronic currents caused by Larmor precession of electrons inside atoms.

Langevin theory of paramagnetism A theory which treats a substance as a classical (non-quantum-mechanical) collection of permanent magnetic dipoles with no interactions between them, having a Boltzmann distribution with respect to energy of interaction with an applied field.

Langmuir-Child equation *See* Child's law.

Langmuir dark space A nonluminous region surrounding a negatively charged probe inserted in the positive column of a glow discharge.

Langmuir effect The ionization of atoms of low ionization potential that come into contact with a hot metal with a high work function.

L antenna An antenna that consists of an elevated horizontal wire having a vertical down-lead connected at one end.

lap dissolve Changeover from one television scene to another so that the new picture appears gradually as the previous picture simultaneously disappears.

Laplace law *See* Ampère law.

lapping Moving a quartz, semiconductor, or other crystal slab over a flat plate on which a liquid abrasive has been poured, to obtain a flat polished surface or to reduce the thickness a carefully controlled amount.

lap winding A two-layer winding in which each coil is connected in series to the adjacent coil.

large-scale integrated circuit A very complex integrated circuit, which contains well over 100 interconnected individual devices, such as basic logic gates and transistors, placed on a single semiconductor chip. Abbreviated LSI circuit. Also known as chip circuit; multiple-function chip.

Larmor formula The rate at which energy is radiated by a nonrelativistic, accelerated charge is $2q^2a^2/3c^3$, where q is the particle's charge in esu (electrostatic units), a is its acceleration, and c is the speed of light.

Larmor frequency The angular frequency of the Larmor precession, equal in esu (electrostatic units) to the negative of a particle's charge times the magnetic induction divided by the product of twice the particle's mass and the speed of light.

Larmor orbit The motion of a charged particle in a uniform magnetic field, which is a superposition of uniform circular motion in a plane perpendicular to the field, and uniform motion parallel to the field.

Larmor precession A common rotation superposed upon the motion of a system of charged particles, all having the same ration of charge to mass, by a magnetic field.

Larmor radius For a charged particle moving transversely in a uniform magnetic field, the radius of curvature of the projection of its path on a plane perpendicular to the field. Also known as gyromagnetic radius.

Larmor's theorem The theorem that for a system of charged particles, all having the same ratio of charge to mass, moving in a central field of force, the motion in a uniform magnetic induction B is, to first order in B, the same as a possible motion in the absence of B except for the superposition of a common precession of angular frequency equal to the Larmor frequency.

LASCR See light-activated silicon controlled rectifier.

LASCS See light-activated silicon controlled switch.

laser amplifier A laser which is used to increase the output of another laser. Also known as light amplifier.

laser flash tube A high-power, air-cooled or water-cooled xenon flash tube designed to produce high-intensity flashes for pumping applications.

laser radiation detector A photodetector that responds primarily to the coherent visible, infrared, or ultraviolet light of a laser beam.

laser threshold The minimum pumping energy required to initiate lasing action in a laser.

laser-triggered switch A high-voltage high-power switch that consists of a spark gap triggered into conduction by a laser beam.

latch-in relay A relay that maintains its contacts in the last position assumed, even without coil energization.

lateral quadrupole An electric or magnetic quadrupole which produces a field equivalent to that of two equal and opposite electric or magnetic dipoles separated by a small distance perpendicular to the direction of the dipoles.

lattice dynamics The study of the thermal vibrations of a crystal lattice. Also known as crystal dynamics.

lattice energy The energy required to separate ions in an ionic crystal an infinite distance from each other.

lattice filter An electric filter consisting of a lattice network whose branches have LC parallel-resonant circuits shunted by quartz crystals.

lattice network A network that is composed of four branches connected in series to form a mesh; two nonadjacent junction points serve as input terminals, and the remaining two junction points serve as output terminals.

lattice polarization Electric polarization of a solid due to displacement of ions from equilibrium positions in the lattice.

lattice scattering Scattering of electrons by collisions with vibrating atoms in a crystal lattice, reducing the mobility of charge carriers in the crystal and thereby affecting its conductivity.

lattice vibration A periodic oscillation of the atoms in a crystal lattice about their equilibrium positions.

lattice wave A disturbance propagated through a crystal lattice in which atoms oscillate about their equilibrium positions.

lattice winding A winding made of lattice coils and used for electric machines.

lattice-wound coil See honeycomb coil.

launching The process of transferring energy from a coaxial cable or transmission line to a waveguide.

Lauritsen electroscope A rugged and sensitive electroscope in which a metallized quartz fiber is the sensitive element.

lawnmower Type of radio-frequency preamplifier used with radar receivers.

law of electric charges The law that like charges repel, and unlike charges attract.

law of electromagnetic induction See Faraday's law of electromagnetic induction.

law of magnetism The law that like poles repel, and unlike poles attract.

Lawrence tube See chromatron.

layer capacitance See cathode interface capacitance.

layer impedance See cathode interface impedance.

layer winding Coil-winding method in which adjacent turns are laid evenly and side by side along the length of the coil form; any number of additional layers may be wound over the first, usually with sheets of insulating material between the layers.

lay ratio The ratio of the axial length of one complete turn of the helix formed by the core of a cable or the wire of a stranded conductor, to the mean diameter of the cable.

lazy H antenna An antenna array in which two or more dipoles are stacked one above the other to obtain greater directivity.

LCD See liquid crystal display.

LC filter See inductive filter.

LC ratio The inductance of a circuit in henrys divided by capacitance in farads.

L display A radarscope display in which the target appears as two horizontal pulses or blips, one extending to the right and one to the left from a central vertical time base; when the radar antenna is correctly aimed in azimuth at the target, both blips are of equal amplitude; when not correctly aimed, the relative blip amplitudes indicate the pointing error; the position of the signal along the baseline indicates target distance; the display may be rotated 90° when used for elevation aiming instead of azimuth aiming.

lead A wire used to connect two points in a circuit.

lead-acid battery A storage battery in which the electrodes are grids of lead containing lead oxides that change in composition during charging and discharging, and the electrolyte is dilute sulfuric acid.

lead-covered cable A cable whose conductors are protected from moisture and mechanical damage by a lead sheath.

lead-l-lead junction A Josephson junction consisting of two pieces of lead separated by a thin insulating barrier of lead oxide. Abbreviated Pb-I-Pb junction.

lead-in A single wire used to connect a single-terminal outdoor antenna to a receiver or transmitter. Also known as down-lead.

leading current An alternating current that reaches its maximum value up to 90° ahead of the voltage that produces it.

leading load Load that is predominately capacitive, so that its current leads the voltage applied to the load.

leading phase In three-phase power measurement, the phase whose voltage is leading upon that of one of the other phases by 120°.

lead-in insulator A tubular insulator inserted in a hole drilled through a wall, through which the lead-in wire can be brought into a building.

lead sulfide cell A cell used to detect infrared radiation; either its generated voltage or its change of resistance may be used as a measure of the intensity of the radiation.

leakage coefficient *See* leakage factor.

leakage conductance The conductance of the path over which leakage current flows; it is normally a low value.

leakage current 1. Undesirable flow of current through or over the surface of an insulating material or insulator. 2. The flow of direct current through a poor dielectric in a capacitor. 3. The alternating current that passes through a rectifier without being rectified.

leakage factor The total magnetic flux in an electric rotating machine or transformer divided by the useful flux that passes through the armature or secondary winding. Also known as leakage coefficient.

leakage flux Magnetic lines of force that go beyond their intended path and do not serve their intended purpose.

leakage indicator An instrument used to measure or detect current leakage from an electric system to earth. Also known as earth detector.

leakage inductance Self-inductance due to leakage flux in a transformer.

leakage radiation In a radio transmitting system, radiation from anything other than the intended radiating system.

leakage reactance Inductive reactance due to leakage flux that links only the primary winding of a transformer.

leakage resistance The resistance of the path over which leakage current flows; it is normally high.

leaky Pertaining to a condition in which the leakage resistance has dropped so much below its normal value that excessive leakage current flows; usually applied to a capacitor.

leaky-wave antenna A wide-band microwave antenna that radiates a narrow beam whose direction varies with frequency; it is fundamentally a perforated waveguide, thin enough to permit flush mounting for aircraft and missile radar applications.

Lecher line *See* Lecher wires.

Lecher wires Two parallel wires that are several wavelengths long and a small fraction of a wavelength apart, used to measure the wavelength of a microwave source that is connected to one end of the wires; a shorting bar which slides along the wires is used to determine the position of standing-wave nodes. Also known as Lecher line; Lecher wire wavemeter.

Lecher wire wavemeter See Lecher wires.

Leclanché cell The common dry cell, which is a primary cell having a carbon positive electrode and a zinc negative electrode in an electrolyte of sal ammoniac and a depolarizer.

LED See light-emitting diode.

Leduc current An asymmetrical alternating current obtained from, or similar to that obtained from, the secondary winding of an induction coil; used in electrobiology.

LEED See low-energy electron diffraction.

left-hand polarization In elementary-particle discussions, circular or elliptical polarization of an electromagnetic wave in which the electric field vector at a fixed point in space rotates in the left-hand sense about the direction of propagation; in optics, the opposite convention is used; in facing the source of the beam, the electric vector is observed to rotate counterclockwise.

left-hand rule 1. For a current-carrying wire, the rule that if the fingers of the left hand are placed around the wire so that the thumb points in the direction of electron flow, the fingers will be pointing in the direction of the magnetic field produced by the wire. 2. For a current-carrying wire in a magnetic field, such as a wire on the armature of a motor, the rule that if the thumb, first, and second fingers of the left hand are extended at right angles to one another, with the first finger representing the direction of magnetic lines of force and the second finger representing the direction of current flow, the thumb will be pointing in the direction of force on the wire. Also known as Fleming's rule.

left-hand taper A taper in which there is greater resistance in the counterclockwise half of the operating range of a rheostat or potentiometer (looking from the shaft end) than in the clockwise half.

Lenard rays Cathode rays produced in air by a Lenard tube.

Lenard tube An early experimental electron-beam tube that had a thin glass or metallic foil window at the end opposite the cathode, through which the electron beam could pass into the atmosphere.

lengthened dipole An antenna element with lumped inductance to compensate an end loss.

lens See electron lens; magnetic lens.

lens antenna A microwave antenna in which a dielectric lens is placed in front of the dipole or horn radiator to concentrate the radiated energy into a narrow beam or to focus received energy on the receiving dipole or horn.

Lenz's law The law that whenever there is an induced electromotive force (emf) in a conductor, it is always in such a direction that the current it would produce would oppose the change which causes the induced emf.

level 1. A single bank of contacts, as on a stepping relay. 2. The difference between a quantity and an arbitrarily specified reference quantity, usually expressed as the logarithm of the ratio of the quantities. 3. A charge value that can be stored in a given storage element of a charge storage tube and distinguished in the output from other charge values.

level compensator 1. Automatic transmission-regulating feature or device used to minimize the effects of variations in amplitude of the received signal. 2. Automatic gain control device used in the receiving equipment of a telegraph circuit.

level converter An amplifier that converts nonstandard positive or negative logic input voltages to standard DTL or other logic levels.

level shifting Changing the logic level at the interface between two different semiconductor logic systems.

lever switch A switch having a lever-shaped operating handle.

Leyden jar An early type of capacitor, consisting simply of metal foil sheets on the inner and outer surfaces of a glass jar.

Lichtenberg figures Patterns produced on a photographic emulsion, or in fine powder spread over the surface of a solid dielectric, by an electric discharge produced by a high transient voltage. Also known as Lichtenberger figures.

Lienard-Wiechert potentials The retarded and advanced electromagnetic scaler and vector potentials produced by a moving point charge, expressed in terms of the (retarded or advanced) position and velocity of the charge.

lifting magnet A type of electromagnet in which a material to be held or moved is initially placed in contact with the magnet, in contrast to a traction magnet. Also known as holding magnet.

light-activated silicon controlled rectifier A silicon-controlled rectifier having a glass window for incident light that takes the place of, or adds to the action of, an electric gate current in providing switching action. Abbreviated LASCR. Also known as photo-SCR; photothyristor.

light-activated silicon controlled switch A semiconductor device that has four layers of silicon alternately doped with acceptor and donor impurities, but with all four of the p and n layers made accessible by terminals; when a light beam hits the active light-sensitive surface, the photons generate electron-hole pairs that make the device turn on; removal of light does not reverse the phenomenon; the switch can be turned off only by removing or reversing its positive bias. Abbreviated LASCS.

light amplifier 1. Any electronic device which, when actuated by a light image, reproduces a similar image of enhanced brightness, and which is capable of operating at very low light levels without introducing spurious brightness variations (noise) into the reproduced image. Also known as image intensifier. 2. *See* laser amplifier.

light-beam oscillograph An oscillograph in which a beam of light, focused to a point by a lens, is reflected from a tiny mirror attached to the moving coil of a galvanometer onto a photographic film moving at constant speed.

light bulb *See* incandescent lamp.

light carrier injection A method of introducing the carrier in a facsimile system by periodic variation of the scanner light beam, the average amplitude of which is varied by the density changes of the subject copy. Also known as light modulation.

light chopper A rotating fan or other mechanical device used to interrupt a light beam that is aimed at a phototube, to permit alternating-current amplification of the phototube output and to make its output independent of strong, steady ambient illumination.

light-emitting diode A semiconductor diode that converts electric energy efficiently into spontaneous and noncoherent electromagnetic radiation at visible and near-infrared wavelengths by electroluminescence at a forward-biased pn junction. Abbreviated LED. Also known as solid-state lamp.

light-gating cathode-ray tube A cathode-ray tube in which the electron beam varies the transmission or reflection properties of a screen that is positioned in the beam of an external light source.

light gun A light pen mounted in a gun-type housing.

lighthouse tube *See* disk-seal tube.

lighting branch circuit A circuit that supplies power to outlets for lighting fixtures only.

light microsecond Distance a light wave travels in free space in one-millionth of a second.

244 light modulation

light modulation *See* light carrier injection.

light modulator The combination of a source of light, an appropriate optical system, and a means for varying the resulting light beam to produce an optical sound track on motion picture film.

light-negative Having negative photoconductivity, hence decreasing in conductivity (increasing in resistance) under the action of light.

lightning arrester A protective device designed primarily for connection between a conductor of an electrical system and ground to limit the magnitude of transient overvoltages on equipment. Also known as arrester; surge arrester.

lightning conductor A conductor designed to carry the current of a lightning discharge from a lightning rod to ground.

lightning generator A high-voltage power supply used to generate surge voltages resembling lightning, for testing insulators and other high-voltage components.

lightning protection Means, such as lightning rods and lightning arresters, of protecting electrical systems, buildings, and other property from lightning.

lightning recorder *See* sferics receiver.

lightning rod A metallic rod set up on an exposed elevation of a structure and connected to a low-resistance ground to intercept lightning discharges and to provide a direct conducting path to ground for them.

lightning surge A transient disturbance in an electric circuit due to lightning.

lightning switch A manually operated switch used to connect a radio antenna to ground during electrical storms, rather than to the radio receiver.

light-operated switch A switch that is operated by a beam or pulse of light, such as a light-activated silicon controlled rectifier.

light panel *See* electroluminescent panel.

light pen A tiny photocell or photomultiplier, mounted with or without fiber or plastic light pipe in a pen-shaped housing; it is held against a cathode-ray screen to make measurements from the screen or to change the nature of the display.

light-positive Having positive photoconductivity; selenium ordinarily has this property.

light relay *See* photoelectric relay.

light-sensitive Having photoconductive, photoemissive, or photovoltaic characteristics. Also known as photosensitive.

light-sensitive cell *See* photodetector.

light-sensitive detector *See* photodetector.

light-sensitive tube *See* phototube.

light sensor photodevice *See* photodetector.

light valve 1. A device whose light transmission can be made to vary in accordance with an externally applied electrical quantity, such as voltage, current, electric field, or magnetic field, or an electron beam. 2. Any direct-view electronic display optimized for reflecting or transmitting an image with an independent collimated light source for projection purposes.

limited integrator A device used in analog computers that has two input signals and one output signal whose value is proportional to the integral of one of the input signals with respect to the other as long as this output signal does not exceed specified limits.

limited signal Radar signal that is intentionally limited in amplitude by the dynamic range of the radar system.

linear distortion 245

limiter An electronic circuit used to prevent the amplitude of an electronic waveform from exceeding a specified level while preserving the shape of the waveform at amplitudes less than the specified level. Also known as amplitude limiter; amplitude-limiting circuit; automatic peak limiter; clipper; clipping circuit; limiter circuit; peak limiter.

limiter circuit See limiter.

limiting A desired or undesired amplitude-limiting action performed on a signal by a limiter. Also known as clipping.

limit ratio Ratio of peak value to limited value, or comparison of such ratios.

limit switch A switch designed to cut off power automatically at or near the limit of travel of a moving object controlled by electrical means.

Lindeck potentiometer A potentiometer in which an unknown potential difference is balanced against a known potential difference derived from a fixed resistance carrying a variable current; the converse of most potentiometers.

Lindemann electrometer A variant of the quadrant electrometer, designed for portability and insensitivity to changes in position, in which the quadrants are two sets of plates about 6 millimeters apart, mounted on insulating quartz pillars; a needle rotates about a taut silvered quartz suspension toward the oppositely charged plates when voltage is applied to it, and its movement is observed through a microscope.

Linde's rule The rule that the increase in electrical resistivity of a monovalent metal produced by a substitutional impurity per atomic percent impurity is equal to $a + b(v-1)^2$, where a and b are constants for a given solvent metal and a given row of the periodic table for the impurity, and v is the valence of the impurity.

L indicator See L scope.

line 1. The path covered by the electron beam of a television picture tube in one sweep from left to right across the screen. 2. One horizontal scanning element in a facsimile system. 3. See trace.

linear amplifier An amplifier in which changes in output current are directly proportional to changes in applied input voltage.

linear array An antenna array in which the dipole or other half-wave elements are arranged end to end on the same straight line. Also known as collinear array.

linear circuit See linear network.

linear collision cascade A sputtering event in which the bombarding projectile collides directly with a small number of target atoms, which collide with others, and the sharing of energy then proceeds through many generations before one or more target atoms are ejected; the density of atoms in motion remains sufficiently small so that collisions between atoms can be ignored.

linear comparator A comparator circuit which operates on continuous, or nondiscrete, waveforms. Also known as continuous comparator.

linear computing element A linear circuit in an analog computer.

linear conductor antenna An antenna consisting of one or more wires which all lie along a straight line.

linear control Rheostat or potentiometer having uniform distribution of graduated resistance along the entire length of its resistance element.

linear detection Detection in which the output voltage is substantially proportional, over the useful range of the detecting device, to the voltage of the input wave.

linear distortion Amplitude distortion in which the output signal envelope is not proportional to the input signal envelope and no alien frequencies are involved.

linear electrical constants of a uniform line Series resistance, series inductance, shunt conductance, and shunt capacitance per unit length of line.

linear integrated circuit An integrated circuit that provides linear amplification of signals.

linearity control A cathode-ray-tube control which varies the distribution of scanning speed throughout the trace interval. Also known as distribution control.

linear-logarithmic intermediate-frequency amplifier Amplifier used to avoid overload or saturation as a protection against jamming in a radar receiver.

linearly graded junction A pn junction in which the impurity concentration does not change abruptly from donors to acceptors, but varies smoothly across the junction, and is a linear function of position.

linear magnetic amplifier A magnetic amplifier employing negative feedback to make its output load voltage a linear function of signal current.

linear motor An electric motor that has in effect been split and unrolled into two flat sheets, so that the motion between rotor and stator is linear rather than rotary.

linear network A network in which the parameters of resistance, inductance, and capacitance are constant with respect to current or voltage, and in which the voltage or current of sources is independent of or directly proportional to other voltages and currents, or their derivatives, in the network. Also known as linear circuit.

linear-phase Pertaining to a filter or other network whose image phase constant is a linear function of frequency.

linear power amplifier A power amplifier in which the signal output voltage is directly proportional to the signal input voltage.

linear rectifier A rectifier, the output current of voltage of which contains a wave having a form identical with that of the envelope of an impressed signal wave.

linear repeater A repeater used in communication satellites to amplify input signals a fixed amount, generally with traveling-wave tubes or solid-state devices operating in their linear region.

linear sweep A cathode-ray sweep in which the beam moves at constant velocity from one side of the screen to the other, then suddenly snaps back to the starting side.

linear-sweep delay circuit A widely used form of linear time-delay circuit in which the input signal initiates action by a linear sawtooth generator, such as the bootstrap or Miller integrator, whose output is then compared with a calibrated direct-current reference voltage level.

linear-sweep generator An electronic circuit that provides a voltage or current that is a linear function of time; the waveform is usually recurrent at uniform periods of time.

linear taper A taper that gives the same change in resistance per degree of rotation over the entire range of a potentiometer.

linear time base A time base that makes the electron beam of a cathode-ray tube move at a constant speed along the horizontal time scale.

linear transducer A transducer for which the pertinent measures of all the waves concerned are linearly related.

linear unit An electronic device used in analog computers in which the change in output, due to any change in one of two or more input signals, is proportional to the change in that input and does not depend upon the values of the other inputs.

linear variable-differential transformer A transformer in which a diaphragm or other transducer sensing element moves an armature linearly inside the coils of a differential transformer, to change the output voltage by changing the inductances of the coils in equal but opposite amounts. Abbreviated LVDT.

line balance 1. Degree of electrical similarity of the two conductors of a transmission line. 2. Matching impedance, equaling the impedance of the line at all frequencies, that is used to terminate a two-wire line.

line-balance converter See balun.

line-building-out network See impedance-matching network.

line circuit 1. Relay equipment associated with each station connected to a dial or manual switchboard. 2. A circuit to interconnect an individual telephone and a channel terminal.

line conductor A metal used as a conductor in a power line; the most frequently used conductors are copper and aluminum.

line-controlled blocking oscillator A circuit formed by combining a monostable blocking oscillator with an open-circuit transmission line in the regenerative circuit; it is capable of generating pulses with large amounts of power.

line cord A two-wire cord terminating in a two-prong plug at one end and connected permanently to a radio receiver or other appliance at the other end; used to make connections to a source of power. Also known as power cord.

line-cord resistor An asbestos-enclosed wire-wound resistor incorporated in a line cord along with the two regular wires.

line driver An integrated circuit that acts as the interface between logic circuits and a two-wire transmission line.

line drop The voltage drop existing between two points on a power line or transmission line, due to the impedance of the line.

line-drop compensator A device that restores the voltage lost when electricity is transmitted along a wire.

line equalizer An equalizer containing inductance or capacitance, inserted in a transmission line to modify the frequency response of the line.

line fault A defect, such as an open circuit, short circuit, or ground, in an electric line for transmission or distribution of power or of speech, music, or other content.

line filter 1. A filter inserted between a power line and a receiver, transmitter, or other unit of electric equipment to prevent passage of noise signals through the power line in either direction. Also known as power-line filter. 2. A filter inserted in a transmission line or high-voltage power line for carrier communication purposes.

line flux A local inductive field of a telephone or power line.

line frequency The number of times per second that the scanning spot sweeps across the screen in a horizontal direction in a television system. Also known as horizontal frequency; horizontal line frequency.

line-frequency blanking pulse See horizontal blanking pulse.

line impedance The impedance measured across the terminals of a transmission line.

line influence The effect of a local inductive field around a telephone line.

line interlace See interlaced scanning.

line lengthener Device for altering the electrical length of a waveguide or transmission line without altering other electrical characteristics, or the physical length.

line location The location of power and communications lines when two or more such lines run along the same route; they should either be used jointly, or located with

248 line loop resistance

respect to each other so as to avoid unnecessary crossings, conflicts, and inductive exposures.

line loop resistance Metallic resistance of the line wires that extend from an individual telephone set to the dial central office.

line loss Total of the various energy losses occurring in a transmission line.

line of electrostatic induction A unit of electric flux equal to the electric flux associated with a charge of 1 statcoulomb.

line of magnetic induction *See* maxwell.

line of sight The straight line for a transmitting radar antenna in the direction of the beam.

line pad Pad inserted between a program amplifier and a transmission line, to isolate the amplifier from impedance variations of the line.

line pulsing Method of pulsing a transmitter in which an artificial line is charged over a relatively long period of time and then discharged through the transmitter tubes in a short interval determined by the line characteristic.

line radiation Electromagnetic radiation from a power line caused mainly by corona pulses; gives rise to radio interference.

line regulation The maximum change in the output voltage or current of a regulated power supply for a specified change in alternating-current line voltage, such as from 105 to 125 volts.

line relay Relay which is controlled over a subscriber line or trunkline.

line side Terminal connections to an external or outstation source, such as data terminal connections to a communications circuit connecting to another data terminal.

line-stabilized oscillator Oscillator in which a section of line is used as a sharply selective circuit element for the purpose of controlling the frequency.

line stretcher Section of waveguide or rigid coaxial line whose physical length is variable to provide impedance matching.

line switching Connecting or disconnecting the line voltage from a piece of electronic equipment.

line synchronizing pulse *See* horizontal synchronizing pulse.

line-to-ground fault A defect in a power or communications line in which faulty insulation allows the conductor to make contact with the earth.

line transducer A special type of electret transducer consisting essentially of a coaxial cable with polarized dielectric, and with the center conductor and shield serving as electrodes; mechanical excitation resulting in a deformation of the shield at any point along the length of the cable produces an electrical output signal.

line transformer Transformer connecting a transmission line to terminal equipment; used for isolation, line balance, impedance matching, or additional circuit connections.

line trap A filter consisting of a series inductance shunted by a tuning capacitor, inserted in series with the power or telephone line for a carrier-current system to minimize the effects of variations in line attenuation and reduce carrier energy loss.

line tuning Adjustment of the frequency of carrier current of a communication system to tune out the reactance of a capacitor with suitable inductance.

line-turn *See* maxwell-turn.

line unit Electric control device used to send, receive, and control the impulses of a teletypewriter.

lithium-sulfur battery 249

line voltage The voltage provided by a power line at the point of use.

line-voltage regulator A regulator that counteracts variations in power-line voltage, so as to provide an essentially constant voltage for the connected load.

linkage See flux linkage.

link circuit Closed loop used for coupling purposes; it generally consists of two coils, each having a few turns of wire, connected by a twisted pair of wires or by other means, with each coil placed over, near, or in one of the two coils that are to be coupled.

link coupling Modification of inductive coupling where the two coils are connected together by a short length of transmission line, with each coil inductively coupled to the coil of a separate tuned circuit.

lin-log amplifier Automatic gain control amplifier that operates in a linear manner for low-amplitude input signals, but responds in a logarithmic manner to high-amplitude input signals.

liquid crystal display A digital display that consists of two sheets of glass separated by a sealed-in, normally transparent, liquid crystal material; the outer surface of each glass sheet has a transparent conductive coating such as tin oxide or indium oxide, with the viewing-side coating etched into character-forming segments that have leads going to the edges of the display; a voltage applied between front and back electrode coatings disrupts the orderly arrangement of the molecules, darkening the liquid enough to form visible characters even though no light is generated. Abbreviated LCD.

liquid-dielectric capacitor A capacitor in which the plate assemblies are mounted in a tank filled with a suitable oil or liquid dielectric.

liquid fuse unit Fuse unit in which the fuse link is immersed in a liquid, or provision is made for drawing the arc into the liquid when the fuse link melts.

liquid-metal fuel cell A fuel cell that uses molten potassium and bismuth as reactants and a molten salt electrolyte; has very high power output, but a relatively short life.

liquid-metal MHD generator A system for generating electric power in which the kinetic energy of a flowing, molten metal is converted to electric energy by magnetohydrodynamic (MHD) interaction.

liquid rheostat A variable-resistance type of voltage regulator in which the variable-resistance element is liquid, usually water; carbon electrodes are raised or lowered in the liquid to change resistance ratings and control voltage flow.

liquid semiconductor An amorphous material in solid or liquid state that possesses the properties of varying resistance induced by charge carrier injection.

lithium battery A solid-state battery with a lithium anode, an iodine-polyvinyl pyridine cathode, and an electrolyte consisting of a layer of lithium iodide; used in cardiac pacemakers.

lithium cell A primary cell for producing electrical energy by using lithium metal for one electrode immersed in usually an organic electrolyte.

lithium-drifted germanium crystal A high-resolution junction detector, used especially for more penetrating gamma-radiation and higher-energy electrons, produced by drifting lithium ions through a germanium crystal to produce an intrinsic region where impurity-based carrier generation centers are deactivated, sandwiched between a p layer and an n layer.

lithium-sulfur battery A storage battery in which the cells use a molten lithium cathode and a molten sulfur anode separated by a molten salt electrolyte that consists of lithium iodide, potassium iodide, and lithium chloride.

Litzendraht wire *See* Litz wire.

Litz wire Wire consisting of a number of separately insulated strands woven together so each strand successively takes up all possible positions in the cross section of the entire conductor, to reduce skin effect and thereby reduce radio-frequency resistance. Derived from Litzendraht wire.

live *See* energized.

live chassis A radio, television, or other chassis that has a direct chassis connection to one side of the alternating-current line.

LLL circuit *See* low-level logic circuit.

L network A network composed of two branches in series, with the free ends connected to one pair of terminals; the junction point and one free end are connected to another pair of terminals.

load 1. A device that consumes electric power. 2. The amount of electric power that is drawn from a power line, generator, or other power source. 3. The material to be heated by an induction heater or dielectric heater. Also known as work. 4. The device that receives the useful signal output of an amplifier, oscillator, or other signal source.

load-break switch An electric switch in a circuit with several hundred thousand volts, designed to carry a large amount of current without overheating the open position, having enough insulation to isolate the circuit in closed position, and equipped with arc interrupters to interrupt the load current.

load cell A device which measures large pressures by applying the pressure to a piezoelectric crystal and measuring the voltage across the crystal; the cell plus a recording mechanism constitutes a strain gage.

load characteristic Relation between the instantaneous values of a pair of variables such as an electrode voltage and an electrode current, when all direct electrode supply voltages are maintained constant. Also known as dynamic characteristic.

load circuit Complete circuit required to transform power from a source such as an electron tube to a load.

load circuit efficiency Ratio between useful power delivered by the load circuit to the load and the load circuit power input.

load curve A graph that plots the power supplied by an electric power system versus time.

load divider Unit for distributing power to various units.

loaded line Wire line in which loading coils have been inserted at regular intervals to reduce attenuation and phase lag at the frequencies within the band used.

loaded motional impedance *See* motional impedance.

loaded Q 1. The Q factor of an impedance which is connected or coupled under working conditions. Also known as working Q. 2. The Q factor of a specific mode of resonance of a microwave tube or resonant cavity when there is external coupling to that mode.

load factor The ratio of average electric load to peak load, usually calculated over a 1-hour period.

load impedance The complex impedance presented to a transducer by its load.

loading The addition of inductance to a transmission line to improve its transmission characteristics throughout a given frequency band. Also known as electrical loading.

loading coil 1. An iron-core coil connected into a telephone line or cable at regular intervals to lessen the effect of line capacitance and reduce distortion. Also known

local oscillator injection   251

as Pupin coil; telephone loading coil. 2. A coil inserted in series with a radio antenna to increase its electrical length and thereby lower the resonant frequency.

loading disk Circular metal piece mounted at the top of a vertical antenna to increase its natural wavelength.

load isolator Waveguide or coaxial device that provides a good energy path from a signal source to a load, but provides a poor energy path for reflections from a mismatched load back to the signal source.

load line A straight line drawn across a series of tube or transistor characteristic curves to show how output signal current will change with input signal voltage when a specified load resistance is used.

load loss The sum of the copper loss of a transformer, due to resistance in the windings, plus the eddy current loss in the winding, plus the stray loss.

load power Of an energy load, the average rate of flow of energy through the terminals of that load when connected to a specified source.

load regulation The maximum change in the output voltage or current of a regulated power supply for a specified change in load conditions.

load shedding A procedure in which parts of an electric power system are disconnected in an attempt to prevent failure of the entire system due to overloading.

lobe A part of the radiation pattern of a directional antenna representing an area of stronger radio-signal transmission. Also known as radiation lobe.

lobe-half-power width In a plane containing the direction of the maximum energy of a lobe, the angle between the two directions in that plane about the maximum in which the radiation intensity is one-half the maximum value of the lobe.

lobe penetration Penetration of the radar coverage of a station which is not limited by pulse repetition frequency, scope limitations, or the screening angle at the azimuth of penetration.

lobe switching *See* beam switching.

lobing Formation of maxima and minima at various angles of the vertical plane antenna pattern by the reflection of energy from the surface surrounding the radar antenna; these reflections reinforce the main beam at some angles and detract from it at other angles, producing fingers of energy.

local action 1. Internal losses of a battery caused by chemical reactions producing local currents between different parts of a plate. 2. Quantitatively, the percentage loss per month in the capacity of a battery on open circuit, or the amount of current needed to keep the battery fully charged.

local battery Battery that actuates the telegraphic station recording instruments, as distinguished from the battery furnishing current to the line.

local battery telephone set Telephone set for which the transmitter current is supplied from a battery, or other current supply circuit, individual to the telephone set; the signaling current may be supplied from a local hand generator or from a centralized power source.

local cell A galvanic cell resulting from differences in potential between adjacent areas on the surface of a metal immersed in an electrolyte.

local oscillator The oscillator in a superheterodyne receiver, whose output is mixed with the incoming modulated radio-frequency carrier signal in the mixer to give the frequency conversions needed to produce the intermediate-frequency signal.

local oscillator injection Adjustment used to vary the magnitude of the local oscillator signal that is coupled into the mixer.

252 local oscillator radiation

local oscillator radiation Radiation of the fundamental or harmonics of the local oscillator of a superheterodyne receiver.

lock To fasten onto and automatically follow a target by means of a radar beam.

locked oscillator A sine-wave oscillator whose frequency can be locked by an external signal to the control frequency divided by an integer.

locked-oscillator detector A frequency-modulation detector in which a local oscillator follows, or is locked to, the input frequency; the phase difference between local oscillator and input signal is proportional to the frequency deviation, and an output voltage is generated proportional to the phase difference.

locked-rotor current The current drawn by a stalled electric motor.

lock-in Shifting and automatic holding of one or both of the frequencies of two oscillating systems which are coupled together, so that the two frequencies have the ratio of two integral numbers.

lock-in amplifier An amplifier that uses some form of automatic synchronization with an external reference signal to detect and measure very weak electromagnetic radiation at radio or optical wavelengths in the presence of very high noise levels.

locking Controlling the frequency of an oscillator by means of an applied signal of constant frequency.

lock-on 1. The procedure wherein a target-seeking system (such as some types of radars) is continuously and automatically following a target in one or more coordinates (for example, range, bearing, elevation). 2. The instant at which radar begins to track a target automatically.

lockout circuit A switching circuit which responds to concurrent inputs from a number of external circuits by responding to one, and only one, of these circuits at any time. Also known as finding circuit; hunting circuit.

lock-up relay A relay that locks in its energized position either by permanent magnetic biasing which can be released only by applying a reverse magnetic pulse or by auxiliary contacts that keep its coil energized until the circuit is interrupted.

logarithmic amplifier An amplifier whose output signal is a logarithmic function of the input signal.

logarithmic diode A diode that has an accurate semilogarithmic relationship between current and voltage over wide and forward dynamic ranges.

logarithmic fast time constant Constant false alarm rate scheme which has a logarithmic intermediate-frequency amplifier followed by a fast time constant circuit.

logarithmic multiplier A multiplier in which each variable is applied to a logarithmic function generator, and the outputs are added together and applied to an exponential function generator, to obtain an output proportional to the product of two inputs.

logic 1. The basic principles and applications of truth tables, interconnections of on/off circuit elements, and other factors involved in mathematical computation in a computer. 2. General term for the various types of gates, flip-flops, and other on/off circuits used to perform problem-solving functions in a digital computer.

logical gate *See* switching gate.

logic card A small fiber chassis on which resistors, capacitors, transistors, magnetic cores, and diodes are mounted and interconnected in such a way as to perform some computer function; computers employing this type of construction may be repaired by removing the faulty card and replacing it with a new card.

logic level One of the two voltages whose values have been arbitrarily chosen to represent the binary numbers 1 and 0 in a particular data-processing system.

long-persistence screen 253

logic swing The voltage difference between the logic levels used for 1 and 0; magnitude is chosen arbitrarily for a particular system and is usually well under 10 volts.

logic switch A diode matrix or other switching arrangement that is capable of directing an input signal to one of several outputs.

log-periodic antenna A broad-band antenna which consists of a sheet of metal with two wedge-shaped cutouts, each with teeth cut into its radii along circular arcs; characteristics are repeated at a number of frequencies that are equally spaced on a logarithmic scale.

loktal base A special base for small vacuum tubes, so designed that it locks the tube firmly in a corresponding special eight-pin loktal socket; the tube pins are sealed directly into the glass envelope.

London equations Equations for the time derivative and the curl of the current in a superconductor in terms of the electric and magnetic field vectors respectively, derived in the London superconductivity theory.

London penetration depth A measure of the depth which electric and magnetic fields can penetrate beneath the surface of a superconductor from which they are otherwise excluded, according to the London superconductivity theory.

London superconductivity theory An extension of the two-fluid model of superconductivity, in which it is assumed that superfluid electrons behave as if the only force acting on them arises from applied electric fields, and that the curl of the superfluid current vanishes in the absence of a magnetic field.

long-conductor antenna See long-wire antenna.

long discharge 1. A capacitor or other electrical charge accumulator which takes a long time to leak off. 2. A gaseous electrical discharge in which the length of the discharge channel is very long compared with its diameter; lightning discharges are natural examples of long discharges. Also known as long spark.

longitudinal circuit Circuit formed by one telephone wire (or by two or more telephone wires in parallel) with return through the earth or through any other conductors except those which are taken with the original wire or wires to form a metallic telephone circuit.

longitudinal current Current which flows in the same direction in the two wires of a parallel pair using the earth or other conductors for a return path.

longitudinal-made delay line A magnetostrictive delay in which signals are propagated by means of longitudinal vibrations in the magnetostrictive material.

longitudinal magnetoresistance The change of electrical resistance produced in a current-carrying metal or semiconductor upon application of a magnetic field parallel to the current flow.

longitudinal quadrupole An electric or magnetic quadrupole which produces a field equivalent to that of two equal and opposite electric or magnetic dipoles separated by a small distance parallel to the direction of the dipoles. Also known as axial quadrupole.

long-line current A current that flows through the earth from an anodic to a cathodic area and returns along an underground pipe or other metal structure, often over a considerable distance and as the result of concentration cell action.

long-line effect An effect occurring when an oscillator is coupled to a transmission line with a bad mismatch; two or more frequencies may then be equally suitable for oscillation, and the oscillator jumps from one of these frequencies to another as its load changes.

long-persistence screen A fluorescent screen containing phosphorescent compounds that increase the decay time, so a pattern may be seen for several seconds after it is produced by the electron beam.

long-range order A tendency for some property of atoms in a lattice (such as spin orientation or type of atom) to follow a pattern which is repeated every few unit cells.

long-tail pair A two-tube or transistor circuit that has a common resistor (tail resistor) which gives strong negative feedback.

long-wire antenna An antenna whose length is a number of times greater than its operating wavelength, so as to give a directional radiation pattern. Also known as long-conductor antenna.

look-through 1. When jamming, a technique whereby the jamming emission is interrupted irregularly for extremely short periods to allow monitoring of the victim signal during jamming operations. 2. When being jammed, the technique of observing or monitoring a desired signal during interruptions in the jamming signal.

loop 1. A closed path or circuit over which a signal can circulate, as in a feedback control system. 2. Commercially, the portion of a connection from central office to subscriber in a telephone system. 3. *See* coupling loop; loop antenna; mesh.

loop antenna A directional-type antenna consisting of one or more complete turns of a conductor, usually tuned to resonance by a variable capacitor connected to the terminals of the loop. Also known as loop.

loop coupling A method of transferring energy between a waveguide and an external circuit, by inserting a conducting loop into the waveguide, oriented so that electric lines of flux pass through it.

loop filter A low-pass filter, which may be a simple RC filter or may include an amplifier, and which passes the original modulating frequencies but removes the carrier-frequency components and harmonics from a frequency-modulated signal in a locked-oscillator detector.

loop gain Total usable power gain of a carrier terminal or two-wire repeater; maximum usable gain is determined by, and may not exceed, the losses in the closed path.

loop-mile Length of wire in a mile of two-wire line.

loopstick antenna *See* ferrite-rod antenna.

loop test A telephone or telegraph line test that is made by connecting a faulty line to good lines in such a way as to form a loop in which measurements can be made to determine the position of the fault.

loose coupling Coupling of a degree less than the critical coupling.

Lorentz electron A model of the electron as a damped harmonic oscillator; used to explain the variation of the real and imaginary parts of the index of refraction of a substance with frequency.

Lorentz equation The equation of motion for a charged particle, which sets the rate of change of its momentum equal to the Lorentz force.

Lorentz force The force on a charged particle moving in electric and magnetic fields, equal to the particle's charge times the sum of the electric field and the cross product of particle's velocity with the magnetic flux density.

Lorentz force density The force per unit volume on a charge density and current density, assuming that these densities arise from large numbers of charged particles experiencing a Lorentz force.

Lorentz gage Any gage in which the sum of the divergence of the vector potential and the partial derivative of the scalar potential divided by the speed of light (in Gaussian units) vanishes identically; it is always possible to find a gage satisfying this condition.

Lorentz gas A model of completely ionized gas in which ions are assumed to be stationary and interactions between electrons are neglected.

Lorentz-Heaviside system See Heaviside-Lorentz system.

Lorentz local field In a theory of electric polarization, the average electric field due to the polarization at a molecular site that is calculated under the assumption that the field due to polarization by molecules inside a small sphere centered at the site may be neglected. Also known as Mossotti field.

Lorentz number The thermal conductivity of a metal divided by the product of its temperature and its electrical conductivity, according to the Wiedemann-Franz law.

Lorentz relation See Wiedemann-Franz law.

loss angle A measure of the power loss in an inductor or a capacitor, equal to the amount by which the angle between the phasors denoting voltage and current across the inductor or capacitor differs from 90°.

loss current 1. The current which passes through a capacitor as a result of the conductivity of the dielectric and results in power loss in the capacitor. 2. The component of the current across an inductor which is in phase with the voltage (in phasor notation) and is associated with power losses in the inductor.

losser circuit Resonant circuit having sufficient high-frequency resistance to prevent sustained oscillation at the resonant frequency.

loss evaluation A method of achieving an economic balance between buyer and seller in adding material to a transformer design to get lower losses, in which one calculates a value in dollars per kilowatt for load loss and for no-load loss.

Lossev effect See injection electroluminescence.

loss factor The power factor of a material multiplied by its dielectric constant; determines the amount of heat generated in a material.

lossless junction A waveguide junction in which all the power incident on the junction is reflected from it.

loss modulation See absorption modulation.

loss of information See walk-down.

lossy attenuator In waveguide technique, a length of waveguide deliberately introducing a transmission loss by the use of some dissipative material.

lossy line 1. Cable used in test measurements which has a large attenuation per unit length. 2. Transmission line designed to have a high degree of attenuation.

loudness analyzer An instrument that produces a cathode-ray display which shows the loudness of airborne sounds at a number of subdivisions of part or all of the audio spectrum.

low-energy electron diffraction A technique for studying the atomic structure of single crystal surfaces, in which electrons of uniform energy in the approximate range 5–500 electronvolts are scattered from a surface, and those scattered electrons that have lost no energy are selected and accelerated to a fluorescent screen where the diffraction pattern from the surface can be observed. Abbreviated LEED.

lower half-power frequency The frequency on an amplifier response curve which is smaller than the frequency for peak response and at which the output voltage is $1/\sqrt{2}$ of its midband or other reference value.

lower-sideband upconverter Parametric amplifier in which the frequency, power, impedance, and gain considerations are the same as for the nondegenerate amplifier; here, however, the output is taken at the difference frequency, or the lower sideband, rather than the signal-input frequency.

low-frequency antenna An antenna designed to transmit or receive radiation at frequencies of less than about 300 kilohertz.

low-frequency compensation

low-frequency compensation Compensation that serves to extend the frequency range of a broad-band amplifier to lower frequencies.

low-frequency current An alternating current having a frequency of less than about 300 kilohertz.

low-frequency cutoff A frequency below which the gain of a system or device decreases rapidly.

low-frequency gain The gain of the voltage amplifier at frequencies less than those frequencies at which this gain is close to its maximum value.

low-frequency impedance corrector Electric network designed to be connected to a basic network, or to a basic network and a building-out network, so that the combination will simulate, at low frequencies, the sending-end impedance, including dissipation, of a line.

low-frequency padder In a superheterodyne receiver, a small adjustable capacitor connected in series with the oscillator tuning coil and adjusted during alignment to obtain correct calibration of the circuit at the low-frequency end of the tuning range.

low-frequency propagation Propagation of radio waves at frequencies between 30 and 300 kilohertz.

low-frequency transconductance The change in the plate current of a vacuum tube divided by the change in the control-grid voltage that produces it, at frequencies small enough for these two quantities to be considered in phase.

low-frequency tube An electron tube operated at frequencies small enough so that the transit time of an electron between electrodes is much smaller than the period of oscillation of the voltage.

low-impedance measurement The measurement of an impedance which is small enough to necessitate use of indirect methods.

low-impedance switching tube A gas tube which has a static impedance on the order of 10,000 ohms, but zero or negative dynamic impedance, and therefore can be used as a relay and transmits information with negligible loss as well.

low level The less positive of the two logic levels or states in a digital logic system.

low-level logic circuit A modification of a diode-transistor logic circuit in which a resistor and capacitor in parallel are replaced by a diode, with the result that a relatively low voltage swing is required at the base of the transistor to switch it on or off. Abbreviated LLL circuit.

low-level modulation Modulation produced at a point in a system where the power level is low compared with the power level at the output of the system.

low-loss Having a small dissipation of electric or electromagnetic power.

low-noise amplifier An amplifier having very low background noise when the desired signal is weak or absent; field-effect transistors are used in audio preamplifiers for this purpose.

low-noise preamplifier A low-noise amplifier placed in a system prior to the main amplifier, sometimes close to the source; used to establish a satisfactory noise figure at an early point in the system.

low-pass filter A filter that transmits alternating currents below a given cutoff frequency and substantially attenuates all other currents.

low-Q filter A filter in which the energy dissipated in each cycle is a fairly large fraction of the energy stored in the filter.

low-reactance grounding Use of grounding connections with a moderate amount of inductance to effect a moderate reduction in the short-circuit current created by a line-to-ground fault.

low voltage 1. Voltage which is small enough to be regarded as safe for indoor use, usually 120 volts in the United States. 2. Voltage which is less than that needed for normal operation; a result of low voltage may be burnout of electric motors due to loss of electromotive force.

low-voltage winding The coil of wire wound around the core of a power transformer which has the smaller number of turns, and therefore the lower voltage.

LSA diode A microwave diode in which a space charge is developed in the semiconductor by the applied electric field and is dissipated during each cycle before it builds up appreciably, thereby limiting transit time and increasing the maximum frequency of oscillation. Derived from limited space-charge accumulation diode.

L scan *See* L scope.

L scope A cathode-ray scope on which a trace appears as a vertical or horizontal range scale, the signals appearing as left and right horizontal (or up and down vertical) deflections as echoes are received by two antennas, the left and right (or up and down) deflections being proportional to the strength of the echoes received by the two antennas. Also known as L indicator; L scan.

luminaire An electric lighting fixture, wall bracket, portable lamp, or other complete lighting unit designed to contain one or more electric lighting sources and associated reflectors, refractors, housing, and such support for those items as necessary.

luminescent cell *See* electroluminescent panel.

luminescent center A point-lattice defect in a transparent crystal that exhibits luminescence.

luminescent screen The screen in a cathode-ray tube, which becomes luminous when bombarded by an electron beam and maintains its luminosity for an appreciable time.

luminous sensitivity of phototube Quotient of the anode current by the incident luminous flux.

lumped constant A single constant that is electrically equivalent to the total of that type of distributed constant existing in a coil or circuit. Also known as lumped parameter.

lumped-constant network An analytical tool in which distributed constants (inductance, capacitance, and resistance) are represented as hypothetical components.

lumped discontinuity An analytical tool in the study of microwave circuits in which the effective values of inductance, capacitance, and resistance representing a discontinuity in a waveguide are shown as discrete components of equivalent value.

lumped element A section of a transmission line designed so that electric or magnetic energy is concentrated in it at specified frequencies, and inductance or capacitance may therefore be regarded as concentrated in it, rather than distributed over the length of the line.

lumped impedance An impedance concentrated in a single component rather than distributed throughout the length of a transmission line.

lumped parameter *See* lumped constant.

Luneberg lens A type of antenna consisting of a dielectric sphere whose index of refraction varies with distance from the center of the sphere so that a beam of parallel rays falling on the lens is focused at a point on the lens surface diametrically opposite from the direction of incidence, and, conversely, energy emanating from a point on the surface is focused into a plane wave. Also spelled Luneburg lens.

Luneburg lens *See* Luneberg lens.

LVDT *See* linear variable-differential transformer.

Lyddane-Sachs-Teller relation For an infinite ionic crystal, the relation $\epsilon(0)/\epsilon(\infty) = \omega_L^2/\omega_T^2$, where $\epsilon(0)$ is the crystal's static dielectric constant, $\epsilon(\infty)$ is the dielectric constant at a frequency at which electronic polarizability is effective but ionic polarizability is not, ω_L is the frequency of longitudinal optical phonons with zero wave vectors, and ω_T is the frequency of transverse optical phonons with large wave vector.

M

mA *See* milliampere.

machine switching system *See* automatic exchange.

MAD *See* magnetic anomaly detector.

Madelung constant A dimensionless constant which determines the electrostatic energy of a three-dimensional periodic crystal lattice consisting of a large number of positive and negative point charges when the number and magnitude of the charges and the nearest-neighbor distance between them is specified.

madistor A cryogenic semiconductor device in which injection plasma can be steered or controlled by transverse magnetic fields, to give the action of a switch.

MADT *See* microalloy diffused transistor.

MAG *See* maximum available gain.

magamp *See* magnetic amplifier.

magic eye *See* cathode-ray tuning indicator.

magic tee *See* hybrid tee.

magn A unit of absolute permeability equal to 1 henry per meter; it has been proposed by the Soviet Union but has not won general acceptance.

magnesium anode Bar of magnesium buried in the earth, connected to an underground cable to prevent cable corrosion due to electrolysis.

magnesium cell A primary cell in which the negative electrode is made of magnesium or one of its alloys.

magnesium–copper sulfide rectifier Dry-disk rectifier consisting of magnesium in contact with copper sulfide.

magnesium–manganese dioxide cell Type of electrochemical (dry) cell battery in which the active elements are magnesium and manganese dioxide.

magnesium–silver chloride cell A reserve primary cell that is activated by adding water; active elements are magnesium and silver chloride.

magnesyn A portion of a repeater unit; a two-pole permanently magnetized rotor within a three-phase two-pole delta-connected stator which carries the indicating pointer and is free to rotate in any direction.

magnet A piece of ferromagnetic or ferrimagnetic material whose domains are sufficiently aligned so that it produces a net magnetic field outside itself and can experience a net torque when placed in an external magnetic field.

magnet coil *See* iron-core coil.

magnetic Pertaining to magnetism or a magnet.

260 magnetic amplifier

magnetic amplifier A device that employs saturable reactors to modulate the flow of alternating-current electric power to a load in response to a lower-energy-level direct-current input signal. Abbreviated magamp. Also known as transductor.

magnetic anisotropy The dependence of the magnetic properties of some materials on direction.

magnetic anomaly detector A sensitive magnetometer carried at the end of a boom projecting from the tail of a patrol plane which can detect very small changes in the earth's magnetic field caused by a ferrous object, such as a submerged submarine; used to pinpoint a submarine's location, for effective deployment of suitable weapons. Abbreviated MAD.

magnetic bias A steady magnetic field applied to the magnetic circuit of a relay or other magnetic device.

magnetic blowout 1. A permanent magnet or electromagnet used to produce a magnetic field that lengthens the arc between opening contacts of a switch or circuit breaker, thereby helping to extinguish the arc. 2. *See* blowout.

magnetic bubble A cylindrical stable (nonvolatile) region of magnetization produced in a thin-film magnetic material by an external magnetic field; direction of magnetization is perpendicular to the plane of the material. Also known as bubble.

magnetic cell One unit of a magnetic memory, capable of storing one bit of information as a zero state or a one state.

magnetic circuit A group of magnetic flux lines each forming a closed path, especially when this circuit is regarded as analogous to an electric circuit because of the similarity of its magnetic field equations to direct-current circuit equations.

magnetic coercive force *See* coercive force.

magnetic constant The absolute permeability of empty space, equal to 1 electromagnetic unit in the centimeter-gram-second system, and to $4\pi \times 10^{-7}$ henry per meter or, numerically, to 1.25664×10^{-6} henry per meter in International System units. Symbolized μ_0.

magnetic core 1. A configuration of magnetic material, usually a mixture of iron oxide or ferrite particles mixed with a binding agent and formed into a tiny doughnutlike shape, that is placed in a spatial relationship to current-carrying conductors, and is used to maintain a magnetic polarization for the purpose of storing data, or for its nonlinear properties as a logic element. Also known as core; memory core. 2. A quantity of ferrous material placed in a coil or transformer to provide a better path than air for magnetic flux, thereby increasing the inductance of the coil and increasing the coupling between the windings of a transformer. Also known as core.

magnetic coupling For a pair of particles or systems, the effect of the magnetic field created by one system on the magnetic moment or angular momentum of the other.

magnetic Curie temperature The temperature below which a magnetic material exhibits ferromagnetism, and above which ferromagnetism is destroyed and the material is paramagnetic.

magnetic damping Damping of a mechanical motion by means of the reaction between a magnetic field and the current generated by the motion of a coil through the magnetic field.

magnetic deflection Deflection of an electron beam by the action of a magnetic field, as in a television picture tube.

magnetic delay line Delay line, used for the storage of data in a computer, consisting essentially of a metallic medium along which the velocity of the propagation of magnetic energy is small compared to the speed of light; storage is accomplished by the recirculation of wave patterns containing information, usually in binary form.

magnetic diffusivity A measure of the tendency of a magnetic field to diffuse through a conducting medium at rest; it is equal to the partial derivative of the magnetic field strength with respect to time divided by the Laplacian of the magnetic field, or to the reciprocal of $4\pi\mu\sigma$, where μ is the magnetic permeability and σ is the conductivity in electromagnetic units.

magnetic dipole An object, such as a permanent magnet, current loop, or particle with angular momentum, which experiences a torque in a magnetic field, and itself gives rise to a magnetic field, as if it consisted of two magnetic poles of opposite sign separated by a small distance.

magnetic dipole antenna Simple loop antenna capable of radiating an electromagnetic wave in response to a circulation of electric current in the loop.

magnetic dipole density See magnetization.

magnetic dipole moment A vector associated with a magnet, current loop, particle, or such, whose cross product with the magnetic induction (or alternatively, the magnetic field strength) of a magnetic field is equal to the torque exerted on the system by the field. Also known as dipole moment; magnetic moment.

magnetic disk storage See disk storage.

magnetic displacement See magnetic induction.

magnetic domain See ferromagnetic domain.

magnetic drum See drum.

magnetic drum receiving equipment Radar developed for detection of targets beyond line of sight using ionospheric reflection and very low power.

magnetic drum storage See drum.

magnetic energy The energy required to set up a magnetic field.

magnetic ferroelectric A substance which possesses both magnetic ordering and spontaneous electric polarization.

magnetic field 1. One of the elementary fields in nature; it is found in the vicinity of a magnetic body or current-carrying medium and, along with electric field, in a light wave; charges moving through a magnetic field experience the Lorentz force. 2. See magnetic field strength.

magnetic field intensity See magnetic field strength.

magnetic field strength An auxiliary vector field, used in describing magnetic phenomena, whose curl, in the case of static charges and currents, equals (in meter-kilogram-second units) the free current density vector, independent of the magnetic permeability of the material. Also known as magnetic field; magnetic field intensity; magnetic force; magnetic intensity; magnetizing force.

magnetic film See magnetic thin film.

magnetic firing circuit A type of firing circuit in which the capacitor is discharged through the igniter by saturating a reactor, which is connected in series with the capacitor; often used in ignitron rectifiers to obtain longer life and greater reliability than is possible with thyratron firing tubes.

magnetic flaw detector A flaw detector in which a ferrous object is magnetized with an electromagnet or permanent magnet and sprayed with magnetic particles or a solution containing fine suspended magnetic particles which outline surface or near-surface flaws.

magnetic flux 1. The integral over a specified surface of the component of magnetic induction perpendicular to the surface. 2. See magnetic lines of force.

magnetic flux density See magnetic induction.

262　magnetic focusing

magnetic focusing　Focusing a beam of electrons or other charged particles by using the action of a magnetic field.

magnetic force　*See* magnetic field strength.

magnetic gap　The space between a magnet's pole faces.

magnetic groups　*See* Shubnikov groups.

magnetic head　The electromagnet used for reading, recording, or erasing signals on a magnetic disk, drum, or tape. Also known as magnetic read/write head.

magnetic hysteresis　Lagging of changes in the magnetization of a substance behind changes in the magnetic field as the magnetic field is varied. Also known as hysteresis.

magnetic induction　A vector quantity that is used as a quantitative measure of magnetic field; the force on a charged particle moving in the field is equal to the particle's charge times the cross product of the particle's velocity with the magnetic induction (mks units). Also known as magnetic displacement; magnetic flux density; magnetic vector.

magnetic intensity　*See* magnetic field strength.

magnetic leakage　Passage of magnetic flux outside the path along which it can do useful work.

magnetic lens　A magnetic field with axial symmetry, capable of converging beams of charged particles of uniform velocity and of forming images of objects placed in the path of such beams; the field may be produced by solenoids, electromagnets, or permanent magnets. Also known as lens.

magnetic lines of flux　*See* magnetic lines of force.

magnetic lines of force　Lines used to represent the magnetic induction in a magnetic field, selected so that they are parallel to the magnetic induction at each point, and so that the number of lines per unit area of a surface perpendicular to the induction is equal to the induction. Also known as magnetic flux; magnetic lines of flux.

magnetic material　A material exhibiting ferromagnetism.

magnetic-memory plate　Magnetic memory consisting of a ferrite plate having a grid of small holes through which the read-in and read-out wires are threaded; printed wiring may be applied directly to the plate in place of conventionally threaded wires, permitting mass production of plates having a high storage capacity.

magnetic modulator　A modulator in which a magnetic amplifier serves as the modulating element for impressing an intelligence signal on a carrier.

magnetic moment　*See* magnetic dipole moment.

magnetic monopole　A hypothetical particle carrying magnetic charge; it would be a source for magnetic field in the same way that a charged particle is a source for electric field. Also known as monopole.

magnetic multipole　One of a series of types of static or oscillating distributions of magnetization, which is a magnetic multipole of order 2; the electric and magnetic fields produced by a magnetic multipole of order 2^n are equivalent to those of two magnetic multipoles of order 2^{n-1} of equal strength but opposite sign, separated from each other by a short distance.

magnetic multipole field　The electric and magnetic fields generated by a static or oscillating magnetic multipole.

magnetic needle　1. A bar magnet or collection of bar magnets which is hung so as to show the direction of the magnetic field. 2. In particular, a slender bar magnet, pointed at both ends, that is pivoted or freely suspended in a magnetic compass.

magnetic rigidity   263

magnetic octupole moment A quantity characterizing a distribution of magnetization; obtained by integrating the product of the divergence of the magnetization, the third power of the distance from the origin, and a spherical harmonic Y^*_{3m} over the magnetization distribution.

magnetic oscillograph An instrument that records a trace measuring one component of the earth's magnetic field.

magnetic pendulum A bar magnet which is hung by a thread or balanced on a pivot so that it oscillates in a horizontal plane when disturbed and released in a magnetic field having a horizontal component.

magnetic pinch See pinch effect.

magnetic polarization See intrinsic induction.

magnetic pole 1. One of two regions located at the ends of a magnet that generate and respond to magnetic fields in much the same way that electric charges generate and respond to electric fields. 2. A particle which generates and responds to magnetic fields in exactly the same way that electric charges generate and respond to electric fields; the particle probably does not have physical reality, but it is often convenient to imagine that a magnetic dipole consists of two magnetic poles of opposite sign, separated by a small distance.

magnetic pole strength The magnitude of a (fictional) magnetic pole, equal to the force exerted on the pole divided by the magnetic induction (or, alternatively, by the magnetic field strength). Also known as pole strength.

magnetic potential See magnetic scalar potential.

magnetic printing The permanent and usually undesired transfer of a recorded signal from one section of a magnetic recording medium to another when these sections are brought together, as on a reel of tape. Also known as crosstalk; magnetic transfer.

magnetic probe A small coil inserted in a magnetic field to measure changes in field strength.

magnetic pumping A method of moving a conducting liquid by applying a magnetic field which varies with time.

magnetic quadrupole lens A magnetic field generated by four magnetic poles of alternating sign arranged in a circle; used to focus beams of charged particles in devices such as electron microscopes and particle accelerators.

magnetic read/write head See magnetic head.

magnetic recorder An instrument that records information, generally in the form of audio-frequency or digital signals, on magnetic tape or magnetic wire as magnetic variations in the medium.

magnetic recording Recording by means of a signal-controlled magnetic field.

magnetic reed switch See reed switch.

magnetic reluctance See reluctance.

magnetic reproducer An instrument which moves a magnetic recording medium, such as a tape, wire, or disk, past an electromagnetic transducer that converts magnetic signals on the medium into electric signals.

magnetic reproducing The conversion of information on magnetic tape or magnetic wire, which was originally produced by electric signals, back into electric signals.

magnetic rigidity A measure of the momentum of a particle moving perpendicular to a magnetic field, equal to the magnetic induction times the particle's radius of curvature.

magnetics The study of magnetic phenomena, comprising magnetostatics and electromagnetism.

magnetic saturation The condition in which, after a magnetic field strength becomes sufficiently large, further increase in the magnetic field strength produces no additional magnetization in a magnetic material. Also known as saturation.

magnetic scalar potential The work which must be done against a magnetic field to bring a magnetic pole of unit strength from a reference point (usually at infinity) to the point in question. Also known as magnetic potential.

magnetic shell Two layers of magnetic charge of opposite sign, separated by an infinitesimal distance.

magnetic shielding See magnetostatic shielding.

magnetic shunt Piece of iron, usually adjustable as to position, used to divert a portion of the magnetic lines of force passing through an air gap in an instrument or other device.

magnetic stepping motor See stepper motor.

magnetic strain energy The potential energy of a magnetic domain, subject to both a tensile stress and a magnetic field, associated with the domain's magnetostriction expansion.

magnetic susceptibility The ratio of the magnetization of a material to the magnetic field strength; it is a tensor when these two quantities are not parallel; otherwise it is a simple number. Also known as susceptibility.

magnetic tape A plastic, paper, or metal tape that is coated or impregnated with magnetizable iron oxide particles; used in magnetic recording.

magnetic tape core Toroidal core formed by winding a strip of thin magnetic-core material around a form.

magnetic tape reader A computer device that is capable of reading information recorded on magnetic tape by transforming this information into electric pulses.

magnetic test coil See exploring coil.

magnetic thermometer A sample of a paramagnetic salt whose magnetic susceptibility is measured and whose temperature is then calculated from the inverse relationship between the two quantities; useful at temperatures below about 1 K.

magnetic thin film A sheet or cylinder of magnetic material less than 5 micrometers thick, usually possessing uniaxial magnetic anisotropy; used mainly in computer storage and logic elements. Also known as ferromagnetic film; magnetic film.

magnetic transducer A device for transforming mechanical into electrical energy, which consists of a magnetic field including a variable-reluctance path and a coil surrounding all or a part of this path, so that variation in reluctance leads to a variation in the magnetic flux through the coil and a corresponding induced emf (electromotive force).

magnetic transfer See magnetic printing.

magnetic vector See magnetic induction.

magnetic vector potential See vector potential.

magnetic wave The spread of magnetization from a small portion of a substance where an abrupt change in the magnetic field has taken place.

magnetic-wave device A device that depends on magnetoelastic or magnetostatic wave propagation through or on the surface of a magnetic or dielectric material.

magnetization 1. The property and in particular, the extent of being magnetized; quantitatively, the magnetic moment per unit volume of a substance. Also known as magnetic dipole density; magnetization intensity. 2. The process of magnetizing a magnetic material.

magnetization curve See B-H curve; normal magnetization curve.

magnetizing current The current that flows through the primary winding of a power transformer when no loads are connected to the secondary winding; this current established the magnetic field in the core and furnished energy for the no-load power losses in the core. Also known as exciting current.

magnetizing force See magnetic field strength.

magnet keeper See keeper.

magneto An alternating-current generator that uses one or more permanent magnets to produce its magnetic field; frequently used as a source of ignition energy on tractor, marine, industrial, and aviation engines. Also known as magnetoelectric generator.

magnetoelastic coupling The interaction between the magnetization and the strain of a magnetic material.

magnetoelasticity Phenomenon in which an elastic strain alters the magnetization of a ferromagnetic material.

magnetoelectric effect A linear coupling between magnetization and polarization found in certain magnetic ferroelectrics, such as $BaMnF_4$ at low temperatures.

magnetoelectric generator See magneto.

magnetoelectricity 1. Magnetic techniques for generating voltages, such as in an ordinary generator. 2. The appearance of an electric field in certain substances, such as chromic oxide (Cr_2O_3), when they are subjected to a static magnetic field.

magnetogalvanic effect See galvanomagnetic effect.

magnetograph A set of three variometers attached to a suitable recording unit, which records the components of the magnetic field vector in each of three perpendicular directions.

magnetohydrodynamic generator A system for generating electric power in which the kinetic energy of a flowing conducting fluid is converted to electric energy by a magnetohydrodynamic interaction. Abbreviated MHD generator.

magneto ignition system An ignition system in which the voltage required to cause a flow of current in the primary winding of the ignition coil is generated by a set of permanent magnets, instead of being supplied by a battery.

magnetomotive force The work that would be required to carry a magnetic pole of unit strength once around a magnetic circuit. Abbreviated mmf.

magnetooptical modulator An arrangement for modulating a beam of light by passing it through a single crystal of yttrium iron garnet, which provides intensity modulation by using a magnetic field to produce optical rotation.

magnetoplasmadynamics The generation of electric current by shooting a beam of ionized gas through a magnetic field, to give the same effect as moving copper bars near a magnet.

magnetoresistance The change in electrical resistance produced in a current-carrying conductor or semiconductor on application of a magnetic field.

magnetoresistivity The change in resistivity produced in a current-carrying conductor or semiconductor on application of a magnetic field.

magnetoresistor Magnetic field–controlled variable resistor.

magnetostatic Pertaining to magnetic properties that do not depend upon the motion of magnetic fields.

magnetostatic mode A spin wave in a magnetic material whose wavelength is greater than about one-tenth the size of the sample.

magnetostatics The study of magnetic fields that remain constant with time. Also known as static magnetism.

magnetostatic shielding The use of an enclosure made of a high-permeability magnetic material to prevent a static magnetic field outside the enclosure from reaching objects inside it, or to confine a magnetic field within the enclosure. Also known as magnetic shielding.

magnetostriction The dependence of the state of strain (dimensions) of a ferromagnetic sample on the direction and extent of its magnetization.

magnetostriction transducer A transducer used with sonar equipment to change an alternating current to sound energy at the same frequency and to form the sound energy into a beam; its operation depends on the interaction between the magnetization and the deformation of a material having magnetostrictive properties.

magnetostrictive delay line A delay line made of nickel or other magnetostrictive material, in which the amount of delay is determined by a shock wave traveling through the length of the line at the speed of sound.

magnetostrictive filter Filter network which uses the magnetostrictive phenomena to form high-pass, low-pass, band-pass, or band-elimination filters; the impedance characteristic is the inverse of that of a crystal.

magnetostrictive oscillator An oscillator whose frequency is controlled by a magnetostrictive element.

magnetostrictive resonator Ferromagnetic rod so designed that it can be excited magnetically into resonant vibration at one or more definite and known frequencies.

magnetostrictor A device for converting electric oscillations to mechanical oscillations by employing the property of magnetostriction.

magneto telephone set Local battery telephone set in which current for signaling by the telephone station is supplied from a local hand generator, usually a magneto.

magnet power The electric power supplied to the coils of an electromagnet.

magnetron One of a family of crossed-field microwave tubes, wherein electrons, generated from a heated cathode, move under the combined force of a radial electric field and an axial magnetic field in such a way as to produce microwave radiation in the frequency range 1–40 gigahertz; a pulsed microwave radiation source for radar, and continuous source for microwave cooking.

magnetron oscillator Oscillator circuit employing a magnetron tube.

magnetron pulling Frequency shift of a magnetron caused by factors which vary the standing waves or the standing-wave ratio on the radio-frequency lines.

magnetron pushing Frequency shift of a magnetron caused by faulty operation of the modulator.

magnetron vacuum gage A vacuum gage that is essentially a magnetron operated beyond cutoff in the vacuum being measured.

magnet wire The insulated copper or aluminum wire used in the coils of all types of electromagnetic machines and devices.

magnistor A device that utilizes the effects of magnetic fields on injection plasmas in semiconductors such as indium antimonide.

magnon A quasi-particle which is introduced to describe small departures from complete ordering of electronic spins in ferro-, ferri-, antiferro-, and helimagnetic arrays. Also known as quantized spin wave.

manual telephone set   267

mag-slip   See synchro.

main   1. One of the conductors extending from the service switch, generator bus, or converter bus to the main distribution center in interior wiring.   2. See power transmission line.

main-and-transfer bus   A substation switching arrangement similar to a single bus but with an additional transfer bus provided.

main bang   Transmitted pulse, within a radar system.

main distributing frame   Frame which terminates the permanent outside line entering the central office building on one side and the subscriber-line multiple cabling, trunk multiple cabling, and so on, used for associating an outside line with any desired terminal on the other side; it usually carries the control-office protective devices, and functions as a test point between line and office. Also known as main frame.

main exciter   Exciter which supplies energy for the field excitation of a principal electric machine.

main frame   See main distributing frame.

main lobe   See major lobe.

main sweep   On certain fire-control radar, the longest range scale available.

majority carrier   The type of carrier, that is, electron or hole, that constitutes more than half the carriers in a semiconductor.

majority emitter   Of a transistor, an electrode from which a flow of minority carriers enters the interelectrode region.

major lobe   Antenna lobe indicating the direction of maximum radiation or reception. Also known as main lobe.

major node   A point in an electrical network at which three or more elements are connected together. Also known as junction.

major relay station   Tape relay station which has two or more trunk circuits connected thereto to provide an alternate route or to meet command requirements.

make   Closing of relay, key, or other contact.

make contact   Contact of a device which closes a circuit upon the operation of the device (normally open).

making current   The peak value attained by the current during the first cycle after a switch, circuit breaker, or similar apparatus is closed.

Malter effect   A phenomenon in which a metal with a nonconducting surface film has a large coefficient of secondary electron emission; this is particularly notable in aluminum whose surface has been oxidized and then coated with cesium oxide.

Manchester plate   A storage battery consisting of a heavy alloy grid with circular openings into which are pressed pure lead buttons that are made from lead tape by crimping and rolling to develop a large surface area and are coated with lead peroxide, PbO_2.

manual rate-aided tracking   Radar circuit which tracks individual targets by computing the velocity from position fixes inserted manually into the circuitry.

manual switchboard   Telephone switchboard in which the connections are made manually, by plugs and jacks, or by keys.

manual switching   Method by which manual connection is made between two or more teletypewriter circuits.

manual telephone set   Telephone set not equipped with a dial.

268 Marconi antenna

Marconi antenna Antenna system of which the ground is an essential part, as distinguished from a Hertz antenna.

marginal checking A preventive-maintenance procedure in which certain operating conditions, such as supply voltage or frequency, are varied about their normal values in order to detect and locate incipient defective units.

marginal relay Relay with a small margin between its nonoperative current value (maximum current applicable without operation) and its operative value (minimum current that operates the relay).

marginal test A test of electronic equipment in which conditions are varied until failures occur or faults can be detected, allowing measurement of permissible operating margins.

marking current Magnitude and polarity of current in the line when the receiving mechanism is in the operating position.

marking pulse In a teletypewriter, the signal interval during which time the teletypewriter selector unit is operated.

marking wave In telegraphic communications, that portion of the emission during which the active portions of the code character are being transmitted. Also known as keying wave.

mark-space multiplier A multiplier used in analog computers in which one input controls the mark-to-space ratio of a square wave while the other input controls the amplitude of the wave, and the output, obtained by a smoothing operation, is proportional to the average value of the signal. Also known as time-division multiplier.

mark-space ratio *See* mark-to-space ratio.

mark-to-space ratio The ratio of the duration of the positive-amplitude part of a square wave to that of the negative-amplitude part. Also known as mark-space ratio.

Marx circuit An electric circuit used in an impulse generator in which capacitors are charged in parallel through charging resistors, and then connected in series and discharged through the test piece by the simultaneous sparkover of spark gaps.

Marx effect The effect wherein the energy of photoelectrons emitted from an illuminated surface is decreased when the surface is simultaneously illuminated by light of lower frequency than that causing the emission.

maser amplifier A maser which is used to increase the power produced by another maser.

mash connection *See* delta connection.

mask A thin sheet of metal or other material containing an open pattern, used to shield selected portions of a semiconductor or other surface during a deposition process.

masking 1. Using a covering or coating on a semiconductor surface to provide a masked area for selective deposition or etching. 2. A programmed procedure for eliminating radar coverage in areas where such transmissions may be of use to the enemy for navigation purposes, by weakening the beam in appropriate directions or by use of additional transmitters on the same frequency at suitable sites to interfere with homing; also used to suppress the beam in areas where it would interfere with television reception.

mass resistivity The product of the electrical resistance of a conductor and its mass, divided by the square of its length; the product of the electrical resistivity and the density.

master gain Control of overall gain of an amplifying system as opposed to varying the gain of several individual inputs.

master multivibrator Master oscillator using a multivibrator unit.

master oscillator An oscillator that establishes the carrier frequency of the output of an amplifier or transmitter.

master-oscillator power amplifier Transmitter using an oscillator followed by one or more stages of radio-frequency amplification.

master plan position indicator In a radar system, a plan position indicator which controls remote indicators or repeaters.

master switch 1. Switch that dominates the operation contactors, relays, or other magnetically operated devices. 2. Switch electrically ahead of a number of individual switches.

matched filter A filter with the property that, when the input consists of noise in addition to a specified desired signal, the signal-to-noise ratio is the maximum which can be obtained in any linear filter.

matched impedance An impedance of a load which is equal to the impedance of a generator, so that maximum power is delivered to the load.

matched load A load having the impedance value that results in maximum absorption of energy from the signal source.

matched transmission line Transmission line terminated with a load equivalent to its characteristic impedance.

matching Connecting two circuits or parts together with a coupling device in such a way that the maximum transfer of energy occurs between the two circuits, and the impedance of either circuit will be terminated in its image.

matching diaphragm Diaphragm consisting of a slit in a thin sheet of metal, placed transversely across a waveguide for matching purposes; the orientation of the slit with respect to the long dimension of the waveguide determines whether the diaphragm acts as a capacitive or inductive reactance.

matching impedance Impedance value that must be connected to the terminals of a signal-voltage source for proper matching.

matching section A section of transmission line, a quarter or half wavelength long, inserted between a transmission line and a load to obtain impedance matching.

matching stub Device placed on a radio-frequency transmission line which varies the impedance of the line; the impedance of the line can be adjusted in this manner.

matrix 1. The section of a color television transmitter that transforms the red, green, and blue camera signals into color-difference signals and combines them with the chrominance subcarrier. Also known as color coder; color encoder; encoder. 2. The section of a color television receiver that transforms the color-difference signals into the red, green, and blue signals needed to drive the color picture tube. Also known as color decoder; decoder.

Matthias' rules Several empirical rules giving the dependence of the transition temperatures of superconducting metals and alloys on the position of the metals in the periodic table and in the composition of the alloys.

Matthiessen's rule An empirical rule which states that the total resistivity of a crystalline metallic specimen is the sum of the resistivity due to thermal agitation of the metal ions of the lattice and the resistivity due to imperfections in the crystal.

mattress array *See* billboard array.

mavar *See* parametric amplifier.

maximum available gain The theoretical maximum power gain available in a transistor stage; it is seldom achieved in practical circuits because it can be approached only when feedback is negligible. Abbreviated MAG.

maximum average power output In television, the maximum of radio-frequency output power that can occur under any combination of signals transmitted, averaged over the longest repetitive modulation cycle.

maximum demand The greatest average value of the power, apparent power, or current consumed by a customer of an electric power system, the averages being taken over successive time periods, usually 15 or 30 minutes in length.

maximum keying frequency In facsimile, the frequency in hertz that is numerically equal to the spot speed divided by twice the horizontal dimension of the spot.

maximum modulating frequency Highest picture frequency required for a facsimile transmission system; the maximum modulating frequency and the maximum keying frequency are not necessarily equal.

maximum retention time Maximum time between writing into and reading an acceptable output from a storage element of a charge storage tube.

maximum signal level In an amplitude-modulated facsimile system, the level corresponding to copy black or copy white, whichever has the highest amplitude.

maximum unambiguous range The range beyond which the echo from a pulsed radar signal returns after generation of the next pulse, and can thus be mistaken as a short-range echo of the next cycle.

maximum undistorted power output Of a transducer, the maximum power delivered under specified conditions with a total harmonic output not exceeding a specified percentage.

maxwell A centimeter-gram-second electromagnetic unit of magnetic flux, equal to the magnetic flux which produces an electromotive force of 1 abvolt in a circuit of one turn linking the flux, as the flux is reduced to zero in 1 second at a uniform rate. Abbreviated Mx. Also known as abweber (abWb); line of magnetic induction.

Maxwell bridge A four-arm alternating-current bridge used to measure inductance (or capacitance) in terms of resistance and capacitance (or inductance); bridge balance is independent of frequency. Also known as Maxwell-Wien bridge; Wien-Maxwell bridge.

Maxwell field equations Four differential equations which relate the electric and magnetic fields to electric charges and currents, and form the basis of the theory of electromagnetic waves. Also known as electromagnetic field equations; Maxwell equations.

Maxwell relation According to Maxwell's electromagnetic theory, that relation wherein the dielectric constant of a substance equals the square of its index of refraction.

Maxwell's displacement current See displacement current.

Maxwell's electromagnetic theory A mathematical theory of electric and magnetic fields which predicts the propagation of electromagnetic radiation, and is valid for electromagnetic phenomena where effects on an atomic scale can be neglected.

Maxwell's law A movable portion of a circuit will always move in such a direction as to give maximum magnetic flux linkages through the circuit.

Maxwell's stress tensor A second-rank tensor whose product with a unit vector normal to a surface gives the force per unit area transmitted across the surface by an electromagnetic field.

maxwell-turn A centimeter-gram-second electromagnetic unit of flux linkage, equal to the flux linkage of a coil consisting of one complete loop of wire through which passes a magnetic flux of one maxwell. Also known as line-turn.

Maxwell-Wagner mechanism A capacitor consisting of two parallel metal plates with two layers of material between them, one with vanishing conductivity, the other with finite conductivity and vanishing electric susceptibility.

Maxwell-Wien bridge See Maxwell bridge.

MBE See molecular beam epitaxy.

M center A color center consisting of an F center combined with two ion vacancies.

McNally tube Reflex klystron tube, the frequency of which may be electrically controlled over a wide range; used as a local oscillator.

M-derived filter A filter consisting of a series of T or pi sections whose impedances are matched at all frequencies, even though the sections may have different resonant frequencies.

M display A modified radarscope A display in which target distance is determined by moving an adjustable pedestal signal along the baseline until it coincides with the horizontal position of the target deflection.

meaconing A system for receiving electromagnetic signals and rebroadcasting them with the same frequency so as, for instance, to confuse navigation; a confusion reflector, such as chaff, is an example.

mean carrier frequency Average carrier frequency of a transmitter corresponding to the resting frequency in a frequency-modulated system.

mean power of a radio transmitter Power supplied to the antenna transmission line by a transmitter during normal operation, averaged over a time sufficiently long compared with the period of the lowest frequency encountered in the modulation; a time of 1/10 second during which the mean power is greatest will be selected normally.

mechanical dialer See automatic dialer.

mechanical filter Filter, used in intermediate-frequency amplifiers of highly selective superheterodyne receivers, consisting of shaped metal rods that act as coupled mechanical resonators when used with piezoelectric input and output transducers. Also known as mechanical wave filter.

mechanical jamming See passive jamming.

mechanical modulator A device that varies a carrier wave by moving some part of a circuit element.

mechanical rectifier A rectifier in which rectification is accomplished by mechanical action, as in a synchronous vibrator.

mechanical stepping motor A device in which a voltage pulse through a solenoid coil causes reciprocating motion by a solenoid plunger, and this is transformed into rotary motion through a definite angle by ratchet-and-pawl mechanisms or other mechanical linkages.

mechanical tilt 1. Vertical tilt of the mechanical axis of a radar antenna. 2. The angle indicated by the tilt indicator dial.

mechanical wave filter See mechanical filter.

medical electronics A branch of electronics in which electronic instruments and equipment are used for such medical applications as diagnosis, therapy, research, anesthesia control, cardiac control, and surgery.

medium-frequency tube An electron tube operated at frequencies between 300 and 3000 kilohertz, at which the transit time of an electron between electrodes is much smaller than the period of oscillation of the voltage.

medium-scale integration Solid-state integrated circuits having more than about 12 gate-equivalent circuits. Abbreviated MSI.

megagauss A unit of magnetic induction equal to 10^6 gauss or 100 tesla.

megatron *See* disk-seal tube.

megavolt A unit of potential difference or emf (electromotive force), equal to 1,000,000 volts. Abbreviated MV.

megawatt year of electricity A unit of electric energy, equal to the energy from a power of 1,000,000 watts over a period of 1 tropical year, or to 3.1557×10^{13} joules. Abbreviated MWYE.

megohm A unit of resistance, equal to 1,000,000 ohms.

megohmmeter An instrument which is used for measuring the high resistance of electrical materials of the order of 20,000 megohms at 1000 volts; one direct-reading type employs a permanent magnet and a moving coil.

Meissner effect The expulsion of magnetic flux from the interior of a piece of superconducting material as the material undergoes the transition to the superconducting phase. Also known as flux jumping; Meissner-Ochsenfeld effect.

Meissner-Ochsenfeld effect *See* Meissner effect.

Meissner oscillator Electron-tube oscillator in which the grid and plate circuits are inductively coupled through an independent tank circuit which determines the frequency.

melodeon Broadband panoramic receiver used for countermeasures reception; all types of received electromagnetic radiation are presented as vertical pips on a frequency-calibrated cathode-ray indicator screen.

meltback transistor A junction transistor in which the junction is made by melting a properly doped semiconductor and allowing it to solidify again.

memistor Nonmagnetic memory device consisting of a resistive substrate in an electrolyte; when used in an adaptive system, a direct-current signal removes copper from an anode and deposits it on the substrate, thus lowering the resistance of the substrate; reversal of the current reverses the process, raising the resistance of the substrate.

memory core *See* magnetic core.

memory switch *See* ovonic memory switch.

memory tube *See* storage tube.

memotron An electrical-visual storage tube which is capable of bistable visual-signal display, controllable in duration from a few milliseconds to infinity, and which is suited to specialized oscillography.

mercury arc An electric discharge through ionized mercury vapor, giving off a brilliant bluish-green light containing strong ultraviolet radiation.

mercury-arc rectifier A gas-filled rectifier tube in which the gas is mercury vapor; small sizes use a heated cathode, while larger sizes rated up to 8000 kilowatts and higher use a mercury-pool cathode. Also known as mercury rectifier; mercury-vapor rectifier.

mercury cell A primary dry cell that delivers an essentially constant output voltage throughout its useful life by means of a chemical reaction between zinc and mercury oxide; widely used in hearing aids. Also known as mercury oxide cell.

mercury delay line An acoustic delay line in which mercury is the medium for sound transmission. Also known as mercury memory; mercury storage.

mercury lamp *See* mercury-vapor lamp.

mercury memory *See* mercury delay line.

mercury-pool cathode A cathode of a gas tube consisting of a pool of mercury; an arc spot on the pool emits electrons.

metadyne 273

mercury-pool rectifier *See* pool-cathode mercury arc rectifier.

mercury rectifier *See* mercury-are rectifier.

mercury storage *See* mercury delay line.

mercury switch A switch that is closed by making a large globule of mercury move up to the contacts and bridge them; the mercury is usually moved by tilting the entire switch.

mercury tank A container of mercury, with pairs of transducers at opposite ends, used in a mercury delay line.

mercury tube *See* mercury-vapor tube; pool tube.

mercury-vapor lamp A lamp in which light is produced by an electric arc between two electrodes in an ionized mercury-vapor atmosphere; it gives off a bluish-green light rich in ultraviolet radiation. Also known as mercury lamp.

mercury-vapor rectifier *See* mercury-arc rectifier.

mercury-vapor tube A gas tube in which the active gas is mercury vapor. Also known as mercury tube.

mercury-wetted reed switch A reed switch containing a pool of mercury at one end and normally operated vertically; the contacts on the reeds are covered with a mercury film by capillary action; each operation of the switch renews this mercury film contact, thereby increasing the operating life of the switch many times.

merged-transistor logic *See* integrated injection logic.

merit A performance rating that governs the choice of a device for a particular application; it must be qualified to indicate type of rating, as in gain-bandwidth merit or signal-to-noise merit.

mesa device Any device produced by diffusing the surface of a germanium or silicon wafer and then etching down all but selected areas, which then appear as physical plateaus or mesas.

mesa diode A diode produced by diffusing the entire surface of a large germanium or silicon wafer and then delineating the individual diode areas by a photoresist-controlled etch that removes the entire diffused area except the island or mesa at each junction site.

mesa transistor A transistor in which a germanium or silicon wafer is etched down in steps so the base and emitter regions appear as physical plateaus above the collector region.

MESFET *See* metal semiconductor field-effect transistor.

mesh A set of branches forming a closed path in a network so that if any one branch is omitted from the set, the remaining branches of the set do not form a closed path. Also known as loop.

mesh analysis A method of electrical circuit analysis in which the mesh currents are taken as independent variables and the potential differences around a mesh are equated to 0.

mesh currents The currents which are considered to circulate around the meshes of an electric network, so that the current in any branch of the network is the algebraic sum of the mesh currents of the meshes to which that branch belongs. Also known as cyclic currents; Maxwell's cyclic currents.

mesh impedance The ratio of the voltage to the current in a mesh when all other meshes are open. Also known as self-impedance.

metadyne A type of rotating magnetic amplifier having more than one brush per pole, used for voltage regulation or transformation.

metal-air battery See air depolarized battery.

metal antenna An antenna which has a relatively small metal surface, in contrast to a slot antenna.

metal-clad substation An electric power substation housed in a metal cabinet, either indoors or outdoors.

metal detector An electronic device for detecting concealed metal objects, such as guns, knives, or buried pipelines, generally by radiating a high-frequency electromagnetic field and detecting the change produced in that field by the ferrous or nonferrous metal object being sought. Also known as electronic locator; metal locator; radio metal locator.

metal-film resistor A resistor in which the resistive element is a thin film of metal or alloy, deposited on an insulating substrate of an integrated circuit.

metal halide lamp A discharge lamp in which metal halide salts are added to the contents of a discharge tube in which there is a high-pressure arc in mercury vapor; the added metals generate different wavelengths, to give substantially white light at an efficiency approximating that of high-pressure sodium lamps.

metal-insulator semiconductor Semiconductor construction in which an insulating layer, generally a fraction of a micrometer thick, is deposited on the semiconducting substrate before the pattern of metal contacts is applied. Abbreviated MIS.

metal-insulator transition The change of certain low-dimensional conductors from metals to insulators as the temperature is lowered through a certain value, due to the lattice distortion and band gap accompanying the onset of a charge-density wave.

metallic circuit Wire circuit of which the ground or earth forms no part.

metallic-disk rectifier See metallic rectifier.

metallic electrode arc lamp A type of arc lamp in which light is produced by luminescent vapor introduced into the arc by evaporation from the cathode; the anode is solid copper, and the cathode is formed of magnetic iron oxide with titanium as the light-producing element and other chemicals to control steadiness and vaporization.

metallic insulator Section of transmission line used as a mechanical support device; the section is an odd number of quarter-wavelengths long at the frequency of interest, and the input impedance becomes high enough so that the section effectively acts as an insulator.

metallic rectifier A rectifier consisting of one or more disks of metal under pressure-contact with semiconductor coatings or layers, such as a copper oxide, selenium, or silicon rectifier. Also known as contact rectifier; dry-disk rectifier; dry-plate rectifier; metallic-disk rectifier; semiconductor rectifier.

metallized capacitor A capacitor in which a film of metal is deposited directly on the dielectric to serve in place of a separate foil strip; has self-healing characteristics.

metallized-paper capacitor A modification of a paper capacitor in which metal foils are replaced by extremely thin films of metal deposited on the paper; if a breakdown occurs, these films burn away in the area of the breakdown.

metallized resistor A resistor made by depositing a thin film of high-resistance metal on the surface of a glass or ceramic rod or tube.

metal locator See metal detector.

metal-nitride-oxide semiconductor A semiconductor structure that has a double insulating layer; typically, a layer of silicon dioxide (SiO_2) is nearest the silicon substrate, with a layer of silicon nitride (Si_3N_4) over it. Abbreviated MNOS.

metal oxide resistor A metal-film resistor in which an oxide of a metal such as tin is deposited as a film onto an insulating substrate.

microcircuitry 275

metal oxide semiconductor A metal insulator semiconductor structure in which the insulating layer is an oxide of the substrate material; for a silicon substrate, the insulating layer is silicon dioxide (SiO_2). Abbreviated MOS.

metal oxide semiconductor field-effect transistor A field-effect transistor having a gate that is insulated from the semiconductor substrate by a thin layer of silicon dioxide. Abbreviated MOSFET; MOST; MOS transistor. Formerly known as insulated-gate field-effect transistor.

metal oxide semiconductor integrated circuit An integrated circuit using metal oxide semiconductor transistors; it can have a higher density of equivalent parts than a bipolar integrated circuit.

metal semiconductor field-effect transistor A field-effect transistor that uses a thin film of gallium arsenide, with a Schottky barrier gate formed by depositing a layer of metal directly onto the surface of the film. Abbreviated MESFET.

metascope An infrared receiver used for converting pulsed invisible infrared rays into visible signals for communication purposes; also used with an infrared source for reading maps in darkness.

meter-type relay A relay that uses a meter movement having a contact-bearing pointer which moves toward or away from a fixed contact mounted on the meter scale.

metric waves Radio waves having wavelengths between 1 and 10 meters, corresponding to frequencies between 30 and 300 megahertz (the very-high-frequency band).

mF *See* millifarad.

mG *See* milligauss.

mH *See* millihenry.

MHD generator *See* magnetohydrodynamic generator.

mho *See* siemens.

MIC *See* microwave integrated circuit.

mica capacitor A capacitor whose dielectric consists of thin rectangular sheets of mica and whose electrodes are either thin sheets of metal foil stacked alternately with mica sheets, or thin deposits of silver applied to one surface of each mica sheet.

micro *See* microcomputer.

microalloy diffused transistor A microalloy transistor in which the semiconductor wafer is first subjected to gaseous diffusion to produce a nonuniform base region. Abbreviated MADT.

microalloy transistor A transistor in which the emitter and collector electrodes are formed by etching depressions, then electroplating and alloying a thin film of the impurity metal to the semiconductor wafer, somewhat as in a surface-barrier transistor.

microammeter An ammeter whose scale is calibrated to indicate current values in microamperes.

microampere A unit of current equal to one-millionth of an ampere. Abbreviated µA.

microcapacitor Any very small capacitor used in microelectronics, usually consisting of a thin film of dielectric material sandwiched between electrodes.

microchannel plate A plate that consists of extremely small cylinder-shaped electron multipliers mounted side by side, to provide image intensification factors as high as 100,000.

microcircuitry Electronic circuit structures that are orders of magnitude smaller and lighter than circuit structures produced by the most compact combinations of discrete components. Also known as microelectronic circuitry; microminiature circuitry.

microcomputer A microprocessor combined with input/output interface devices, some type of external memory, and the other elements required to form a working computer system; it is smaller, lower in cost, and usually slower than a minicomputer. Also known as micro.

microcontroller A microcomputer, microprocessor, or other equipment used for precise process control in data handling, communication, and manufacturing.

microcoulomb A unit of electric charge equal to one-millionth of a coulomb. Abbreviated μC.

microelectronic circuitry See microcircuitry.

microelectronics The technology of constructing circuits and devices in extremely small packages by various techniques. Also known as microminiaturization; microsystem electronics.

microelement Resistor, capacitor, transistor, diode, inductor, transformer, or other electronic element or combination of elements mounted on a ceramic wafer 0.025 centimeter thick and about 0.75 centimeter square; individual microelements are stacked, interconnected, and potted to form micromodules.

microfarad A unit of capacitance equal to one-millionth of a farad. Abbreviated μF.

microhm A unit of resistance, reactance, and impedance equal to one-millionth of an ohm.

microhysteresis effect Hysteresis that results from the motion of domain walls lagging behind an applied magnetic or elastic stress when these walls are held up by dislocations and other imperfections in the material.

microlock 1. Satellite telemetry system that uses phase-lock techniques in the ground receiving equipment to achieve extreme sensitivity. 2. A lock by a tracking station upon a minitrack radio transmitter. 3. The system by which this lock is effected.

micromicrofarad See picofarad.

microminiature circuitry See microcircuitry.

microminiaturization See microelectronics.

micromodule Cube-shaped, plug-in, miniature circuit composed of potted microelements; each microelement can consist of a resistor, capacitor, transistor, or other element, or a combination of elements.

microphonics Noise caused by mechanical vibration of the elements of an electron tube, component, or system. Also known as microphonism.

microphonism See microphonics.

microprocessing unit A microprocessor with its external memory, input/output interface devices, and buffer, clock, and driver circuits. Abbreviated MPU.

microprocessor A single silicon chip on which the arithmetic and logic functions of a computer are placed.

micropump See electroosmotic driver.

microradiometer A radiometer used for measuring weak radiant power, in which a thermopile is supported on and connected directly to the moving coil of a galvanometer. Also known as radiomicrometer.

microstrip A strip transmission line that consists basically of a thin-film strip in intimate contact with one side of a flat dielectric substrate, with a similar thin-film ground-plane conductor on the other side of the substrate.

microsystem electronics *See* microelectronics.

microvolt A unit of potential difference equal to one-millionth of a volt. Abbreviated μV.

microvoltmeter A voltmeter whose scale is calibrated to indicate voltage values in microvolts.

microvolts per meter Field strength of antenna which is the ratio of the antenna voltage in microvolts to the antenna length in meters, as measured at a given point.

microwave An electromagnetic wave which has a wavelength between about 0.3 and 30 centimeters, corresponding to frequencies of 1–100 gigahertz; however, there are no sharp boundaries distinguishing microwaves from infrared and radio waves.

microwave amplifier A device which increases the power of microwave radiation.

microwave antenna A combination of an open-end waveguide and a parabolic reflector or horn, used for receiving and transmitting microwave signal beams at microwave repeater stations.

microwave attenuator A device that causes the field intensity of microwaves in a waveguide to decrease by absorbing part of the incident power; usually consists of a piece of lossy material in the waveguide along the direction of the electric field vector.

microwave bridge A microwave circuit equivalent to an ordinary electrical bridge and used to measure impedance; consists of six waveguide sections arranged to form a multiple junction.

microwave cavity *See* cavity resonator.

microwave circuit Any particular grouping of physical elements, including waveguides, attenuators, phase changers, detectors, wavemeters, and various types of junctions, which are arranged or connected together to produce certain desired effects on the behavior of microwaves.

microwave circulator *See* circulator.

microwave detector A device that can demonstrate the presence of a microwave by a specific effect that the wave produces, such as a bolometer, or a semiconductor crystal making a pinpoint contact with a tungsten wire.

microwave device Any device capable of generating, amplifying, modifying, detecting, or measuring microwaves, or voltages having microwave frequencies.

microwave filter A device which passes microwaves of certain frequencies in a transmission line or waveguide while rejecting or absorbing other frequencies; consists of resonant cavity sections or other elements.

microwave generator *See* microwave oscillator.

microwave gyrator *See* gyrator.

microwave heating Heating of food by means of electromagnetic energy in or just below the microwave spectrum for cooking, dehydration, sterilization, thawing, and other purposes.

microwave integrated circuit A microwave circuit that uses integrated-circuit production techniques involving such features as thin or thick films, substrates, dielectrics, conductors, resistors, and microstrip lines, to build passive assemblies on a dielectric. Abbreviated MIC.

microwave optics The study of those properties of microwaves which are analogous to the properties of light waves in optics.

microwave oscillator A type of electron tube or semiconductor device used for generating microwave radiation or voltage waveforms with microwave frequencies. Also known as microwave generator.

278 microwave radiometer

microwave radiometer *See* radiometer.

microwave receiver Complete equipment that is needed to convert modulated microwaves into useful information.

microwave reflectometer A pair of single-detector couplers on opposite sides of a waveguide, one of which is positioned to monitor transmitted power, and the other to measure power reflected from a single discontinuity in the line.

microwave refractometer An instrument that measures the index of refraction of the atmosphere by measuring the travel time of microwave signals through each of two precision microwave transmission cavities, one of which is hermetically sealed to serve as a reference.

microwave resonance cavity *See* cavity resonator.

microwave solid-state device A semiconductor device for the generation or amplification of electromagnetic energy at microwave frequencies.

microwave spectrum The range of wavelengths or frequencies of electromagnetic radiation that are designated microwaves.

microwave transmission line A material structure forming a continuous path from one place to another and capable of directing the transmission of electromagnetic energy along this path.

microwave tube A high-vacuum tube designed for operation in the frequency region from approximately 3000 to 300,000 megahertz.

microwave waveguide *See* waveguide.

microwave wavemeter Any device for measuring the free-space wavelengths (or frequencies) of microwaves; usually made of a cavity resonator whose dimensions can be varied until resonance with the microwaves is achieved.

middle-ultraviolet lamp A mercury-vapor lamp designed to produce radiation in the wavelength band from 2800 to 3200 angstrom units (280 to 320 nanometers) such as sunlamps and photochemical lamps.

mid-frequency gain The maximum gain of an amplifier, when this gain depends on the frequency; for an RC-coupled voltage amplifier the gain is essentially equal to this value over a large range of frequencies.

migration 1. The movement of charges through a semiconductor material by diffusion or drift of charge carriers or ionized atoms. 2. The movement of crystal defects through a semiconductor crystal under the influence of high temperature, strain, or a continuously applied electric field.

Miller bridge Type of bridge circuit for measuring amplification factors of vacuum tubes.

Miller effect The increase in the effective grid-cathode capacitance of a vacuum tube due to the charge induced electrostatically on the grid by the anode through the grid-anode capacitance.

Miller generator *See* bootstrap integrator.

Miller integrator A resistor-capacitor charging network having a high-gain amplifier paralleling the capacitor; used to produce a linear time-base voltage. Also known as Miller time-base.

Miller time-base *See* Miller integrator.

milliammeter An ammeter whose scale is calibrated to indicate current values in milliamperes.

milliampere A unit of current equal to one-thousandth of an ampere. Abbreviated mA.

millifarad A unit of capacitance equal to one-thousandth of a farad. Abbreviated mF.

minority emitter 279

milligauss A unit of magnetic flux density equal to one-thousandth of a gauss. Abbreviated mG.

millihenry A unit of inductance equal to one-thousandth of a henry. Abbreviated mH.

Millikan meter An integrating ionization chamber in which a gold-leaf electroscope is charged a known amount and ionizing events reduce this charge, so that the resulting angle through which the gold leaf is repelled at any given time indicates the number of ionizing events that have occurred.

millimeter wave An electromagnetic wave having a wavelength between 1 millimeter and 1 centimeter, corresponding to frequencies between 30 and 300 gigahertz. Also known as millimetric wave.

millimetric wave *See* millimeter wave.

millivolt A unit of potential difference or emf equal to one-thousandth of a volt. Abbreviated mV.

millivoltmeter A voltmeter whose scale is calibrated to indicate voltage values in millivolts.

M indicator *See* M scope.

miniature electron tube A small electron tube having no base, with tube electrode leads projecting through the glass bottom in positions corresponding to those of pins for either a seven-pin or nine-pin tube base.

miniaturization Reduction in the size and weight of a system, package, or component by using small parts arranged for maximum utilization of space.

minimum detectable signal *See* threshold signal.

minimum discernible signal Receiver input power level that is just sufficient to produce a discernible signal in the receiver output; a receiver sensitivity test.

minimum firing current The limit below which firing will not occur in electric blasting caps.

minimum-loss attenuator A section linking two unequal resistive impedances which is designed to introduce the smallest attenuation possible. Also known as minimum-loss pad.

minimum-loss matching Design of a network linking two resistive impedances so that it introduces a loss which is as small as possible.

minimum-loss pad *See* minimum-loss attenuator.

minimum signal level In facsimile, level corresponding to the copy white or copy black signal, whichever is the lower.

minitrack A subminiature radio transmitter capable of sending data over 4000 miles (6500 kilometers) on extremely low power.

Minkowski electrodynamics An electromagnetic theory, compatible with the special theory of relativity, which takes into account the presence of matter with electric and magnetic polarization.

minor bend Rectangular waveguide bent so that throughout the length of a bend a longitudinal axis of the guide lies in one plane which is parallel to the narrow side of the waveguide.

minority carrier The type of carrier, electron, or hole that constitutes less than half the total number of carriers in a semiconductor.

minority emitter Of a transistor, an electrode from which a flow of minority carriers enters the interelectrode region.

minor lobe Any lobe except the major lobe of an antenna radiation pattern. Also known as secondary lobe; side lobe.

minor relay station A tape relay station which has tape relay responsibility but does not provide an alternate route.

minor switch Single-motion stepping switch mounted atop the telephone connectors and most commonly used for the party-line selection.

mirror galvanometer A galvanometer having a small mirror attached to the moving element, to permit use of a beam of light as an indicating pointer. Also known as reflecting galvanometer.

MIS *See* metal-insulator semiconductor.

misfire Failure to establish an arc between the main anode and the cathode of an ignitron or other mercury-arc rectifier during a scheduled conducting period.

mismatch The condition in which the impedance of a source does not match or equal the impedance of the connected load or transmission line.

mismatch factor *See* reflection factor.

mismatch loss Loss of power delivered to a load as a result of failure to make an impedance match of a transmission line with its load or with its source.

mismatch slotted line A slotted line linking two waveguides which is not properly designed to minimize the power reflected or transmitted by it.

missile plume The region of electromagnetic and other disturbances that follow a missile during reentry and make the missile more readily detectable.

mixed reflection *See* spread reflection.

mixer 1. A device having two or more inputs, usually adjustable, and a common output; used to combine separate audio or video signals linearly in desired proportions to produce an output signal. 2. The stage in a superheterodyne receiver in which the incoming modulated radio-frequency signal is combined with the signal of a local r-f oscillator to produce a modulated intermediate-frequency signal. Also known as first detector; heterodyne modulator; mixer-first detector.

mixer-first detector *See* mixer.

mixer tube A multigrid electron tube, used in a superheterodyne receiver, in which control voltages of different frequencies are impressed upon different control grids, and the nonlinear properties of the tube cause the generation of new frequencies equal to the sum and difference of the impressed frequencies.

mixing Combining two or more signals, such as the outputs of several microphones.

mmf *See* magnetomotive force.

MNOS *See* metal-nitride-oxide semiconductor.

mobility *See* drift mobility.

Möbius resistor A nonreactive resistor made by placing strips of aluminum or other metallic tape on opposite sides of a length of dielectric ribbon, twisting the strip assembly half a turn, joining the ends of the metallic tape, then soldering leads to opposite surfaces of the resulting loop.

mode A form of propagation of guided waves that is characterized by a particular field pattern in a plane transverse to the direction of propagation. Also known as transmission mode.

mode converter *See* mode transducer.

mode filter A waveguide filter designed to separate waves of the same frequency but of different transmission modes.

modulator 281

mode jump Change in mode of magnetron operation from one pulse to the next; each mode represents a different frequency and power level.

modem A combination modulator and demodulator at each end of a telephone line to convert binary digital information to audio tone signals suitable for transmission over the line, and vice versa. Also known as dataset. Derived from modulator-demodulator.

mode number 1. The number of complete cycles during which an electron of average speed is in the drift space of a reflex klystron. 2. The number of radians of phase in the microwave field of a magnetron divided by 2π as one goes once around the anode.

mode shift Change in mode of magnetron operation during a pulse.

mode skip Failure of a magnetron to fire on each successive pulse.

mode switch A microwave control device, often consisting of a waveguide section of special cross section, which is used to change the mode of microwave power transmission in the waveguide.

mode transducer Device for transforming an electromagnetic wave from one mode of propagation to another. Also known as mode converter; mode transformer.

mode transformer See mode transducer.

modified constant-voltage charge Charging of a storage battery in which the voltage of the charging circuit is held substantially constant, but a fixed resistance is inserted in the battery circuit producing a rising voltage characteristic at the battery terminals as the charge progresses.

moding Defect of magnetron oscillation in which it oscillates in one or more undesired modes.

modular circuit Any type of circuit assembled to form rectangular or cubical blocks that perform one or more complete circuit functions.

modular structure 1. An assembly involving the use of integral multiples of a given length for the dimensions of electronic components and electronic equipment, as well as for spacings of holes in a chassis or printed wiring board. 2. An assembly made from modules.

modulate To vary the amplitude, frequency, or phase of a wave, or vary the velocity of the electrons in an electron beam in some characteristic manner.

modulated amplifier Amplifier stage in a transmitter in which the modulating signal is introduced and modulates the carrier.

modulated stage Radio-frequency stage to which the modulator is coupled and in which the continuous wave (carrier wave) is modulated according to the system of modulation and the characteristics of the modulating wave.

modulating electrode Electrode to which a potential is applied to control the magnitude of the beam current.

modulation capability Of an aural transmitter, the maximum percentage modulation that can be obtained without exceeding a given distortion figure.

modulation-doped structure An epitaxially grown crystal structure in which successive semiconductor layers contain different types of electrical dopants.

modulation rise Increase of the modulation percentage caused by nonlinearity of any tuned amplifier, usually the last intermediate-frequency stage of a receiver.

modulator 1. The transmitter stage that supplies the modulating signal to the modulated amplifier stage or that triggers the modulated amplifier stage to produce pulses at desired instants as in radar. 2. A device that produces modulation by any means, such as by virtue of a nonlinear characteristic or by controlling some circuit quantity in accordance with the waveform of a modulating signal. 3. One of the electrodes of a spacistor.

modulator-demodulator *See* modem.

modulator glow tube Cold cathode recorder tube that is used for facsimile and sound-on-film recording; provides a modulated high-intensity point source of light.

module A packaged assembly of wired components, built in a standardized size and having standardized plug-in or solderable terminations.

molded capacitor Capacitor, usually mica, that has been encased in a molded plastic insulating material.

molectronics *See* molecular electronics.

molecular beam epitaxy A technique of growing single crystals in which beams of atoms or molecules are made to strike a single-crystalline substrate in a vacuum, giving rise to crystals whose crystallographic orientation is related to that of the substrate. Abbreviated MBE.

molecular binding The force which holds a molecule at some site on the surface of a crystal.

molecular circuit A circuit in which the individual components are physically indistinguishable from each other.

molecular electronics The branch of electronics that deals with the production of complex electronic circuits in microminiature form by producing semiconductor devices and circuit elements integrally while growing multizoned crystals in a furnace. Also known as molectronics.

molecular engineering The use of solid-state techniques to build, in extremely small volumes, the components necessary to provide the functional requirements of overall equipments, which when handled in more conventional ways are vastly bulkier.

molecular field theory *See* Weiss theory.

monitoring amplifier A power amplifier used primarily for evaluation and supervision of a program.

monitoring key Key which, when operated, makes it possible for an attendant or operator to listen on a telephone circuit without appreciably impairing transmission on the circuit.

monobrid circuit Integrated circuit using a combination of monolithic and multichip techniques by means of which a number of monolithic circuits, or a monolithic device in combination with separate diffused or thin-film components, are interconnected in a single package.

monocharge electret A type of foil electret that carries electrical charge of the same sign on both surfaces.

monochromatic radiation Electromagnetic radiation having wavelengths confined to an extremely narrow range.

monochrome channel In a color television system, any path which is intended to carry the monochrome signal; the monochrome channel may also carry other signals.

monochrome signal 1. A signal wave used for controlling luminance values in monochrome television. 2. The portion of a signal wave that has major control of the luminance values in a color television system, regardless of whether the picture is displayed in color or in monochrome. Also known as M signal.

monocord switchboard Local battery switchboard in which each telephone line terminates in a single jack and plug.

motional induction 283

monofier Complete master oscillator and power amplifier system in a single evacuated tube envelope; electrically, it is equivalent to a stable low-noise oscillator, an isolator, and a two- or three-cavity klystron amplifier.

monolithic ceramic capacitor A capacitor that consists of thin dielectric layers interleaved with staggered metal-film electrodes; after leads are connected to alternate projecting ends of the electrodes, the assembly is compressed and sintered to form a solid monolithic block.

monolithic integrated circuit An integrated circuit having elements formed in place on or within a semiconductor substrate, with at least one element being formed within the substrate.

monopinch Antijam application of the monopulse technique where the error signal is used to provide discrimination against jamming signals.

monopole *See* magnetic monopole.

monopole antenna An antenna, usually in the form of a vertical tube or helical whip, on which the current distribution forms a standing wave, and which acts as one part of a dipole whose other part is formed by its electrical image in the ground or in an effective ground plane. Also known as spike antenna.

monoscope A signal-generating electron-beam tube in which a picture signal is produced by scanning an electrode that has a predetermined pattern of secondary-emission response over its surface. Also known as monotron; phasmajector.

monostable Having only one stable state.

monostable blocking oscillator A blocking oscillator in which the electron tube or other active device carries no current unless positive voltage is applied to the grid. Also known as driven blocking oscillator.

monostable circuit A circuit having only one stable condition, to which it returns in a predetermined time interval after being triggered.

monostable multivibrator A multivibrator with one stable state and one unstable state; a trigger signal is required to drive the unit into the unstable state, where it remains for a predetermined time before returning to the stable state. Also known as one-shot multivibrator; single-shot multivibrator; start-stop multivibrator; univibrator.

monotonicity In an analog-to-digital converter, the condition wherein there is an increasing output for every increasing value of input voltage over the full operating range.

monotron *See* monoscope.

mosaic A light-sensitive surface used in television camera tubes, consisting of a thin mica sheet coated on one side with a large number of tiny photosensitive silver-cesium globules, insulated from each other.

MOSFET *See* metal oxide semiconductor field-effect transistor.

Mossotti field *See* Lorentz local field.

MOST *See* metal oxide semiconductor field-effect transistor.

MOS transistor *See* metal oxide semiconductor field-effect transistor.

motional electromotive force An electromotive force in a circuit that results from the motion of all or part of the circuit through a magnetic field.

motional impedance Of a transducer, the complex remainder after the blocked impedance has been subtracted from the loaded impedance. Also known as loaded motional impedance.

motional induction The production of an electromotive force in a circuit by motion of all or part of the circuit through a magnetic field in such a way that the circuit cuts across the magnetic flux.

284 motion picture pickup

motion picture pickup Use of a television camera to pick up scenes directly from motion picture film.

motor A machine that converts electric energy into mechanical energy by utilizing forces produced by magnetic fields on current-carrying conductors. Also known as electric motor.

motorboating Undesired oscillation in an amplifying system or transducer, usually of a pulse type, occurring at a subaudio or low-audio frequency.

motor branch circuit A branch circuit that terminates at a motor; it must have conductors with current-carrying capacity at least 125% of the motor full-load current rating, and overcurrent protection capable of carrying the starting current of the motor.

motor control *See* electronic motor control.

motor-converter Induction motor and a synchronous converter with their rotors mounted on the same shaft and with their rotor windings connected in series; such converter operates synchronously at a speed corresponding to the sum of the numbers of poles of the two machines.

motor effect The mutually repulsive force exerted by neighboring conductors that carry current in opposite directions.

motor-generator set A motor and one or more generators that are coupled mechanically for use in changing one power-source voltage to other desired voltages or frequencies.

mount The flange or other means by which a switching tube, or tube and cavity, is connected to a waveguide.

mountain effect The effect of rough terrain on radio-wave propagation, causing reflections that produce errors in radio direction-finder indications.

movable contact The relay contact that is mechanically displaced to engage or disengage one or more stationary contacts. Also known as armature contact.

moving-coil instrument Any instrument in which current is sent through one or more coils suspended or pivoted in a magnetic field, and the motion of the coils is used to measure either the current in the coils or the strength of the field.

moving-coil meter A meter in which a pivoted coil is the moving element.

moving-coil pickup *See* dynamic pickup.

moving-target indicator A device that limits the display of radar information primarily to moving targets; signals due to reflections from stationary objects are canceled by a memory circuit. Abbreviated MTI.

MPU *See* microprocessing unit.

M scan *See* M scope.

M scope A modified form of A scope on which part of the time base is slightly displaced in a vertical direction by insertion of an adjustable step which serves as a range marker. Also known as M indicator; M scan.

MSI *See* medium-scale integration.

M signal *See* monochrome signal.

MTI *See* moving-target indicator.

M-type backward-wave oscillator A backward-wave oscillator in which focusing and interaction are through magnetic fields, as in a magnetron. Also known as M-type carcinotron; type-M carcinotron.

M-type carcinoron *See* M-type backward-wave oscillator.

mu factor Ratio of the change in one electrode voltage to the change in another electrode voltage under the conditions that a specified current remains unchanged and that all other electrode voltages are maintained constant; a measure of the relative effect of the voltages on two electrodes upon the current in the circuit of any specified electrode.

multianode tube Electron tube having two or more main anodes and a single cathode.

multiaperture reluctance switch Two-aperture ferrite storage core which may be used to provide a nondestructive readout computer memory.

multicavity klystron A klystron in which there is at least one cavity between the input and output cavities, each of which remodulates the beam so that electrons are more closely bunched.

multicavity magnetron A magnetron in which the circuit includes a plurality of cavities, generally cut into the solid cylindrical anode so that the mouths of the cavities face the central cathode.

multicellular horn A cluster of horn antennas having mouths that lie in a common surface and that are fed from openings spaced one wavelength apart in one face of a common waveguide.

multichannel field-effect transistor A field-effect transistor in which appropriate voltages are applied to the gate to control the space within the current flow channels.

multichip microcircuit Microcircuit in which discrete, miniature, active electronic elements (transistor or diode chips) and thin-film or diffused passive components or component clusters are interconnected by thermocompression bonds, alloying, soldering, welding, chemical deposition, or metallization.

multicollector electron tube An electron tube in which electrons travel to more than one electrode.

multicoupler A device for connecting several receivers to one antenna and properly matching the impedances of the receivers to the antenna.

multielectrode tube Electron tube containing more than three electrodes associated with a single electron stream.

multielement array An antenna array having a large number of antennas.

multielement parasitic array Antennas consisting of an array of driven dipoles and parasitic elements, arranged to produce a beam of high directivity.

multielement vacuum tube A vacuum tube which has one or more grids in addition to the cathode and plate electrodes.

multigrid tube An electron tube having two or more grids between cathode and anode, as a tetrode or pentode.

multigun tube A cathode-ray tube having more than one electron gun.

multipactor A high-power, high-speed microwave switching device in which a thin electron cloud is driven back and forth between two parallel plane surfaces in a vacuum by a radio-frequency electric field.

multipath *See* multipath transmission.

multipath transmission The propagation phenomenon that results in signals reaching a radio receiving antenna by two or more paths, causing distortion in radio and ghost images in television. Also known as multipath.

multiple 1. Group of terminals arranged to make a circuit or group of circuits accessible at a number of points at any one of which connection can be made. 2. To connect in parallel. 3. *See* parallel.

multiple appearance Jack arrangement in telephone switchboards whereby a single-line circuit appears before two or more operators.

multiple-beam antenna An antenna or antenna array which radiates several beams in different directions.

multiple-contact switch *See* selector switch.

multiple jacks Series of jacks with tip, ring, and sleeve, respectively connected in parallel, and appearing in different panels of the face equipment of a telephone exchange.

multiple lamp holder A device that can be inserted in a lamp holder to act as two or more lamp holders. Also known as current tap.

multiple reflection echoes Radar echoes returned from a real target by reflection from some object in the radar beam; such echoes appear at a false bearing and false range.

multiple resonance Two or more resonances at different frequencies in a circuit consisting of two or more coupled circuits which are resonant at slightly different frequencies.

multiple switchboard Manual telephone switchboard in which each subscriber line is attached to two or more jacks, to be within reach of several operators.

multiple target generator An electronic countermeasures device that produces several false responses in a hostile radar set.

multiple-tuned antenna Low-frequency antenna having a horizontal section with a multiplicity of tuned vertical sections.

multiple twin quad Quad cable in which the four conductors are arranged in two twisted pairs, and the two pairs twisted together.

multiple-unit semiconductor device Semiconductor device having two or more seats of electrodes associated with independent carrier streams.

multiple-unit steerable antenna *See* musa.

multiple-unit tube *See* multiunit tube.

multiplexer A device for combining two or more signals, as for multiplex, or for creating the composite color video signal from its components in color television. Also spelled multiplexor.

multiplexor *See* multiplexer.

multiplication An increase in current flow through a semiconductor because of increased carrier activity.

multiplier 1. A resistor used in series with a voltmeter to increase the voltage range. Also known as multiplier resistor. 2. A device that has two or more inputs and an output that is a representation of the product of the quantities represented by the input signals; voltages are the quantities commonly multiplied. 3. *See* electron multiplier; frequency multiplier.

multiplier phototube A phototube with one or more dynodes between its photocathode and the output electrode; the electron stream from the photocathode is reflected off each dynode in turn, with secondary emission adding electrons to the stream at each reflection. Also known as electron-multiplier phototube; photoelectric electron-multiplier tube; photomultiplier; photomultiplier tube.

multiplier resistor *See* multiplier.

multiplier traveling-wave photodiode Photodiode in which the construction of a traveling-wave tube is combined with that of a multiplier phototube to give increased sensitivity.

multiplier tube Vacuum tube using secondary emission from a number of electrodes in sequence to obtain increased output current; the electron stream is reflected, in turn, from one electrode of the multiplier to the next.

multipolar Having more than one pair of magnetic poles.

multipolar machine An electric machine that has a field magnet with more than one pair of poles.

multipole One of a series of types of static or oscillating distributions of charge or magnetization; namely, an electric multipole or a magnetic multipole.

multipole fields The electric and magnetic fields generated by static or oscillating electric or magnetic multipoles.

multisegment magnetron Magnetron with an anode divided into more than two segments, usually by slots parallel to its axis.

multistable circuit A circuit having two or more stable operating conditions.

multistage amplifier *See* cascade amplifier.

multistator watt-hour meter An induction type of watt-hour meter in which several stators exert a torque on the rotor.

multistrip coupler A series of parallel metallic strips placed on a surface acoustic wave filter between identical apodized interdigital transducers; it converts the spatially nonuniform surface acoustic wave generated by one transducer into a spatially uniform wave received at the other transducer, and helps to reject spurious bulk acoustic modes.

multiturn potentiometer A precision wire-wound potentiometer in which the resistance element is formed into a helix, generally having from 2 to 10 turns.

multiunit tube Electron tube containing within one glass or metal envelope, two or more groups of electrodes, each associated with separate electron streams. Also known as multiple-unit tube.

multivator 1. Type of dc-to-dc up-converter. 2. An automatic device for analyzing a number of dust samples that might be collected by spacecraft on the moon, Mars, and other planets, to detect the presence of microscopic organisms with a multiplier phototube that measures the fluorescence given off.

multivibrator A relaxation oscillator using two tubes, transistors, or other electron devices, with the output of each coupled to the input of the other through resistance-capacitance elements or other elements to obtain in-phase feedback voltage.

Murray loop test A method of localizing a fault in a cable by replacing two arms of a Wheatstone bridge with a loop formed by the cable under test and a good cable connected to the far end of the defective cable.

musa An electrically steerable receiving antenna whose directional pattern can be rotated by varying the phases of the contributions of the individual units. Derived from multiple-unit steerable antenna.

muting circuit 1. Circuit which cuts off the output of a receiver when no radio-frequency carrier greater than a predetermined intensity is reaching the first detector. 2. Circuit for making a receiver insensitive during operation of its associated transmitter.

muting switch 1. A switch used in connection with automatic tuning systems to silence the receiver while tuning from one station to another. 2. A switch used to ground the output of a phonograph pickup automatically while a record changer is in its change cycle.

mutual admittance For two meshes of a network carrying alternating current, the ratio of the complex current in one mesh to the complex voltage in the other, when the voltage in all meshes besides these two is 0.

mutual capacitance The accumulation of charge on the surfaces of conductors of each of two circuits per unit of potential difference between the circuits.

mutual conductance *See* transconductance.

mutual impedance For two meshes of a network carrying alternating current, the ratio of the complex voltage in one mesh to the complex current in the other, when all meshes besides the latter one carry no current.

mutual inductance Property of two neighboring circuits, equal to the ratio of the electromotive force induced in one circuit to the rate of change of current in the other circuit.

mutual induction The generation of a voltage in one circuit by a varying current in another.

mV *See* millivolt.

MV *See* megavolt.

MWYE *See* megawatt year of electricity.

Mx *See* maxwell.

Mylar capacitor A capacitor that uses Mylar film as a dielectric between rolled strips of metal foil.

myriametric waves Electromagnetic waves having wavelengths between 10 and 100 kilometers, corresponding to the very low frequency band.

N

nancy receiver *See* infrared receiver.

NAND circuit A logic circuit whose output signal is a logical 1 if any of its inputs is a logical 0, and whose output signal is a logical 0 if all of its inputs are logical 1.

narrow-band amplifier An amplifier which increases the magnitude of signals over a band of frequencies whose bandwidth is small compared to the average frequency of the band.

narrow-band-pass filter A band-pass filter in which the band of frequencies transmitted by the filter has a bandwidth which is small compared to the average frequency of the band.

narrow-beam antenna An antenna which radiates most of its power in a cone having a radius of only a few degrees.

narrow-sector recorder A radio direction finder with which atmospherics are received from a limited sector related to the position of the antenna; this antenna is usually rotated continuously and the bearings of the atmospherics recorded automatically.

natural antenna frequency Lowest resonant frequency of an antenna without added inductance or capacitance.

natural frequency The lowest resonant frequency of an antenna, circuit, or component.

natural function generator *See* analytical function generator.

natural law function generator *See* analytical function generator.

natural wavelength Wavelength corresponding to the natural frequency of an antenna or circuit.

navigation receiver An electronic device that determines a ship's position by receiving and comparing radio signals from transmitters at known locations.

n-channel A conduction channel formed by electrons in an n-type semiconductor, as in an n-type field-effect transistor.

N curve A plot of voltage against current for a negative-resistance device; its slope is negative for some values of current or voltage.

N display Radar display in which the target appears as a pair of vertical deflections from a horizontal time base; direction is indicated by relative amplitude of the blips; target distance is determined by moving an adjustable pedestal signal along the base line until it coincides with the horizontal position of the blips; the pedestal control is calibrated in distance.

NEA material *See* negative electron affinity material.

near field The electromagnetic field that exists within one wavelength of a source of electromagnetic radiation, such as a transmitting antenna.

near-infrared radiation Infrared radiation having a relatively short wavelength, between 0.75 and about 2.5 micrometers (some scientists place the upper limit from 1.5 to 3 micrometers), at which radiation can be detected by photoelectric cells, and which corresponds in frequency range to the lower electronic energy levels of molecules and semiconductors. Also known as photoelectric infrared radiation.

near-ultraviolet radiation Ultraviolet radiation having relatively long wavelength, in the approximate range from 300 to 400 nanometers.

needle gap Spark gap in which the electrodes are needle points.

needle scratch *See* surface noise.

needle test point A sharp steel probe connected to a test cord for making contact with a conductor.

Néel point *See* Néel temperature.

Néel temperature A temperature, characteristic of certain metals, alloys, and salts, below which spontaneous nonparalleled magnetic ordering takes place so that they become antiferromagnetic, and above which they are paramagnetic. Also known as Néel point.

Néel's theory A theory of the behavior of antiferromagnetic and other ferrimagnetic materials in which the crystal lattice is divided into two or more sublattices; each atom in one sublattice responds to the magnetic field generated by nearest neighbors in other sublattices, with the result that magnetic moments of all the atoms in any sublattice are parallel, but magnetic moments of two different sublattices can be different.

Néel wall The boundary between two magnetic domains in a thin film in which the magnetization vector remains parallel to the faces of the film in passing through the wall.

negative Having a negative charge.

negative booster Booster used in connection with a ground-return system to reduce the difference of potential between two points on the grounded return.

negative charge The type of charge which is possessed by electrons in ordinary matter, and which may be produced in a resin object by rubbing with wool. Also known as negative electricity.

negative effective mass amplifiers and generators Class of solid-state devices for broad-band amplification and generation of electrical waves in the microwave region; these devices use the property of the effective masses of charge carriers in semiconductors becoming negative with sufficiently high kinetic energies.

negative electricity *See* negative charge.

negative electrode *See* cathode; negative plate.

negative electron affinity material A material, such as gallium phosphide, whose surface has been treated with a substance, such as cesium, so that the surface barrier is reduced, band-bending occurs so that the top of the conduction band lies above the vacuum level, and the electron affinity of the substance is negative. Abbreviated NEA material.

negative glow The luminous flow in a glow-discharge cold-cathode tube occurring between the cathode dark space and the Faraday dark space.

negative-grid generator Conventional oscillator circuit in which oscillation is produced by feedback from the plate circuit to a grid which is normally negative with respect to the cathode, and which is designed to operate without drawing grid current at any time.

negative-grid thyratron A thyratron with only one grid, which serves to prevent the flow of current until its potential relative to the cathode is made less negative than a certain critical value.

negative impedance An impedance such that when the current through it increases, the voltage drop across the impedance decreases.

negative-impedance repeater A telephone repeater that provides an effective gain for voice-frequency signals by insertion into the line of a negative impedance that cancels out line impedances responsible for transmission losses.

negative modulation 1. Modulation in which an increase in brightness corresponds to a decrease in amplitude-modulated transmitter power; used in United States television transmitters and in some facsimile systems. 2. Modulation in which an increase in brightness corresponds to a decrease in the frequency of a frequency-modulated facsimile transmitter. Also known as negative transmission.

negative phase sequence The phase sequence that corresponds to the reverse of the normal order of phases in a polyphase system.

negative-phase-sequence relay Relay which functions in conformance with the negative-phase-sequence component of the current, voltage, or power of the circuit.

negative picture phase The video signal phase in which the signal voltage swings in a negative direction for an increase in brilliance.

negative plate The internal plate structure that is connected to the negative terminal of a storage battery. Also known as negative electrode.

negative potential An electrostatic potential which is lower than that of the ground, or of some conductor or point in space that is arbitrarily assigned to have zero potential.

negative resistance The resistance of a negative-resistance device.

negative-resistance device A device having a range of applied voltages within which an increase in this voltage produces a decrease in the current.

negative-resistance oscillator An oscillator in which a parallel-tuned resonant circuit is connected to a vacuum tube so that the combination acts as the negative resistance needed for continuous oscillation.

negative-resistance repeater Repeater in which gain is provided by a series negative resistance or a shunt negative resistance, or both.

negative terminal The terminal of a battery or other voltage source that has more electrons than normal; electrons flow from the negative terminal through the external circuit to the positive terminal.

negative thermion *See* thermoelectron.

negative-transconductance oscillator Electron-tube oscillator in which the output of the tube is coupled back to the input without phase shift, the phase condition for oscillation being satisfied by the negative transconductance of the tube.

negative transmission *See* negative modulation.

negatron *See* dynatron.

neon glow lamp A glow lamp containing neon gas, usually rated between 1/25 and 3 watts, and producing a characteristic red glow; used as an indicator light and electronic circuit component.

neon oscillator Relaxation oscillator in which a neon tube or lamp serves as the switching element.

neon tube An electron tube in which neon gas is ionized by the flow of electric current through long lengths of gas tubing, to produce a luminous red glow discharge; used chiefly in outdoor advertising signs.

Nernst bridge A four-arm bridge containing capacitors instead of resistors, used for measuring capacitance values at high frequencies.

Nernst glower See Nernst lamp.

Nernst lamp An electric lamp consisting of a short, slender rod of zirconium oxide in open air, heated to brilliant white incandescence by current. Also known as Nernst glower.

nesistor A negative-resistance semiconductor device that is basically a bipolar field-effect transistor.

net power flow The difference between the power carried by electromagnetic waves traveling in a given direction along a waveguide and the power carried by waves traveling in the opposite direction.

network A collection of electric elements, such as resistors, coils, capacitors, and sources of energy, connected together to form several interrelated circuits. Also known as electric network.

network admittance The admittance between two terminals of a network under specified conditions.

network analysis Derivation of the electrical properties of a network, from its configuration, element values, and driving forces.

network constant One of the resistance, inductance, mutual inductance, or capacitance values involved in a circuit or network; if these values are constant, the network is said to be linear.

network filter A combination of electrical elements (for example, interconnected resistors, coils, and capacitors) that represents relatively small attenuation to signals of a certain frequency, and great attenuation to all other frequencies.

network flow Flow of current in a network.

network input impedance The impedance between the input terminals of a network under specified conditions.

network master relay Relay that performs the chief functions of closing and tripping an alternating-current low-voltage network protector.

network phasing relay Relay which functions in conjunction with a master relay to limit closure of the network protector to a predetermined relationship between the voltage and the network voltage.

network relay Form of voltage, power, or other type of relay used in the protection and control of alternating-current low-voltage networks.

network synthesis Derivation of the configuration and element values of a network with given electrical properties.

network theory The systematizing and generalizing of the relations between the currents, voltages, and impedances associated with the elements of an electrical network.

network transfer admittance The current that would flow through a short circuit between one pair of terminals in a network if a unit voltage were applied across the other pair.

Neumann's formula A formula for the mutual inductance M_{12} between two closed circuits C_1 and C_2; it is

$$M_{12} = \frac{\mu_0}{4\pi} \int_{C_1} \int_{C_2} \frac{ds_1\, ds_2}{r}$$

nickel delay line 293

where r is the distance between line elements ds_1 and ds_2, and μ_0 is the permeability of the empty space.

neuristor A device that behaves like a nerve fiber in having attenuationless propagation of signals; one goal of research is development of a complete artificial nerve cell, containing many neuristors, that could duplicate the function of the human eye and brain in recognizing characters and other visual images.

neutral conductor A conductor of a polyphase circuit or of a single-phase, three-wire circuit which is intended to have a potential such that the potential differences between it and each of the other conductors are approximately equal in magnitude and are also equally spaced in phase.

neutral ground Ground connected to the neutral point or points of an electric circuit, transformer, rotating machine, or system.

neutralize To nullify oscillation-producing voltage feedback from the output to the input of an amplifier through tube interelectrode capacitances; an external feedback path is used to produce at the input a voltage that is equal in magnitude but opposite in phase to that fed back through the interelectrode capacitance.

neutralized radio-frequency stage Stage having an additional circuit connected to feed back, in the opposite phase, an amount of energy equivalent to what is causing the oscillation, thus neutralizing any tendency to oscillate and making the circuit function strictly as an amplifier.

neutralizing capacitor Capacitor, usually variable, employed in a radio receiving or transmitting circuit to feed a portion of the signal voltage from the plate circuit of a stage back to the grid circuit.

neutralizing circuit Portion of an amplifier circuit which provides an intentional feedback path from plate to grid to prevent regeneration.

neutralizing transformer A transformer installed on a communication line which produces counter electromotive forces which largely cancel out induced longitudinal voltage and allow normal operation of the line during inductive disturbances.

neutralizing voltage Voltage developed in the plate circuit (Hazeltine neutralization) or in the grid circuit (Rice neutralization), used to nullify or cancel the feedback through the tube.

neutral point Point which has the same potential as the point of junction of a group of equal nonreactive resistances connected at their free ends to the appropriate main terminals or lines of the system.

neutral relay Relay in which the movement of the armature does not depend upon the direction of the current in the circuit controlling the armature. Also known as nonpolarized relay.

neutral return path A route from the load back to the power source, completing a circuit in an electric power distribution system, which is grounded, usually by connections to water pipes.

neutral temperature The temperature of the hot junction of a thermocouple at which the electromotive force of the thermocouple attains its maximum value, when the cold junction is maintained at a constant temperature of 0°C.

nicad battery See nickel-cadmium battery.

nickel-cadmium battery A sealed storage battery having a nickel anode, a cadmium cathode, and an alkaline electrolyte; widely used in cordless appliances; without recharging, it can serve as a primary battery. Also known as cadmium-nickel storage cell; nicad battery.

nickel delay line An acoustic delay line in which nickel is used to transmit sound signals.

nickel-iron battery *See* Edison battery.

N indicator *See* N scope.

NIXIE display indicator *See* NIXIE Tube.

NIXIE Tube Trademark of Burroughs Corporation for a cold-cathode gas readout tube having a common anode and 10 different metallic cathodes, each formed in the shape of a different numeral, alphabetic character, or special symbol; the desired character is surrounded with a brilliant glow when the corresponding cathode is energized. Also known as NIXIE display indicator.

N-level logic An arrangement of gates in a digital computer in which not more than N gates are connected in series.

NMRR *See* normal-mode rejection ratio.

nn junction In a semiconductor, a region of transition between two regions having different properties in n-type semiconducting material.

no-break power Power system designed to fulfill load requirements during the interval between the failure of the primary power and the time the auxiliary power can be made available.

nodal analysis A method of electrical circuit analysis in which potential differences are taken as independent variables and the sum of the currents flowing into a node is equated to 0.

nodal points Junction points in a transmission system; the automatic switches and switching centers are the nodal points in automated systems.

node 1. A junction point within a network. 2. *See* branch point.

noise Interfering and unwanted currents or voltages in an electrical device or system.

noise analyzer A device used for noise analysis.

noise factor The ratio of the total noise power per unit bandwidth at the output of a system to the portion of the noise power that is due to the input termination, at the standard noise temperature of 290 K. Also known as noise figure.

noise field strength *See* radio-noise field strength.

noise figure *See* noise factor.

noise filter 1. A filter that is inserted in an alternating-current power line to block noise interference that would otherwise travel through the line in either direction and affect the operation of receivers. 2. A filter used in a radio receiver to reduce noise, usually an auxiliary low-pass filter which can be switched in or out of the audio system.

noise generator A device which produces (usually random) electrical noise, for use in tests of the response of electrical systems to noise, and in measurements of noise intensity. Also known as noise source.

noise jammer 1. An electronic jammer that emits a carrier modulated with recordings or synthetic reproductions of natural atmospheric noise; the radio-frequency carrier may be suppressed; used to discourage the enemy by simulating naturally adverse communications conditions. 2. During World War II, a powerful transmitter modulated with white noise tuned to the approximate frequency of an enemy transmitter and used to obscure intelligible output at the receiver.

noise killer 1. Device installed in a circuit to reduce its interference to other circuits. 2. *See* noise suicide circuit.

noise limiter A limiter circuit that cuts off all noise peaks that are stronger than the highest peak in the desired signal being received, thereby reducing the effects of atmospheric or human-caused interference. Also known as noise silencer; noise suppressor.

noise measurement Any of a wide range of measurements of random and nonrandom electrical noise, but usually noise-power measurement.

noise-metallic In telephone communications, weighted noise current in a metallic circuit at a given point when the circuit is terminated at that point in the nominal characteristic impedance of the circuit.

noise-modulated jamming Random electronic noise that appears at the radar receiver as background noise and tends to mask the desired radar echo or radio signal.

noise-power measurement Measurement of the power carried by electrical noise averaged over some brief interval of time, usually by amplifying noise from the source in a linear amplifier and then using a quadratic detector followed by a low-pass filter and an indicating device.

noise-reducing antenna system Receiving antenna system so designed that only the antenna proper can pick up signals; it is placed high enough to be out of the noise-interference zone, and is connected to the receiver with a shielded cable or twisted transmission line that is incapable of picking up signals.

noise silencer *See* noise limiter.

noise source *See* noise generator.

noise suicide circuit A circuit which reduces the gain of an amplifier for a short period whenever a sufficiently large noise pulse is received. Also known as noise killer.

noise suppression Any method of reducing or eliminating the effects of undesirable electrical disturbances, as in frequency modulation whenever the signal carrier level is greater than the noise level.

noise suppressor 1. A circuit that blocks the audio-frequency amplifier of a radio receiver automatically when no carrier is being received, to eliminate background noise. Also known as squelch circuit. 2. A circuit that reduces record surface noise when playing phonograph records, generally by means of a filter that blocks out the higher frequencies where such noise predominates. 3. *See* noise limiter.

noise temperature The temperature at which the thermal noise power of a passive system per unit bandwidth would be equal to the actual noise at the actual terminals; the standard reference temperature for noise measurements is 290 K.

noise testing The measurement of the power dissipated in a resistance termination of given value joined to one end of a telephone or telegraph circuit when no test power is applied to the circuit.

noise tube A gas tube used as a source of white noise.

noise weighting Use of an electrical network to obtain a weighted average over frequency of the noise power, which is representative of the relative disturbing effects of noise in a communications system at various frequencies.

no-load current The current which flows in a network when the output is open-circuited.

no-load loss The power loss of a device that is operated at rated voltage and frequency but is not supplying power to a load.

no-load voltage *See* open-circuit voltage.

nominal impedance Impedance of a circuit under conditions at which it was designed to operate; normally specified at center of operating frequency range.

nominal value The value of some property (such as resistance, capacitance, or impedance) of a device at which it is supposed to operate, under normal conditions, as opposed to actual value.

296 noncontacting piston

noncontacting piston See choke piston.

noncontacting plunger See choke piston.

nondegenerate amplifier Parametric amplifier that is characterized by a pumping frequency considerably higher than twice the signal frequency; the output is taken at the signal input frequency; the amplifier exhibits negative impedance characteristics, indicative of infinite gain, and is therefore capable of oscillation.

nondestructive breakdown Breakdown of the barrier between the gate and channel of a field-effect transistor without causing failure of the device; in a junction field-effect transistor, avalanche breakdown occurs at the *pn* junction.

nondirectional See omnidirectional.

nondirectional antenna See omnidirectional antenna.

nondissipative stub Nondissipative length of waveguide or transmission line.

nonhoming tuning system Motor-driven automatic tuning system in which the motor starts up in the direction of previous rotation; if this direction is incorrect for the new station, the motor reverses, after turning to the end of the dial, then proceeds to the desired station.

noninductive Having negligible or zero inductance.

noninductive capacitor A capacitor constructed so it has practically no inductance; foil layers are staggered during winding, so an entire layer of foil projects at either end for contact-making purposes; all currents then flow laterally rather than spirally around the capacitor.

noninductive resistor A wire-wound resistor constructed to have practically no inductance, either by using a hairpin winding or be reversing connections to adjacent sections of the winding.

noninductive winding A winding constructed so that the magnetic field of one turn or section cancels the field of the next adjacent turn or section.

noninverting amplifier An operational amplifier in which the input signal is applied to the ungrounded positive input terminal to give a gain greater than unity and make the output voltage change in phase with the input voltage.

noninverting parametric device Parametric device whose operation depends essentially upon three frequencies, a harmonic of the pump frequency and two signal frequencies, of which one is the sum of the other plus the pump harmonic.

nonlinear amplifier An amplifier in which a change in input does not produce a proportional change in output.

nonlinear capacitor Capacitor having a mean charge characteristic or a peak charge characteristic that is not linear, or a reversible capacitance that varies with bias voltage.

nonlinear circuit A circuit in which the current and voltage in any element that results from two sources of energy acting together is not equal to the sum of the currents or voltages that result from each of the sources acting alone.

nonlinear circuit component An electrical device for which a change in applied voltage does not produce a proportional charge in current. Also known as nonlinear device; nonlinear element.

nonlinear coil Coil having an easily saturable core, possessing high impedance at low or zero current and low impedance when current flows and saturates the core.

nonlinear coupler A type of frequency multiplier which uses the nonlinear capacitance of a junction diode to couple energy from the input circuit, which is tuned to the fundamental, to the output circuit, which is tuned to the desired harmonic.

nonshorting contact switch   297

nonlinear crystal A crystal in which some influence (such as stress, electric field, or magnetic field) produces a response (such as strain, electric polarization, or magnetization) which is not proportional to the influence.

nonlinear detection Detection based on the curvature of a tube characteristic, such as square-law detection.

nonlinear device See nonlinear circuit component.

nonlinear dielectric A dielectric whose polarization is not proportional to the applied electric field.

nonlinear distortion Distortion in which the output of a system or component does not have the desired linear relation to the input.

nonlinear element See nonlinear circuit component.

nonlinear inductance The behavior of an inductor for which the voltage drop across the inductor is not proportional to the rate of change of current, such as when the inductor has a core of magnetic material in which magnetic induction is not proportional to magnetic field strength.

nonlinear network A network in which the current or voltage in any element that results from two sources of energy acting together is not equal to the sum of the currents or voltages that result from each of the sources acting alone.

nonlinear oscillator A radio-frequency oscillator that changes frequency in response to an audio signal; it is the basic circuit used in eavesdropping devices.

nonlinear reactance The behavior of a coil or capacitor whose voltage drop is not proportional to the rate of change of current through the coil, or the charge on the capacitor.

nonlinear resistance The behavior of a substance (usually a semiconductor) which does not obey Ohm's law but has a voltage drop across it that is proportional to some power of the current.

nonlinear taper Nonuniform distribution of resistance throughout the element of a potentiometer or rheostat.

nonloaded Q Of an electric impedance, the Q value of the impedance without external coupling or connection. Also known as basic Q.

nonmagnetic Not magnetizable, and therefore not affected by magnetic fields.

nonmetallic sheathed cable Assembly of two or more rubber-covered conductors in an outer sheath of nonconducting fibrous material that has been treated to make it flame-resistant and moisture-repellent.

nonmultiple switchboard Manual telephone switchboard in which each subscriber line is attached to only one jack.

nonreactive Pertaining to a circuit, component, or load that has no capacitance or impedance, so that an alternating current is in phase with the corresponding voltage.

nonreactive load See resistive load.

nonrenewable fuse unit Fuse unit that cannot be readily restored for service after operation.

nonresonant antenna A long-wire or traveling-wave antenna which does not have natural frequencies of oscillation, and responds equally well to radiation over a broad range of frequencies.

nonresonant line Transmission line having no reflected waves, and neither current nor voltage standing waves.

nonshorting contact switch Selector switch in which the width of the movable contact is less than the distance between contact clips, so that the old circuit is broken before the new circuit is completed.

298 nonsinusoidal waveform

nonsinusoidal waveform The representation of a wave which does not vary in a sinusoidal manner, and which therefore contains harmonics.

nonstorage camera tube Television camera tube in which the picture signal is, at each instant, proportional to the intensity of the illumination on the corresponding area of the scene.

nonsynchronous Not related in phase, frequency, or speed to other quantities in a device or circuit.

nonsynchronous timer A circuit at the receiving end of a communications link which restores the time relationship between pulses when no timing pulses are transmitted.

nonsynchronous transmission A data transmission process in which a clock is not used to control the unit intervals within a block or a group of data signals.

nonsynchronous vibrator Vibrator that interrupts a direct-current circuit at a frequency unrelated to the other circuit constants and does not rectify the resulting stepped-up alternating voltage.

nonthermal decimetric emission A radio-wave emission above the 4-centimeter wavelength from the planet Jupiter that has a nearly constant flux between 5-centimeter and 1-meter wavelength. Also known as DIM.

NOR circuit A circuit in which output voltage appears only when signal is absent from all of its input terminals.

Nordheim's rule The rule that the residual resistivity of a binary alloy that contains mole fraction x of one element and $1 - x$ of the other is proportional to $x(1 - x)$.

normal electrode Standard electrode used for measuring electrode potentials.

normal impedance *See* free impedance.

normal incidence reflectivity The ratio of the energy of electromagnetic radiation reflected from the interface between two media to the energy of the incident radiation when the incident radiation travels in a direction perpendicular to the surface.

normal induction Limiting induction, either positive or negative, in a magnetic material that is under the influence of a magnetizing force which varies between two specific limits.

normalized admittance The reciprocal of the normalized impedance.

normalized coupling coefficient Mutual inductance, expressed on a scale running from zero to one.

normalized current The current divided by the square root of the characteristic admittance of a waveguide or transmission line.

normalized impedance An impedance divided by the characteristic impedance of a transmission line or waveguide.

normalized Q The ratio of the reactive component of the impedance of a filter section to the resistive component.

normalized susceptance The susceptance of an element of a waveguide or transmission line divided by the characteristic admittance.

normalized voltage The voltage divided by the square root of the characteristic impedance of a waveguide or transmission line.

normal magnetization curve Curve traced on a graph of magnetic induction versus magnetic field strength in an originally unmagnetized specimen, as the magnetic field strength is increased from zero. Also known as magnetization curve.

normal-mode helix A type of helical antenna whose diameter and electrical length are considerably less than a wavelength, and which has a radiation pattern with greatest intensity normal to the helix axis.

n-type germanium   299

normal-mode rejection ratio The ability of an amplifier to reject spurious signals at the power-line frequency or at harmonics of the line frequency. Abbreviated NMRR.

north pole The pole of a magnet at which magnetic lines of force are considered as leaving the magnet; the lines enter the south pole; if the magnet is freely suspended, its north pole points toward the north geomagnetic pole.

Norton's theorem The theorem that the voltage across an element that is connected to two terminals of a linear network is equal to the short-circuit current between these terminals in the absence of the element, divided by the sum of the admittances between the terminals associated with the element and the network respectively.

notch Rectangular depression extending below the sweep line of the radar indicator in some types of equipment.

notch antenna Microwave antenna in which the radiation pattern is determined by the size and shape of a notch or slot in a radiating surface.

notch filter A band-rejection filter that produces a sharp notch in the frequency response curve of a system; used in television transmitters to provide attenuation at the low-frequency end of the channel, to prevent possible interference with the sound carrier of the next lower channel.

notching Term indicating that a predetermined number of separate impulses are required to complete operation of a relay.

NOT circuit A logic circuit with one input and one output that inverts the input signal at the output; that is, the output signal is a logical 1 if the input signal is a logical 0, and vice versa. Also known as inverter circuit.

novar Beam-power tube having a nine-pin base.

npin transistor An npn transistor which has a layer of high-purity germanium between the base and collector to extend the frequency range.

npnp diode See pnpn diode.

npnp transistor An npn-junction transistor having a transition or floating layer between p and n regions, to which no ohmic connection is made. Also known as $pnpn$ transistor.

npn semiconductor Double junction formed by sandwiching a thin layer of p-type material between two layers of n-type material of a semiconductor.

npn transistor A junction transistor having a p-type base between an n-type emitter and an n-type collector; the emitter should then be negative with respect to the base, and the collector should be positive with respect to the base.

NPO-body Referring to a series of temperature-compensating capacitors that have an invariant dielectric constant over a specified temperature range.

np semiconductor Region of transition between n- and p-type material.

N scan See N scope.

N scope A cathode-ray scope combining the features of K and M scopes. Also known as N indicator; N scan.

n-type conduction The electrical conduction associated with electrons, as opposed to holes, in a semiconductor.

n-type crystal rectifier Crystal rectifier in which forward current flows when the semiconductor is negative with respect to the metal.

n-type germanium Germanium to which more impurity atoms of donor type (with valence 5, such as antimony) than of acceptor type (with valence 3, such as indium) have been added, with the result that the conduction electron density exceeds the hole density.

n-type semiconductor An extrinsic semiconductor in which the conduction electron density exceeds the hole density.

nuclear electric power generation Large-scale generation of electric power in which the source of energy is nuclear fission, generally in a nuclear reactor, or nuclear fusion.

nuclear triode detector A type of junction detector that has two outputs which together determine the precise location on the detector where the ionizing radiation was incident, as well as the energy of the ionizing particle.

null-current circuit A circuit used to measure current, in which the unknown current is opposed by a current resulting from applying a voltage controlled by a slide wire across a series resistor, and the slide wire is continuously adjusted so that the resulting current, as measured by a direct-current detector amplifier, is equal to zero.

null-current measurement Measurement of current using a null-current circuit.

null detection Altering of adjustable bridge circuit components, to obtain zero current.

numerical display device Any device for visually displaying numerical figures, such as a numerical indicator tube, a device utilizing electroluminescence, or a device in which any one of a stack of transparent plastic strips engraved with digits can be illuminated by a small light at the edge of the strip.

numerical indicator tube An electron tube capable of visually displaying numerical figures; some varieties also display alphabetical characters and commonly used symbols.

nuvistor Electron tube in which all electrodes are cylindrical, placed one inside the other with close spacing, in a ceramic envelope.

Nyquist's theorem The mean square noise voltage across a resistance in thermal equilibrium is four times the product of the resistance, Boltzmann's constant, the absolute temperature, and the frequency range within which the voltage is measured.

O

O attenuator A dissipative attenuator in which the circuit has the form of a ladder with two rungs, and the resistances across the rungs are unequal, so that the impedances across the two pairs of terminals are unequal.

octal base Tube base having a central aligning key and positioned for eight equally spaced pins.

octave-band oscillator An oscillator that can be tuned over a frequency range of 2 to 1, so that its highest frequency is twice its lowest frequency.

octode An eight-electrode electron tube containing an anode, a cathode, a control electrode, and five additional electrodes that are ordinarily grids.

odoriferous homing Homing on the ionized air produced by the exhaust gases of a snorkeling submarine.

oersted The unit of magnetic field strength in the centimeter-gram-second electromagnetic system of units, equal to the field strength at the center of a plane circular coil of one turn and 1-centimeter radius, when there is a current of $1/2 \pi$ abampere in the coil.

off-center plan position indicator A plan position indicator in which the center of the display that represents the location of the radar can be moved from the center of the screen to any position on the face of the PPI.

ohm The unit of electrical resistance in the rationalized meter-kilogram-second system of units, equal to the resistance through which a current of 1 ampere will flow when there is a potential difference of 1 volt across it. Symbolized Ω.

ohmic contact A region where two materials are in contact, which has the property that the current flowing through it is proportional to the potential difference across it.

ohmic dissipation Loss of electric energy when a current flows through a resistance due to conversion into heat. Also known as ohmic loss.

ohmic loss See ohmic dissipation.

ohmic resistance Property of a substance, circuit, or device for which the current flowing through it is proportional to the potential difference across it.

Ohm's law The law that the direct current flowing in an electric circuit is directly proportional to the voltage applied to the circuit; it is valid for metallic circuits and many circuits containing an electrolytic resistance.

oil-break Property of an electrical switch, circuit breaker, or similar apparatus whose contacts separate in oil.

oil circuit breaker A high-voltage circuit breaker in which the arc is drawn in oil to dissipate the heat and extinguish the arc; the intense heat of the arc decomposes the oil, generating a gas whose high pressure produces a flow of fresh fluid through the arc that furnishes the necessary insulation to prevent a restrike of the arc.

oil-filled cable Cable having insulation impregnated with an oil which is fluid at all operating temperatures and provided with facilities such as longitudinal ducts or channels and with reservoirs; by this means positive oil pressure can be maintained within the cable at all times, incipient voids are promptly filled during periods of expansion, and all surplus oil is adequately taken care of during periods of contraction.

oil-immersed Property of a transformer, reactor, regulator, or similar apparatus whose coils are immersed in an insulating liquid that is usually, but not necessarily, oil.

oil switch A switch whose contacts are immersed in oil in order to suppress the arc and prevent the contacts from being damaged.

O indicator *See* O scope.

olivette Standing floodlight used in the wings for lighting stage entrances and acting areas at fairly close range; bulb wattage ranges from 500 to 1500 watts.

omegatron A miniature mass spectrograph, about the size of a receiving tube, that can be sealed to another tube and used to identify the residual gases left after evacuation.

omnidirectional Radiating or receiving equally well in all directions. Also known as nondirectional.

omnidirectional antenna An antenna that has an essentially circular radiation pattern in azimuth and a directional pattern in elevation. Also known as nondirectional antenna.

OMS *See* ovonic memory switch.

ondograph An instrument that draws the waveform of an alternating-current voltage step by step; a capacitor is charged momentarily to the amplitude of a point on the voltage wave, then discharged into a recording galvanometer, with the action being repeated a little further along on the waveform at intervals of about 0.01 second.

ondoscope A glow-discharge tube used to detect high-frequency radiation, as in the vicinity of a radar transmitter; the radiation ionizes the gas in the tube and produces a visible glow.

one-digit subtracter *See* half-subtracter.

one-quadrant multiplier Of an analog computer, a multiplier in which operation is restricted to a single sign of both input variables.

one-shot multivibrator *See* monostable multivibrator.

one-sided abrupt junction An abrupt junction that is realized by giving one side of the junction a high doping level compared with the other; that is, an n^+p or p^+n junction.

O network Network composed of four impedance branches connected in series to form a closed circuit, two adjacent junction points serving as input terminals, the remaining two junction points serving as output terminals.

one-way trunk Trunk between two central offices, used for calls that originate at one of those offices, but not for calls that originate at the other. Also known as outgoing trunk.

on-line The state in which a piece of equipment or a subsystem is connected and powered to deliver its proper output to the system.

on-off switch A switch used to turn a receiver or other equipment on or off; often combined with a volume control in radio and television receivers.

on-off tests Tests conducted to determine the source of interference by switching various suspected sources on and off while observing the victim receiver.

Onsager theory of dielectrics A theory for calculating the dielectric constant of a material with polar molecules in which the local field at a molecule is calculated for an actual spherical cavity of molecular size in the dielectric using Laplace's equation, and the polarization catastrophe of the Lorentz field theory is thereby avoided.

on the beam Centered on a beam of, or on an equisignal zone of, radiant energy, as a radio range.

open 1. Condition in which conductors are separated so that current cannot pass. 2. Break or discontinuity in a circuit which can normally pass a current.

open circuit An electric circuit that has been broken, so that there is no complete path for current flow.

open-circuited line A microwave discontinuity which reflects an infinite impedance.

open-circuit impedance Of a line or four-terminal network, the driving-point impedance when the far end is open.

open-circuit jack Jack that normally leaves its circuit open; the circuit can be closed only through a circuit connected to the plug that is inserted in the jack.

open-circuit voltage The voltage at the terminals of a source when no appreciable current is flowing. Also known as no-load voltage.

open-delta connection An unsymmetrical transformer connection which is employed when one transformer of a bank of three single-phase delta-connected units must be cut out, because of failure. Also known as V connection.

open-fuse cutout Enclosed fuse cutout in which the fuse support and fuse holder are exposed.

open-link fuse A simple type of fuse that consists of a strip of fuse material bolted to open terminal blocks.

open-phase protection Effect of a device operating on the loss of current in one phase of a polyphase circuit to cause and maintain the interruption of power in the circuit.

open-phase relay Relay which functions by reason of the opening of one or more phases of a polyphase circuit, when sufficient current is flowing in the remaining phase or phases.

open plug Plug designed to hold jack springs in their open position.

open wire A conductor supported above the ground, separate from other conductors.

open-wire feeder *See* open-wire transmission line.

open-wire loop Branch line on a main open-wire line.

open-wire transmission line A transmission line consisting of two spaced parallel wires supported by insulators, at the proper distance to give a desired value of surge impedance. Also known as open-wire feeder.

operate time Total elapsed time from application of energizing current to a relay coil to the time the contacts have opened or closed.

operating angle Electrical angle of the input signal (for example, portion of a cycle) during which plate current flows in a vacuum tube amplifier.

operating point Point on a family of characteristic curves of a vacuum tube or transistor where the coordinates of the point represent the instantaneous values of the electrode voltages and currents for the operating conditions under study or consideration.

operating power Power that is actually supplied to a radio transmitter antenna.

operating range The frequency range over which a reversible transducer is operable.

operational amplifier An amplifier having high direct-current stability and high immunity to oscillation, generally achieved by using a large amount of negative feedback; used to perform analog-computer functions such as summing and integrating.

optical branch The vibrations of an optical mode plotted on a graph of frequency versus wave number; it is separated from, and has higher frequencies than, the acoustic branch.

optical coupler See optoisolator.

optical coupling Coupling between two circuits by means of a light beam or light pipe having transducers at opposite ends, to isolate the circuits electrically.

optical Doppler effect A change in the observed frequency of light or other electromagnetic radiation caused by relative motion of the source and observer.

optical electronic reproducer See optical sound head.

optical encoder An encoder that converts positional information into corresponding digital data by interrupting light beams directed on photoelectric devices.

optical harmonic Light, generated by passing a laser beam with a power density on the order of 10^{10} watts/cm^2 or more through certain transparent materials, which has a frequency which is an integral multiple of that of the incident laser light.

optical isolator See optoisolator.

optically coupled isolator See optoisolator.

optical mask A thin sheet of metal or other substance containing an open pattern, used to suitably expose to light a photoresistive substance overlaid on a semiconductor or other surface to form an integrated circuit.

optical mode A type of vibration of a crystal lattice whose frequency varies with wave number only over a limited range, and in which neighboring atoms or molecules in different sublattices move in opposition to each other.

optical phenomena Phenomena associated with the generation, transmission, and detection of electromagnetic radiation in the visible, infrared, or ultraviolet regions.

optical phonon A quantum of an optical mode of vibration of a crystal lattice.

optical properties The effects of a substance or medium on light or other electromagnetic radiation passing through it, such as absorption, scattering, refraction, and polarization.

optical relay An optoisolator in which the output device is a light-sensitive switch that provides the same on and off operations as the contacts of a relay.

optical scanner See flying-spot scanner.

optical sound head The assembly in motion picture projection which reproduces photographically recorded sound; light from an incandescent lamp is focused on a slit, light from the slit is in turn focused on the optical sound track of a film, and the light passing through the film is detected by a photoelectric cell. Also known as optical electronic reproducer.

optical sound recorder See photographic sound recorder.

optical sound reproducer See photographic sound reproducer.

optical waveguide A waveguide in which a light-transmitting material such as a glass or plastic fiber is used for transmitting information from point to point at wavelengths somewhere in the ultraviolet, visible-light, or infrared portions of the spectrum. Also known as fiber waveguide; optical-fiber cable.

orthogonal antennas 305

optimum array current The current distribution in a broadside antenna array which is such that for a specified side-lobe level the beam width is as narrow as possible, and for a specified first null the side-lobe level is as small as possible.

optimum bunching Bunching condition required for maximum output in a velocity modulation tube.

optimum coupling *See* critical coupling.

optimum filter An electric filter in which the mean square value of the error between a desired output and the actual output is at a minimum.

optocoupler *See* optoisolator.

optoelectronic isolator *See* optoisolator.

optoelectronics The branch of electronics that deals with solid-state and other electronic devices for generating, modulating, transmitting, and sensing electromagnetic radiation in the ultraviolet, visible-light, and infrared portions of the spectrum.

optoelectronic scanner A scanner in which lenses, mirrors, or other optical devices are used between a light source or image and a photodiode or other photoelectric device.

optoisolator A coupling device in which a light-emitting diode, energized by the input signal, is optically coupled to a photodetector such as a light-sensitive output diode, transistor, or silicon controlled rectifier. Also known as optical coupler; optical isolator; optically coupled isolator; optocoupler; optoelectronic isolator; photocoupler; photoisolator.

OR circuit *See* OR gate.

order-disorder transition The transition of an alloy or other solid solution between a state in which atoms of one element occupy certain regular positions in the lattice of another element, and a state in which this regularity is not present.

ordering A solid-state transformation in certain solid solutions, in which a random arrangement in the lattice is transformed into a regular ordered arrangement of the atoms with respect to one another; a so-called superlattice is formed.

organic electrolyte cell A type of wet cell that is based on the use of particularly reactive metals such as lithium, calcium, or magnesium in conjunction with organic electrolytes; the best-known type is the lithium–cupric fluoride cell.

OR gate A multiple-input gate circuit whose output is energized when any one or more of the inputs is in a prescribed state; performs the function of the logical inclusive-or; used in digital computers. Also known as OR circuit.

orientation The physical positioning of a directional antenna or other device having directional characteristics.

orientation effect Those bulk properties of a material which result from orientation polarization.

orientation polarization Polarization arising from the orientation of molecules which have permanent dipole moments arising from an asymmetric charge distribution. Also known as dipole polarization.

orifice Opening or window in a side or end wall of a waveguide or cavity resonator through which energy is transmitted.

orthicon A camera tube in which a beam of low-velocity electrons scans a photoemissive mosaic that is capable of storing a pattern of electric charges; has higher sensitivity than the iconoscope.

orthogonal antennas In radar, a pair of transmitting and receiving antennas, or a single transmitting-receiving antenna, designed for the detection of a difference in polarization between the transmitted energy and the energy returned from the target.

O scan *See* O scope.

osciducer Transducer in which information pertaining to the stimulus is provided in the form of deviation from the center frequency of an oscillator.

oscillating magnetic field A magnetic field which varies periodically in time.

oscillator 1. An electronic circuit that converts energy from a direct-current source to a periodically varying electric output. 2. The stage of a superheterodyne receiver that generates a radio-frequency signal of the correct frequency to mix with the incoming signal and produce the intermediate-frequency value of the receiver. 3. The stage of a transmitter that generates the carrier frequency of the station or some fraction of the carrier frequency.

oscillator harmonic interference Interference occurring in a superheterodyne receiver due to the interaction of incoming signals with harmonics (usually the second harmonic) of the local oscillator.

oscillator-mixer-first detector *See* converter.

oscillatory circuit Circuit containing inductance or capacitance, or both, and resistance, connected so that a voltage impulse will produce an output current which periodically reverses or oscillates.

oscillatory discharge Alternating current of gradually decreasing amplitude which, under certain conditions, flows through a circuit containing inductance, capacitance, and resistance when a voltage is applied.

oscillatory surge Surge which includes both positive and negative polarity values.

oscillistor A bar of semiconductor material, such as germanium, that will oscillate much like a quartz crystal when it is placed in a magnetic field and is carrying direct current that flows parallel to the magnetic field.

oscillograph tube Cathode-ray tube used to produce a visible pattern, which is the graphical representation of electric signals, by variations of the position of the focused spot or spots according to these signals.

oscilloscope *See* cathode-ray oscilloscope.

O scope An A scope modified by the inclusion of an adjustable notch for measuring range. Also known as O indicator; O scan.

OTS *See* ovonic threshold switch.

O-type backward-wave oscillator A backward-wave tube in which an electron gun produces an electron beam focused longitudinally throughout the length of the tube, a slow-wave circuit interacts with the beam, and at the end of the tube a collector terminates the beam. Also known as O-type carcinotron; type-O carcinotron.

O-type carcinotron *See* O-type backward-wave oscillator.

outage A failure in an electric power system.

outer-shell electron *See* conduction electron.

outgoing trunk *See* one-way trunk.

outlet A power line termination from which electric power can be obtained by inserting the plug of a line cord. Also known as convenience receptacle; electric outlet; receptacle.

outlet box A box at which lines in an electric wiring system terminate, so that electric appliances or fixtures may be connected.

out-of-service jack Jack associated with a test jack which removes the circuit from service when a shorted plug is inserted.

overload protection 307

output 1. The current, voltage, power, driving force, or information which a circuit or device delivers. 2. Terminals or other places where a circuit or device can deliver current, voltage, power, driving force, or information.

output bus driver A device that power-amplifies output signals from a computer to allow them to drive heavy circuit loads.

output capacitance Of an n-terminal electron tube, the short-circuit transfer capacitance between the output terminal and all other terminals, except the input terminal, connected together.

output gap An interaction gap by means of which usable power can be abstracted from an electron stream in a microwave tube.

output impedance The impedance presented by a source to a load.

output power Power delivered by a system or transducer to its load.

output resistance The resistance across the output terminals of a circuit or device.

output stage The final stage in any electronic equipment.

output transformer The iron-core audio-frequency transformer used to match the output stage of a radio receiver or an amplifier to its loudspeaker or other load.

output tube Power-amplifier tube designed for use in an output stage.

output winding Of a saturable reactor, a winding, other than a feedback winding, which is associated with the load, and through which power is delivered to the load.

overall response The ratio between system input and output.

overbunching In velocity-modulated streams of electrons, the bunching condition produced by the continuation of the bunching process beyond the optimum condition.

overcompound To use sufficiently many series turns in a compound-wound generator so that the terminal voltage at rated load is greater than at no load, usually to compensate for increased line drop.

overcoupled circuits Two resonant circuits which are tuned to the same freqency but coupled so closely that two response peaks are obtained; used to attain broad-band response with substantially uniform impedance.

overcurrent An abnormally high current, usually resulting from a short circuit.

overcurrent protection *See* overload protection.

overdriven amplifier Amplifier stage which is designed to distort the input-signal waveform by permitting the grid signal to drive the stage beyond cutoff or plate-current saturation.

overlay transistor Transistor containing a large number of emitters connected in parallel to provide maximum power amplification at extremely high frequencies.

overload A load greater than that which a device is designed to handle; may cause overheating of power-handling components and distortion in signal circuits.

overload capacity Current, voltage, or power level beyond which permanent damage occurs to the device considered.

overload current A current greater than that which a circuit is designed to carry; may melt wires or damage elements of the circuit.

overload level Level above which operation ceases to be satisfactory as a result of signal distortion, overheating, damage, and so forth.

overload protection Effect of a device operative on excessive current, but not necessarily on short circuit, to cause and maintain the interruption of current flow to the device governed. Also known as overcurrent protection.

overload relay A relay that opens a circuit when the load in the circuit exceeds a preset value, in order to provide overload protection; usually responds to excessive current, but may respond to excessive values of power, temperature, or other quantities. Also known as overload release.

overload release See overload relay.

overpotential See overvoltage.

overshoot The reception of microwave signals where they were not intended, due to an unusual atmospheric condition that sets up variations in the index of refraction.

over-the-horizon propagation See scatter propagation.

over-the-horizon radar Long-range radar in which the transmitted and reflected beams are bounced off the ionosphere layers to achieve ranges far beyond line of sight.

overtone crystal Quartz crystal cut in such a manner that it will operate at a higher order than its fundamental frequency, or operate at two frequencies simultaneously as in a synthesizer.

overvoltage 1. A voltage greater than that at which a device or circuit is designed to operate. Also known as overpotential. 2. The amount by which the applied voltage exceeds the Geiger threshold in a radiation counter tube.

overvoltage crowbar A circuit that monitors the output of a power supply and prevents the output voltage from exceeding a preset voltage, under any failure condition, by having a low resistance (crowbar) placed across the output terminals when an overvoltage occurs.

ovonic device See glass switch.

ovonic memory switch A glass switch which, after being brought from the highly resistive state to the conducting state, remains in the conducting state until a current pulse returns it to its highly resistive state. Abbreviated OMS. Also known as memory switch.

ovonic threshold switch A glass switch which, after being brought from the highly resistive state to the conducting state, returns to the highly resistive state when the current falls below a holding current value. Abbreviated OTS.

Ovshinsky effect The characteristic of a special thin-film solid-state switch that responds identically to both positive and negative polarities so that current can be made to flow in both directions equally.

Owen bridge A four-arm alternating-current bridge used to measure self-inductance in terms of capacitance and resistance; bridge balance is independent of frequency.

oxide-coated cathode A cathode that has been coated with oxides of alkaline-earth metals to improve electron emission at moderate temperatures. Also known as Wehnelt cathode.

oxide isolation Isolation of the elements of an integrated circuit by forming a layer of silicon oxide around each element.

oxide passivation Passivation of a semiconductor surface by producing a layer of an insulating oxide on the surface.

P

P *See* permeance; primary winding.

pA *See* picoampere.

packaged magnetron Integral structure comprising a magnetron, its magnetic circuit, and its output matching device.

packaging The process of physically locating, connecting, and protecting devices or components.

packaging density The number of components per unit volume in a working system or subsystem.

packing density The number of devices or gates per unit area of an integrated circuit.

pad 1. An arrangement of fixed resistors used to reduce the strength of a radio-frequency or audio-frequency signal by a desired fixed amount without introducing appreciable distortion. Also known as fixed attenuator. 2. *See* terminal area.

padder A trimmer capacitor inserted in series with the oscillator tuning circuit of a superheterodyne receiver to control calibration at the low-frequency end of a tuning range.

paint Vernacular for a target image on a radarscope.

pair Two like conductors employed to form an electric circuit.

paired cable Cable in which the single conductors are twisted together in groups of two, none of which is arranged with others to form quads.

pairing In television, imperfect interlace of lines composing the two fields of one frame of the picture; instead of having the proper equal spacing, the lines appear in groups of two.

Palmer scan Combination of circular or raster and conical radar scans; the beam is swung around the horizon, and at the same time a conical scan is performed.

panadapter *See* panoramic adapter.

pancake coil A coil having the shape of a pancake, usually with the turns arranged in the form of a flat spiral.

panel board *See* control board.

panel display An unconventional method of displaying color television pictures in which luminescent conversion devices, such as light-emitting diodes or electroluminescent devices, are arranged in a matrix array, forming a flat-panel screen, and are controlled by signals sent over vertical and horizontal wires connected to both electrodes of the devices.

panoramic adapter A device designed to operate with a search receiver to provide a visual presentation on an oscilloscope screen of a band of frequencies extending

310 panoramic display

above and below the center frequency to which the search receiver is tuned. Also known as panadapter.

panoramic display A display that simultaneously shows the relative amplitudes of all signals received at different frequencies.

panoramic receiver Radio receiver that permits continuous observation on a cathode-ray-tube screen of the presence and relative strength of all signals within a wide frequency range.

pan-range Intensity-modulated, A-type radar indication with a slow vertical sweep applied to video; stationary targets give solid vertical deflection, and moving targets give broken vertical deflection.

paper capacitor A capacitor whose dielectric material consists of oiled paper sandwiched between two layers of metallic foil.

paraballoon Air-inflated radar antenna.

parabolic antenna Antenna with a radiating element and a parabolic reflector that concentrates the radiated power into a beam.

parabolic reflector An antenna having a concave surface which is generated either by translating a parabola perpendicular to the plane in which it lies (in a cylindrical parabolic reflector), or rotating it about its axis of symmetry (in a paraboloidal reflector). Also known as dish.

paraboloidal antenna *See* paraboloidal reflector.

paraboloidal reflector An antenna having a concave surface which is a paraboloid of revolution; it concentrates radiation from a source at its focal point into a beam. Also known as paraboloidal antenna.

parallel Connected to the same pair of terminals. Also known as multiple; shunt.

parallel buffer Electronic device (magnetic core or flip-flop) used to temporarily store digital data in parallel, as opposed to series storage.

parallel circuit An electric circuit in which the elements, branches (having elements in series), or components are connected between two points, with one of the two ends of each component connected to each point.

parallel feed Application of a direct-current voltage to the plate or grid of a tube in parallel with an alternating-current circuit, so that the direct-current and the alternating-current components flow in separate paths. Also known as shunt feed.

parallel impedance One of two or more impedances that are connected to the same pair of terminals.

paralleling reactor Reactor for correcting the division of load between parallel-connected transformers which have unequal impedance voltages.

parallel operation The connecting together of the outputs of two or more batteries or other power supplies so that the sum of their output currents flows to a common load.

parallel padding Method of parallel operation for two or more power supplies in which their current limiting or automatic crossover output characteristic is employed so that each supply regulates a portion of the total current, each parallel supply adding to the total and padding the output only when the load current demand exceeds the capability, or limit setting, of the first supply.

parallel-plate capacitor A capacitor consisting of two parallel metal plates, with a dielectric filling the space between them.

parallel-plate waveguide Pair of parallel conducting planes used for propagating uniform circularly cylindrical waves having their axes normal to the plane.

parametric amplifier 311

parallel programming Method of parallel operation for two or more power supplies in which their feedback terminals (voltage control terminals) are also paralleled; these terminals are often connected to a separate programming source.

parallel rectifier One of two or more rectifiers that are connected to the same pair of terminals, generally in series with small resistors or inductors, when greater current is desired than can be obtained with a single rectifier.

parallel resonance Also known as antiresonance. 1. The frequency at which the inductive and capacitive reactances of a parallel resonant circuit are equal. 2. The frequency at which the parallel impedance of a parallel resonant circuit is a maximum. 3. The frequency at which the parallel impedance of a parallel resonant circuit has a power factor of unity.

parallel resonant circuit A circuit in which an alternating-current voltage is applied across a capacitor and a coil in parallel. Also known as antiresonant circuit.

parallel resonant interstage A coupling between two amplifier stages achieved by means of a parallel-tuned LC circuit.

parallel-rod oscillator Ultra-high-frequency oscillator circuit in which parallel rods or wires of required length and dimensions form the tank circuits.

parallel series Circuit in which two or more parts are connected together in parallel to form parallel circuits, and in which these circuits are then connected together in series so that both methods of connection appear.

parallel-T network A network used in capacitance measurements at radio frequencies, having two sets of three impedances, each in the form of the letter T, with the arms of the two Ts joined to common terminals, and the source and detector each connected between two of these terminals. Also known as twin-T network.

parallel-tuned circuit A circuit with two parallel branches, one having an inductance and a resistance in series, the other a capacitance and a resistance in series.

parallel wires Two conductors which are parallel to each other; often used in transmission lines.

paramagnetic Exhibiting paramagnetism.

paramagnetic crystal A crystal whose permeability is slightly greater than that of vacuum and is independent of the magnetic field strength.

paramagnetic material A material within which an applied magnetic field is increased by the alignment of electron orbits.

paramagnetic relaxation The approach of a system, which displays paramagnetism because of electronic magnetic moments of atoms or ions, to an equilibrium or steady-state condition over a period of time, following a change in the magnetic field.

paramagnetic salt A salt whose permeability is slightly greater than that of vacuum and is independent of magnetic field strength; used in adiabatic demagnetization.

paramagnetic susceptibility The susceptibility of a paramagnetic substance, which is a positive number and is, in general, much smaller than unity.

paramagnetism A property exhibited by substances which, when placed in a magnetic field, are magnetized parallel to the field to an extent proportional to the field (except at very low temperatures or in extremely large magnetic fields).

parameter 1. The resistance, capacitance, inductance, or impedance of a circuit element. 2. The value of a transistor or tube characteristic.

parametric amplifier A highly sensitive ultra-high-frequency or microwave amplifier having as its basic element an electron tube or solid-state device whose reactance

parametric converter

can be varied periodically by an alternating-current voltage at a pumping frequency. Also known as mavar; paramp; reactance amplifier.

parametric converter Inverting or noninverting parametric device used to convert an input signal at one frequency into an output signal at a different frequency.

parametric device Electronic device whose operation depends essentially upon the time variation of a characteristic parameter usually understood to be a reactance.

parametric down-converter Parametric converter in which the output signal is at a lower frequency than the input signal.

parametric oscillator An oscillator in which the reactance parameter of an energy-storage device is varied to obtain oscillation.

parametric phase-locked oscillator See parametron.

parametric up-converter Parametric converter in which the output signal is at a higher frequency than the input signal.

parametron A resonant circuit in which either the inductance or capacitance is made to vary periodically at one-half the driving frequency; used as a digital computer element, in which the oscillation represents a binary digit. Also known as parametric phase-locked oscillator; phase-locked oscillator; phase-locked subharmonic oscillator.

paramp See parametric amplifier.

paraphase amplifier An amplifier that provides two equal output signals 180° out of phase.

parasite Current in a circuit, due to some unintentional cause, such as inequalities of temperature or of composition; particularly troublesome in electrical measurements.

parasitic An undesired and energy-wasting signal current, capacitance, or other parameter of an electronic circuit.

parasitic antenna See parasitic element.

parasitic current An eddy current in a piece of electrical machinery; gives rise to energy losses.

parasitic element An antenna element that serves as part of a directional antenna array but has no direct connection to the receiver or transmitter and reflects or reradiates the energy that reaches it, in a phase relationship such as to give the desired radiation pattern. Also known as parasitic antenna; parasitic reflector; passive element.

parasitic oscillation An undesired self-sustaining oscillation or a self-generated transient impulse in an oscillator or amplifier circuit, generally at a frequency above or below the correct operating frequency.

parasitic reflector See parasitic element.

parasitic suppressor A suppressor, usually in the form of a coil and resistor in parallel, inserted in a circuit to suppress parasitic high-frequency oscillations.

paraxial trajectory A trajectory of a charged particle in an axially symmetric electric or magnetic field in which both the distance of the particle from the axis of symmetry and the angle between this axis and the tangent to the trajectory are small for all points on the trajectory.

Parker-Washburn boundary A surface which separates two regions in a solid in which the crystal axes point in different directions, and which is made up of a single array of dislocations.

partial-read pulse Current pulse that is applied to a magnetic memory to select a specific magnetic cell for reading.

partial-select output The voltage response produced by applying partial-read or partial-write pulses to an unselected magnetic cell.

partition noise Noise that arises in an electron tube when the electron beam is divided between two or more electrodes, as between screen grid and anode in a pentode.

parylene capacitor A highly stable fixed capacitor using parylene film as the dielectric; it can be operated at temperatures up to 170°C, as well as at cryogenic temperatures.

Paschen's law The law that the sparking potential between two parallel plate electrodes in a gas is a function of the product of the gas density and the distance between the electrodes. Also known as Paschen's rule.

Paschen's rule *See* Paschen's law.

passband A frequency band in which the attenuation of a filter is essentially zero.

pass element Controlled variable resistance device, either a vacuum tube or power transistor, in series with the source of direct-current power; the pass element is driven by the amplified error signal to increase its resistance when the output needs to be lowered or to decrease its resistance when the output must be raised.

passivation Growth of an oxide layer on the surface of a semiconductor to provide electrical stability by isolating the transistor surface from electrical and chemical conditions in the environment; this reduces reverse-current leakage, increases breakdown voltage, and raises power dissipation rating.

passive AND gate *See* AND gate.

passive antenna An antenna which influences the directivity of an antenna system but is not directly connected to a transmitter or receiver.

passive component *See* passive element.

passive corner reflector A corner reflector that is energized by a distant transmitting antenna; used chiefly to improve the reflection of radar signals from objects that would not otherwise be good radar targets.

passive double reflector A combination of two passive reflectors positioned to bend a microwave beam over the top of a mountain or ridge, generally without appreciably changing the general direction of the beam.

passive electronic countermeasures Electronic countermeasures that do not radiate energy, including reconnaissance or surveillance equipment that detects and analyzes electromagnetic radiation from radar and communications transmitters, and devices such as chaff which return spurious echoes to enemy radar.

passive element 1. An element of an electric circuit that is not a source of energy, such as a resistor, inductor, or capacitor. Also known as passive component. 2. *See* parasitic element.

passive filter An electric filter composed of passive elements, such as resistors, inductors, or capacitors, without any active elements, such as vacuum tubes or transistors.

passive jamming Use of confusion reflectors to return spurious and confusing signals to enemy radars. Also known as mechanical jamming.

passive junction A waveguide junction that does not have a source of energy.

passive network A network that has no source of energy.

passive reflector A flat reflector used to change the direction of a microwave or radar beam; often used on microwave relay towers to permit placement of the transmitter, repeater, and receiver equipment on the ground, rather than at the tops of towers. Also known as plane reflector.

passive system Electronic system which emits no energy, and does not give away its position or existence.

passive transducer A transducer containing no internal source of power.

paste In batteries, the medium in the form of a paste or jelly, containing an electrolyte; it is positioned adjacent to the negative electrode of a dry cell; in an electrolytic cell, the paste serves as one of the conducting plates.

pasted-plate storage battery *See* Faure storage battery.

patch A temporary connection between jacks or other terminations on a patch board.

patch board A board or panel having a number of jacks at which circuits are terminated; patch cords are plugged into the jacks to connect various circuits temporarily as required in broadcast, communication, and computer work.

patch cord A cord equipped with plugs at each end, used to connect two jacks on a patch board.

path plotting In laying out a microwave system, the plotting of the path followed by the microwave beam on a profile chart which indicates the earth's curvature.

pattern generator A signal generator used to generate a test signal that can be fed into a television receiver to produce on the screen a pattern of lines having usefulness for servicing purposes.

Patterson function A function of three spatial coordinates, constructed in the Patterson-Harker method, which has peaks at all vectors between two atoms in a crystal, the heights of the peaks being approximately proportional to the product of the atomic numbers of the corresponding atoms.

Patterson-Harker method A method of analyzing the structure of a crystal from x-ray diffraction results; a Fourier series involving squares of the absolute values of the structure factors, which are directly observable, is used to construct a vectorial representation of interatomic distances in the crystal (Patterson map).

Patterson map A contour chart of the Patterson function.

Patterson projection A projection of the Patterson function on a section through a crystal.

Patterson vectors In analysis of crystal structure, the vectors of peaks relative to the origin in a Patterson function or Patterson projection.

Pauling rule A rule governing the number of ions of opposite charge in the neighborhood of a given ion in an ionic crystal, in accordance with the requirement of local electrical neutrality of the structure.

Pauli spin susceptibility The susceptibility of free electrons in a metal due to the tendency of their spins to align with a magnetic field.

Pb-I-Pb junction *See* lead-I-lead junction.

PD *See* potential difference.

PDA *See* postacceleration.

P display *See* plan position indicator.

4PDT *See* four-pole double-throw.

peak cathode current 1. Maximum instantaneous value of a periodically recurring cathode current. 2. Highest instantaneous value of a randomly recurring pulse of cathode current. 3. Highest instantaneous value of a nonrecurrent pulse of cathode current occurring under fault conditions.

peak detector A detector whose output voltage approximates the true peak value of an applied signal; the detector tracks the signal in its sample mode and preserves the highest input signal in its hold mode.

peak envelope power Of a radio transmitter, the average power supplied to the antenna transmission line by a transmitter during one radio-frequency cycle at the highest crest of the modulation envelope, taken under conditions of normal operation.

peaker A small fixed or adjustable inductance used to resonate with stray and distributed capacitances in a broad-band amplifier to increase the gain at the higher frequencies.

peak forward voltage The maximum instantaneous voltage applied to an electronic device in the direction of lesser resistance to current flow.

peaking circuit A circuit used to improve the high-frequency response of a broad-band amplifier; in shunt peaking, a small coil is placed in series with the anode load; in series peaking, the coil is placed in series with the grid of the following stage.

peaking network Type of interstage coupling network in which an inductance is effectively in series (series-peaking network), or in shunt (shunt-peaking network), with the parasitic capacitance to increase the amplification at the upper end of the frequency range.

peaking transformer A transformer in which the number of ampere-turns in the primary is high enough to produce many times the normal flux density values in the core; the flux changes rapidly from one direction of saturation to the other twice per cycle, inducing a highly peaked voltage pulse in a secondary winding.

peak inverse anode voltage Maximum instantaneous anode voltage in the direction opposite to that in which the tube or other device is designed to pass current.

peak inverse voltage Maximum instantaneous anode-to-cathode voltage in the reverse direction which is actually applied to the diode in an operating circuit.

peak limiter *See* limiter.

peak load The maximum instantaneous load or the maximum average load over a designated interval of time. Also known as peak power.

peak power 1. The maximum instantaneous power of a transmitted radar pulse. 2. *See* peak load.

peak signal level Expression of the maximum instantaneous signal power or voltage as measured at any point in a facsimile transmission system; this includes auxiliary signals.

peak value The maximum instantaneous value of a varying current, voltage, or power during the time interval under consideration. Also known as crest value.

pedestal *See* blanking level.

pedestal level *See* blanking level.

Peierls-Nabarro force The force required to displace a dislocation along its slip plane.

pel *See* pixel.

pencil beam A beam of radiant energy concentrated in an approximately conical or cylindrical portion of space of relatively small diameter; this type of beam is used for many revolving navigational lights and radar beams.

pencil beam antenna Unidirectional antenna designed so that cross sections of the major lobe formed by planes perpendicular to the direction of maximum radiation are approximately circular.

pencil tube A small tube designed especially for operation in the ultra-high-frequency band; used as an oscillator or radio-frequency amplifier.

penetration depth In induction heating, the thickness of a layer, extending inward from a conductor's surface, whose resistance to direct current equals the resistance of the whole conductor to alternating current of a given frequency.

316 penetration phosphors

penetration phosphors A system for creating color cathode-ray-tube displays, in which phosphors of two different colors are placed on the screen of a cathode-ray tube in separate layers, and a high-energy beam penetrates the first layer and excites the second, while a low-energy beam is stopped by the first layer and excites it.

Penning gage See Philips ionization gage.

pentagrid See heptode.

pentode A five-electrode electron tube containing an anode, a cathode, a control electrode, and two additional electrodes that are ordinarily grids.

pentode transistor Point-contact transistor with four-point-contact electrodes; the body serves as a base with three emitters and one collector.

percentage differential relay Differential relay which functions when the difference between two quantities of the same nature exceeds a fixed percentage of the smaller quantity. Also known as biased relay; ratio-balance relay; ratio-differential relay.

percentage ripple Ratio of the effective value of the ripple voltage to the average value of the total voltage, expressed as a percentage.

percent make 1. In pulse testing, the length of time a circuit stands closed compared to the length of the test signal. 2. Percentage of time during a pulse period that telephone dial pulse springs are making contact.

percolation limit In a disordered crystalline alloy having one constituent with a magnetic moment, the concentration of the magnetic element above which the spin-glass phase is replaced by the ferromagnetic state.

perfect dielectric See ideal dielectric.

periodic antenna An antenna in which the input impedance varies as the frequency is altered.

periodic duty Intermittent duty in which the load conditions are regularly recurrent.

periodic field focusing Focusing of an electron beam where the electrons follow a trochoidal path and the focusing field interacts with them at selected points.

periodic line Line consisting of successive and identical sections, similarly oriented, the electrical properties of each section not being uniform throughout; the periodicity is in space and not in time; an example of a periodic line is the loaded line with loading coils uniformly spaced.

peristaltic charge-coupled device A high-speed charge-transfer integrated circuit in which the movement of the charges is similar to the peristaltic contractions and dilations of the digestive system.

permanent echo A signal reflected from an object that is fixed with respect to a radar site.

permanent magnet A piece of hardened steel or other magnetic material that has been strongly magnetized and retains its magnetism indefinitely. Abbreviated PM.

permanent-magnet focusing Focusing of the electron beam in a television picture tube by means of the magnetic field produced by one or more permanent magnets mounted around the neck of the tube.

permanent-magnet stepper motor A stepper motor in which the rotor is a powerful permanent magnet and each stator coil is energized independently in sequence; the rotor aligns itself with the stator coil that is energized.

permanent-split capacitor motor A capacitor motor in which the starting capacitor and the auxiliary winding remain in the circuit for both starting and running. Abbreviated PSC motor. Also known as capacitor start-run motor.

phantom group 317

Permasyn motor A synchronous motor which has permanent magnets embedded in the squirrel-cage rotor to provide an equivalent direct-current field.

permatron Thermionic gas-discharge diode in which the start of conduction is controlled by an external magnetic field.

permeability A factor, characteristic of a material, that is proportional to the magnetic induction produced in a material divided by the magnetic field strength; it is a tensor when these quantities are not parallel.

permeability tuning Process of tuning a resonant circuit by varying the permeability of an inductor; it is usually accomplished by varying the amount of magnetic core material of the inductor by slug movement.

permeance A characteristic of a portion of a magnetic circuit, equal to magnetic flux divided by magnetomotive force; the reciprocal of reluctance. Symbolized P.

permittivity The dielectric constant multiplied by the permittivity of empty space, where the permittivity of empty space (ϵ_0) is a constant appearing in Coulomb's law, having the value of 1 in centimeter-gram-second electrostatic units, and of 8.854×10^{-12} farad/meter in rationalized meter-kilogram-second units. Symbolized ϵ.

persistence A measure of the length of time that the screen of a cathode-ray tube remains luminescent after excitation is removed; ranges from 1 for short persistence to 7 for long persistence.

persistent-image device An optoelectronic amplifier capable of retaining an image for a definite length of time.

persistron A device in which electroluminescence and photoconductivity are used in a single panel capable of producing a steady or persistent display with pulsed signal input.

persuader Element of storage tube which directs secondary emission to electron multiplier dynodes.

perveance The space-charge-limited cathode current of a diode divided by the 3/2 power of the anode voltage.

Petersen coil *See* arc-suppression coil.

petticoat insulator Insulator having an outward-flaring lower part that is hollow inside to increase the length of the surface leakage path and keep part of the path dry at all times.

pF *See* picofarad.

PFE *See* photoferroelectric effect.

phanotron A hot-filament diode rectifier tube utilizing an arc discharge in mercury vapor or an inert gas, usually xenon.

phantastran A solid-state phantastron.

phantastron A monostable pentode circuit used to generate sharp pulses at an adjustable and accurately timed interval after receipt of a triggering signal.

phantom-circuit loading coil Loading coil for introducing a desired amount of inductance into a phantom circuit, and a minimum amount of inductance into its constituent circuits.

phantom-circuit repeating coil Repeating coil used at a terminal of a phantom circuit, in the terminal circuit extending from the midpoints of the associated side-circuit repeating coils.

phantom group 1. Group of four open-wire conductors suitable for the derivation of a phantom circuit. 2. Three circuits which are derived from simplexing two physical circuits to form a phantom circuit.

phantom repeating coil A side-circuit repeating coil or a phantom-circuit repeating coil when discrimination between these two types is not necessary.

phantom signals Signals appearing on the screen of a cathode-ray-tube indicator, the cause of which cannot readily be determined and which may be caused by circuit fault, interference, propagation anomalies, jamming, and so on.

phantom target See echo box.

phase advancer Phase modifier which supplies leading reactive volt-amperes to the system to which it is connected; may be either synchronous or asynchronous.

phase-balance relay Relay which functions by reason of a difference between two quantities associated with different phases of a polyphase circuit.

phase-change coefficient See phase constant.

phase-comparison relaying A method of detecting faults in an electric power system in which signals are transmitted from each of two terminals every half cycle so that a continuous signal is received at an intermediate point if there is no fault between the terminals, while a periodic signal is received if there is a fault.

phase conductor In a polyphase circuit, any conductor other than the neutral conductor.

phase constant A rating for a line or medium through which a plane wave of a given frequency is being transmitted; it is the imaginary part of the propagation constant, and is the space rate of decrease of phase of a field component (or of the voltage or current) in the direction of propagation, in radians per unit length. Also known as phase-change coefficient; wavelength constant.

phase control 1. A control that changes the phase angle at which the alternating-current line voltage fires a thyratron, ignitron, or other controllable gas tube. Also known as phase-shift control. 2. See hue control.

phase converter A converter that changes the number of phases in an alternating-current power source without changing the frequency.

phase-correcting network See phase equalizer.

phased array An array of dipoles on a radar antenna in which the signal feeding each dipole is varied so that antenna beams can be formed in space and scanned very rapidly in azimuth and elevation.

phase detector A circuit that provides a direct-current output voltage which is related to the phase difference between an oscillator signal and a reference signal, for use in controlling the oscillator to keep it in synchronism with the reference signal. Also known as phase discriminator.

phase discriminator See phase detector.

phase equalizer A network designed to compensate for phase-frequency distortion within a specified frequency band. Also known as phase-correcting network.

phase factor 1. The argument (phase) of a structure factor; it cannot be directly observed. 2. See power factor.

phase generator An instrument that accepts single-phase input signals over a given frequency range, or generates its own signal, and provides continuous shifting of the phase of this signal by one or more calibrated dials.

phase inversion Production of a phase difference of 180° between two similar wave shapes of the same frequency.

phase inverter A circuit or device that changes the phase of a signal by 180°, as required for feeding a push-pull amplifier stage without using a coupling transformer, or for

phase-shift oscillator 319

changing the polarity of a pulse; a triode is commonly used as a phase inverter. Also known as inverter.

phase jitter Jitter that undesirably shortens or lengthens pulses intermittently during data processing or transmission.

phase lock Technique of making the phase of an oscillator signal follow exactly the phase of a reference signal by comparing the phases between the two signals and using the resultant difference signal to adjust the frequency of the reference oscillator.

phase-locked loop A circuit that consists essentially of a phase detector which compares the frequency of a voltage-controlled oscillator with that of an incoming carrier signal or reference-frequency generator; the output of the phase detector, after passing through a loop filter, is fed back to the voltage-controlled oscillator to keep it exactly in phase with the incoming or reference frequency. Abbreviated PLL.

phase-locked oscillator See parametron.

phase-locked subharmonic oscillator See parametron.

phase modifier Machine whose chief purpose is to supply leading or lagging reactive volt-amperes to the system to which it is connected; may be either synchronous or asynchronous.

phase-modulation detector A device which recovers or detects the modulating signal from a phase-modulated carrier.

phase-modulation transmitter A radio transmitter used to broadcast a phase-modulated signal.

phase modulator An electronic circuit that causes the phase angle of a modulated wave to vary (with respect to an unmodulated carrier) in accordance with a modulating signal.

phaser Microwave ferrite phase shifter employing a longitudinal magnetic field along one or more rods of ferrite in a waveguide.

phase response A graph of the phase shift of a network as a function of frequency.

phase-sequence relay Relay which functions according to the order in which the phase voltages successively reach their maximum positive values. Also known as phase-rotation relay.

phase shift The phase angle between the input and output signals of a network or system.

phase-shift circuit A network that provides a voltage component which is shifted in phase with respect to a reference voltage.

phase-shift control See phase control.

phase-shift discriminator A discriminator that uses two similarly connected diodes, fed by a transformer that is tuned to the center frequency; when the frequency-modulated or phase-modulated input signal swings away from this center frequency, one diode receives a stronger signal than the other; the net output of the diodes is then proportional to the frequency displacement. Also known as Foster-Seely discriminator.

phase shifter A device used to change the phase relation between two alternating-current values.

phase-shifting transformer A transformer which produces a difference in phase angle between two circuits.

phase-shift oscillator An oscillator in which a network having a phase shift of 180° per stage is connected between the output and the input of an amplifier.

phase splitter A circuit that takes a single input signal voltage and produces two output signal voltages 180° apart in phase.

phase transformation A change of polyphase power from three-phase to six-phase, from three-phase to twelve-phase, and so forth, by use of transformers.

phase undervoltage relay Relay which functions by reason of the reduction of one phase voltage in a polyphase circuit.

phasing *See* framing.

phasing line That portion of the length of scanning line set aside for the phasing signal in a television or facsimile system.

phasing signal A signal used to adjust the picture position along the scanning line in a facsimile system.

phasitron An electron tube used to frequency-modulate a radio-frequency carrier; internal electrodes are designed to produce a rotating disk-shaped corrugated sheet of electrons; audio input is applied to a coil surrounding the glass envelope of the tube, to produce a varying axial magnetic field that gives the desired phase or frequency modulation of the rf carrier input to the tube.

phasmajector *See* monoscope.

phasor A low-energy collective excitation of the conduction electrons in a metal, corresponding to a slowly varying phase modulation of a charge-density wave.

Philips ionization gage An ionization gage in which a high voltage is applied between two electrodes, and a strong magnetic field deflects the resulting electron stream, increasing the length of the electron path and thus increasing the chance for ionizing collisions of electrons with gas molecules. Abbreviated pig. Also known as cold-cathode ionization gage; Penning gage.

phone patch A device connecting an amateur or citizens'-band transceiver temporarily to a telephone system.

phone plug A standard plug having a ¾-inch-diameter (19-millimeter-diameter) shank, used with headphones, microphones, and other audio equipment; usually designed for use with either two or three conductors. Also known as telephone plug.

phonic motor A small synchronous motor which is driven by the current of an accurate oscillator, such as a crystal oscillator, and whose frequency is thus constant to a high degree of accuracy; used in astronomical instruments where a driving speed of great accuracy is required.

phono jack A jack designed to accept a phono plug and provide a ground connection for the shield of the conductor connected to the plug.

phonon A quantum of an acoustic mode of thermal vibration in a crystal lattice.

phonon-electron interaction An interaction between an electron and a vibration of a lattice, resulting in a change in both the momentum of the particle and the wave vector of the vibration.

phonon emission The production of a phonon in a crystal lattice, which may result from the interaction of other phonons via anharmonic lattice forces, from scattering of electrons in the lattice, or from scattering of x-rays or particles which bombard the crystal.

phonon wind A stream of nonthermal phonons that is effective in propelling electron-hole droplets through a crystal.

phono plug A plug designed for attaching to the end of a shielded conductor, for feeding audio-frequency signals from a phonograph or other a-f source to a mating phono jack on a preamplifier or amplifier.

phosphor dot One of the tiny dots of phosphor material that are used in groups of three, one group for each primary color, on the screen of a color television picture tube.

photocathode A photosensitive surface that emits electrons when exposed to light or other suitable radiation; used in phototubes, television camera tubes, and other light-sensitive devices.

photocell A solid-state photosensitive electron device whose current-voltage characteristic is a function of incident radiation. Also known as electric eye; photoelectric cell.

photocell relay A relay actuated by a signal received when light falls on, or is prevented from falling on, a photocell.

photoconduction An increase in conduction of electricity resulting from absorption of electromagnetic radiation.

photoconductive cell A device for detecting or measuring electromagnetic radiation by variation of the conductivity of a substance (called a photoconductor) upon absorption of the radiation by this substance. Also known as photoresistive cell; photoresistor.

photoconductive device A photoelectric device which utilizes the photoinduced change in electrical conductivity to provide an electrical signal.

photoconductive film A film of material whose current-carrying ability is enhanced when illuminated.

photoconductive gain factor The ratio of the number of electrons per second flowing through a circuit containing a cube of semiconducting material, whose sides are of unit length, to the number of photons per second absorbed in this volume.

photoconductive meter An exposure meter in which a battery supplies power through a photoconductive cell to a milliammeter.

photoconductivity The increase in electrical conductivity displayed by many nonmetallic solids when they absorb electromagnetic radiation.

photoconductivity gain The number of charge carriers that circulate through a circuit involving a photoconductor for each charge carrier generated by light.

photoconductor A nonmetallic solid whose conductivity increases when it is exposed to electromagnetic radiation.

photoconductor diode *See* photodiode.

photocoupler *See* optoisolator.

photodarlington A Darlington amplifier in which the input transistor is a phototransistor.

photodetector A detector that responds to radiant energy; examples include photoconductive cells, photodiodes, photoresistors, photoswitches, phototransistors, phototubes, and photovoltaic cells. Also known as light-sensitive cell; light-sensitive detector; light sensor photodevice; photoelectric detector; photosensor.

photodiffusion effect *See* Dember effect.

photodiode A semiconductor diode in which the reverse current varies with illumination; examples include the alloy-junction photocell and the grown-junction photocell. Also known as photoconductor diode.

photoelectret An electret produced by the removal of light from an illuminated photoconductor in an electric field.

photoelectric Pertaining to the electrical effects of light, such as the emission of electrons, generation of voltage, or a change in resistance when exposed to light.

photoelectric absorption Absorption of photons in one of the several photoelectric effects.

photoelectric cell See photocell.

photoelectric constant The ratio of the frequency of radiation causing emission of photoelectrons to the voltage corresponding to the energy absorbed by a photoelectron; equal to Planck's constant divided by the electron charge.

photoelectric control Control of a circuit or piece of equipment by changes in incident light.

photoelectric counter A photoelectrically actuated device used to record the number of times a given light path is intercepted by an object.

photoelectric cutoff register control Use of a photoelectric control system as a longitudinal position regulator to maintain the position of the point of cutoff with respect to a repetitive pattern of moving material.

photoelectric detector See photodetector.

photoelectric device A device which gives an electrical signal in response to visible, infrared, or ultraviolet radiation.

photoelectric effect See photoelectricity.

photoelectric electron-multiplier tube See multiplier phototube.

photoelectric infrared radiation See near-infrared radiation.

photoelectric intrusion detector A burglar-alarm system in which interruption of a light beam by an intruder reduces the illumination on a phototube and thereby closes an alarm circuit.

photoelectricity The liberation of an electric charge by electromagnetic radiation incident on a substance; includes photoemission, photoionization, photoconduction, the photovoltaic effect, and the Auger effect (an internal photoelectric process). Also known as photoelectric effect; photoelectric process.

photoelectric lighting control Use of a photoelectric relay actuated by a change in illumination in a given area or at a given point.

photoelectric process See photoelectricity.

photoelectric relay A relay combined with a phototube and amplifier, arranged so changes in incident light on the phototube make the relay contacts open or close. Also known as light relay.

photoelectric tube See phototube.

photoelectromagnetic effect The effect whereby, when light falls on a flat surface of an intermetallic semiconductor located in a magnetic field that is parallel to the surface, excess hole-electron pairs are created, and these carriers diffuse in the direction of the light but are deflected by the magnetic field to give a current flow through the semiconductor that is at right angles to both the light rays and the magnetic field.

photoelectromotive force Electromotive force caused by photovoltaic action.

photoelectron An electron emitted by the photoelectric effect.

photoemission The ejection of electrons from a solid (or less commonly, a liquid) by incident electromagnetic radiation. Also known as external photoelectric effect.

photoemission threshold The energy of a photon which is just sufficient to eject an electron from a solid or liquid in photoemission.

photoemissive cell A device which detects or measures radiant energy by measurement of the resulting emission of electrons from the surface of a photocathode.

photonegative 323

photoemissivity The property of a substance that emits electrons when struck by light.

photoemitter A material that emits electrons when sufficiently illuminated.

photoferroelectric effect An effect observed in ferroelectric ceramics such as PLZT materials, in which light at or near the band-gap energy of the material has an effect on the electric field in the material created by an applied voltage, and, at a certain value of the voltage, also influences the degree of ferroelectric remanent polarization. Abbreviated PFE.

photoflash lamp A lamp consisting of a glass bulb filled with finely shredded aluminum foil in an atmosphere of oxygen; when the foil is ignited by a low-voltage dry cell, it burns with a burst of high-intensity light of short time duration and with definitely regulated time characteristics.

photoflash unit A portable electronic light source for photographic use, consisting of a capacitor-discharge power source, a flash tube, a battery for charging the capacitor, and sometimes also a high-voltage pulse generator to trigger the flash.

photoflood lamp An incandescent lamp used in photography which has a high-temperature filament, so that it gives high illumination and high color temperature for a short lifetime.

photoglow tube Gas-filled phototube used as a relay by making the operating voltage sufficiently high so that ionization and a flow discharge occur, with considerable current flow, when a certain illumination is reached.

photographic sound recorder A sound recorder having means for producing a modulated light beam and means for moving a light-sensitive medium relative to the beam to give a photographic recording of sound signals. Also known as optical sound recorder.

photographic sound reproducer A sound reproducer in which an optical sound record on film is moved through a light beam directed at a light-sensitive device, to convert the recorded optical variations back into audio signals. Also known as optical sound reproducer.

photoisland grid Photosensitive surface in the storage-type, Farnsworth dissector tube for television cameras.

photoisolator *See* optoisolator.

photomagnetoelectric effect The generation of a voltage when a semiconductor material is positioned in a magnetic field and one face is illuminated.

photomask A film or glass negative that has many high-resolution images, used in the production of semiconductor devices and integrated circuits.

photomultiplier *See* multiplier phototube.

photomultiplier cell A transistor whose *pn*-junction is exposed so that it conducts more readily when illuminated.

photomultiplier counter A scintillation counter that has a built-in multiplier phototube.

photomultiplier tube *See* multiplier phototube.

photon coupled isolator Circuit coupling device, consisting of an infrared emitter diode coupled to a photon detector over a short shielded light path, which provides extremely high circuit isolation.

photon coupling Coupling of two circuits by means of photons passing through a light pipe.

photonegative Having negative photoconductivity, hence decreasing in conductivity (increasing in resistance) under the action of light; selenium sometimes exhibits photonegativity.

photopositive Having positive photoconductivity, hence increasing in conductivity (decreasing in resistance) under the action of light; selenium ordinarily has photopositivity.

photoresistive cell See photoconductive cell.

photoresistor See photoconductive cell.

photo-SCR See light-activated silicon controlled rectifier.

photosensitive See light-sensitive.

photosensor See photodetector.

photothyristor See light-activated silicon controlled rectifier.

phototransistor A junction transistor that may have only collector and emitter leads or also a base lead, with the base exposed to light through a tiny lens in the housing; collector current increases with light intensity, as a result of amplification of base current by the transistor structure.

phototronic photocell See photovoltaic cell.

phototropism A reversible change in the structure of a solid exposed to light or other radiant energy, accompanied by a change in color. Also known as phototropy.

phototropy See phototropism.

phototube An electron tube containing a photocathode from which electrons are emitted when it is exposed to light or other electromagnetic radiation. Also known as electric eye; light-sensitive tube; photoelectric tube.

phototube cathode The photoemissive surface which is the most negative element of a phototube.

phototube relay A photoelectric relay in which a phototube serves as the light-sensitive device.

photovaristor Varistor in which the current-voltage relation may be modified by illumination, for example, one in which the semiconductor is cadmium sulfide or lead telluride.

photovoltaic Capable of generating a voltage as a result of exposure to visible or other radiation.

photovoltaic cell A device that detects or measures electromagnetic radiation by generating a potential at a junction (barrier layer) between two types of material, upon absorption of radiant energy. Also known as barrier-layer cell; barrier-layer photocell; boundary-layer photocell; photronic photocell.

photovoltaic effect The production of a voltage in a nonhomogeneous semiconductor, such as silicon, or at a junction between two types of material, by the absorption of light or other electromagnetic radiation.

photovoltaic meter An exposure cell in which a photovoltaic cell produces a current proportional to the light falling on the cell, and this current is measured by a sensitive microammeter.

photox cell Type of photovoltaic cell in which a voltage is generated between a copper base and a film of cuprous oxide during exposure to visible or other radiation.

photronic cell Type of photovoltaic cell in which a voltage is generated in a layer of selenium during exposure to visible or other radiation.

photronic photocell See photovoltaic cell.

physical electronics The study of physical phenomena basic to electronics, such as discharges, thermionic and field emission, and conduction in semiconductors and metals.

piezoelectric hysteresis 325

pi attenuator An attenuator consisting of a pi network whose impedances are all resistances.

pickoff A device used to convert mechanical motion into a proportional electric signal.

pickup 1. A device that converts a sound, scene, measurable quantity, or other form of intelligence into corresponding electric signals, as in a microphone, phonograph pickup, or television camera. 2. The minimum current, voltage, power, or other value at which a relay will complete its intended function. 3. Interference from a nearby circuit or system.

pickup voltage Of a magnetically operated device, the voltage at which the device starts to operate.

picoampere A unit of current equal to 10^{-12} ampere, or one-millionth of a microampere. Abbreviated pA.

picofarad A unit of capacitance equal to 10^{-12} farad, or one-millionth of a microfarad. Also known as micromicrofarad (deprecated usage); puff (British usage). Abbreviated pF.

picture element 1. That portion, in facsimile, of the subject copy which is seen by the scanner at any instant; it can be considered a square area having dimensions equal to the width of the scanning line. 2. In television, any segment of a scanning line, the dimension of which along the line is exactly equal to the nominal line width; the area which is being explored at any instant in the scanning process. Also known as critical area; elemental area; recording spot; scanning spot.

picture frequency See frame frequency.

picture synchronizing pulse See vertical synchronizing pulse.

picture transmitter See visual transmitter.

picture tube A cathode-ray tube used in television receivers to produce an image by varying the electron-beam intensity as the beam is deflected from side to side and up and down to scan a raster on the fluorescent screen at the large end of the tube. Also known as kinescope; television picture tube.

picture-tube brightener A small step-up transformer that can be inserted between the socket and base of a picture tube to increase the heater voltage and thereby increase picture brightness to compensate for normal aging of the tubes.

Pierce oscillator Oscillator in which a piezoelectric crystal unit is connected between the grid and the plate of an electron tube, in what is basically a Colpitts oscillator, with voltage division provided by the grid-cathode and plate-cathode capacitances of the circuit.

piezoelectric Having the ability to generate a voltage when mechanical force is applied, or to produce a mechanical force when a voltage is applied, as in a piezoelectric crystal.

piezoelectric crystal A crystal which exhibits the piezoelectric effect; used in crystal loudspeakers, crystal microphones, and crystal cartridges for phono pickups.

piezoelectric effect 1. The generation of electric polarization in certain dielectric crystals as a result of the application of mechanical stress. 2. The reverse effect, in which application of a voltage between certain faces of the crystal produces a mechanical distortion of the material.

piezoelectric element A piezoelectric crystal used in an electric circuit, for example, as a transducer to convert mechanical or acoustical signals to electric signals, or to control the frequency of a crystal oscillator.

piezoelectric hysteresis Behavior of a piezoelectric crystal whose electric polarization depends not only on the mechanical stress to which the crystal is subjected, but also on the previous history of this stress.

piezoelectricity Electricity or electric polarization resulting from the piezoelectric effect.

piezoelectric resonator *See* crystal resonator.

piezoelectric semiconductor A semiconductor exhibiting the piezoelectric effect, such as quartz, Rochelle salt, and barium titanate.

piezoelectric transducer A piezoelectric crystal used as a transducer, either to convert mechanical or acoustical signals to electric signals, as in a microphone, or vice versa, as in ultrasonic metal inspection.

piezoelectric vibrator An element cut from piezoelectric material, usually in the form of a plate, bar, or ring, with electrodes attached to or supported near the element to excite one of its resonant frequencies.

piezomagnetism Stress dependence of magnetic properties.

pi filter A filter that has a series element and two parallel elements connected in the shape of the Greek letter pi (π).

pig 1. An ion source based on the same principle as the Philips ionization gage. 2. *See* Philips ionization gage.

piggyback twistor Electrically alterable nondestructive-readout storage device that uses a thin narrow tape of magnetic material wound spirally around a fine copper conductor to store information; another similar tape is wrapped on top of the first, piggyback fashion, to sense the stored information; a binary digit or bit is stored at the intersection of a copper strap and a pair of these twistor wires.

pigtail A short, flexible wire, usually stranded or braided, used between a stationary terminal and a terminal having a limited range of motion, as in relay armatures.

pigtail splice A splice made by twisting together the bared ends of parallel conductors.

pileup A set of moving and fixed contacts, insulated from each other, formed as a unit for incorporation in a relay or switch. Also known as stack.

pill A microwave stripline termination.

pillbox antenna Cylindrical parabolic reflector enclosed by two plates perpendicular to the cylinder, spaced to permit the propagation of only one mode in the desired direction of polarization.

pilot cell Selected cell of a storage battery whose temperature, voltage, and specific gravity are assumed to indicate the condition of the entire battery.

pilot lamp A small lamp used to indicate that a circuit is energized. Also known as pilot light.

pilot light *See* pilot lamp.

pilot motor A small motor used in the automatic control of an electric current.

pilot relaying A system for protecting transmission consisting of protective relays at line terminals and a communication channel between relays which is used by the relays to determine if a fault is within the protected line section, in which case all terminals are tripped simultaneously at high speed, or outside it, in which case tripping is blocked.

pi mode Of a magnetron, the mode of operation for which the phases of the fields of successive anode openings facing the interaction space differ by pi radians.

pin A terminal on an electron tube, semiconductor, integrated circuit, plug, or connector. Also known as base pin; prong.

pinch effect Manifestation of the magnetic self-attraction of parallel electric currents, such as constriction of ionized gas in a discharge tube, or constriction of molten

planar array   **327**

metal through which a large current is flowing. Also known as cylindrical pinch; magnetic pinch; rheostriction.

pinch-off voltage Of a field-effect transistor, the voltage at which the current flow between source and drain is blocked because the channel between these electrodes is completely depleted.

pinch resistor A silicon integrated-circuit resistor produced by diffusing an n-type layer over a p-type resistor; this narrows or pinches the resistive channel, thereby increasing the resistance value.

pincushion distortion Distortion in which all four sides of a received television picture are concave (curving inward).

pin diode A diode consisting of a silicon wafer containing nearly equal p-type and n-type impurities, with additional p-type impurities diffused from one side and additional n-type impurities from the other side; this leaves a lightly doped intrinsic layer in the middle, to act as a dielectric barrier between the n-type and p-type regions. Also known as power diode.

pine-tree array Array of dipole antennas aligned in a vertical plane known as the radiating curtain, behind which is a parallel array of dipole antennas forming a reflecting curtain.

pi network An electrical network which has three impedance branches connected in series to form a closed circuit, with the three junction points forming an output terminal, an input terminal, and a common output and input terminal.

ping A sonic or ultrasonic pulse sent out by an echo-ranging sonar.

pin jack Single conductor jack having an opening for the insertion of a plug of very small diameter.

pin junction A semiconductor device having three regions: p-type impurity, intrinsic (electrically pure), and n-type impurity.

pinning The hindering of motion of dislocations in a solid, and the consequent hardening of the solid, by impurities which collect near the dislocations, resulting in a large energy barrier being imposed against the motion of the dislocations.

pip *See* blip.

pipe-to-soil potential The voltage potential (emf) generated between a buried pipe and its surrounding soil, the result of electrolytic action and a cause of electrolytic corrosion of the pipe.

pi point Frequency at which the insertion phase shift of an electric structure is 180° or an integral multiple of 180°.

pi section filter An electric filter made of several pi networks connected in series.

piston A sliding metal cylinder used in waveguides and cavities for tuning purposes or for reflecting essentially all of the incident energy. Also known as plunger; waveguide plunger.

piston attenuator A microwave attenuator inserted in a waveguide to introduce an amount of attenuation that can be varied by moving an output coupling device along its longitudinal axis.

pi-T transformation *See* Y-delta transformation.

pixel The smallest addressable element in an electronic display; a short form for picture element. Also known as pel.

PLA *See* programmed logic array.

planar array An array of ultrasonic transducers that can be mounted in a single plane or sheet, to permit closer conformation with the hull design of a sonar-carrying ship.

328 planar-array antenna

planar-array antenna An array antenna in which the centers of the radiating elements are all in the same plane.

planar ceramic tube Electron tube having parallel planar electrodes and a ceramic envelope.

planar device A semiconductor device having planar electrodes in parallel planes, made by alternate diffusion of p- and n-type impurities into a substrate.

planar diode A diode having planar electrodes in parallel planes.

planar photodiode A vacuum photodiode consisting simply of a photocathode and an anode; light enters through a window sealed into the base, behind the photocathode.

planar transistor A transistor constructed by an etching and diffusion technique in which the junction is never exposed during processing, and the junctions reach the surface in one plane; characterized by very low leakage current and relatively high gain.

plane Screen of magnetic cores; planes are combined to form stacks.

plane earth Earth that is considered to be a plane surface as used in ground-wave calculations.

plane-earth attenuation Attenuation of an electromagnetic wave over an imperfectly conducting plane earth in excess of that over a perfectly conducting plane.

plane of polarization Plane containing the electric vector and the direction of propagation of electromagnetic wave.

plane-polarized wave An electromagnetic wave whose electric field vector at all times lies in a fixed plane that contains the direction of propagation through a homogeneous isotropic medium.

plane reflector *See* passive reflector.

planigraphy *See* sectional radiography.

planoconvex spotlight A light that can be used as a sharply defined spotlight or for soft-edged lighting; ranges in power from 100 to 2000 watts.

plan position indicator A radarscope display in which echoes from various targets appear as bright spots at the same locations as they would on a circular map of the area being scanned, the radar antenna being at the center of the map. Abbreviated PPI. Also known as P display.

plan position indicator repeater Unit which repeats a plan position indicator (PPI) at a location remote from the radar console. Also known as remote plan position indicator.

Plante cell A type of lead-acid cell in which the active material is formed on the plates by electrochemical means during repeated charging and discharging, instead of being applied as a prepared paste.

plant factor The ratio of the average power load of an electric power plant to its rated capacity. Also known as capacity factor.

plasma cathode A cathode in which the source of electrons is a gas plasma rather than a solid.

plasma diode A diode used for converting heat directly into electricity; it consists of two closely spaced electrodes serving as cathode and anode, mounted in an envelope in which a low-pressure cesium vapor fills the interelectrode space; heat is applied to the cathode, causing emission of electrons.

plasma display A display in which sets of parallel conductors at right angles to each other are deposited on glass plates, with the very small space between the plates filled with a gas; each intersection of two conductors defines a single cell that can be energized to produce a gas discharge forming one element of a dot-matrix display.

plasma generator Any device that produces a high-velocity plasma jet, such as a plasma accelerator, engine, oscillator, or torch.

plasma gun 1. A machine, such as an electric-arc chamber, that will generate very high heat fluxes to convert neutral gases into plasma. 2. An electromagnetic device which creates and accelerates bursts of plasma.

plasma sheath An envelope of ionized gas that surrounds a spacecraft or other body moving through an atmosphere at hypersonic velocities; affects transmission, reception, and diffraction of radio waves.

plasmatron A gas-discharge tube in which independently generated plasma serves as a conductor between a hot cathode and an anode; the anode current is modulated by varying either the conductivity or the effective cross section of the plasma.

plasmon A quantum of a collective longitudinal wave in the electron gas of a solid.

plastic film capacitor A capacitor constructed by stacking, or forming into a roll, alternate layers of foil and a dielectric which consists of a plastic, such as polystyrene or Mylar, either alone or as a laminate with paper.

plastic plate A plate of plastic dielectric material used as a base for a semiconductor device.

plate 1. One of the conducting surfaces in a capacitor. 2. One of the electrodes in a storage battery. 3. *See* anode.

plateau The portion of the plateau characteristic of a counter tube in which the counting rate is substantially independent of the applied voltage.

plateau characteristic The relation between counting rate and voltage for a counter tube when radiation is constant, showing a plateau after the rise from the starting voltage to the Geiger threshold. Also known as counting rate–voltage characteristic.

plate circuit *See* anode circuit.

plate-circuit detector *See* anode-circuit detector.

plate current *See* anode current.

plated circuit A printed circuit produced by electrodeposition of a conductive pattern on an insulating base. Also known as plated printed circuit.

plate detector *See* anode detector.

plated printed circuit *See* plated circuit.

plate efficiency *See* anode efficiency.

plate impedance *See* anode impedance.

plate input power *See* anode input power.

plate-load impedance *See* anode impedance.

plate modulation *See* anode modulation.

plate neutralization *See* anode neutralization.

plate pulse modulation *See* anode pulse modulation.

plate resistance *See* anode resistance.

plate saturation *See* anode saturation.

platinotron A microwave tube that may be used as a high-power saturated amplifier or oscillator in pulsed radar applications; requires permanent magnet just as does a magnetron.

playback head

playback head A head that converts a changing magnetic field on a moving magnetic tape into corresponding electric signals. Also known as reproduce head.

pliotron General term for any hot-cathode vacuum tube having one or more grids.

PLL See phase-locked loop.

plug The half of a connector that is normally movable and is generally attached to a cable or removable subassembly; inserted in a jack, outlet, receptacle, or socket.

plug adapter lamp holder A device that can be inserted in a lamp holder to act as a lamp holder and one or more receptacles. Also known as current tap.

plug fuse A fuse designed for use in a standard screw-base lamp socket.

plugging Braking an electric motor by reversing its connections, so it tends to turn in the opposite direction; the circuit is opened automatically when the motor stops, so the motor does not actually reverse.

plug-in unit A component or subassembly having plug-in terminals so all connections can be made simultaneously by pushing the unit into a suitable socket.

plumbing Slang term for the pipelike waveguide circuit elements used in microwave radio and radar equipment.

plunger See piston.

PM See permanent magnet.

pneumatic transmission lag The time delay in a pneumatic transmission line between the generation of an impulse at one end and the resultant reaction at the other end.

pnip transistor An intrinsic junction transistor in which the intrinsic region is sandwiched between the n-type base and the p-type collector.

pn junction The interface between two regions in a semiconductor crystal which have been treated so that one is a p-type semiconductor and the other is an n-type semiconductor; it contains a permanent dipole charge layer.

pnpn diode A semiconductor device consisting of four alternate layers of p-type and n-type semiconductor material, with terminal connections to the two outer layers. Also known as $npnp$ diode.

pnp transistor A junction transistor having an n-type base between a p-type emitter and a p-type collector.

Pockels readout optical modulator A device for storing data in the form of images; it consists of bismuth silicon oxide crystal coated with an insulating layer of parylene and transparent electrodes evaporated on the surfaces; a blue laser is used for writing and a red laser is used for nondestructive readout or processing. Abbreviated PROM.

Poggendorff's first method See constant-current dc potentiometer.

Poggendorff's second method See constant-resistance dc potentiometer.

Poincaré electron A classical model of the electron in which nonelectromagnetic forces hold the electron together so that it has zero self-stress; it is unstable and has infinite self-energy in the case of a point electron.

point contact A contact between a specially prepared semiconductor surface and a metal point, usually maintained by mechanical pressure but sometimes welded or bonded.

point-contact diode A semiconductor rectifier that uses the barrier formed between a specially prepared semiconductor surface and a metal point to produce the rectifying action.

point-contact transistor A transistor having a base electrode and two or more point contacts located near each other on the surface of an n-type semiconductor.

polaron 331

point jammer Any electronic jammer directed against a specific enemy installation operating on a specific frequency.

point-junction transistor Transistor having a base electrode and both point-contact and junction electrodes.

point-source light A special lamp in which the radiating element is concentrated in a small physical area.

point target In radar, an object which returns a target signal by reflection from a relatively simple discrete surface; such targets are ships, aircraft, projectiles, missiles, and buildings.

point transposition Transposition, usually in an open-wire line, which is executed within a distance comparable to the wire separation, without material distortion of the normal wire configuration outside this distance.

poison A material which reduces the emission of electrons from the surface of a cathode.

polar crystal See ferroelectric crystal.

polarity effect An effect for which the breakdown voltage across a vacuum separating two electrodes, one of which is pointed, is much higher when the pointed electrode is the anode.

polarizability The electric dipole moment induced in a system, such as an atom or molecule, by an electric field of unit strength.

polarizability catastrophe According to a theory using the Lorentz field concept, the phenomenon where, at a certain temperature, the dielectric constant of a material becomes infinite.

polarization 1. The process of producing a relative displacement of positive and negative bound charges in a body by applying an electric field. 2. A vector quantity equal to the electric dipole moment per unit volume of a material. Also known as dielectric polarization; electric polarization. 3. A chemical change occurring in dry cells during use, increasing the internal resistance of the cell and shortening its useful life.

polarization charge See bound charge.

polarization potential One of two vectors from which can be derived, by differentiation, an electric scalar potential and magnetic vector potential satisfying the Lorentz condition. Also known as Hertz vector.

polarized electrolytic capacitor An electrolytic capacitor in which the dielectric film is formed adjacent to only one metal electrode; the impedance to the flow of current is then greater in one direction than in the other.

polarized electromagnetic radiation Electromagnetic radiation in which the direction of the electric field vector is not random.

polarized ion source A device that generates ion beams in such a manner that the spins of the ions are aligned in some direction.

polarized plug A plug that can be inserted in its receptacle only when in a predetermined position.

polarized receptacle A receptacle designed for use with a polarized plug, to ensure that the grounded side of an alternating-current line or the positive side of a direct-current line is always connected to the same terminal on a piece of equipment.

polarized relay Relay in which the movement of the armature depends upon the direction of the current in the circuit controlling the armature. Also known as polar relay.

polaron An electron in a crystal lattice together with a cloud of phonons that result from the deformation of the lattice produced by the interaction of the electron with ions or atoms in the lattice.

polar radiation pattern Diagram showing the relative strength of the radiation from an antenna in all directions in a given plane.

polar relay See polarized relay.

pole 1. One of the electrodes in an electric cell. 2. An output terminal on a switch; a double-pole switch has two output terminals.

pole-changing control A method of obtaining two or more running speeds of a three-phase motor by making changes in the number of magnetic poles, usually by making changes in the coil connections at the winding terminals.

pole face The end of a magnetic core that faces the air gap in which the magnetic field performs useful work.

pole-face winding Winding in the pole face of a motor or generator used to neutralize the cross-magnetizing armature reaction under the pole faces, which would otherwise cause a nonuniform distribution of voltage between commutator segments. Also known as compensated winding.

pole horn The part of a pole piece or pole shoe in an electrical machine that projects circumferentially beyond the pole core.

pole piece A piece of magnetic material forming one end of an electromagnet or permanent magnet, shaped to control the distribution of magnetic flux in the adjacent air gap.

pole shoe Portion of a field pole facing the armature of the machine; it may be separable from the body of the pole.

pole strength See magnetic pole strength.

poling Adjustment of polarity; specifically, in wire-line practice, the use of transpositions between transposition sections of open wire or between lengths of cable, to cause the residual cross-talk couplings in individual sections or lengths to oppose one another.

polychromatic radiation Electromagnetic radiation that is spread over a range of frequencies.

polygonization A phenomenon observed during the annealing of plastically bent crystals in which the edge dislocations created by cold working organize themselves vertically above each other so that polygonal domains are formed.

polyphase Having or utilizing two or more phases of an alternating-current power line.

polyphase circuit Group of alternating-current circuits (usually interconnected) which enter (or leave) a delimited region at more than two points of entry; they are intended to be so energized that, in the steady state, the alternating currents through the points of entry, and the alternating potential differences between them, all have exactly equal periods, but have differences in phase, and may have differences in waveform.

polyphase rectifier A rectifier which utilizes two or more diodes (usually three), each of which operates during an equal fraction of an alternating-current cycle to achieve an output current which varies less than that in an ordinary half-wave or full-wave rectifier.

polyphase synchronous generator Generator whose alternating-current circuits are so arranged that two or more symmetrical alternating electromotive forces with definite phase relationships are produced at its terminals.

polyrod antenna End-fire directional dielectric antenna consisting of a polystyrene rod energized by a section of waveguide.

polystyrene capacitor A capacitor that uses film polystyrene as a dielectric between rolled strips of metal foil.

polystyrene dielectric Polystyrene used in applications where its very high resistivity, good dielectric strength, and other electrical properties are important, such as for electrical insulation or in dielectrics.

pool cathode A cathode at which the principal source of electron emission is a cathode spot on a liquid-metal electrode, usually mercury.

pool-cathode mercury-arc rectifier A pool tube connected in an electric circuit; its rectifying properties result from the fact that only the mercury-pool cathode, and not the anode, can emit electrons. Also known as mercury-pool rectifier.

pool-cathode tube See pool tube.

pool tube A gas-discharge tube having a mercury-pool cathode. Also known as mercury tube; pool-cathode tube.

popcorn noise Noise produced by erratic jumps of bias current between two levels at random intervals in operational amplifiers and other semiconductor devices.

porcelain capacitor A fixed capacitor in which the dielectric is a high grade of porcelain, molecularly fused to alternate layers of fine silver electrodes to form a monolithic unit that requires no case or hermetic seal.

port 1. An entrance or exit for a network. 2. An opening in a waveguide component, through which energy may be fed or withdrawn, or measurements made.

posistor A thermistor having a large positive resistance-temperature characteristic.

positive Having fewer electrons than normal, and hence having ability to attract electrons.

positive bias A bias such that the control grid of an electron tube is positive with respect to the cathode.

positive charge The type of charge which is possessed by protons in ordinary matter, and which may be produced in a glass object by rubbing with silk.

positive column The luminous glow, often striated, that occurs between the Faraday dark space and the anode in a glow-discharge tube. Also known as positive glow.

positive electrode See anode.

positive glow See positive column.

positive-grid oscillator See Barkhausen-Kurz oscillator; retarding-field oscillator.

positive-ion sheath Collection of positive ions on the control grid of a gas-filled triode tube.

positive logic Pertaining to a logic circuit in which the more positive voltage (or current level) represents the 1 state; the less positive level represents the 0 state.

positive modulation In an amplitude-modulated television system, that form of television modulation in which an increase in brightness corresponds to an increase in transmitted power.

positive phase sequence The phase sequence that corresponds to the normal order of phases in a polyphase system.

positive-phase-sequence relay Relay which functions in conformance with the positive-phase-sequence component of the current, voltage, or power of the circuit.

positive ray A stream of positively charged atoms or molecules, produced by a suitable combination of ionizing agents, accelerating fields, and limiting apertures.

positive terminal The terminal of a battery or other voltage source toward which electrons flow through the external circuit.

postaccelerating electrode See intensifier electrode.

postacceleration Acceleration of beam electrons after deflection in an electron-beam tube. Also known as postdeflection acceleration (PDA).

postdeflection accelerating electrode *See* intensifier electrode.

postdeflection acceleration *See* postacceleration.

posttuning drift In a frequency-agile source such as the fast-tuning oscillators used in set-on jammers for electronic warfare equipment, the increase in frequency brought about by the drop in temperature of the varactor after warm-up time, settling time, and the time when the oscillator has reached a new frequency. Abbreviated PTD.

pot *See* potentiometer.

pot core A ferrite magnetic core that has the shape of a pot, with a magnetic post in the center and a magnetic plate as a cover; the coils for a choke or transformer are wound on the center post.

potential *See* electric potential.

potential difference Between any two points, the work which must be done against electric forces to move a unit charge from one point to the other. Abbreviated PD.

potential divider *See* voltage divider.

potential drop The potential difference between two points in an electric circuit.

potential gradient Difference in the values of the voltage per unit length along a conductor or through a dielectric.

potential transformer phase angle Angle between the primary voltage vector and the secondary voltage vector rsed; this angle is conveniently considered as positive when the reversed, secondary voltage vector leads the primary voltage vector.

potentiometer A resistor having a continuously adjustable sliding contact that is generally mounted on a rotating shaft; used chiefly as a voltage divider. Also known as pot (slang).

potentiometry Use of a potentiometer to measure electromotive forces, and the applications of such measurements.

Potier diagram Vector diagram showing the voltage and current relations in an alternating-current generator.

potted circuit A pulse-forming network immersed in oil and enclosed in a metal container.

potted line Pulse-forming network immersed in oil and enclosed in a metal container.

potting Process of filling a complete electronic assembly with a thermosetting compound for resistance to shock and vibration, and for exclusion of moisture and corrosive agents.

powdered-iron core *See* ferrite core.

powder method A method of x-ray diffraction analysis in which a collimated, monochromatic beam of x-rays is directed at a sample consisting of an enormous number of tiny crystals having random orientation, producing a diffraction pattern that is recorded on film or with a counter tube. Also known as x-ray powder method.

powder pattern The pattern created by very fine powders or colloidal particles, spread over the surface of a magnetic material; reveals the magnetic domains in a single crystal of such material.

power amplification *See* power gain.

power amplifier The final stage in multistage amplifiers, such as audio amplifiers and radio transmitters, designed to deliver maximum power to the load, rather than maximum voltage gain, for a given percent of distortion.

power pack 335

power amplifier tube *See* power tube.

power component *See* active component.

power density The amount of power per unit area in a radiated microwave or other electromagnetic field, usually expressed in units of watts per square centimeter.

power detection Form of detection in which the power output of the detecting device is used to supply a substantial amount of power directly to a device such as a loudspeaker or recorder.

power detector Detector capable of handling strong input signals without appreciable distortion.

power diode *See* pin-diode.

power divider A device used to produce a desired distribution of power at a branch point in a waveguide system.

power factor The ratio of the average (or active) power to the apparent power (root-mean-square voltage times rms current) of an alternating-current circuit. Abbreviated pf. Also known as phase factor.

power factor controller A solid-state electronic device that reduces excessive energy waste in alternating-current induction motors by holding constant the phase angle between current and voltage.

power factor regulator Regulator which functions to maintain the power factor of a line or an apparatus at a predetermined value, or to vary it according to a predetermined plan.

power flow The rate at which energy is transported across a surface by an electromagnetic field.

power frequency The frequency at which electric power is generated and distributed; in most of the United States it is 60 hertz.

power gain 1. The ratio of the power delivered by a transducer to the power absorbed by the input circuit of the transducer. Also known as power amplification. 2. An antenna ratio equal to 4π (12.57) times the ratio of the radiation intensity in a given direction to the total power delivered to the antenna.

power generator A device for producing electric energy, such as an ordinary electric generator or a magnetohydrodynamic, thermionic, or thermoelectric power generator.

power level The ratio of the amount of power being transmitted past any point in an electric system to a reference power value; usually expressed in decibels.

power line Two or more wires conducting electric power from one location to another. Also known as electric power line.

power-line carrier The use of transmission lines to transmit speech, metering indications, control impulses, and other signals from one station to another, without interfering with the lines' normal function of transmitting power.

power-line filter *See* line filter.

power loss The ratio of the power absorbed by the input circuit of a transducer to the power delivered to a specified load; usually expressed in decibels. Also known as power attenuation.

power output The alternating-current power in watts delivered by an amplifier to a load.

power output tube *See* power tube.

power pack Unit for converting power from an alternating- or direct-current supply into an alternating- or direct-current power at voltages suitable for supplying an electronic device.

power rating The power available at the output terminals of a component or piece of equipment that is operated according to the manufacturer's specifications.

power ratio The ratio of the maximum power to the minimum power in a waveguide that is improperly terminated.

power rectifier A device which converts alternating current to direct current and operates at high power loads.

power relay Relay that functions at a predetermined value of power; may be an overpower relay, an underpower relay, or a combination of both.

power resistor A resistor used in electric power systems, ranging in size from 5 watts to many kilowatts, and cooled by air convection, air blast, or water.

power semiconductor A semiconductor device capable of dissipating appreciable power (generally over 1 watt) in normal operation; may handle currents of thousands of amperes or voltages up into thousands of volts, at frequencies up to 10 kilohertz.

power supply A source of electrical energy, such as a battery or power line, employed to furnish the tubes and semiconductor devices of an electronic circuit with the proper electric voltages and currents for their operation. Also known as electronic power supply.

power supply circuit An electrical network used to convert alternating current to direct current.

power switch An electric switch which energizes or deenergizes an electric load; ranges from ordinary wall switches to load-break switches and disconnecting switches in power systems operating at voltages of hundreds of thousands of volts.

power switchboard Part of a switch gear which consists of one or more panels upon which are mounted the switching control, measuring, protective, and regulatory equipment; the panel or panel supports may also carry the main switching and interrupting devices together with their connection.

power switching Switching between supplies of electrical energy at high levels of current and voltage.

power transfer theorem The theorem that, in an electrical network which carries direct or sinusoidal alternating current, the greatest possible power is transferred from one section to another when the impedance of the section that acts as a load is the complex conjugate of the impedance of the section that acts as a source, where both impedances are measured across the pair of terminals at which the power is transferred, with the other part of the network disconnected.

power transformer An iron-core transformer having a primary winding that is connected to an alternating-current power line and one or more secondary windings that provide different alternating voltage values.

power transistor A junction transistor designed to handle high current and power; used chiefly in audio and switching circuits.

power transmission line The facility in an electric power system used to transfer large amounts of power from one location to a distant location; distinguished from a subtransmission or distribution line by higher voltage, greater power capability, and greater length. Also known as electric main; main (both British usages).

power transmission tower A rigid steel tower supporting a high-voltage electric power transmission line, having a large enough spacing between conductors, and between conductors and ground, to prevent corona discharge.

power tube An electron tube capable of handling more current and power than an ordinary voltage-amplifier tube; used in the last stage of an audio-frequency amplifier or in high-power stages of a radio-frequency amplifier. Also known as power amplifier tube; power output tube.

power winding In a saturable reactor, a winding to which is supplied the power to be controlled; commonly the functions of the output and power windings are accomplished by the same winding, which is ten termed the output winding.

Poynting theorem A theorem, derived from Maxwell's equations, according to which the rate of loss of energy stored in electric and magnetic fields within a region of space is equal to the sum of the rate of dissipation of electrical energy as heat and the rate of flow of electromagnetic energy outward through the surface of the region.

Poynting vector A vector, equal to the cross product of the electric-field strength and the magnetic-field strength (mks units) whose outward normal component, when integrated over a closed surface, gives the outward flow of electromagnetic energy through that surface.

PPI *See* plan position indicator.

pp junction A region of transition between two regions having different properties in p-type semiconducting material.

p$^+$-type semiconductor A p-type semiconductor in which the excess mobile hole concentration is very large.

practical units The units of the meter-kilogram-second-ampere system.

preamplifier An amplifier whose primary function is boosting the output of a low-level audio-frequency, radio-frequency, or microwave source to an intermediate level so that the signal may be further processed without appreciable degradation of the signal-to-noise ratio of the system. Also known as preliminary amplifier.

precipitation attenuation Loss of radio energy due to the passage through a volume of the atmosphere containing precipitation; part of the energy is lost by scattering, and part by absorption.

precipitation clutter suppression Technique of reducing, by one of the various devices integral to the radar system, clutter caused by rain in the radar range.

precipitation noise Noise generated in an antenna circuit, generally in the form of a relaxation oscillation, caused by the periodic discharge of the antenna or conductors in the vicinity of the antenna into the atmosphere.

precision-balanced hybrid circuit Circuit used to interconnect a four-wire telephone circuit to a particular two-wire circuit, in which the impedance of the balancing network is adjusted to give a relatively high degree of balance.

precision net In a four-wire terminating set or similar device employing a hybrid coil, an artificial line designed and adjusted to provide an accurate balance for the loop and subscribers set or line impedance.

precision sweep Delayed expanded radar sweep for high resolution and range accuracy.

preconduction current Low value of plate current flowing in a thyratron or other grid-controlled gas tube prior to the start of conduction.

predetection combining Method used to produce an optimum signal from multiple receivers involved in diversity reception of signals.

preece A unit of electrical resistivity equal to 10^{13} times the product of 1 ohm and 1 meter.

preemphasis A process which increases the magnitude of some frequency components with respect to the magnitude of others to reduce the effects of noise introduced in subsequent parts of the system.

preemphasis network An RC (resistance-capacitance) filter inserted in a system to emphasize one range of frequencies with respect to another. Also known as emphasizer.

338 preferred numbers

preferred numbers A series of numbers adopted by the Electronic Industries Association and the military services for use as nominal values of resistors and capacitors, to reduce the number of different sizes that must be kept in stock for replacements. Also known as preferred values.

preferred values See preferred numbers.

prefocus lamp A light bulb whose filaments are precisely positioned with respect to the lamp socket.

preheat fluorescent lamp A fluorescent lamp in which a manual switch or thermal starter is used to preheat the cathode for a few seconds before high voltage is applied to strike the mercury arc.

preliminary amplifier See preamplifier.

preplumbed system Fixed nontunable waveguides or coaxial transmission lines.

prescaler A scaler that extends the upper frequency limit of a counter by dividing the input frequency by a precise amount, generally 10 or 100.

preselector 1. Device in automatic switching which performs its selecting operation before seizing an idle trunk. 2. A tuned radio-frequency amplifier stage used ahead of the frequency converter in a superheterodyne receiver to increase the selectivity and sensitivity of the receiver.

presentation See radar display.

press-to-talk switch A switch mounted directly on a microphone to provide a convenient means for switching two-way radiotelephone equipment or electronic dictating equipment to the talk position.

pressure cable A cable in which a fluid such as oil or gas, at greater than atmospheric pressure, surrounds the conductors and insulation and keeps their temperature down.

pressure pickup A device that converts changes in the pressure of a gas or liquid into corresponding changes in some more readily measurable quantity such as inductance or resistance.

pressure switch A switch that is actuated by a change in pressure of a gas or liquid.

pre-transmit-receive tube See pre-TR tube.

pretrigger Trigger used to initiate sweep ahead of transmitted pulse.

pre-TR tube Gas-filled radio-frequency switching tube used in some radar systems to protect the TR tube from excessively high power and the receiver from frequencies other than the fundamental. Derived from pre-transmit-receive tube.

prewhitening filter See whitening filter.

PRF See pulse repetition rate.

pri See primary winding.

primary 1. One of the high-voltage conductors of a power distribution system. 2. See primary winding.

primary battery A battery consisting of one or more primary cells.

primary cell A cell that delivers electric current as a result of an electrochemical reaction that is not efficiently reversible, so that the cell cannot be recharged efficiently.

primary coil The input coil in an induction coil or transformer.

primary electron An electron which bombards a solid surface, causing secondary emission.

primary emission Emission of electrons due to primary causes, such as heating of a cathode, and not to secondary effects, such as electron bombardment.

primary extinction A weakening of the stronger beams produced in x-ray diffraction by a very perfect crystal, as compared with the weaker.

primary fault In an electric circuit, the initial breakdown of the insulation of a conductor, usually followed by a flow of power current.

primary flow The current flow that is responsible for the major properties of a semiconductor device.

primary fuel cell A fuel cell in which the fuel and oxidant are continuously consumed.

primary photocurrent A photocurrent resulting from nonohmic contacts unable to replenish charge carriers which pass out of the opposite contact, and whose maximum gain is unity.

primary power cable Power service cables connecting the outside power source to the main-office switch and metering equipment.

primary relay Relay that produces the initial action in a sequence of operations.

primary skip zone Area around a transmitter beyond the ground wave but within the skip distance.

primary voltage The voltage applied to the terminals of the primary winding of a transformer.

primary winding The transformer winding that receives signal energy or alternating-current power from a source. Also known as primary. Abbreviated pri. Symbolized P.

principal E plane Plane containing the direction of radiation of electromagnetic waves and arranged so that the electric vector everywhere lies in the plane.

principal H plane Plane that contains the direction of radiation and the magnetic vector, and is everywhere perpendicular to the E plane.

principal mode *See* fundamental mode.

principle of reciprocity *See* reciprocity theorem.

principle of superposition 1. The principle that the total electric field at a point due to the combined influence of a distribution of point charges is the vector sum of the electric field intensities which the individual point charges would produce at that point if each acted alone. 2. The principle that, in a linear electrical network, the voltage or current in any element resulting from several sources acting together is the sum of the voltages or currents resulting from each source acting alone. Also known as superposition theorem.

printed circuit A conductive pattern that may or may not include printed components, formed in a predetermined design on the surface of an insulating base in an accurately repeatable manner.

printed circuit board A flat board whose front contains slots for integrated circuit chips and connections for a variety of electronic components, and whose back is printed with electrically conductive pathways between the components.

printed-wiring armature An armature in which the conductors consist of printed-wiring strips on both sides of a thin insulating disk, to give a low-inertia armature for servomotors and other variable high-speed applications.

printthrough Transfer of signals from one recorded layer of magnetic tape to the next on a reel.

private branch exchange access line Circuit that connects a main private branch exchange (PBX) to a switching center.

probe A metal rod that projects into but is insulated from a waveguide or resonant cavity; used to provide coupling to an external circuit for injection or extraction of energy or to measure the standing-wave ratio. Also known as waveguide probe.

probe coil In eddy-current nondestructive tests, a type of test coil which is placed on the surface of an object.

prod *See* test prod.

product demodulator A receiver demodulator whose output is the product of the input signal voltage and a local oscillator signal voltage at the input frequency. Also known as product detector.

product detector *See* product demodulator.

product modulator Modulator whose modulated output is substantially equal to the carrier and the modulating wave; the term implies a device in which intermodulation between components of the modulating wave does not occur.

profile chart A vertical cross-section drawing of a microwave path between two stations, indicating terrain, obstructions, and antenna height requirements.

programmable counter A counter that divides an input frequency by a number which can be programmed into decades of synchronous down counters; these decades, with additional decoding and control logic, give the equivalent of a divide-by-N counter system, where N can be made equal to any number.

programmable decade resistor A decade box designed so that the value of its resistance can be remotely controlled by programming logic as required for the control of load, time constant, gain, and other parameters of circuits used in automatic test equipment and automatic controls.

programmable logic array *See* field-programmable logic array.

programmable power supply A power supply whose output voltage can be changed by digital control signals.

programmable read-only memory A large-scale integrated-circuit chip for storing digital data; it can be erased with ultraviolet light and reprogrammed, or it can be programmed only once either at the factory or in the field. Abbreviated PROM.

programmed logic array An array of AND/OR logic gates that provides logic functions for a given set of inputs programmed during manufacture and serves as a read-only memory. Abbreviated PLA.

progressive-wave antenna *See* traveling-wave antenna.

projection cathode-ray tube A television cathode-ray tube designed to produce an intensely bright but relatively small image that can be projected onto a large viewing screen by an optical system.

projection display An electronic system in which an image is generated on a high-brightness cathode-ray tube or similar electronic image generator and then optically projected onto a larger screen.

projection plan position indicator Unit in which the image of a 4-inch (10-centimeter) dark-trace cathode-ray tube is projected on a 24-inch (61-centimeter) horizontal plotting surface; the echoes appear as magenta-colored arcs on white background.

PROM *See* Pockels readout optical modulator; programmable read-only memory.

prong *See* pin.

propagation constant A rating for a line or medium along or through which a wave of a given frequency is being transmitted; it is a complex quantity; the real part is the attenuation constant in nepers per unit length, and the imaginary part is the phase constant in radians per unit length.

propagation delay The time required for a signal to pass through a given complete operating circuit; it is generally of the order of nanoseconds, and is of extreme importance in computer circuits.

propagation velocity Velocity of electromagnetic wave propagation in the medium under consideration.

proportional ionization chamber An ionization chamber in which the initial ionization current is amplified by electron multiplication in a region of high electric-field strength, as in a proportional counter; used for measuring ionization currents or charges over a period of time, rather than for counting.

proportioning reactor A saturable-core reactor used for regulation and control; increasing the input control current from zero to rated value makes output current increase in proportion from cutoff up to full load value.

protective device *See* electric protective device.

protective grounding Grounding of the neutral conductor of a secondary power-distribution system, and of all metal enclosures for conductors, to protect persons from dangerous currents.

protective relay A relay whose principal function is to protect service from interruption or to prevent or limit damage to apparatus.

protective resistance Resistance used in series with a gas tube or other device to limit current flow to a safe value.

protector Device to protect equipment or personnel from high voltages or currents.

protector block Rectangular piece of carbon with an insulated metal insert, or porcelain with a carbon insert, constituting an element of a protector; it forms a gap which will break down and provide a path to ground for voltages over 350 volts.

protector gap A device designed to limit or equalize voltage in order to protect telephone and telegraph equipment; consists of two carbon blocks with an air gap between them, which are brought into contact when there is a steady-state discharge across the gap. Also known as gap.

protector tube A glow-discharge cold-cathode tube that becomes conductive at a predetermined voltage, to protect a circuit against overvoltage.

proton magnetometer A highly sensitive magnetometer which measures the frequency of the proton resonance in ordinary water.

proton microscope A microscope that is similar to the electron microscope but uses protons instead of electrons as the charged particles.

proton scattering microscope A microscope in which protons produced in a cold-cathode discharge are accelerated and focused on a crystal in a vacuum chamber; protons reflected from the crystal strike a fluorescent screen to give a visual and photographable display that is related to the structure of the target crystal.

proton vector magnetometer 1. A type of proton magnetometer with a system of auxiliary coils that permits measurement of horizontal intensity or vertical intensity as well as total intensity. **2.** A type of proton magnetometer with a system of auxiliary coils that permits measurement of horizontal intensity or vertical intensity as well as total intensity.

proximity effect Redistribution of current in a conductor brought about by the presence of another conductor.

PRR *See* pulse repetition rate.

psophometric electromotive force The true noise voltage that exists in a circuit.

psophometric voltage The noise voltage as actually measured in a circuit under specified conditions.

PTD *See* posttuning drift.

p-type conductivity The conductivity associated with holes in a semiconductor, which are equivalent to positive charges.

p-type crystal rectifier Crystal rectifier in which forward current flows when the semiconductor is positive with respect to the metal.

p-type semiconductor An extrinsic semiconductor in which the hole density exeeds the conduction electron density.

p-type silicon Silicon to which more impurity atoms of acceptor type (with valence of 3, such as boron) than of donor type (with valence of 5, such as phosphorus) have been added, with the result that the hole density exceeds the conduction electron density.

puff See picofarad.

pulling An effect that forces the frequency of an oscillator to change from a desired value; causes include undesired coupling to another frequency source or the influence of changes in the oscillator load impedance.

pulling figure The total frequency change of an oscillator when the phase angle of the reflection coefficient of the load impedance varies through 360°, the absolute value of this reflection coefficient being constant at 0.20.

pulsating current Periodic direct current.

pulsating electromotive force Sum of a direct electromotive force and an alternating electromotive force. Also known as pulsating voltage.

pulsating voltage See pulsating electromotive force.

pulse amplifier An amplifier designed specifically to amplify electric pulses without appreciably changing their waveforms.

pulse analyzer An instrument used to measure pulse widths and repetition rates, and to display on a cathode-ray screen the waveform of a pulse.

pulse circuit An active electrical network designed to respond to discrete pulses of current or voltage.

pulse compression A matched filter technique used to discriminate against signals which do not correspond to the transmitted signal.

pulse counter A device that indicates or records the total number of pulses received during a time interval.

pulse-delay network A network consisting of two or more components such as resistors, coils, and capacitors, used to delay the passage of a pulse.

pulse discriminator A discriminator circuit that responds only to a pulse having a particular duration or amplitude.

pulsed oscillator An oscillator that generates a carrier-frequency pulse or a train of carrier-frequency pulses as the result of self-generated or externally applied pulses.

pulse droop A distortion of an otherwise essentially flat-topped rectangular pulse, characterized by a decline of the pulse top.

pulse-duration discriminator A circuit in which the sense and magnitude of the output are a function of the deviation of the pulse duration from a reference.

pulse-forming network A network used to shape the leading or trailing edge of a pulse.

pulse generator 1. A generator that produces repetitive pulses or signal-initiated pulses. 2. See impulse generator.

pulse height The strength or amplitude of a pulse, measured in volts.

pulse-height discriminator A circuit that produces a specified output pulse when and only when it receives an input pulse whose amplitude exceeds an assigned value. Also known as amplitude discriminator.

pulse-width discriminator 343

pulse-height selector A circuit that produces a specified output pulse only when it receives an input pulse whose amplitude lies between two assigned values. Also known as amplitude selector; diffractional pulse-height discriminator.

pulse integrator An RC (resistance-capacitance) circuit which stretches in time duration a pulse applied to it.

pulse interference eliminator Device which removes pulsed signals which are not precisely on the radar operating frequency.

pulse interference separator and blanker Automatic interference blanker that will blank all video signals not synchronous with the radar pulse-repetition frequency.

pulse-link repeater Arrangement of apparatus used in telephone signaling systems for receiving pulses from one E and M signaling circuit, and retransmitting corresponding pulses into another E and M signaling circuit.

pulse operation For microwave tubes, a method of operation in which the energy is delivered in pulses.

pulser A generator used to produce high-voltage, short-duration pulses, as required by a pulsed microwave oscillator or a radar transmitter.

pulse-rate telemetering Telemetering in which the number of pulses per unit time is proportional to the magnitude of the measured quantity.

pulse recurrence rate *See* pulse repetition rate.

pulse regeneration The process of restoring pulses to their original relative timings, forms, and magnitudes.

pulse repeater Device used for receiving pulses from one circuit and transmitting corresponding pulses into another circuit; it may also change the frequencies and waveforms of the pulses and perform other functions.

pulse repetition frequency *See* pulse repetition rate.

pulse repetition rate The number of times per second that a pulse is transmitted. Abbreviated PRR. Also known as pulse recurrence rate; pulse repetition frequency (PRF).

pulse scaler A scaler that produces an output signal when a prescribed number of input pulses has been received.

pulse selector A circuit or device for selecting the proper pulse from a sequence of telemetering pulses.

pulse shaper A transducer used for changing one or more characteristics of a pulse, such as a pulse regenerator or pulse stretcher.

pulse stretcher A pulse shaper that produces an output pulse whose duration is greater than that of the input pulse and whose amplitude is proportional to the peak amplitude of the input pulse.

pulse synthesizer A circuit used to supply pulses that are missing from a sequence due to interference or other causes.

pulse transformer A transformer capable of operating over a wide range of frequencies, used to transfer nonsinusoidal pulses without materially changing their waveforms.

pulse transmitter A pulse-modulated transmitter whose peak-power-output capabilities are usually large with respect to the average-power-output rating.

pulse voltage *See* impulse voltage.

pulse-width discriminator Device that measures the pulse length of video signals and passes only those whose time duration falls into some predetermined design tolerance.

pulse-width modulated static inverter A variation of the quasi-square-wave static inverter, operating at high frequency, in which the pulse width, and not the amplitude, of the square wave is adjusted to approximate the sine wave.

pulsing transformer Transformer designed to supply pulses of voltage or current.

pump Of a parametric device, the source of alternating-current power which causes the nonlinear reactor to behave as a time-varying reactance.

pumped hydroelectric storage A method of energy storage in which excess electrical energy produced at times of low demand is used to pump water into a reservoir, and this water is released at times of high demand to operate hydroelectric generators.

pumped tube An electron tube that is continuously connected to evacuating equipment during operation; large pool-cathode tubes are often operated in this manner.

pumping frequency Frequency at which pumping is provided in a maser, quadrupole amplifier, or other amplifier requiring high-frequency excitation.

pump oscillator Alternating-current generator that supplies pumping energy for maser and parametric amplifiers; operates at twice or some higher multiple of the signal frequency.

punch-through An emitter-to-collector breakdown which can occur in a junction transistor with very narrow base region at sufficiently high collector voltage when the space-charge layer extends completely across the base region.

puncture Disruptive discharge through insulation involving a sudden and large increase in current through the insulation due to complete failure under electrostatic stress.

puncture voltage The voltage at which a test specimen is electrically punctured.

Pupin coil *See* loading coil.

pup jack *See* tip jack.

purity coil A coil mounted on the neck of a color picture tube, used to produce the magnetic field needed for adjusting color purity; the direct current through the coil is adjusted to a value that makes the magnetic field orient the three individual electron beams so each strikes only its assigned color of phosphor dots.

purity control A potentiometer or rheostat used to adjust the direct current through the purity coil.

purity magnet An adjustable arrangement of one or more permanent magnets used in place of a purity coil in a color television receiver.

purple plague A compound formed by intimate contact of gold and aluminum, which appears on silicon planar devices and integrated circuits using gold leads bonded to aluminum thin-film contacts and interconnections, and which seriously degrades the reliability of semiconductor devices.

push-button dialing Dialing a number by pushing buttons on the telephone rather than turning a circular wheel; each depressed button causes a transistor oscillator to oscillate simultaneously at two different frequencies, generating a pair of audio tones which are recognized by central-office (or PBX) switching equipment as digits of a telephone number. Also known as tone dialing; touch call.

push-button switch A master switch that is operated by finger pressure on the end of an operating button.

push-button tuner A device that automatically tunes a radio receiver or other piece of equipment to a desired frequency when the button assigned to that frequency is pressed.

pushing A change in the resonant frequency of a circuit due to changes in the applied voltages.

pyrone detector   345

push-pull amplifier A balanced amplifier employing two similar electron tubes or equivalent amplifying devices working in phase opposition.

push-pull currents *See* balanced currents.

push-pull electret transducer A type of transducer in which a foil electret is sandwiched between two electrodes and is specially treated or arranged so that the electrodes exert forces in opposite directions on the diaphragm, and the net force is a linear function of the applied voltage.

push-pull magnetic amplifier A realization of a push-pull amplifier using magnetic amplifiers.

push-pull oscillator A balanced oscillator employing two similar electron tubes or equivalent amplifying devices in phase opposition.

push-pull transformer An audio-frequency transformer having a center-tapped winding and designed for use in a push-pull amplifier.

push-pull transistor 1. A realization of a push-pull amplifier using transistors. 2. A Darlington circuit in which the two transistors required for a push-pull amplifier exist in a single substrate.

push-pull voltages *See* balanced voltages.

push-push amplifier An amplifier employing two similar electron tubes with grids connected in phase opposition and with anodes connected in parallel to a common load; usually used as a frequency multiplier to emphasize even-order harmonics; transistors may be used in place of tubes.

push-to-talk circuit Simplex circuit in which changeover from the receive to transmit state is accomplished by depressing a single spring-return switch, and releasing the switch returns the circuit to the receive state; the push-to-talk switch is located on microphones and telephone handsets; it is most often applied to radio circuits.

pyroconductivity Electrical conductivity that develops in a material only at high temperature, chiefly at fusion, in solids that are practically nonconductive at atmospheric temperatures.

pyroelectric crystal A crystal exhibiting pyroelectricity, such as tourmaline, lithium sulfate monohydrate, cane sugar, and ferroelectric barium titanate.

pyroelectricity The property of certain crystals to produce a state of electrical polarity by a change of temperature.

pyrone detector Crystal detector in which rectification occurs between iron pyrites and copper or other metallic points.

Q

Q band A radio-frequency band of 36 to 46 gigahertz.

Q multiplier A filter that gives a sharp response peak or a deep rejection notch at a particular frequency, equivalent to boosting the Q of a tuned circuit at that frequency.

QPSK *See* quarternary phase-shift keying.

Q signal The quadrature component of the chrominance signal in color television, having a bandwidth of 0 to 0.5 megahertz; it consists of $+0.48(R-Y)$ and $+0.41(B-Y)$, where Y is the luminance signal, R is the red camera signal, and B is the blue camera signal.

quad 1. A series of four separately insulated conductors, generally twisted together in pairs. 2. A series-parallel combination of transistors; used to obtain increased reliability through double redundancy, because the failure of one transistor will not disable the entire circuit.

quadded cable Cable in which at least some of the conductors are arranged in the form of quads.

quad in-line An integrated-circuit package that has two rows of staggered pins on each side, spaced closely enough together to permit 48 or more pins per package. Abbreviated QUIL.

quadrant *See* international henry.

quadrature amplifier An amplifier that shifts the phase of a signal 90°; used in a color television receiver to amplify the 3.58-megahertz chrominance subcarrier and shift its phase 90° for use in the Q demodulator.

quadrature component 1. A vector representing an alternating quantity which is in quadrature (at 90°) with some reference vector. 2. *See* reactive component.

quadrature current *See* reactive current.

quadruplex circuit Telegraph circuit designed to carry two messages in each direction at the same time.

quadrupole A distribution of charge or magnetization which produces an electric or magnetic field equivalent to that produced by two electric or magnetic dipoles whose dipole moments have the same magnitude but point in opposite directions, and which are separated from each other by a small distance.

quadrupole amplifier A low-noise parametric amplifier consisting of an electron-beam tube in which quadrupole fields act on the fast cyclotron wave of the electron beam to produce high amplification at frequencies in the range of 400–800 megahertz.

quadrupole field 1. An electric or magnetic field equivalent to that produced by two electric or magnetic dipoles whose dipole moments have the same magnitude but point in opposite directions, and which are separated from each other by a small distance. 2. The field produced by a quadrupole lens.

quadrupole lens A device for focusing beams of charged particles which has four electrodes or magnetic poles of alternating sign arranged in a circle about the beam; used in instruments such as electron microscopes and particle accelerators.

quadrupole moment A quantity characterizing a distribution of charge or magnetization; it is given by integrating the product of the charge density or divergence of magnetization density, the second power of the distance from the origin, and a spherical harmonic Y^*_{2m} over the charge or magnetization distribution.

quantized spin wave *See* magnon.

quantizer A device that measures the magnitude of a time-varying quantity in multiples of some fixed unit, at a specified instant or specified repetition rate, and delivers a proportional response that is usually in pulse code or digital form.

quantum efficiency The average number of electrons photoelectrically emitted from a photocathode per incident photon of a given wavelength in a phototube.

quantum electronics The branch of electronics associated with the various energy states of matter, motions within atoms or groups of atoms, and various phenomena in crystals; examples of practical applications include the atomic hydrogen maser and the cesium atomic-beam resonator.

quarternary phase-shift keying Modulation of a microwave carrier with two parallel streams of nonreturn-to-zero data in such a way that the data is transmitted as 90° phase shifts of the carrier; this gives twice the message channel capacity of binary phase-shift keying in the same bandwidth. Abbreviated QPSK.

quarter-wave Having an electrical length of one quarter-wavelength.

quarter-wave antenna An antenna whose electrical length is equal to one quarter-wavelength of the signal to be transmitted or received.

quarter-wave attenuator Arrangement of two wire gratings, spaced an odd number of quarter-wavelengths apart in a waveguide, used to attenuate waves traveling through in one direction.

quarter-wave line *See* quarter-wave stub.

quarter-wave matching section *See* quarter-wave transformer.

quarter-wave stub A section of transmission line that is one quarter-wavelength long at the fundamental frequency being transmitted; when shorted at the far end, it has a high impedance at the fundamental frequency and all odd harmonics, and a low impedance for all even harmonics. Also known as quarter-wave line; quarter-wave transmission line.

quarter-wave termination Metal plate and a wire grating spaced about one quarter-wavelength apart in a waveguide, with the plate serving as the termination of the guide; waves reflected from the metal plate are canceled by waves reflected from the grating so that all energy is absorbed (none is reflected) by the quarter-wave termination.

quarter-wave transformer A section of transmission line approximately one quarter-wavelength long, used for matching a transmission line to an antenna or load. Also known as quarter-wave matching section.

quarter-wave transmission line *See* quarter-wave stub.

quartz crystal A natural or artificially grown piezoelectric crystal composed of silicon dioxide, from which thin slabs or plates are carefully cut and ground to serve as a crystal plate.

quick-make switch 349

quartz-crystal filter A filter which utilizes a quartz crystal; it has a small bandwidth, a high rate of cutoff, and a higher unloaded Q than can be obtained in an ordinary resonator.

quartz-crystal resonator A quartz plate whose natural frequency of vibration is used to control the frequency of an oscillator. Also known as quartz resonator.

quartz delay line An acoustic delay line in which quartz is used as the medium of sound transmission.

quartz-fiber electroscope Electroscope in which a gold-plated quartz fiber serves the same function as the gold leaf of a conventional electroscope.

quartz-iodine lamp An electric lamp having a tungsten filament and a quartz envelope filled with iodine vapor.

quartz lamp A mercury-vapor lamp having a transparent envelope made from quartz instead of glass; quartz resists heat, permitting higher currents, and passes ultraviolet rays that are absorbed by ordinary glass.

quartz oscillator An oscillator in which the frequency of the output is determined by the natural frequency of vibration of a quartz crystal.

quartz plate *See* crystal plate.

quartz resonator *See* quartz-crystal resonator.

quartz strain gage A device used to measure small deformations of a substance by determining the resulting voltage that develops in a quartz attached to it.

quasi-free-electron theory A modification of the free-electron theory of metals to take into account the periodic variation of the potential acting on a conduction electron, in which these electrons are assigned an effective scalar mass which differs from their real mass.

quasi-square-wave static inverter A static inverter that generates two square waves superimposed on one another to approximate an ac sine wave, using a silicon-controlled rectifier bridge and control circuit to control the pulse width and amplitude of the resulting wave, thereby achieving regulation.

quenched spark gap A spark gap having provisions for rapid deionization; one form consists of many small gaps between electrodes that have relatively large mass and are good radiators of heat; the electrodes serve to cool the gaps rapidly and thereby stop conduction.

quench frequency Number of times per second that a circuit is caused to go in and out of oscillation.

quenching 1. The process of terminating a discharge in a gas-filled radiation-counter tube by inhibiting reignition. 2. Reduction of the intensity of resonance radiation resulting from deexcitation of atoms, which would otherwise have emitted this radiation, in collisions with electrons or other atoms in a gas. 3. Reduction in the intensity of sensitized luminescence radiation when energy migrating through a crystal by resonant transfer is dissipated in crystal defects or impurities rather than being reemitted as radiation.

quench oscillator Circuit in a superregenerative receiver which produces the frequency signal.

quick-break fuse A fuse designed to draw out the arc and break the circuit rapidly when the fuse wire melts, generally by separating the broken ends with a spring.

quick-break switch A switch that breaks a circuit rapidly, independently of the rate at which the switch handle is moved, to minimize arcing.

quick-make switch Switch or circuit breaker which has a high contact-closing speed, independent of the operator.

quiescent Condition of a circuit element which has no input signal, so that it does not perform its active function.

quiescent point The point on the characteristic curve of an amplifier representing the conditions that exist when the input signal equals zero.

quiescent push-pull Push-pull output stage so arranged in a radio receiver that practically no current flows when no signal is being received.

quiescent value The voltage or current value for an electron-tube electrode when no signals are present.

quiet automatic volume control *See* delayed automatic gain control.

quiet battery Source of energy of special design or with added filters which is sufficiently quiet and free from interference that it may be used for speech transmission. Also known as talking battery.

quieting sensitivity Minimum signal input to a frequency-modulated receiver which is required to give a specified output signal-to-noise ratio under specified conditions.

quiet tuning Circuit arrangement for silencing the output of a radio receiver, except when it is accurately tuned to an incoming carrier wave.

QUIL *See* quad in-line.

R

race condition An ambiguous condition occurring in control counters when one flip-flop changes to its next state before a second one has had sufficient time to latch.

raceway A channel used to hold and protect wires, cables, or busbars. Also known as electric raceway.

rack panel A panel designed for mounting on a relay rack; its width is 19 inches (48.26 centimeters), height is a multiple of 1¾ inches (4.445 centimeters), and the mounting notches are standardized as to size and position.

radar antenna A device which radiates radio-frequency energy in a radar system, concentrating the transmitted power in the direction of the target, and which provides a large area to collect the echo power of the returning wave.

radar antijamming Measures taken to counteract radar jamming.

radar attenuation Ratio of the power delivered by the transmitter to the transmission line connecting it with the transmitting antenna, to the power reflected from the target which is delivered to the receiver by the transmission line connecting it with the receiving antenna.

radar beam The movable beam of radio-frequency energy produced by a radar transmitting antenna; its shape is commonly defined as the loci of all points at which the power has decreased to one-half of that at the center of the beam.

radar cell Volume whose dimensions are one radar pulse length by one radar beam width.

radar clutter *See* clutter.

radar conspicuous object An object which returns a strong radar echo.

radar constant One of those terms of the radar equation or radar storm-detection equation which are functions of the particular radar to which the equations are applied; these include peak power, antenna gain or aperture, beam width, pulse length, pulse repetition frequency, wavelength, polarization, and noise level of the receiver.

radar control Guidance, direction, or employment exercised over an aircraft, guided missile, gun battery, or the like, by means of, or with the aid of, radar.

radar countermeasure An electronic countermeasure used against enemy radar, such as jamming and confusion reflectors. Abbreviated RCM.

radar cross section *See* echo area.

radar data filtering Quality analysis process that causes the computer to reject certain radar data and to alert personnel of mapping and surveillance consoles to the rejection.

radar display The pattern representing the output data of a radar set, generally produced on the screen of a cathode-ray tube. Also known as presentation; radar presentation.

352 radar distribution switchboard

radar distribution switchboard Switching panel for connecting video, trigger, and bearing from any one of five systems, to any or all of 20 repeaters; also contains order lights, bearing cutouts, alarms, test equipment, and so forth.

radar echo See echo.

radar equation An equation that relates the transmitted and received powers and antenna gains of a primary radar system to the echo area and distance of the radar target.

radar frequency band A frequency band of microwave radiation in which radar operates.

radar image The image of an object which is produced on a radar screen.

radar indicator A cathode-ray tube and associated equipment used to provide a visual indication of the echo signals picked up by a radar set.

radar intelligence item A feature which is radar significant but which cannot be identified exactly at the moment of its appearance as homogeneous.

radar jamming Radiation, reradiation, or reflection of electromagnetic waves so as to impair the usefulness of radar used by the enemy.

radar mile The time for a radar pulse to travel from the radar to a target 1 mile (1.61 kilometers) distant and return, equal to 10.75 microseconds.

radar nautical mile The time interval of approximately 12.355 microseconds that is required for the radio-frequency energy of a radar pulse to travel 1 nautical mile (1852 meters) and return.

radar netting unit Optional electronic equipment that converts the operations central of certain air defense fire distribution systems to a radar netting station.

radar presentation See radar display.

radar pulse Radio-frequency radiation emitted with high power by a pulse radar installation for a period of time which is brief compared to the interval between such pulses.

radar range The maximum distance at which a radar set is ordinarily effective in detecting objects.

radar range equation An equation which expresses radar range in terms of transmitted power, minimum detectable signal, antenna gain, and the target's radar cross section.

radar receiver A high-sensitivity radio receiver that is designed to amplify and demodulate radar echo signals and feed them to a radarscope or other indicator.

radar receiver-transmitter A single component having the dual functions of generating electromagnetic energy for transmission, and of receiving, demodulating, and sometimes presenting intelligence from the reflected electromagnetic energy.

radar reflection The return of electromagnetic waves, generated by a radar installation, from an object on which the waves are incident.

radar reflection interval The time required for a radar pulse to travel from the source to the target and return to the source, taking the velocity of radio propagation to be equal to the velocity of light.

radar reflectivity The fraction of electromagnetic energy generated by a radar installation which is reflected by an object.

radar reflector A device that reflects or deflects radar waves.

radar repeater A cathode-ray indicator used to reproduce the visible intelligence of a radar display at a remote position; when used with a selector switch, the visible intelligence of any one of several radar systems can be reproduced.

radiating curtain 353

radarscope Cathode-ray tube, serving as an oscilloscope, the face of which is the radar viewing screen. Also known as scope.

radar selector switch Manual or motor-driven switch which transfers a plan-position indicator repeater from one system to another, switching video, trigger, and bearing data.

radar shadow A region shielded from radar illumination by an intervening reflecting or absorbing medium such as a hill.

radar signal spectrograph An electronic device in the form of a scanning filter which provides a frequency analysis of the amplitude-modulated back-scattered signal.

radar target An object belonging to a desired class which reflects back a signal sufficient to produce a fluorescent mark on the radar screen.

radar transmitter The transmitter portion of a radar set.

radar volume The volume in space that is irradiated by a given radar; for a continuous-wave radar it is equivalent to the antenna radiation pattern; for a pulse radar it is a function of the cross-section area of the beam of the antenna and the pulse length of the transmitted pulse.

radechon A storage tube having a single electron gun and a dielectric storage medium consisting of a sheet of mica sandwiched between a continuous metal backing plate and a fine-mesh screen; used in simple delay schemes, signal-to-noise improvement, signal comparison, and conversion of signal-time bases. Also known as barrier-grid storage tube.

radial-beam tube A vacuum tube in which a radial beam of electrons is rotated past circumferentially arranged anodes by an external rotating magnetic field; used chiefly as a high-speed switching tube or commutator.

radial Doppler effect The part of the optical Doppler effect which depends on the direction of the relative velocity of source and observer, and is analogous to the acoustical Doppler effect, in contrast to the transverse Doppler effect.

radial grating Conformal wire grating consisting of wires arranged radially in a circular frame, like the spokes of a wagon wheel, and placed inside a circular waveguide to obstruct E waves of zero order while passing the corresponding H waves.

radial lead A wire lead coming from the side of a component rather than axially from the end.

radiancy *See* radiant emittance.

radiant emittance The radiant flux per unit area that emerges from a surface. Also known as radiancy; radiant exitance.

radiant exitance *See* radiant emittance.

radiant flux density The amount of radiant power per unit area that flows across or onto a surface. Also known as irradiance.

radiant intensity The energy emitted per unit time per unit solid angle about the direction considered; usually expressed in watts per steradian.

radiant power The energy carried across or onto a surface by electromagnetic radiation per unit time, or the total radiant energy emitted by a source of electromagnetic radiation per unit time.

radiant reflectance Ratio of reflected radiant power to incident radiant power.

radiant transmittance Ratio of transmitted radiant power to incident radiant power.

radiated power The total power emitted by a transmitting antenna.

radiating curtain Array of dipoles in a vertical plane, positioned to reinforce each other; it is usually placed ¼ wavelength ahead of a reflecting curtain of corresponding half-wave reflecting antennas.

radiating element Basic subdivision of an antenna which in itself is capable of radiating or receiving radio-frequency energy.

radiating guide Waveguide designed to radiate energy into free space; the waves may emerge through slots or gaps in the guide, or through horns inserted in the wall of the guide.

radiation angle The vertical angle between the line of radiation emitted by a directional antenna and the horizon.

radiation cooling Cooling of an electrode resulting from its emission of heat radiation.

radiation damping Damping of a system which loses energy by electromagnetic radiation.

radiation efficiency Of an antenna, the ratio of the power radiated to the total power supplied to the antenna at a given frequency.

radiation field The electromagnetic field that breaks away from a transmitting antenna and radiates outward into space as electromagnetic waves; the other type of electromagnetic field associated with an energized antenna is the induction field.

radiation filter Selectively transparent body, which transmits only certain wavelength ranges.

radiation intensity The power radiated from an antenna per unit solid angle in a given direction.

radiation lobe *See* lobe.

radiation noise *See* electromagnetic noise.

radiation pattern Directional dependence of the radiation of an antenna. Also known as antenna pattern; directional pattern; field pattern.

radiation pressure The pressure exerted by electromagnetic radiation on objects on which it impinges.

radiation resistance The total radiated power of an antenna divided by the square of the effective antenna current measured at the point where power is supplied to the antenna.

radiation thermocouple An infrared detector consisting of several thermocouples connected in series, arranged so that the radiation falls on half of the junctions, causing their temperature to increase so that a voltage is generated.

radiation zone *See* Fraunhofer region.

radiator 1. The part of an antenna or transmission line that radiates electromagnetic waves either directly into space or against a reflector for focusing or directing. 2. A body that emits radiant energy.

radio 1. A prefix denoting the use of radiant energy, particularly radio waves. 2. *See* radio receiver.

radio aid to navigation An aid to navigation which utilizes the propagation characteristics of radio waves to furnish navigation information.

radio antenna *See* antenna.

radio attenuation For one-way propagation, the ratio of the power delivered by the transmitter to the transmission line connecting it with the transmitting antenna to the power delivered to the receiver by the transmission line connecting it with the receiving antenna.

radio B battery A B-type battery used in a radio set, usually consisting of 15 to 30 permanently connected cells.

radio-frequency measurement 355

radio beam A concentrated stream of radio-frequency energy as used in radio ranges, microwave relays, and radar.

radio command A radio control signal to which a guided missile or other remote-controlled vehicle or device responds.

radio compass *See* automatic direction finder.

radio control The control of stationary or moving objects by means of signals transmitted through space by radio.

radio countermeasures Electrical or other techniques depriving the enemy of the benefits which would ordinarily accrue to him through the use of any technique employing the radiation of radio waves; it includes benefits derived from radar and intercept services.

radio emission The emission of radio-frequency electromagnetic radiation by oscillating charges or currents.

radio energy The energy carried by radio-frequency electromagnetic radiation.

radio field intensity Electric or magnetic field intensity at a given location associated with the passage of radio waves.

radio field-to-noise ratio Ratio, at a given location, of the radio field intensity of the desired wave to the noise field intensity.

radio frequency A frequency at which coherent electromagnetic radiation of energy is useful for communication purposes; roughly the range from 10 kilohertz to 100 gigahertz. Abbreviated rf.

radio-frequency alternator A rotating-type alternator designed to produce high power at frequencies above power-line values but generally lower than 100,000 hertz; used chiefly for high-frequency heating.

radio-frequency amplifier An amplifier that amplifies the high-frequency signals commonly used in radio communications.

radio-frequency cable A cable having electric conductors separated from each other by a continuous homogeneous dielectric or by touching or interlocking spacer beads; designed primarily to conduct radio-frequency energy with low losses. Also known as RG line.

radio-frequency cavity preselector A tunable cavity resonator in an ultra-high-frequency circuit, which is similar in function to a tuned resonant circuit.

radio-frequency choke A coil designed and used specifically to block the flow of radio-frequency current while passing lower frequencies or direct current.

radio-frequency current Alternating current having a frequency higher than 10,000 hertz.

radio-frequency filter An electric filter which enhances signals at certain radio frequencies or attenuates signals at undesired radio frequencies.

radio-frequency generator A generator capable of supplying sufficient radio-frequency energy at the required frequency for induction or dielectric heating.

radio-frequency line *See* radio-frequency transmission line.

radio-frequency measurement The precise measurement of frequencies above the audible range by any of various techniques, such as a calibrated oscillator with some means of comparison with the unknown frequency, a digital counting or scaling device which measures the total number of events occurring during a given time interval, or an electronic circuit for producing a direct current proportional to the frequency of its input signal.

radio-frequency oscillator An oscillator that generates alternating current at radio frequencies.

radio-frequency power supply A high-voltage power supply in which the output of a radio-frequency oscillator is stepped up by an air-core transformer to the high voltage required for the second anode of a cathode-ray tube, then rectified to provide the required high direct-current voltage; used in some television receivers.

radio-frequency reactor A reactor used in electronic circuits to pass direct current and offer high impedance at high frequencies.

radio-frequency resistance See high-frequency resistance.

radio-frequency shift See frequency shift.

radio-frequency signal generator A test instrument that generates the various radio frequencies required for alignment and servicing of radio, television, and electronic equipment. Also known as service oscillator.

radio-frequency SQUID A type of SQUID which has only one Josephson junction in a superconducting loop; its state is determined from radio-frequency measurements of the impedance of the ring.

radio-frequency transformer A transformer having a tapped winding or two or more windings designed to furnish inductive reactance or to transfer radio-frequency energy from one circuit to another by means of a magnetic field; may have an air core or some form of ferrite core. Also known as radio transformer.

radio-frequency transmission line A transmission line designed primarily to conduct radio-frequency energy, consisting of two or more conductors supported in a fixed spatial relationship along their own length. Also known as radio-frequency line.

radiogoniometer A goniometer used as part of a radio direction finder.

radio guidance Guidance of a flight-borne missile or other vehicle from a ground station by means of radio signals.

radiolucent Transparent to x-rays and radio waves.

radiometer A receiver for detecting microwave thermal radiation and similar weak wide-band signals that resemble noise and are obscured by receiver noise; examples include the Dicke radiometer, subtraction-type radiometer, and two-receiver radiometer. Also known as microwave radiometer; radiometer-type receiver.

radiometer-type receiver See radiometer.

radiomicrometer See microradiometer.

radio mirage The detection of radar targets at phenomenally long range due to radio ducting.

radio noise Electromagnetic noise having radio frequencies.

radio-noise field strength A quantity which is proportional, or related in a known manner, to the field strength of electromagnetic waves of an interfering character at a point, such as a radio receiver. Also known as noise field strength.

radiopaque Not appreciably penetrable by x-rays or other forms of radiation.

radio pill A device used in biotelemetry for monitoring the physiologic activity of an animal, such as pH values of stomach acid; an example is the Heidelberg capsule.

radio pulse An intense burst of radio-frequency energy lasting for a fraction of a second.

radio receiver A device that converts radio waves into intelligible sounds or other perceptible signals. Also known as radio; radio set; receiving set.

radio scanner See scanning radio.

range calibrator 357

radio scattering *See* scattering.

radio set *See* radio receiver; radio transmitter.

radio sextant An antenna with a high-resolution beam pattern that measures the angle between local direction references and an astronomical radio signal source such as an artificial satellite, the sun, the moon, or a radio star.

radio shielding Metallic covering over all electric wiring and ignition apparatus, which is grounded at frequent intervals for the purpose of eliminating electric interference with radio communications.

radiosonde commutator A component of a radiosonde consisting of a series of alternate electrically conducting and insulating strips; as these are scanned by a contact, the radiosonde transmits temperature and humidity signals alternately.

radio transformer *See* radio-frequency transformer.

radio transmitter The equipment used for generating and amplifying a radio-frequency carrier signal, modulating the carrier signal with intelligence, and feeding the modulated carrier to an antenna for radiation into space as electromagnetic waves. Also known as radio set; transmitter.

radio transponder A transponder which receives and transmits radio waves, in contrast to a sonar transponder, which receives and transmits acoustic waves.

radio tube *See* electron tube.

radio wave An electromagnetic wave produced by reversal of current in a conductor at a frequency in the range from about 10 kilohertz to about 300,000 megahertz.

radio wavefront distortion Change in the direction of advance of a radio wave.

radio-wave propagation The transfer of energy through space by electromagnetic radiation at radio frequencies.

radome A strong, thin shell, made from a dielectric material that is transparent to radio-frequency radiation, and used to house a radar antenna, or a space communications antenna of similar structure.

railing Radar pulse jamming at high recurrence rates (50 to 150 kilohertz); it results in an image on a radar indicator resembling fence railing.

rainbow Technique which applies pulse-to-pulse frequency changing to identifying and discriminating against decoys and chaff.

ramp generator A circuit that generates a sweep voltage which increases linearly in value during one cycle of sweep, then returns to zero suddenly to start the next cycle.

randomized jitter Jitter by means of noise modulation.

random winding A coil winding in which the turns are positioned haphazardly rather than in layers.

range-amplitude display Radar display in which a time base provides the range scale from which echoes appear as deflections normal to the base.

range attenuation In radar terminology, the decrease in power density (flux density) caused by the divergence of the flux lines with distance, this decrease being in accordance with the inverse-square law.

range-bearing display *See* B display.

range calibrator 1. A device with which the operator of a transmitter calculates the distance over which the signal will extend intelligibly. 2. A device for adjusting radar range indications by use of known range targets or delayed signals, so when on target the set will indicate the correct range.

range comprehension In an FM sonar system, valves between the maximum and the minimum ranges.

range delay A control used in radars which permits the operator to present on the radarscope only those echoes from targets which lie beyond a certain distance from the radar; by using range delay, undesired echoes from nearby targets may be eliminated while the indicator range is increased.

rangefinder A device which determines the distance to an object by measuring the time it takes for a radio wave to travel to the object and return.

range gate capture Electronic countermeasure technique using a spoofer radar transmitter to produce a false target echo that can make a fire-control tracking radar move off the real target and follow the false one.

range gating The process of selecting those radar echoes that lie within a small range interval.

range-height indicator display A radar display that presents visually the scalar distance between a reference point and a target, along with the vertical distance between a reference plane and the target. Abbreviated RHI display.

range mark offset Displacement of range mark on a type B indicator.

range rate The rate at which the distance from the measuring equipment to the target or signal source that is being tracked is changing with respect to time.

range ring Accurate, adjustable, ranging mark on a plan position indicator corresponding to a range step on a type M indicator.

range selection Control on a radar indicator for selection of range scale.

range step Vertical displacement on M-indicator sweep to measure range.

range strobe An index mark which may be displayed on various types of radar indicators to assist in the determination of the exact range of a target.

range sweep A sweep intended primarily for measurement of range.

range-tracking element An element in a radar set which measures range and its time derivative; by means of the latter, a range gate is actuated slightly before the predicted instant of signal reception.

range unit Radar system component used for control and indication (usually counters) of range measurements.

range zero Alignment of start sweep trace with zero range.

ranging oscillator Oscillator circuit containing an LC (inductor-capacitor) resonant combination in the cathode circuit, usually used in radar equipment to provide range marks.

rare-earth magnet Any of several types of magnets made with rare-earth elements, such as rare-earth-cobalt magnets, which have coercive forces up to ten times that of ordinary magnets; used for computers and signaling devices.

raster A predetermined pattern of scanning lines that provides substantially uniform coverage of an area; in television the raster is seen as closely spaced parallel lines, most evident when there is no picture.

raster scanning Radar scan very similar to electron-beam scanning in an ordinary television set; horizontal sector scan that changes in elevation.

rate effect The phenomenon of a *pnpn* device switching to a high-conduction mode when anode voltage is applied suddenly or when high-frequency transients exist.

rate feedback The return of a signal, proportional to the rate of change of the output of a device, from the output to the input.

rate-grown transistor A junction transistor in which both impurities (such as gallium and antimony) are placed in the melt at the same time and the temperature is suddenly raised and lowered to produce the alternate p-type and n-type layers of rate-grown junctions. Also known as graded-junction transistor.

rate receiver A guidance antenna that receives a signal from a launched missile as to its rate of speed.

rate transmitter A transmitter in a missile being launched, used with a ground receiver to indicate the rate of speed increase.

ratio arm circuit Two adjacent arms of a Wheatstone bridge, designed so they can be set to provide a variety of indicated resistance ratios.

ratio-balance relay *See* percentage differential relay.

ratio detector A frequency-modulation detector circuit that uses two diodes and requires no limiter at its input; the audio output is determined by the ratio of two developed intermediate-frequency voltages whose relative amplitudes are a function of frequency.

ratio-differential relay *See* percentage differential relay.

rationalized units A system of electrical units, such as occurs in the International System, in which the factor of 4π is removed from the field equations and appears instead in the explicit expressions for the fields of a point charge and current element.

ratio of transformation Ratio of the secondary voltage of a transformer to the primary voltage under no-load conditions, or the corresponding ratio of currents in a current transformer.

ratio of transformer Ratio of the number of turns in one winding of a transformer to the number of turns in the other, unless otherwise specified.

ratio resistor One of the resistors in a Wheatstone or Kelvin bridge whose resistances appear in a pair of ratios which are equal in a balanced bridge.

rat race A particular radar waveguide configuration which allows the handling of greater power.

rawin target A special type of radar target tied beneath a free balloon, and designed to be an efficient reflector of radio energy; such targets usually consist of a corner reflector and are made of some reflecting material stretched over light wooden or metal struts.

Rayleigh balance An apparatus for assigning the value of the ampere in which the force exerted on a movable circular coil by larger circular coils above and below, but coaxial with, the movable coil is compared with the gravitational force on a known mass.

Rayleigh cycle A cycle of magnetization that does not extend beyond the initial portion of the magnetization curve, between zero and the upward bend.

Rayleigh loop A parabolic approximation to a magnetic hysteresis loop.

Rayleigh reciprocity theorem Reciprocal relationship for an antenna when it is transmitting or receiving; the effective heights, radiation resistance, and the radiation pattern are alike, whether the antenna is transmitting or receiving.

Rayleigh scattering Scattering of electromagnetic radiation by independent particles which are much smaller than the wavelength of the radiation.

Raysistor A device which contains a photosensitive semiconductor and a light source; light source can be used to control the conductivity of the semiconductor.

R-C amplifier *See* resistance-capacitance coupled amplifer.

360 R-C circuit

R-C circuit See resistance-capacitance circuit.

R-C constant See resistance-capacitance constant.

R-C coupled amplifier See resistance-capacitance coupled amplifier.

R-C coupling See resistance coupling.

RCM See radar countermeasure.

R-C network See resistance-capacitance network.

R-C oscillator See resistance-capacitance oscillator.

R display Radar display, essentially an expanded A display, in which an echo can be expanded for more detailed examination.

reactance The imaginary part of the impedance of an alternating-current circuit.

reactance amplifier See parametric amplifier.

reactance drop The component of the phasor representing the voltage drop across a component or conductor of an alternating-current circuit which is perpendicular to the current.

reactance frequency multiplier Frequency multiplier whose essential element is a nonlinear reactor.

reactance grounded Grounded through a reactance.

reactance relay Form of impedance relay, the operation of which is a function of the reactance of a circuit.

reactance tube Vacuum tube operated in a way that it presents almost a pure reactance to the circuit.

reactance-tube modulator An electron-tube circuit, used to produce phase or frequency modulation, in which the reactance is varied in accordance with the instantaneous amplitude of the modulating voltage.

reaction motor A synchronous motor whose rotor contains salient poles but which has no windings and no permanent magnets.

reactive Pertaining to either inductive or capacitance reactance; a reactive circuit has a high value of reactance in comparison with resistance.

reactive component In the phasor representation of quantities in an alternating-current circuit, the component of current, voltage, or apparent power which does not contribute power, and which results from inductive or capacitive reactance in the circuit, namely, the reactive current, reactive voltage, or reactive power. Also known as idle component; quadrature component; wattless component.

reactive current In the phasor representation of alternating current, the component of the current perpendicular to the voltage, which contributes no power but increases the power losses of the system. Also known as idle current; quadrature current; wattless current.

reactive factor The ratio of reactive power to apparent power.

reactive ion etching A directed chemical etching process used in integrated circuit fabrication in which chemically active ions are accelerated along electric field lines to meet a substrate perpendicular to its surface.

reactive load A load having inductive or capacitive reactance.

reactive power The power value obtained by multiplying together the effective value of current in amperes, the effective value of voltage in volts, and the sine of the angular phase difference between current and voltage. Also known as wattless power.

reciprocal ferrite switch   361

reactive voltage In the phasor representation of alternating current, the voltage component that is perpendicular to the current.

reactive volt-ampere See volt-ampere reactive.

reactive volt-ampere hour See var hour.

reactor A device that introduces either inductive or capacitive reactance into a circuit, such as a coil or capacitor. Also known as electric reactor.

read To generate an output corresponding to the pattern stored in a charge storage tube.

Read diode A high-frequency semiconductor diode consisting of an avalanching pn-junction, biased to fields of several hundred thousand volts per centimeter, at one end of a high-resistance carrier serving as a drift space for the charge carriers.

real power The component of apparent power that represents true work; expressed in watts, it is equal to volt-amperes multiplied by the power factor.

rear-projection Pertaining to television system in which the picture is projected on a ground-glass screen for viewing from the opposite side of the screen.

received power 1. The total power received at an antenna from a signal, such as a radar target signal. 2. In a mobile communications system, the root-mean-square value of power delivered to a load which properly terminates an isotropic reference antenna.

receiver The complete equipment required for receiving modulated radio waves and converting them into the original intelligence, such as into sounds or pictures, or converting to desired useful information as in a radar receiver.

receiver bandwidth Spread, in frequency, between the halfpower points on the receiver response curve.

receiver gating Application of operating voltages to one or more stages of a receiver only during that part of a cycle of operation when reception is desired.

receiver incremental tuning Control feature to permit receiver tuning (of a transceiver) up to 3 kilohertz to either side of the transmitter frequency.

receiver noise threshold External noise appearing at the front end of a receiver, plus the noise added by the receiver itself, determines a noise threshold that has to be exceeded by the minimum discernible signal.

receiver radiation Radiation of interfering electromagnetic fields by the oscillator of a receiver.

receiver synchro See synchro receiver.

receiving antenna An antenna used to convert electromagnetic waves to modulated radio-frequency currents.

receiving area The factor by which the power density must be multiplied to obtain the received power of an antenna, equal to the gain of the antenna times the square of the wavelength divided by 4π.

receiving set See radio receiver.

receiving tube A low-voltage and low-power vacuum tube used in radio receivers, computers, and sensitive control and measuring equipment.

receptacle See outlet.

rechargeable battery See storage battery.

reciprocal ferrite switch A ferrite switch that can be inserted in a waveguide to switch an input signal to either of two output waveguides; switching is done by a Faraday rotator when acted on by an external magnetic field.

reciprocal impedance Two impedances Z_1 and Z_2 are said to be reciprocal impedances with respect to an impedance Z (invariably a resistance) if they are so related as to satisfy the equation $Z_1 Z_2 = Z^2$.

reciprocal junction A waveguide junction in which the transmission coefficient from the ith port to the jth port is the same as that from the jth port to the ith port; that is, the S matrix is symmetrical.

reciprocal ohm *See* siemens.

reciprocal ohm meter *See* rom.

reciprocal space *See* wave-vector space.

reciprocal transducer Transducer which satisfies the principle of reciprocity.

reciprocation In electronics, a process of deriving a reciprocal impedance from a given impedance, or finding a reciprocal network for a given network.

reciprocity theorem Also known as principle of reciprocity. **1.** The electric potentials V_1 and V_2 produced at some arbitrary point, due to charge distributions having total charges of q_1 and q_2 respectively, are such that $q_1 V_2 = q_2 V_1$. **2.** In an electric network consisting of linear passive impedances, the ratio of the electromotive force introduced in any branch to the current in any other branch is equal in magnitude and phase to the ratio that results if the positions of electromotive force and current are exchanged. **3.** Given two loop antennas, a and b, then $I_{ab}/V_a = I_{ba}/V_b$, where I_{ab} denotes the current received in b when a is used as transmitter, and V_a denotes the voltage applied in a; I_{ba} and V_b are the corresponding quantities when b is the transmitter, a the receiver; it is assumed that the frequency and impedances remain unchanged.

reclosing relay Form of voltage, current, power, or other type of relay which functions to reclose a circuit.

recombination coefficient The rate of recombination of positive ions with electrons or negative ions in a gas, per unit volume, divided by the product of the number of positive ions per unit volume and the number of electrons or negative ions per unit volume.

recombination electroluminescence *See* injection electroluminescence.

recombination radiation The radiation emitted in semiconductors when electrons in the conduction band recombine with holes in the valence band.

recombination velocity On a semiconductor surface, the ratio of the normal component of the electron (or hole) current density at the surface to the excess electron (or hole) charge density at the surface.

reconditioned carrier reception Method of reception in which the carrier is separated from the sidebands to eliminate amplitude variations and noise, and is then added at an increased level to the sideband, to obtain a relatively undistorted output.

reconstruction A process in which atoms at the surface of a solid displace and form bands different from those existing in the bulk solid.

recontrol time *See* deionization time.

record head *See* recording head.

recording-completing trunk Trunk for extending a connection from a local line to a toll operator, used for recording the call and for completing the toll connection.

recording head A magnetic head used only for recording. Also known as record head.

recording lamp A lamp whose intensity can be varied at an audio-frequency rate, for exposing variable-density sound tracks on motion picture film and for exposing paper or film in photographic facsimile recording.

rectifier rating 363

recording level Amplifier output level required to secure a satisfactory recording.

recording noise Noise that is introduced during a recording process.

recording spot See picture element.

recording storage tube Type of cathode-ray tube in which the electric equivalent of an image can be stored as an electrostatic charge pattern on a storage surface; there is no visual display, but the stored information can be read out at any later time as an electric output signal.

recording trunk Trunk extending from a local central office or private branch exchange to a toll office, which is used only for communications with toll operators and not for completing toll connections.

recovery time 1. The time required for the control electrode of a gas tube to regain control after anode-current interruption. 2. The time required for a fired TR (transmit-receive) or pre-TR tube to deionize to such a level that the attenuation of a low-level radio-frequency signal transmitted through the tube is decreased to a specified value. 3. The time required for a fired ATR (anti-transmit-receive) tube to deionize to such a level that the normalized conductance and susceptance of the tube in its mount are within specified ranges. 4. The interval required, after a sudden decrease in input signal amplitude to a system or component, to attain a specified percentage (usually 63%) of the ultimate change in amplification or attenuation due to this decrease. 5. The time required for a radar receiver to recover to half sensitivity after the end of the transmitted pulse, so it can receive a return echo.

rectangular cavity A resonant cavity having the shape of a rectangular parallelepiped.

rectangular pulse A pulse in which the wave amplitude suddenly changes from zero to another value at which it remains constant for a short period of time, and then suddenly changes back to zero.

rectangular scanning Two-dimensional sector scanning in which a slow sector scanning in one direction is superimposed on a rapid sector scanning in a perpendicular direction.

rectangular wave A periodic wave that alternately and suddenly changes from one to the other of two fixed values. Also known as rectangular wave train.

rectangular waveguide A waveguide having a rectangular cross section.

rectangular wave train See rectangular wave.

Rectenna A device that converts microwave energy in direct-current power; consists of a number of small dipoles, each having its own diode rectifier network, which are connected to direct-current buses.

rectification The process of converting an alternating current to a unidirectional current.

rectification factor Quotient of the change in average current of an electrode by the change in amplitude of the alternating sinusoidal voltage applied to the same electrode, the direct voltages of this and other electrodes being maintained constant.

rectified value of an alternating quantity Average of all the positive (or negative) values of the quantity during an integral number of periods.

rectifier A nonlinear circuit component that allows more current to flow in one direction than the other; ideally, it allows current to flow in one direction unimpeded but allows no current to flow in the other direction.

rectifier filter An electric filter used in smoothing out the voltage fluctuation of an electron tube rectifier, and generally placed between the rectifier's output and the load resistance.

rectifier rating A performance rating for a semiconductor rectifier, usually on the basis of the root-mean-square value of sinusoidal voltage that it can withstand in the reverse direction and the average current density that it will pass in the forward direction.

rectifier stack A dry-disk rectifier made up of layers or stacks of disks of individual rectifiers, as in a selenium rectifier or copper-oxide rectifier.

rectifier transformer Transformer whose secondary supplies energy to the main anodes of a rectifier.

rectilinear scanning Process of scanning an area in a predetermined sequence of narrow parallel strips.

recycling Returning to an original condition, as to 0 or 1 in a counting circuit.

redistribution The alteration of charges on an area of a storage surface by secondary electrons from any other area of the surface in a charge storage tube or television camera tube.

redox cell Cell designed to convert the energy of reactants to electrical energy; an intermediate reductant, in the form of liquid electrolyte, reacts at the anode in a conventional manner; it is then regenerated by reaction with a primary fuel.

reed relay A relay having contacts mounted on magnetic reeds sealed into a length of small glass tubing; an actuating coil is wound around the tubing or wound on an auxiliary ferrite-core structure, to provide the magnetic field required for relay operation.

reed switch A switch that has contacts mounted on ferromagnetic reeds sealed in a glass tube, designed for actuation by an external magnetic field. Also known as magnetic reed switch.

reentrant winding Armature winding that returns to its starting point, thus forming a closed circuit.

reference angle Angle formed between the center line of a radar beam as it strikes a reflecting surface and the perpendicular drawn to that reflecting surface.

reference burst *See* color burst.

reference dipole Straight half-wave dipole tuned and matched for a given frequency, and used as a unit of comparison in antenna measurement work.

reference mark One of the marks used in a design of a printed circuit, giving scale dimensions and indicating the edges of the circuit board.

reference noise The power level used as a basis of comparison when designating noise power expressed in decibels above reference noise (dBrn); the reference usually used is 10^{-12} watt (-90 decibels above 1 milliwatt; dBm) at 1000 hertz.

reference supply A source of stable and constant voltage, such as a Zener diode, used in analog computers, regulated power supplies, and a variety of other circuits for comparison with a varying voltage.

reference voltage An alternating-current voltage used for comparison, usually to identify an in-phase or out-of-phase condition in an ac circuit.

reference white level In television, the level at the point of observation corresponding to the specified maximum excursion of the picture signal in the white direction.

reflectance *See* reflection factor.

reflected impedance 1. Impedance value that appears to exist across the primary of a transformer due to current flowing in the secondary. 2. Impedance which appears at the input terminals as a result of the characteristics of the impedance at the output terminals.

reflected resistance Resistance value that appears to exist across the primary of a transformer when a resistive load is across the secondary.

regeneration 365

reflecting antenna An antenna used to achieve greater directivity or desired radiation patterns, in which a dipole, slot, or horn radiates toward a larger reflector which shapes the radiated wave to produce the desired pattern; the reflector may consist of one or two plane sheets, a parabolic or paraboloidal sheet, or a paraboloidal horn.

reflecting curtain A vertical array of half-wave reflecting antennas, generally used one quarter-wavelength behind a radiating curtain of dipoles to form a high-gain antenna.

reflecting electrode Tabular outer electrode or the repeller plate in a microwave oscillator tube, corresponding in construction but not in function to the plate of an ordinary triode; used for generating extremely high frequencies.

reflecting galvanometer See mirror galvanometer.

reflecting grating Arrangement of wires placed in a waveguide to reflect one desired wave while allowing one or more other waves to pass freely.

reflection factor Ratio of the load current that is delivered to a particular load when the impedances are mismatched to that delivered under conditions of matched impedances. Also known as mismatch factor; reflectance; transition factor.

reflection lobes Three-dimensional sections of the radiation pattern of a directional antenna, such as a radar antenna, which results from reflection of radiation from the earth's surface.

reflection loss 1. Reciprocal of the ratio, expressed in decibels, of the scalar values of the volt-amperes delivered to the load to the volt-amperes that would be delivered to a load of the same impedance as the source. 2. Apparent transmission loss of a line which results from a portion of the energy being reflected toward the source due to a discontinuity in the transmission line.

reflectometer See microwave reflectometer.

reflector 1. A single rod, system of rods, metal screen, or metal sheet used behind an antenna to increase its directivity. 2. A metal sheet or screen used as a mirror to change the direction of a microwave radio beam. 3. See repeller.

reflector characteristic A chart of power output and frequency deviation of a reflex klystron as a function of reflector voltage.

reflector voltage Voltage between the reflector electrode and the cathode in a reflex klystron.

reflex bunching The bunching that occurs in an electron stream which has been made to reverse its direction in the drift space.

reflex circuit A circuit in which the signal is amplified twice by the same amplifier tube or tubes, once as an intermediate-frequency signal before detection and once as an audio-frequency signal after detection.

reflex klystron A single-cavity klystron in which the electron beam is reflected back through the cavity resonator by a repelling electrode having a negative voltage; used as a microwave oscillator. Also known as reflex oscillator.

reflex oscillator See reflex klystron.

refraction loss Portion of the transmission loss that is due to refraction resulting from nonuniformity of the medium.

refractivity 1. Some quantitative measure of refraction, usually a measure of the index of refraction. 2. The index of refraction minus 1.

regenerate 1. To restore pulses to their original shape. 2. To restore stored information to its original form in a storage tube in order to counteract fading and disturbances.

regeneration Replacement or restoration of charges in a charge storage tube to overcome decay effects, including loss of charge by reading.

regenerative amplifier An amplifier that uses positive feedback to give increased gain and selectivity.

regenerative braking A system of dynamic braking in which the electric drive motors are used as generators and return the kinetic energy of the motor armature and load to the electric supply system.

regenerative clipper A type of monostable multivibrator which is a modification of a Schmitt trigger; used for pulse generation.

regenerative detector A vacuum-tube detector circuit in which radio-frequency energy is fed back from the anode circuit to the grid circuit to give positive feedback at the carrier frequency, thereby increasing the amplification and sensitivity of the circuit.

regenerative divider Frequency divider which employs modulation, amplification, and selective feedback to produce the output wave.

regenerative fuel cell A fuel cell in which the reaction product is processed to regenerate the reactants.

regenerative receiver A radio receiver that uses a regenerative detector.

register circuit A switching circuit with memory elements that can store from a few to millions of bits of coded information; when needed, the information can be taken from the circuit in the same code as the input, or in a different code.

regular In a definite direction; not diffused or scattered, when applied to reflection, refraction, or transmission.

regulated power supply A power supply containing means for maintaining essentially constant output voltage or output current under changing load conditions.

regulating transformer Transformer having one or more windings excited from the system circuit or a separate source and one or more windings connected in series with the system circuit for adjusting the voltage or the phase relation or both in steps, usually without interrupting the load.

regulating winding Of a transformer, a supplementary winding connected in series with one of the main windings to change the ratio of transformation or the phase relation, or both, between circuits.

regulation 1. The change in output voltage that occurs between no load and full load in a transformer, generator, or other source. 2. The difference between the maximum and minimum tube voltage drops within a specified range of anode current in a gas tube.

regulation of constant-current transformer Maximum departure of the secondary current from its rated value expressed in percent of the rated secondary current, with rated primary voltage and frequency applied.

Reinartz crystal oscillator Crystal-controlled vacuum-tube oscillator in which the crystal current is kept low by placing a resonant circuit in the cathode lead tuned to half the crystal frequency; the resulting regeneration at the crystal frequency improves efficiency without the danger of uncontrollable oscillation at other frequencies.

reinserter *See* direct-current restorer.

reinsertion of carrier Combining a locally generated carrier signal in a receiver with an incoming signal of the suppressed carrier type.

rejection band Also known as stop band. The band of frequencies below the cutoff frequency in a uniconductor waveguide.

rejector *See* trap.

rel Unit of reluctance equal to 1 ampere-turn per magnetic line of force.

relative attenuation The ratio of the peak output voltage of an electric filter to the voltage at the frequency being considered.

relative bandwidth For an electric filter, the ratio of the bandwidth being considered to a specified reference bandwidth, such as the bandwidth between frequencies at which there is an attenuation of 3 decibels.

relative gain of an antenna Gain of an antenna in a given direction when the reference antenna is a half-wave, loss-free dipole isolated in space whose equatorial plane contains the given direction.

relative power gain Of one transmitting or receiving antenna over another, the measured ratio of the signal power one produces at the receiver input terminals to that produced by the other, the transmitting power level remaining fixed.

relative resistance The ratio of the resistance of a piece of a material to the resistance of a piece of specified material, such as annealed copper, having the same dimensions and temperature.

relative response In a transducer, the amount (in decibels) by which the response under some particular condition exceeds the response under a reference condition.

relativistic electrodynamics The study of the interaction between charged particles and electric and magnetic fields when the velocities of the particles are comparable with that of light.

relaxation circuit Circuit arrangement, usually of vacuum tubes, reactances, and resistances, which has two states or conditions, one, both, or neither of which may be stable; the transient voltage produced by passing from one to the other, or the voltage in a state of rest, can be used in other circuits.

relaxation inverter An inverter that uses a relaxation oscillator circuit to convert direct-current power to alternating-current.

relaxation oscillator An oscillator whose fundamental frequency is determined by the time of charging or discharging a capacitor or coil through a resistor, producing waveforms that may be rectangular or sawtooth.

relaxation time The travel time of an electron in a metal before it is scattered and loses its momentum.

relay A device that is operated by a variation in the conditions in one electric circuit and serves to make or break one or more connections in the same or another electric circuit. Also known as electric relay.

relay contact One of the pair of contacts that are closed or opened by the movement of the armature of a relay.

relay selector Relay circuit associated with a selector, consisting of a magnetic impulse counter, for registering digits and holding a circuit.

relay system Dial switching equipment that does not use mechanical switches, but is made up principally of relays.

relieving anode Of a pool-cathode tube, an auxiliary anode which provides an alternative conducting path for reducing the current to another electrode.

reluctance A measure of the opposition presented to magnetic flux in a magnetic circuit, analogous to resistance in an electric circuit; it is equal to magnetomotive force divided by magnetic flux. Also known as magnetic reluctance.

reluctance motor A synchronous motor, similar in construction to an induction motor, in which the member carrying the secondary circuit has salient poles but no direct-current excitation; it starts as an induction motor but operates normally at synchronous speed.

remanence The magnetic flux density that remains in a magnetic circuit after the removal of an applied magnetomotive force; if the magnetic circuit has an air gap, the remanence will be less than the residual flux density.

remember condition Condition of a flip-flop circuit in which no change takes place between a given internal state and the next state.

remodulator A circuit that converts amplitude modulation to audio frequency-shift modulation for transmission of facsimile signals over a voice-frequency radio channel. Also known as converter.

remote-cutoff tube *See* variable-mu tube.

remote indicator 1. An indicator located at a distance from the data-gathering sensing element, with data being transmitted to the indicator mechanically, electrically over wires, or by means of light, radio, or sound waves. 2. *See* repeater.

remote plan position indicator *See* plan position indicator repeater.

remote sensing Sensing, by a power supply, of voltage directly at the load, so that variations in the load lead drop do not affect load regulation.

repeater 1. An amplifier or other device that receives weak signals and delivers corresponding stronger signals with or without reshaping of waveforms; may be either a one-way or two-way repeater. 2. An indicator that shows the same information as is shown on a master indicator. Also known as remote indicator. 3. *See* repeating coil.

repeater jammer A jammer that intercepts an enemy radar signal and reradiates the signal after modifying it to incorporate erroneous data on azimuth, range, or number of targets.

repeating coil A transformer used to provide inductive coupling between two sections of a telephone line when a direct connection is undesirable. Also known as repeater.

repeating-coil bridge cord In telephony, a method of connecting the common office battery to the cord circuits by connecting the battery to the midpoints of a repeating coil, bridged across the cord circuit.

repeller An electrode whose primary function is to reverse the direction of an electron stream in an electron tube. Also known as reflector.

reproduce head *See* playback head.

repulsion-induction motor A repulsion motor that has a squirrel-cage winding in the rotor in addition to the repulsion-motor winding.

repulsion motor An alternating-current motor having stator windings connected directly to the source of ac power and rotor windings connected to a commutator; brushes on the commutator are short-circuited and are positioned to produce the rotating magnetic field required for starting and running.

repulsion-start induction motor An alternating-current motor that starts as a repulsion motor; at a predetermined speed the commutator bars are short-circuited to give the equivalent of a squirrel-cage winding for operation as an induction motor with constant-speed characteristics.

rescap A capacitor and resistor assembly manufactured as a packaged encapsulated circuit. Also known as capacitor-resistor unit; capristor; packaged circuit; resistor-capacitor unit.

reserve battery A battery which is inert until an operation is performed which brings all the cell components into the proper state and location to become active.

reset condition Condition of a flip-flop circuit in which the internal state of the flip-flop is reset to zero.

reset pulse 1. A drive pulse that tends to reset a magnetic cell in the storage section of a digital computer. 2. A pulse used to reset an electronic counter to zero or to some predetermined position.

resettability The ability of the tuning element of an oscillator to retune the oscillator to the same operating frequency for the same set of input conditions.

residual charge The charge remaining on the plates of a capacitor after initial discharge.

residual current Current flowing through a thermionic diode when there is no anode voltage, due to the velocity of the electrons emitted by the heated cathode.

residual field The magnetic field left in an iron core after excitation has been removed.

residual flux density The magnetic flux density at which the magnetizing force is zero when the material is in a symmetrically and cyclically magnetized condition. Also known as residual induction; residual magnetic induction; residual magnetism.

residual resistance The value to which the electrical resistance of a metal drops as the temperature is lowered to near absolute zero, caused by imperfections and impurities in the metal rather than by lattice vibrations.

residual voltage Vector sum of the voltages to ground of the several phase wires of an electric supply circuit.

resistance 1. The opposition that a device or material offers to the flow of direct current, equal to the voltage drop across the element divided by the current through the element. Also known as electrical resistance. 2. In an alternating-current circuit, the real part of the complex impedance.

resistance box A box containing a number of precision resistors connected to panel terminals or contacts so that a desired resistance value can be obtained by withdrawing plugs (as in a post-office bridge) or by setting multicontact switches.

resistance bridge *See* Wheatstone bridge.

resistance-capacitance circuit A circuit which has a resistance and a capacitance in series, and in which inductance is negligible. Abbreviated *R-C* circuit.

resistance-capacitance constant Time constant of a resistive-capacitive circuit; equal in seconds to the resistance value in ohms multiplied by the capacitance value in farads. Abbreviated *R-C* constant.

resistance-capacitance coupled amplifier An amplifier in which a capacitor provides a path for signal currents from one stage to the next, with resistors connected from each side of the capacitor to the power supply or to ground; it can amplify alternating-current signals but cannot handle small changes in direct currents. Also known as *R-C* amplifier; *R-C* coupled amplifier; resistance-coupled amplifier.

resistance-capacitance coupling *See* resistance coupling.

resistance-capacitance network Circuit containing resistances and capacitances arranged in a particular manner to perform a specific function. Abbreviated *R-C* network.

resistance-capacitance oscillator Oscillator in which the frequency is determined by resistance and capacitance elements. Abbreviated *R-C* oscillator.

resistance commutation Commutation of an electric rotating machine in which brushes with relatively high resistance span at least one commutator segment, in order to achieve a linear variation of current with time, and thereby minimize self-inductive voltage in the coils.

resistance-coupled amplifier *See* resistance-capacitance coupled amplifier.

resistance coupling Coupling in which resistors are used as the input and output impedances of the circuits being coupled; a coupling capacitor is generally used

370 resistance drop

between the resistors to transfer the signal from one stage to the next. Also known as *R-C* coupling; resistance-capacitance coupling; resistive coupling.

resistance drop The voltage drop occurring between two points on a conductor due to the flow of current through the resistance of the conductor; multiplying the resistance in ohms by the current in amperes gives the voltage drop in volts. Also known as *IR* drop.

resistance element An element of resistive material in the form of a grid, ribbon, or wire, used singly or built into groups to form a resistor for heating purposes, as in an electric soldering iron.

resistance grounding Electrical grounding in which lines are connected to ground by a resistive (totally dissipative) impedance.

resistance heating The generation of heat by electric conductors carrying current; degree of heating is proportional to the electrical resistance of the conductor; used in electrical home appliances, home or space heating, and heating ovens and furnaces.

resistance lamp Electric lamp used to prevent the current in a circuit from exceeding a desired limit.

resistance loss Power loss due to current flowing through resistance; its value in watts is equal to the resistance in ohms multiplied by the square of the current in amperes.

resistance material Material having sufficiently high resistance per unit length or volume to permit its use in the construction of resistors.

resistance measurement The quantitative determination of that property of an electrically conductive material, component, or circuit called electrical resistance.

resistance noise *See* thermal noise.

resistance-start motor A split-phase motor having a resistance connected in series with the auxiliary winding; the auxiliary circuit is opened when the motor attains a predetermined speed.

resistance strain gage A strain gage consisting of a strip of material that is cemented to the part under test and that changes in resistance with elongation or compression.

resistive coupling *See* resistance coupling.

resistive load A load whose total reactance is zero, so that the alternating current is in phase with the terminal voltage. Also known as nonreactive load.

resistive unbalance Unequal resistance in the two wires of a transmission line.

resistivity *See* electrical resistivity.

resistor A device designed to have a definite amount of resistance; used in circuits to limit current flow or to provide a voltage drop. Also known as electrical resistor.

resistor-capacitor-transistor logic A resistor-transistor logic with the addition of capacitors that are used to enhance switching speed.

resistor color code Code adopted by the Electronic Industries Association to mark the values of resistance on resistors in a readily recognizable manner; the first color represents the first significant figure of the resistor value, the second color the second significant figure, and the third color represents the number of zeros following the first two figures; a fourth color is sometimes added to indicate the tolerance of the resistor.

resistor core Insulating support on which a resistor element is wound or otherwise placed.

resistor element That portion of a resistor which possesses the property of electric resistance.

resonant chamber 371

resistor network An electrical network consisting entirely of resistances.

resistor termination A thick-film conductor pad overlapping and contacting a thick-film resistor area.

resistor-transistor logic One of the simplest logic circuits, having several resistors, a transistor, and a diode. Abbreviated RTL.

resnatron A microwave-beam tetrode containing cavity resonators, used chiefly for generating large amounts of continuous power at high frequencies.

resolution 1. In television, the maximum number of lines that can be discerned on the screen at a distance equal to tube height; this ranges from 350 to 400 for most receivers. 2. In radar, the minimum separation between two targets, in angle or range, at which they can be distinguished on a radar screen. Also known as resolving power.

resolver 1. A synchro or other device whose rotor is mechanically driven to translate rotor angle into electrical information corresponding to the sine and cosine of rotor angle; used for interchanging rectangular and polar coordinates. Also known as sine-cosine generator; synchro resolver. 2. A synchro or other device whose input is the angular position of an object, such as the rotor of an electric machine, and whose output is electric signals, usually proportional to the sine and cosine of an angle, and often in digital form; used to interchange rectangular and polar coordinates, and in servomechanisms to report the orientation of controlled objects. Also known as angular resolver. 3. A device that accepts a single vector-valued analog input and produces for output either analog or digital signals proportional to two or three orthogonal components of the vector. Also known as vector resolver.

resolving cell In radar, volume in space whose diameter is the product of slant range and beam width, and whose length is the pulse length.

resolving power 1. The reciprocal of the beam width of a unidirectional antenna, measured in degrees. 2. See resolution.

resonance A phenomenon exhibited by an alternating-current circuit in which there are relatively large currents near certain frequencies, and a relatively unimpeded oscillation of energy from a potential to a kinetic form; a special case of the physics definition.

resonance bridge A four-arm alternating-current bridge used to measure inductance, capacitance, or frequency; the inductor and the capacitor, which may be either in series or in parallel, are tuned to resonance at the frequency of the source before the bridge is balanced.

resonance curve Graphical representation illustrating the manner in which a tuned circuit responds to the various frequencies in and near the resonant frequency.

resonance method A method of determining the impedance of a circuit element, in which resonance frequency of a resonant circuit containing the element is measured.

resonance transformer 1. A high-voltage transformer in which the secondary circuit is tuned to the frequency of the power supply. 2. An electrostatic particle accelerator, used principally for acceleration of electrons, in which the high-voltage terminal oscillates between voltages which are equal in magnitude and opposite in sign.

resonant antenna An antenna for which there is a sharp peak in the power radiated or intercepted by the antenna at a certain frequency, at which electric currents in the antenna form a standing-wave pattern.

resonant capacitor A tubular capacitor that is wound to have inductance in series with its capacitance.

resonant cavity See cavity resonator.

resonant chamber See cavity resonator.

resonant-chamber switch Waveguide switch in which a tuned cavity in each waveguide branch serves the functions of switch contacts; detuning of a cavity blocks the flow of energy in the associated waveguide.

resonant circuit A circuit that contains inductance, capacitance, and resistance of such values as to give resonance at an operating frequency.

resonant coupling Coupling between two circuits that reaches a sharp peak at a certain frequency.

resonant diaphragm Diaphragm, in waveguide technique, so proportioned as to introduce no reactive impedance at the design frequency.

resonant element *See* cavity resonator.

resonant gate transistor Surface field-effect transistor incorporating a cantilevered beam which resonates at a specific frequency to provide high-Q-frequency discrimination.

resonant helix An inner helical conductor in certain types of transmission lines and resonant cavities, which carries currents with the same frequency as the rest of the line or cavity.

resonant iris A resonant window in a circular waveguide; it resembles an optical iris.

resonant line A transmission line having values of distributed inductance and distributed capacitance so as to make the line resonant at the frequency it is handling.

resonant line oscillator Oscillator in which one or more sections of transmission lines are employed as resonant elements.

resonant-line tuner A television tuner in which resonant lines are used to tune the antenna, radio-frequency amplifier, and radio-frequency oscillator circuits; tuning is achieved by moving shorting contacts that change the electrical lengths of the lines.

resonant-reed relay A reed relay in which the reed switch closes only when the required frequency is applied to the operating coil, to make one of the reeds vibrate until its amplitude is sufficient to make contact with the other reed; used in selective paging systems.

resonant resistance Resistance value to which a resonant circuit is equivalent.

resonant voltage step-up Ability of an inductor and a capacitor in a series resonant circuit to deliver a voltage several times greater than the input voltage of the circuit.

resonant wavelength The wavelength in free space of electromagnetic radiation having a frequency equal to a natural resonance frequency of a cavity resonator.

resonant window A parallel combination of inductive and capacitive diaphragms, used in a waveguide structure to provide transmission at the resonant frequency and reflection at other frequencies.

resonate To bring to resonance, as by tuning.

resonating cavity Short piece of waveguide of adjustable length, terminated at either or both ends by a metal piston, an iris diaphragm, or some other wave-reflecting device; it is used as a filter, as a means of coupling between guides of different diameters, and as impedance networks corresponding to those used in radio circuits.

resonator grid Grid that is attached to a cavity resonator in velocity-modulated tubes to provide coupling between the resonator and the electron beam.

resonator wavemeter Any resonant circuit used to determine wavelength, such as a cavity-resonator frequency meter.

responder The transmitter section of a radar beacon.

responder beacon The radar beacon that serves to emit the signals of the responder in a transponder.

response time The time it takes for the pointer of an electrical or electronic instrument to come to rest at a new value, after the quantity it measures has been abruptly changed.

responsor The receiving section of an interrogator-responsor.

restore Periodic charge regeneration of volatile computer storage systems.

restorer pulses In computers, pairs of complement pulses, applied to restore the coupling-capacitor charge in an alternating-current flip-flop.

rest potential Residual potential difference remaining between an electrode and an electrolyte after the electrode has become polarized.

retardation coil A high-inductance coil used in telephone circuits to permit passage of direct current or low-frequency ringing current while blocking the flow of audio-frequency currents.

retarded field An electric or magnetic field strength as found from the retarded potentials.

retarded potentials The electromagnetic potentials at an instant in time t and a point in space r as a function of the charges and currents that existed at earlier times at points on the past light cone of the event r,t.

retarding-field oscillator An oscillator employing an electron tube in which the electrons oscillate back and forth through a grid that is maintained positive with respect to both the cathode and anode; the field in the region of the grid exerts a retarding effect through the grid in either direction. Also known as positive-grid oscillator.

retard transmitter Transmitter in which a delay period is introduced between the time of actuation and the time of transmission.

retention time The maximum time between writing into a storage tube and obtaining an acceptable output by reading. Also known as storage time.

retentivity The residual flux density corresponding to the saturation induction of a magnetic material.

Retgers' law The law that the properties of crystalline mixtures of isomorphous substances are continuous functions of the percentage composition.

retrace See flyback.

retrace blanking Blanking a television picture tube during vertical retrace intervals to prevent retrace lines from showing on the screen.

retrace line The line traced by the electron beam in a cathode-ray tube in going from the end of one line or field to the start of the next line or field. Also known as return line.

retransmission unit Control unit used at an intermediate station for feeding one radio receiver-transmitter unit for two-way communication.

return See echo.

return interval Interval corresponding to the direction of sweep not used for delineation.

return trace See flyback.

return wire The ground wire, common wire, or negative wire of a direct-current power circuit.

reverse bias A bias voltage applied to a diode or a semiconductor junction with polarity such that little or no current flows; the opposite of forward bias.

reverse-blocking tetrode thyristor See silicon controlled switch.

reverse-blocking triode thyristor *See* silicon controlled rectifier.

reverse current Small value of direct current that flows when a semiconductor diode has reverse bias.

reverse-current protection A device which senses when there is a reversal in the normal direction of current in an electric power system, indicating an abnormal condition of the system, and which initiates appropriate action to prevent damage to the system.

reverse-current relay Relay that operates whenever current flows in the reverse direction.

reverse direction *See* inverse direction.

reverse key Key used in a circuit to reverse the polarity of that circuit.

reverse voltage In the case of two opposing voltages, voltage of that polarity which produces the smaller current.

reversible booster Booster capable of adding to and subtracting from the voltage of a circuit.

reversible capacitance Limit, as the amplitude of an applied sinusoidal capacitor voltage approaches zero, of the ratio of the amplitude of the resulting in-phase fundamental-frequency component of transferred charge to the amplitude of the applied voltage, for a given constant bias voltage superimposed on the sinusoidal voltage.

reversible motor A motor in which the direction of rotation can be reversed by means of a switch that changes motor connections when the motor is stopped.

reversible transducer Transducer whose loss is independent of transmission direction.

reversing motor A motor for which the direction of rotation can be reversed by changing electric connections or by other means while the motor is running at full speed; the motor will then come to a stop, reverse, and attain full speed in the opposite direction.

reversing switch A switch intended to reverse the connections of one part of a circuit.

rewind 1. The components on a magnetic tape recorder that serve to return the tape to the supply reel at high speed. 2. To return a magnetic tape to its starting position.

rf *See* radio frequency.

RG line *See* radio-frequency cable.

rheostat A resistor constructed so that its resistance value may be changed without interrupting the circuit to which it is connected. Also known as variable resistor.

rheostatic control A method of controlling the speed of electric motors that involves varying the resistance or reactance in the armature or field circuit; used in motors that drive elevators.

rheostriction *See* pinch effect.

RHI display *See* range-height indicator display.

rhombic antenna A horizontal antenna having four conductors forming a diamond or rhombus; usually fed at one apex and terminated with a resistance or impedance at the opposite apex. Also known as diamond antenna.

rhumbatron *See* cavity resonator.

ribbon cable A cable made of normal, round, insulated wires arranged side by side and fastened together by a cohesion process to form a flexible ribbon.

Riblet coupler *See* three-decibel coupler.

ringing circuit 375

rice neutralization Development of voltage in the grid circuit of a vacuum tube in order to nullify or cancel feedback through the tube.

rice neutralizing circuit Radio-frequency amplifier circuit that neutralizes the grid-to-plate capacitance of an amplifier tube.

Richardson-Dushman equation An equation for the current density of electrons that leave a heated conductor in thermionic emission. Also known as Dushman equation.

Richardson effect *See* thermionic emission.

Richardson plot A graph of log (J/T^2) against $1/T$, where J is the current density of electrons leaving a heated conductor in thermionic emission, and T is the temperature of the conductor; according to the Richardson-Dushman equation, this is a straight line.

ridge waveguide A circular or rectangular waveguide having one or more longitudinal internal ridges that serve primarily to increase transmission bandwidth by lowering the cutoff frequency.

Rieke diagram A chart showing contours of constant power output and constant frequency for a microwave oscillator, drawn on a Smith chart or other polar diagram whose coordinates represent the components of the complex reflection coefficient at the oscillator load.

right-hand polarization In elementary particle discussions, circular or elliptical polarization of an electromagnetic wave in which the electric field vector at a fixed point in space rotates in the right-hand sense about the direction of propagation; in optics, the opposite convention is used; in facing the source of the beam, the electric vector is observed to rotate clockwise.

right-hand rule 1. For a current-carrying wire, the rule that if the fingers of the right hand are placed around the wire so that the thumb points in the direction of current flow, the fingers will be pointing in the direction of the magnetic field produced by the wire. 2. For a moving wire in a magnetic field, such as the wire on the armature of a generator, if the thumb, first, and second fingers of the right hand are extended at right angles to one another, with the first finger representing the direction of magnetic lines of force and the second finger representing the direction of current flow induced by the wire's motion, the thumb will be pointing in the direction of motion of the wire. Also known as Fleming's rule.

right-hand taper Taper in which there is greater resistance in the clockwise half of the operating range of a rheostat or potentiometer (looking from the shaft end) than in the counterclockwise half.

rigid copper coaxial line A coaxial cable in which the central conductor and outer conductor are formed by joining rigid pieces of copper.

rigid insulation Electrical insulation that is part of a rigid structure, and must provide mechanical strength and stability of form as well as a dielectric barrier; mica, glass, porcelain, and thermosetting resins are the principal materials used.

R indicator *See* R scope.

ring bus A substation switching arrangement that may consist of four, six, or more breakers connected in a closed loop, with the same number of connection points.

ring circuit In waveguide practice, a hybrid T junction having the physical configuration of a ring with radial branches.

ring counter A loop of binary scalers or other bistable units so connected that only one scaler is in a specified state at any given time; as input signals are counted, the position of the one specified state moves in an ordered sequence around the loop.

ringing circuit A circuit which has a capacitance in parallel with a resistance and inductance, with the whole in parallel with a second resistance; it is highly underdamped and is supplied with a step or pulse input.

ring modulator A modulator in which four diode elements are connected in series to form a ring around which current flows readily in one direction; input and output connections are made to the four nodal points of the ring; used as a balanced modulator, demodulator, or phase detector.

ring power transmission line A power transmission line that is closed upon itself to form a ring; provides two paths between the power station and any customer, and enables a faulty section of the line to be disconnected without interrupting service to customers.

ring time The length of time in microseconds required for a pulse of energy transmitted into an echo box to die out; a measurement of the performance of the radar.

ripple The alternating-current component in the output of a direct-current power supply, arising within the power supply from incomplete filtering or from commutator action in a dc generator.

ripple filter A low-pass filter designed to reduce ripple while freely passing the direct current obtained from a rectifier or direct-current generator. Also known as smoothing circuit; smoothing filter.

ripple voltage The alternating component of the unidirectional voltage from a rectifier or generator used as a source of direct-current power.

rise time The time for the pointer of an electrical instrument to make 90% of the change to its final value when electric power suddenly is applied from a source whose impedance is high enough that it does not affect damping.

rising-sun magnetron A multicavity magnetron in which resonators having two different resonant frequencies are arranged alternately for the purpose of mode separation; the cavities appear as alternating long and short radial slots around the perimeter of the anode structure, resembling the rays of the sun.

roc A unit of electrical conductivity equal to the conductivity of a material in which an electric field of 1 volt per centimeter gives rise to a current density of 1 ampere per square centimeter. Derived from reciprocal ohm centimeter.

rocket antenna An antenna carried on a rocket, to receive signals controlling the rocket or to transmit measurements made by instruments aboard the rocket.

rocky point effect Transient but violent discharges between electrodes in high-voltage transmitting tubes.

rod gap 1. A device that is usually formed of two ½-square-inch (3-square-centimeter) rods, one grounded and the other connected to the line conductor, but may also have the shape of rings or horns, used to limit the magnitude of transient overvoltages on an electrical system as a result of lightning strikes. 2. Spark gap in which the electrodes are two coaxial rods, with ends between which the discharge takes place, cut perpendicularly to the axis.

rod thermistor A type of thermistor that has high resistance, long time constant, and moderate power dissipation; it is extruded as a long vertical rod 0.250–2.0 inches (0.63–5.1 centimeters) long and 0.050–0.110 inch (0.13–0.28 centimeter) in diameter, of oxide-binder mix and sintered; ends are coated with conducting paste and leads are wrapped on the coated area.

roentgen optics *See* x-ray optics.

rolling transposition Transposition in which the conductors of an open wire circuit are physically rotated in a substantially helical manner; with two wires, a complete transposition is usually executed in two consecutive spans.

roll-off Gradually increasing loss or attenuation with increase or decrease of frequency beyond the substantially flat portion of the amplitude-frequency response characteristic of a system or transducer.

rom A unit of electrical conductivity, equal to the conductivity of a material in which an electric field of 1 volt per meter gives rise to a current density of 1 ampere per square meter. Derived from reciprocal ohm meter.

roof filter Low-pass filter used in carrier telephone systems to limit the frequency response of the equipment to frequencies needed for normal transmission, thereby blocking unwanted higher frequencies induced in the circuit by external sources; improves runaround cross-talk suppression and minimizes high-frequency singing.

root-mean-square current *See* effective current.

rope-lay conductor Cable composed of a central core surrounded by one or more layers of helically laid groups of wires.

Rosa and Dorsey method A method of measuring the speed of light by comparing the capacitance of a capacitor in electromagnetic units, as measured experimentally, with values of currents determined from a current balance, to the capacitance of the same capacitor in electrostatic units, as determined from its geometrical dimensions.

Rosenberg crossed-field generator A type of dynamoelectric amplifier which is self-regulating and can operate while the rotor varies in speed, the current never rising above a certain value.

rosin joint A soldered joint in which one of the wires is surrounded by an almost invisible film of insulating rosin, making the joint intermittently or continuously open even though it looks good.

rotary beam Short-wave antenna system highly directional in azimuth and altitude, mounted in such a manner that it can be rotated to any desired position, either manually or by an electric motor drive.

rotary converter *See* dynamotor.

rotary coupler *See* rotating joint.

rotary gap *See* rotary spark gap.

rotary joint *See* rotating joint.

rotary phase converter Machine which converts power from an alternating-current system of one or more phases to an alternating-current system of a different number of phases, but of the same frequency.

rotary power source An uninterruptible power system in which a battery driven dc motor mechanically drives an ac generator in the event of a power outage.

rotary solenoid A solenoid in which the armature is rotated when actuated; the rotary stroke, ranging from 25 to 95°, is usually converted to linear motion to give a longer stroke than is possible with a conventional plunger-type solenoid.

rotary spark gap A spark gap in which sparks occur between one or more fixed electrodes and a number of electrodes projecting outward from the circumference of a motor-driven metal disk. Also known as rotary gap.

rotary stepping relay *See* stepping relay.

rotary stepping switch *See* stepping relay.

rotary switch A switch that is operated by rotating its shaft.

rotary transformer A rotating machine used to transform direct-current power from one voltage to another.

rotary-vane attenuator Device designed to introduce attenuation into a waveguide circuit by varying the angular position of a resistive material in the guide.

rotating-anode tube An x-ray tube in which the anode rotates continuously to bring a fresh area of its surface into the beam of electrons, allowing greater output without melting the target.

rotating crystal method Any method of studying crystalline structures by x-ray or neutron diffraction in which a monochromatic, collimated beam of x-rays or neutrons falls on a single crystal that is rotated about an axis perpendicular to the beam.

rotating joint A joint that permits one section of a transmission line or waveguide to rotate continuously with respect to another while passing radio-frequency energy. Also known as rotary coupler; rotary joint.

rotating magnetic amplifier A prime-mover-driven direct-current generator whose power output can be controlled by small field input powers, to give power gain as high as 10,000. Also known as rotary amplifier; rotating amplifier.

rotation camera An instrument for studying crystalline structure by x-ray or neutron diffraction, in which a monochromatic, collimated beam of x-rays or neutrons falls on a single crystal which is rotated about an axis perpendicular to the beam and parallel to one of the crystal axes, and the various diffracted beams are registered on a cylindrical film concentric with the axis of rotation.

rotator A device that rotates the plane of polarization of a plane-polarized electromagnetic wave, such as a twist in a waveguide.

rotoflector In radar, elliptically shaped, rotating reflector used to reflect a vertically directed radar beam at right angles so that it radiates in a horizontal direction.

rotor The rotating member of an electrical machine or device, such as the rotating armature of a motor or generator, or the rotating plates of a variable capacitor.

rotor plate One of the rotating plates of a variable capacitor, usually directly connected to the metal frame.

round trip echoes Multiple reflection echoes produced when a radar pulse is reflected from a target strongly enough so that the echo is reflected back to the target where it produces a second echo.

Rowland ring A ring-shaped sample of magnetic material, generally surrounded by a coil of wire carrying a current.

R scan *See* R scope.

R scope An A scope presentation with a segment of the horizontal trace expanded near the target spot (pip) for greater accuracy in range measurement. Also known as R indicator; R scan.

ruggedization Making electronic equipment and components resistant to severe shock, temperature changes, high humidity, or other detrimental environmental influences.

runaway effect The phenomenon whereby an increase in temperature causes an increase in a collector-terminal current in a transistor, which in turn results in a higher temperature and, ultimately, failure of the transistor; the effect limits the power output of the transistor.

runaway electron An electron, in an ionized gas to which an electric field is applied, that gains energy from the field faster than it loses energy by colliding with other particles in the gas.

run motor In facsimile equipment, a motor which supplies the power to drive the scanning or recording mechanisms; a synchronous motor is used to limit the speed.

S

S *See* secondary winding; siemens.

safety factor The amount of load, above the normal operating rating, that a device can handle without failure.

sag Slack introduced in an aerial cable or openwire line to compensate for contraction during cold weather.

Saint Elmo's fire A visible electric discharge, sometimes seen on the mast of a ship, on metal towers, and on projecting parts of aircraft, due to concentration of the atmospheric electric field at such projecting parts.

salammoniac cell Cell in which the electrolyte consists primarily of a solution of ammonium chloride.

salient pole A structure of magnetic material on which is mounted a field coil of a generator, motor, or similar device.

salient-pole field winding A type of field winding in electric machinery where the winding turns are concentrated around the pole core.

Salisbury dark box Isolating chamber used for test work in connection with radar equipment; the walls of the chamber are specially constructed to absorb all impinging microwave energy at a certain frequency.

samarium-cobalt magnet A rare-earth permanent magnet that is more efficient, has lower leakage and greater resistance to demagnetization, and can be magnetized to higher levels than conventional permanent magnets.

sample-and-hold circuit A circuit that measures an input signal at a series of definite points in time, and whose output remains constant at a value corresponding to the most recent measurement until the next measurement is made.

sampling gate A gate circuit that extracts information from the input waveform only when activated by a selector pulse.

sampling switch *See* commutator switch.

sanatron circuit A variable time-delay circuit having two pentodes and two diodes, used to produce very short gate waveforms having time durations that vary linearly with a reference voltage.

sand load An attenuator used as a power-dissipating terminating section for a coaxial line or waveguide; the dielectric space in the line is filled with a mixture of sand and graphite that acts as a matched-impedance load, preventing standing waves.

SANTA *See* systematic analog network testing approach.

satellite antenna Antenna on an artificial satellite to receive command signals from earth, act as a beacon for tracking, or transmit scientific or other data to earth.

saturable-core reactor *See* saturable reactor.

saturable reactor An iron-core reactor having an additional control winding that carries direct current whose value is adjusted to change the degree of saturation of the core, thereby changing the reactance that the alternating-current winding offers to the flow of alternating current; with appropriate external circuits, a saturable reactor can serve as a magnetic amplifier. Also known as saturable-core reactor; transductor.

saturable transformer A saturable reactor having additional windings to provide voltage transformation or isolation from the alternating-current supply.

saturated diode A diode that is passing the maximum possible current, so further increases in applied voltage have no effect on current.

saturating signal In radar, a signal of an amplitude greater than the dynamic range of the receiving system.

saturation 1. The condition that occurs when a transistor is driven so that it becomes biased in the forward direction (the collector becomes positive with respect to the base, for example, in a *pnp* type of transistor). 2. *See* anode saturation; magnetic saturation; temperature saturation.

saturation current 1. In general, the maximum current which can be obtained under certain conditions. 2. In a vacuum tube, the space-charge-limited current, such that further increase in filament temperature produces no specific increase in anode current. 3. In a vacuum tube, the temperature-limited current, such that a further increase in anode-cathode potential difference produces only a relatively small increase in current. 4. In a gaseous-discharge device, the maximum current which can be obtained for a given mode of discharge. 5. In a semiconductor, the maximum current which just precedes a change in conduction mode.

saturation flux density *See* saturation induction.

saturation induction The maximum intrinsic induction possible in a material. Also known as saturation flux density.

saturation limiting Limiting the minimum output voltage of a vacuum-tube circuit by operating the tube in the region of plate-current saturation (not to be confused with emission saturation).

saturation magnetization The maximum possible magnetization of a material.

saturation signal A radio signal (or radar echo) which exceeds a certain power level fixed by the design of the receiver equipment; when a receiver or indicator is "saturated," the limit of its power output has been reached.

sawtooth generator A generator whose output voltage has a sawtooth waveform; used to produce sweep voltages for cathode-ray tubes.

sawtooth modulated jamming Electronic countermeasure technique when a high level jamming signal is transmitted, thus causing large automatic gain control voltages to be developed at the radar receiver that, in turn, cause target pip and receiver noise to completely disappear.

sawtooth pulse An electric pulse having a linear rise and a virtually instantaneous fall, or conversely, a virtually instantaneous rise and a linear fall.

sawtooth waveform A waveform characterized by a slow rise time and a sharp fall, resembling a tooth of a saw.

saxophone Vertex-fed linear array antenna giving a cosecant-squared radiation pattern.

S-band hiran *See* shiran.

scale-of-ten circuit *See* decade scaler.

scale-of-two circuit *See* binary scaler.

scaler A circuit that produces an output pulse when a prescribed number of input pulses is received. Also known as counter; scaling circuit.

scaling Counting pulses with a scaler when the pulses occur too fast for direct counting by conventional means.

scaling circuit *See* scaler.

scaling factor The number of input pulses per output pulse of a scaling circuit. Also known as scaling ratio.

scaling ratio *See* scaling factor.

scan The motion, usually periodic, given to the major lobe of an antenna; the process of directing the radio-frequency beam successively over all points in a given region of space.

scan converter 1. Equipment that converts radar data images at a 3 kilohertz to 10 kilohertz sampling rate that can be sent over telephone line or narrow bandwidth radio circuits and converted into a slow-scan image, through a similar converter, at the receiving end. 2. A cathode-ray tube that is capable of storing radar, television, and data displays for nondestructive readout over prolonged periods of time.

scanistor Integrated semiconductor optical-scanning device that converts images into electrical signals; the output analog signal represents both amount and position of light shining on its surface.

scanning electron microscope A type of electron microscope in which a beam of electrons, a few hundred angstroms in diameter, systematically sweeps over the specimen; the intensity of secondary electrons generated at the point of impact of the beam on the specimen is measured, and the resulting signal is fed into a cathode-ray-tube display which is scanned in synchronism with the scanning of the specimen.

scanning head Light source and phototube combined as a single unit for scanning a moving strip of paper, cloth, or metal in photoelectric side-register control systems.

scanning linearity In television, the uniformity of scanning speed during the trace interval.

scanning loss In a radar system employing a scanning antenna, the reduction in sensitivity (usually expressed in decibels) due to scanning across the target, compared with that obtained when the beam is directed constantly at the target.

scanning radio A radio receiver that automatically scans across public service, emergency service, or other radio bands and stops at the first preselected station which is on the air. Also known as radio scanner.

scanning spot *See* picture element.

scanning switch *See* commutator switch.

scanning transmission electron microscope A type of electron microscope which scans with an extremely narrow beam that is transmitted through the sample; the detection apparatus produces an image whose brightness depends on atomic number of the sample. Abbreviated STEM.

scanning yoke *See* deflection yoke.

scattering Diffusion of electromagnetic waves in a random manner by air masses in the upper atmosphere, permitting long-range reception, as in scatter propagation. Also known as radio scattering.

scattering coefficient One of the elements of the scattering matrix of a waveguide junction; that is, a transmission or reflection coefficient of the junction.

scattering cross section The power of electromagnetic radiation scattered by an antenna divided by the incident power.

scattering function The intensity of scattered radiation in a given direction per lumen of flux incident upon the scattering material.

scattering loss The portion of the transmission loss that is due to scattering within the medium or roughness of the reflecting surface.

scattering matrix A square array of complex numbers consisting of the transmission and reflection coefficients of a waveguide junction.

scatter propagation Transmission of radio waves far beyond line-of-sight distances by using high power and a large transmitting antenna to beam the signal upward into the atmosphere and by using a similar large receiving antenna to pick up the small portion of the signal that is scattered by the atmosphere. Also known as beyond-the-horizon communication; forward-scatter propagation; over-the-horizon propagation.

scatter reflections Reflections from portions of the ionosphere having different virtual heights, which mutually interfere and cause rapid fading.

schematic circuit diagram *See* circuit diagram.

Schering bridge A four-arm alternating-current bridge used to measure capacitance and dissipation factor; bridge balance is independent of frequency.

Schmitt circuit A bistable pulse generator in which an output pulse of constant amplitude exists only as long as the input voltage exceeds a certain value. Also known as Schmitt limiter; Schmitt trigger.

Schmitt limiter *See* Schmitt circuit.

Schmitt trigger *See* Schmitt circuit.

Schottky barrier A transition region formed within a semiconductor surface to serve as a rectifying barrier at a junction with a layer of metal.

Schottky barrier diode A semiconductor diode formed by contact between a semiconductor layer and a metal coating; it has a nonlinear rectifying characteristic; hot carriers (electrons for n-type material or holes for p-type material) are emitted from the Schottky barrier of the semiconductor and move to the metal coating that is the diode base; since majority carriers predominate, there is essentially no injection or storage of minority carriers to limit switching speeds. Also known as hot-carrier diode; Schottky diode.

Schottky defect 1. A defect in an ionic crystal in which a single ion is removed from its interior lattice site and relocated in a lattice site at the surface of the crystal. 2. A defect in an ionic crystal consisting of the smallest number of positive-ion vacancies and negative-ion vacancies which leave the crystal electrically neutral.

Schottky diode *See* Schottky barrier diode.

Schottky effect The enhancement of the thermionic emission of a conductor resulting from an electric field at the conductor surface.

Schottky line A graph of the logarithm of the saturation current from a thermionic cathode as a function of the square root of anode voltage; it is a straight line according to the Schottky theory.

Schottky noise *See* shot noise.

Schottky theory A theory describing the rectification properties of the junction between a semiconductor and a metal that result from formation of a depletion layer at the surface of contact.

Schottky transistor-transistor logic A transistor-transistor logic circuit in which a Schottky diode with forward diode voltage is placed across the base-collector junction of the output transistor in order to improve the speed of the circuit.

Schrage motor A type of alternating-current commutator motor whose speed is controlled by varying the position of sets of brushes on the commutator.

search coil 383

scintillation 1. A rapid apparent displacement of a target indication from its mean position on a radar display; one cause is shifting of the effective reflection point on the target. Also known as target glint; target scintillation; wander. 2. Random fluctuation, in radio propagation, of the received field about its mean value, the deviations usually being relatively small.

scoop *See* ellipsoidal floodlight.

scope *See* cathode-ray oscilloscope; radarscope.

scotoscope A telescope which employs an image intensifier to see in the dark.

Scott connection A type of transformer which transmits power from two-phase to three-phase systems, or vice versa.

Scott top Transformers arranged in the Scott connection for converting electrical power from two-phase to three-phase, or vice versa.

SCR *See* silicon controlled rectifier.

scrambler A circuit that divides speech frequencies into several ranges by means of filters, then inverts and displaces the frequencies in each range so that the resulting reproduced sounds are unintelligible; the process is reversed at the receiving apparatus to restore intelligible speech. Also known as speech inverter; speech scrambler.

screen 1. The surface on which a television, radar, x-ray, or cathode-ray oscilloscope image is made visible for viewing; it may be a fluorescent screen with a phosphor layer that converts the energy of an electron beam to visible light, or a translucent or opaque screen on which the optical image is projected. Also known as viewing screen. 2. Metal partition or shield which isolates a device from external magnetic or electric fields. 3. *See* screen grid.

screen angle Vertical angle bounded by a straight line from the radar antenna to the horizon and the horizontal at the antenna assuming a $\frac{4}{3}$ earth's radius.

screen dissipation Power dissipated in the form of heat on the screen grid as the result of bombardment by the electron stream.

screened trailing cable A flexible cable provided with a protective screen of conducting material, so applied as to enclose each power core separately or to enclose together all the cores of the cable.

screen grid A grid placed between a control grid and an anode of an electron tube, and usually maintained at a fixed positive potential, to reduce the electrostatic influence of the anode in the space between the screen grid and the cathode. Also known as screen.

screening *See* electric shielding.

scribing Cutting a grid pattern of deep grooves with a diamond-tipped tool in a slice of semiconductor material containing a number of devices, so that the slice can be easily broken into individual chips.

SCS *See* silicon controlled switch.

sea clutter A clutter on an airborne radar due to reflection of signals from the sea. Also known as sea return; wave clutter.

sealed tube Electron tube which is hermetically sealed.

sealing compound A compound used in dry batteries, capacitor blocks, transformers, and other components to keep out air and moisture.

search antenna A radar antenna or antenna system designed for search.

search coil *See* exploring coil.

search gate A gate pulse used to search back and forth over a certain range.

sea return *See* sea clutter.

seasoning Overcoming a temporary unsteadiness of a component that may appear when it is first installed.

sec *See* secondary winding.

SEC *See* secondary electron conduction.

secohm *See* international henry.

secondary 1. Low-voltage conductors of a power distributing system. 2. *See* secondary winding.

secondary battery *See* storage battery.

secondary cell *See* storage cell.

secondary electron 1. An electron emitted as a result of bombardment of a material by an incident electron. 2. An electron whose motion is due to a transfer of momentum from primary radiation.

secondary electron conduction Transport of charge by secondary electrons moving through the interstices of a porous material under the influence of an externally applied electric field. Abbreviated SEC.

secondary emission The emission of electrons from the surface of a solid or liquid into a vacuum as a result of bombardment by electrons or other charged particles.

secondary grid emission Electron emission from a grid resulting directly from bombardment of its surface by electrons or other charged particles.

secondary lobe *See* minor lobe.

secondary photocurrent A photocurrent resulting from ohmic contacts that are able to replenish charge carriers which pass out of the opposite contact in order to maintain charge neutrality, and whose maximum gain is much greater than unity.

secondary radar Radar which receives pulses transmitted by an interrogator and makes a return transmission (usually on a different frequency) by its transponder, as opposed to a primary radar which receives pulses returned from illuminated objects.

secondary voltage The voltage across the secondary winding of a transformer.

secondary winding A transformer winding that receives energy by electromagnetic induction from the primary winding; a transformer may have several secondary windings, and they may provide alternating-current voltages that are higher, lower, or the same as that applied to the primary winding. Abbreviated sec. Also known as secondary.

second breakdown Destructive breakdown in a transistor, wherein structural imperfections cause localized current concentrations and uncontrollable generation and multiplication of current carriers; reaction occurs so suddenly that the thermal time constant of the collector regions is exceeded, and the transistor is irreversibly damaged.

second detector The detector that separates the intelligence signal from the intermediate-frequency signal in a superheterodyne receiver.

second-time-around echo A radar echo received after an interval exceeding the pulse interval. Also known as second-trip echo.

second-trip echo *See* second-time-around echo.

sectionalized vertical antenna Vertical antenna that is insulated at one or more points along its length; the insertion of suitable reactances or applications of a driving voltage across the insulated points results in a modified current distribution giving a more desired radiation pattern in the vertical plane.

sectional radiography The technique of making radiographs of plane sections of a body or an object; its purpose is to show detail in a predetermined plane of the body, while blurring the images of structures in other planes. Also known as laminography; planigraphy; tomography.

sector Coverage of a radar as measured in azimuth.

sectoral horn Horn with two opposite sides parallel and the two remaining sides which diverge.

sector display A display in which only a sector of the total service area of a radar system is shown; usually the sector is selectable.

sector scan A radar scan through a limited angle, as distinguished from complete rotation.

Seebeck coefficient The ratio of the open-circuit voltage to the temperature difference between the hot and cold junctions of a circuit exhibiting the Seebeck effect.

Seebeck effect The development of a voltage due to differences in temperature between two junctions of dissimilar metals in the same circuit.

seed A small, single crystal of semiconductor material used to start the growth of a large, single crystal for use in cutting semiconductor wafers.

seeing The introduction of atoms with a low ionization potential into a hot gas to increase electrical conductivity.

seignette-electric *See* ferroelectric.

selecting circuit A simple switching circuit that receives the identity (the address) of a particular item and selects that item from among a number of similar ones.

selective absorption A greater absorption of electromagnetic radiation at some wavelengths (or frequencies) than at others.

selective circuit A circuit that transmits certain types of signals and fails to transmit or attenuates others.

selective identification feature Airborne pulse-type transponder which provides automatic selective identification of aircraft in which it is installed to ground, shipboard, or airborne identification, friend or foe–selective identification feature recognition installations.

selective jamming Jamming in which only a single radio channel is jammed.

selective reflection Reflection of electromagnetic radiation more strongly at some wavelengths (or frequencies) than at others.

selective scattering Scattering of electromagnetic radiation more strongly at some wavelengths than at others.

selectivity The ability of a radio receiver to separate a desired signal frequency from other signal frequencies, some of which may differ only slightly from the desired value.

selector An automatic or other device for making connections to any one of a number of circuits, such as a selector relay or selector switch.

selector switch A manually operated multiposition switch. Also called multiple-contact switch.

selenium cell A photoconductive cell in which a thin film of selenium is used between suitable electrodes; the resistance of the cell decreases when the illumination is increased.

selenium diode A small area selenium rectifier which has characteristics similar to those of selenium rectifiers used in power systems.

selenium rectifier A metallic rectifier in which a thin layer of selenium is deposited on one side of an aluminum plate and a conductive metal coating is deposited on the selenium.

self-bias A grid bias provided automatically by the resistor in the cathode or grid circuit of an electron tube; the resulting voltage drop across the resistor serves as the grid bias. Also known as automatic C bias; automatic grid bias.

self-bias transistor circuit A transistor with a resistance in the emitter lead that gives rise to a voltage drop which is in the direction to reverse-bias the emitter junction; the circuit can be used even if there is zero direct-current resistance in series with the collector terminal.

self-cleaning contact *See* wiping contact.

self-diffusion The spontaneous movement of an atom to a new site in a crystal of its own species.

self-excited Operating without an external source of alternating-current power.

self-excited oscillator An oscillator that depends on its own resonant circuits for initiation of oscillation and frequency determination.

self-fields The electric and magnetic fields generated by an intense beam of charged particles, which act on the beam itself; they limit the beam intensities which can be achieved in storage rings.

self-inductance 1. The property of an electric circuit whereby an electromotive force is produced in the circuit by a change of current in the circuit itself. 2. Quantitatively, the ratio of the electromotive force produced to the rate of change of current in the circuit.

self-induction The production of a voltage in a circuit by a varying current in that same circuit.

self-pulsing Special type of grid pulsing which automatically stops and starts the oscillations at the pulsing rate by a special circuit.

self-quenched detector Superregenerative detector in which the time constant of the grid leak and grid capacitor is sufficiently large to cause intermittent oscillation above audio frequencies, serving to stop normal regeneration each time just before it spills over into a squealing condition.

self-quenching oscillator Oscillator producing a series of short trains of radio-frequency oscillations separated by intervals of quietness.

self-reset Automatically returning to the original position when normal conditions are resumed; applied chiefly to relays and circuit breakers.

self-saturation The connection of half-wave rectifiers in series with the output windings of the saturable reactors of a magnetic amplifier, to give higher gain and faster response.

self-scanned image sensor A solid-state device, still in the early stages of development, which converts an optical image into a television signal without the use of an electron beam; it consists of an array of photoconductor diodes, each located at the intersection of mutually perpendicular address strips respectively connected to horizontal and vertical scan generators and video coupling circuits.

self-screening range Range at which a target can be detected by a radar in the midst of its jamming mask, with a certain specified probability.

self-starting synchronous motor A synchronous motor provided with the equivalent of a squirrel-cage winding, to permit starting as an induction motor.

self-steering microwave array An antenna array used with electronic circuitry that senses the phase of incoming pilot signals and positions the antenna beam in their direction of arrival.

self-synchronous device *See* synchro.

self-synchronous repeater *See* synchro.

Sellmeier's equation An equation for the index of refraction of electromagnetic radiation as a function of wavelength in a medium whose molecules have oscillators of different frequencies.

selsyn *See* synchro.

selsyn generator *See* synchro transmitter.

selsyn motor *See* synchro receiver.

selsyn receiver *See* synchro receiver.

selsyn system *See* synchro system.

selysn transmitter *See* synchro transmitter.

semiconducting compound A compound which is a semiconductor, such as copper oxide, mercury indium telluride, zinc sulfide, cadmium selenide, and magnesium iodide.

semiconducting crystal A crystal of a semiconductor, such as silicon, germanium, or gray tin.

semiconductor A solid crystalline material whose electrical conductivity is intermediate between that of a conductor and an insulator, ranging from about 10^5 mhos to 10^{-7} mho per meter, and is usually strongly temperature-dependent.

semiconductor device Electronic device in which the characteristic distinguishing electronic conduction takes place within a semiconductor.

semiconductor diode Also known as crystal diode; crystal rectifier; diode. 1. A two-electrode semiconductor device that utilizes the rectifying properties of a pn junction or a point contact. 2. More generally, any two-terminal electronic device that utilizes the properties of the semiconductor from which it is constructed.

semiconductor-diode parametric amplifier Parametric amplifier using one or more varactors.

semiconductor doping *See* doping.

semiconductor heterostructure A structure of two different semiconductors in junction contact having useful electrical or electrooptical characteristics not achievable in either conductor separately; used in certain types of lasers and solar cells.

semiconductor intrinsic properties Properties of a semiconductor that are characteristic of the ideal crystal.

semiconductor junction Region of transition between semiconducting regions of different electrical properties, usually between p-type and n-type material.

semiconductor rectifier *See* metallic rectifier.

semiconductor thermocouple A thermocouple made of a semiconductor, which offers the prospect of operation with high-temperature gradients, because semiconductors are good electrical conductors but poor heat conductors.

semiconductor trap *See* trap.

semimagnetic controller Electrical controller having only part of its basic functions performed by devices that are operated by electromagnets.

semitransparent photocathode Photocathode in which radiant flux incident on one side produces photoelectric emission from the opposite side.

sending-end impedance Ratio of an applied potential difference to the resultant current at the point where the potential difference is applied; the sending-end impedance of a line is synonymous with the driving-point impedance of the line.

sense amplifier Circuit used to determine either a phase or voltage change in communications-electronics equipment and to provide automatic control function.

sense antenna An auxiliary antenna used with a directional receiving antenna to resolve a 180° ambiguity in the directional indication. Also known as sensing antenna.

sensing antenna *See* sense antenna.

sensistor Silicon resistor whose resistance varies with temperature, power, and time.

sensitive switch *See* snap-action switch.

sensitivity 1. The minimum input signal required to produce a specified output signal, for a radio receiver or similar device. 2. Of a camera tube, the signal current developed per unit incident radiation, that is, per watt per unit area.

sensitivity time control In a radar receiver a circuit which greatly reduces the gain at the time that the transmitter emits a pulse; following the pulse, the circuit increases the sensitivity; thus reflection from distant objects will be received and those from nearby objects will be prevented from saturating the receiver.

sensitization *See* activation.

separately excited Obtaining excitation from a source other than the machine or device itself.

separation filter Combination of filters used to separate one band of frequencies from another.

separator 1. A porous insulating sheet used between the plates of a storage battery. 2. A circuit that separates one type of signal from another by clipping, differentiating, or integrating action.

septate coaxial cavity Coaxial cavity having a vane or septum, added between the inner and outer conductors, so that it acts as a cavity of a rectangular cross section bent transversely.

septate waveguide Waveguide with one or more septa placed across it to control microwave power transmission.

septum A metal plate placed across a waveguide and attached to the walls by highly conducting joints; the plate usually has one or more windows, or irises, designed to give inductive, capacitive, or resistive characteristics.

sequential circuit A switching circuit whose output depends not only upon the present state of its input, but also on what its input conditions have been in the past.

sequential logic element A circuit element having at least one input channel, at least one output channel, and at least one internal state variable, so designed and constructed that the output signals depend on the past and present states of the inputs.

series An arrangement of circuit components end to end to form a single path for current.

series circuit A circuit in which all parts are connected end to end to provide a single path for current.

series connection A connection that forms a series circuit.

series excitation The obtaining of field excitation in a motor or generator by allowing the armature current to flow through the field winding.

series-fed vertical antenna Vertical antenna which is insulated from the ground and energized at the base.

series feed Application of the direct-current voltage to the plate or grid of a vacuum tube through the same impedance in which the alternating-current flows.

series generator A generator whose armature winding and field winding are connected in series. Also known as series-wound generator.

series loading Loading in which reactances are inserted in series with the conductors of a transmission circuit.

series modulation Modulation in which the plate circuits of a modulating tube and a modulated amplifier tube are in series with the same plate voltage supply.

series motor A commutator-type motor having armature and field windings in series; characteristics are high starting torque, variation of speed with load, and dangerously high speed on no-load. Also known as series-wound motor.

series multiple Type of switchboard jack arrangement in which a single line circuit appears before two or more operators, all appearances being connected in series.

series-parallel circuit A circuit in which some of the components or elements are connected in parallel, and one or more of these parallel combinations are in series with other components of the circuit.

series-parallel control A method of controlling the speed of electric motors in which the motors, or groups of motors, are connected in series at some times and in parallel at other times.

series-parallel switch A switch used to change the connections of lamps or other devices from series to parallel, or vice versa.

series peaking Use of a peaking coil and resistor in series as the load for a video amplifier to produce peaking at some desired frequency in the passband, such as to compensate for previous loss of gain at the high-frequency end of the passband.

series reactor A reactor used in alternating-current power systems for protection against excessively large currents under short-circuit or transient conditions; it consists of coils of heavy insulated cable either cast in concrete columns or supported in rigid frames and mounted on insulators. Also known as current-limiting reactor.

series regulator A regulator that controls output voltage or current by automatically varying a resistance in series with the voltage source.

series repeater A type of negative impedance telephone repeater which is stable when terminated in an open circuit and oscillates when it is connected to a low impedance, in contrast to a shunt repeater.

series resonance Resonance in a series resonant circuit, wherein the inductive and capacitive reactances are equal at the frequency of the applied voltage; the reactances then cancel each other, reducing the impedance of the circuit to a minimum, purely resistive value.

series resonant circuit A resonant circuit in which the capacitor and coil are in series with the applied alternating-current voltage.

series-shunt network *See* ladder network.

series T junction *See* E-plane T junction.

series transistor regulator A voltage regulator whose circuit has a transistor in series with the output voltage, a Zener diode, and a resistor chosen so that the Zener diode is approximately in the middle of its operating range.

series-tuned circuit A simple resonant circuit consisting of an inductance and a capacitance connected in series.

series winding A winding in which the armature circuit and the field circuit are connected in series with the external circuit.

serrated pulse Vertical and horizontal synchronizing pulse divided into a number of small pulses, each of which acts for the duration of half a line in a television system.

serrodyne Phase modulator using transit time modulation of a traveling-wave tube or klystron.

service oscillator *See* radio-frequency signal generator.

service wires The conductors that bring the electric power into a building.

serving A covering, such as thread or tape, that protects a winding from mechanical damage. Also known as coil serving.

servo amplifier An amplifier used in a servomechanism.

servomultiplier An electromechanical multiplier in which one variable is used to position one or more ganged potentiometers across which the other variable voltages are applied.

set The placement of a storage device in a prescribed state, for example, a binary storage cell in the high or 1 state.

set composite Signaling circuit in which two signaling or telegraph legs may be superimposed on a two-wire, interoffice trunk by means of one of a balanced pair of high-impedance coils connected to each side of the line with an associated capacitor network.

set condition Condition of a flip-flop circuit in which the internal state of the flip-flop is set to 1.

set pulse An electronic pulse designed to place a memory cell in a specified state.

setup The ratio between the reference black level and the reference white level in television, both measured from the blanking level; usually expressed as a percentage.

sferics receiver An instrument which measures, electronically, the direction of arrival, intensity, and rate of occurrence of atmospherics; in its simplest form, the instrument consists of two orthogonally crossed antennas, whose output signals are connected to an oscillograph so that one loop measures the north-south component while the other measures the east-west component; the signals are combined vertically to give the azimuth. Also known as lightning recorder.

shaded-pole motor A single-phase induction motor having one or more auxiliary short-circuited windings acting on only a portion of the magnetic circuit; generally, the winding is a closed copper ring embedded in the face of a pole; the shaded pole provides the required rotating field for starting purposes.

shading Television process of compensating for the spurious signal generated in a camera tube during trace intervals.

shading coil *See* shading ring.

shading ring The copper ring used in a shaded-pole motor to produce a rotating magnetic field for starting purposes, or used around a part of the core of an alternating-current relay to prevent contact chatter. Also known as shading coil.

shading signal Television camera signal that serves to increase the gain of the amplifier in the camera during those intervals of time when the electron beam is on an area corresponding to a dark portion of the scene being televised.

shadow attenuation Attenuation of radio waves over a sphere in excess of that over a plane when the distance over the surface and other factors are the same.

shadow factor The ratio of the electric-field strength that would result from propagation of waves over a sphere to that which would result from propagation over a plane under comparable conditions.

shadow mask A thin, perforated metal mask mounted just back of the phosphor-dot faceplate in a three-gun color picture tube; the holes in the mask are positioned to ensure that each of the three electron beams strikes only its intended color phosphor dot. Also known as aperture mask.

shadow region Region in which, under normal propagation conditions, the field strength from a given transmitter is reduced by some obstruction which renders effective radio reception of signals or radar detection of objects in this region improbable.

shaft-position encoder An analog-to-digital converter in which the exact angular position of a shaft is sensed and converted to digital form.

shaker An electromagnetic device capable of imparting known and usually controlled vibratory acceleration to a given object. Also known as electrodynamic shaker; shake table.

shake table *See* shaker.

shaped-beam antenna Antenna with a directional pattern which, over a certain angular range, is of special shape for some particular use.

shape factor *See* form factor.

shaping The adjustment of a plan-position- indicator pattern set up by a rotating magnetic field.

shaping circuit *See* corrective network.

shaping network *See* corrective network.

sharp-cutoff tube An electron tube in which the control-grid openings are uniformly spaced; the anode current then decreases linearly as the grid voltage is made more negative, and cuts off sharply at a particular grid voltage.

sharpness of resonance The narrowness of the frequency band around the resonance at which the response of an electric circuit exceeds an arbitrary fraction of its maximum response, often 70.7%.

sharp tuning Having high selectivity; responding only to a desired narrow range of frequencies.

sheath 1. A protective outside covering on a cable. 2. A space charge formed by ions near an electrode in a gas tube. 3. The metal wall of a waveguide.

sheath-reshaping converter In a waveguide, a mode converter in which the change of wave pattern is achieved by gradual reshaping of the sheath of the waveguide and of conducting metal sheets mounted longitudinally in the guide.

sheet grating Three-dimensional grating consisting of thin, longitudinal, metal sheets extending along the inside of a waveguide for a distance of about a wavelength, and used to stop all waves except one predetermined wave that passes unimpeded.

shell-type transformer Transformer in which the magnetic circuit completely surrounds the windings.

shielded-conductor cable Cable in which the insulated conductor or conductors are enclosed in a conducting envelope or envelopes, constructed so that substantially every point on the surface of the insulation is at ground potential or at some predetermined potential with respect to ground.

shielded joint Cable joint having its insulation so enveloped by a conducting shield that substantially every point on the surface of the insulation is at ground potential, or at some predetermined potential with respect to ground.

shielded line Transmission line, the elements of which confine the propagated waves to an essentially finite space; the external conducting surface is called the sheath.

shielded wire Insulated wire covered with a metal shield, usually of tinned braided copper wire.

shield grid A grid that shields the control grid of a gas tube from electrostatic fields, thermal radiation, and deposition of thermionic emissive material; it may also be used as an additional control electrode.

shield-grid thyratron A thyratron having a shield grid, usually operated at cathode potential.

shielding *See* electric shielding.

shielding ratio The ratio of a field in a specified region when electrical shielding is in place to the field in that region when the shielding is removed.

shiran Specially designed frequency-modulation continuous-wave distance-measuring equipment used for performing distance measurements of an accuracy comparable to first-order triangulation. Derived from S-band hiran.

shock excitation Excitation produced by a voltage or current variation of relatively short duration; used to initiate oscillation in the resonant circuit of an oscillator. Also known as impulse excitation.

Shockley diode A *pnpn* silicon controlled switch having characteristics that permit operation as a unidirectional diode switch.

Shockley partial dislocation A partial dislocation in which the Burger's vector lies in the fault plane, so that it is able to glide, in contrast to a Frank partial dislocation. Also known as glissile dislocation.

shore effect Bending of waves toward the shoreline when traveling over water near a shoreline, due to the slightly greater velocity of radio waves over water than over land; this effect causes errors in radio-direction-finder indications.

short *See* short circuit.

short antenna An antenna shorter than about one-tenth of a wavelength, so that the current may be assumed to have constant magnitude along its length, and the antenna may be treated as an elementary dipole.

short circuit A low-resistance connection across a voltage source or between both sides of a circuit or line, usually accidental and usually resulting in excessive current flow that may cause damage. Also known as short.

short-circuit impedance Of a line or four-terminal network, the driving point impedance when the far-end is short-circuited.

short-circuit transition *See* shunt transition.

short-contact switch Selector switch in which the width of the movable contact is greater than the distance between contact clips, so that the new circuit is contacted before the old one is broken; this avoids noise during switching.

short-gate gain Video gain on short-range gate.

short-path principle *See* Hittorf principle.

short-slot coupler *See* three-decibel coupler.

short-time rating A rating defining the load that a machine, apparatus, or device can carry for a specified short time.

shortwave converter Electronic unit designed to be connected between a receiver and its antenna system to permit reception of frequencies higher than those the receiver ordinarily handles.

shortwave radiation A term used loosely to distinguish radiation in the visible and near-visible portions of the electromagnetic spectrum (roughly 0.4 to 1.0 micrometer in wavelength) from long-wave radiation (infrared radiation).

shot effect *See* shot noise.

shot-firing cable A two-conductor cable which leads from the exploder to the detonator wires. Also known as firing cable.

shot-firing circuit The path taken by the electric current from the exploder along the shot-firing cable, the detonator wires, and finally the detonator when a shot is detonated.

shot noise Noise voltage developed in a thermionic tube because of the random variations in the number and the velocity of electrons emitted by the heated cathode; the effect causes sputtering or popping sounds in radio receivers and snow effects in television pictures. Also known as Schottky noise; shot effect.

Shubnikov–de Haas effect Oscillations of the resistance or Hall coefficient of a metal or semiconductor as a function of a strong magnetic field, due to the quantization of the electron's energy.

Shubnikov groups The point groups and space groups of crystals having magnetic moments. Also known as black-and-white groups; magnetic groups.

shunt 1. A precision low-value resistor placed across the terminals of an ammeter to increase its range by allowing a known fraction of the circuit current to go around the meter. Also known as electric shunt. 2. To place one part in parallel with another. 3. A piece of iron that provides a parallel path for magnetic flux around an air gap in a magnetic circuit. 4. *See* parallel.

shunt-excited Having field windings connected across the armature terminals, as in a direct-current generator.

shunt-excited antenna A tower antenna, not insulated from the ground at the base, whose feeder is connected at a point about one-fifth of the way up the antenna and usually slopes up to this point from a point some distance from the antenna's base.

shunt-fed vertical antenna Vertical antenna connected to the ground at the base and energized at a point suitably positioned above the grounding point.

shunt feed *See* parallel feed.

shunt generator A generator whose field winding and armature winding are connected in parallel, and in which the armature supplies both the load current and the field current.

shunting The act of connecting one device to the terminals of another so that the current is divided between the two devices in proportion to their respective admittances.

shunt loading Loading in which reactances are applied in shunt across the conductors.

shunt motor A direct-current motor whose field circuit and armature circuit are connected in parallel.

shunt neutralization *See* inductive neutralization.

shunt peaking The use of a peaking coil in a parallel circuit branch connecting the output load of one stage to the input load of the following stage, to compensate for high-frequency loss due to the distributed capacitances of the two stages.

shunt reactor A reactor that has a relatively high inductance and is wound on a magnetic core containing an air gap; used to neutralize the charging current of the line to which it is connected.

shunt regulator A regulator that maintains a constant output voltage by controlling the current through a dropping resistance in series with the load.

shunt repeater A type of negative impedance telephone repeater which is stable when it is short-circuited, but oscillates when terminated by a high impedance, in contrast to a series repeater; it can be thought of as a negative admittance.

shunt T junction *See* H-plane T junction.

shunt transition A method of changing the connection of motors from series to parallel in which one motor, or group of motors, is first short-circuited, then disconnected, and finally connected in parallel with the other motor or motors. Also known as short-circuit transition.

shunt wound Having armature and field windings in parallel, as in a direct-current generator or motor.

sideband 1. The frequency band located either above or below the carrier frequency, within which fall the frequency components of the wave produced by the process of modulation. 2. The wave components lying within such bands.

side echo Echo due to a side lobe of an antenna.

side lobe *See* minor lobe.

side-lobe blanking Radar technique which compares relative signal strengths between an omnidirectional antenna and the radar antenna.

siemens A unit of conductance, admittance, and susceptance, equal to the conductance between two points of a conductor such that a potential difference of 1 volt between these points produces a current of 1 ampere; the conductance of a conductor in siemens is the reciprocal of its resistance in ohms. Formerly known as mho ($\mho$); reciprocal ohm. Symbolized S.

Siemens' electrodynamometer An early type of electromagnetic instrument in which current flows through all the coils in series.

signal distortion generator Instrument designed to apply known amounts of distortion on a signal for the purpose of testing and adjusting communications equipment such as teletypewriters.

signaling key *See* key.

signal-shaping network Network inserted in a telegraph circuit, usually at the receiving end, to improve the waveform of the code signals.

signal strength The strength of the signal produced by a radio transmitter at a particular location, usually expressed as microvolts or millivolts per meter of effective receiving antenna height.

signal-strength meter A meter that is connected to the automatic volume-control circuit of a communication receiver and calibrated in decibels or arbitrary S units to read the strength of a received signal. Also known as S meter; S-unit meter.

signal-to-interference ratio The relative magnitude of signal waves and waves which interfere with signal-wave reception.

signal-to-noise ratio The ratio of the amplitude of a desired signal at any point to the amplitude of noise signals at that same point; often expressed in decibels; the peak value is usually used for pulse noise, while the root-mean-square (rms) value is used for random noise. Abbreviated S/N; SNR.

signal tracer An instrument used for tracing the progress of a signal through a radio receiver or an audio amplifier to locate a faulty stage.

signal voltage Effective (root-mean-square) voltage value of a signal.

signal winding Control winding, of a saturable reactor, to which the independent variable (signal wave) is applied.

signature The characteristic pattern of a target as displayed by detection and classification equipment.

silicide resistor A thin-film resistor that uses a silicide of molybdenum or chromium, deposited by direct-current sputtering in an integrated circuit when radiation hardness or high resistance values are required.

silicon capacitor A capacitor in which a pure silicon-crystal slab serves as the dielectric; when the crystal is grown to have a p zone, a depletion zone, and an n zone, the capacitance varies with the externally applied bias voltage, as in a varactor.

silicon controlled rectifier A semiconductor rectifier that can be controlled; it is a *pnpn* four-layer semiconductor device that normally acts as an open circuit, but switches rapidly to a conducting state when an appropriate gate signal is applied to the gate terminal. Abbreviated SCR. Also known as reverse-blocking triode thyristor.

silicon controlled switch A four-terminal switching device having four semiconductor layers, all of which are accessible; it can be used as a silicon controlled rectifier, gate-turnoff switch, complementary silicon controlled rectifier, or conventional silicon transistor. Abbreviated SCS. Also known as reverse-blocking tetrode thyristor.

silicon detector *See* silicon diode.

silicon diode A crystal diode that uses silicon as a semiconductor; used as a detector in ultra-high- and super-high-frequency circuits. Also known as silicon detector.

silicon image sensor A solid-state television camera in which the image is focused on an array of individual light-sensitive elements formed from a charged-coupled-device semiconductor chip. Also known as silicon imaging device.

silicon imaging device *See* silicon image sensor.

silicon-on-sapphire A semiconductor manufacturing technology in which metal oxide semiconductor devices are constructed in a thin single-crystal silicon film grown on an electrically insulating synthetic sapphire substrate. Abbreviated SOS.

silicon rectifier A metallic rectifier in which rectifying action is provided by an alloy junction formed in a high-purity silicon slab.

silicon resistor A resistor using silicon semiconductor material as a resistance element, to obtain a positive temperature coefficient of resistance that does not appreciably change with temperature; used as a temperature-sensing element.

silicon solar cell A solar cell consisting of p and n silicon layers placed one above the other to form a pn junction at which radiant energy is converted into electricity.

silicon-symmetrical switch Thyristor modified by adding a semiconductor layer so that the device becomes a bidirectional switch; used as an alternating-current phase control, for synchronous switching and motor speed control.

silicon transistor A transistor in which silicon is used as the semiconducting material.

silver battery A solid-state battery based on an Ag_4RbI_5 electrolyte that conducts positive silver ions.

silver-cadmium storage battery A storage battery that combines the excellent space and weight characteristics of silver-zinc batteries with long shelf life and other desirable properties of nickel-cadmium batteries.

silvered mica capacitor A mica capacitor in which a coating of silver is deposited directly on the mica sheets to serve in place of conducting metal foil.

silver migration A process, causing reduction in insulation resistance and dielectric failure; silver, in contact with an insulator, at high humidity, and subjected to an electrical potential, is transported ionically from one location to another.

silver oxide cell A primary cell in which depolarization is accomplished by an oxide of silver.

silverstat regulator Multitapped resistor, the taps of which are connected to single-leaf silver contacts; variation of voltage causes a solenoid to open or close these contacts, shorting out more or less of the resistance in the exciter circuit as a means of regulating the output voltage to the desired value.

396 silver-zinc storage battery

silver-zinc storage battery A storage battery that gives higher current output and greater watt-hour capacity per unit of weight and volume than most other types, even at high discharge rates; used in missiles and torpedoes, where its high cost can be tolerated.

simple harmonic current Alternating current, the instantaneous value of which is equal to the product of a constant, and the cosine of an angle varying linearly with time. Also known as sinusoidal current.

simple harmonic electromotive force An alternating electromotive force which is equal to the product of a constant and the cosine or sine of an angle which varies linearly with time.

simultaneous color television A color television system in which the phosphors for the three primary colors are excited at the same time, not one after another; the shadow-mask color picture tube gives a simultaneous display.

simultaneous lobing A radar direction-finding technique in which the signals received by two partly overlapping antenna lobes are compared in phase or power to obtain a measure of the angular displacement of a target from the equisignal direction.

sine-cosine encoder A shaft-position encoder having a special type of angle-reading code disk that gives an output which is a binary representation of the sine of the shaft angle.

sine-cosine generator *See* resolver.

sine potentiometer A potentiometer whose direct-current output voltage is proportional to the sine of the shaft angle; used as a resolver in computer and radar systems.

sine-wave modulated jamming Jamming signal produced by modulating a continuous wave signal with one or more sine waves.

sine-wave oscillator *See* sinusoidal oscillator.

singing-stovepipe effect Reception and reproduction of radio signals by ordinary pieces of metal in contact with each other, such as sections of stovepipe; it occurs when rusty bolts, faulty welds, or mechanically loose connections within strong radiated fields near transmitters produce intermodulation interference; the mechanically poor connections serve as nonlinear diodes.

single bus A substation switching arrangement that involves one common bus for all connections and one breaker per connection.

single-carrier theory A theory of the behavior of a rectifying barrier which assumes that conduction is due to the motion of carriers of only one type; it can be applied to the contact between a metal and a semiconductor.

single-channel multiplier A type of photomultiplier tube in which electrons travel down a cylindrical channel coated on the inside with a resistive secondary-emitting layer, and gain is achieved by multiple electron impacts on the inner surface as the electrons are directed down the channel by an applied voltage over the length of the channel.

single-edged push-pull amplifier circuit Amplifier circuit having two transmission paths designed to operate in a complementary manner and connected to provide a single unbalanced output without the use of an output transformer.

single-end amplifier Amplifier stage which normally employs only one tube or semiconductor or, if more than one tube or semiconductor is used, they are connected in parallel so that operation is asymmetric with respect to ground. Also known as single-sided amplifier.

single-ended Unbalanced, as when one side of a transmission line or circuit is grounded.

single-gun color tube A color television picture tube having only one electron gun and one electron beam; the beam is sequentially deflected across phosphors for the three primary colors to form each color picture element, as in the chromatron.

single in-line package A packaged resistor network or other assembly that has a single row of terminals or lead wires along one edge of the package. Abbreviated SIP.

single knock-on A sputtering event in which target atoms are ejected either directly by the bombarding projectiles or after a small number of collisions.

single-layer solenoid A solenoid which has only one layer of wire, wound in a cylindrical helix.

single-phase Energized by a single alternating voltage.

single-phase circuit Either an alternating-current circuit which has only two points of entry, or one which, having more than two points of entry, is intended to be so energized that the potential differences between all pairs of points of entry are either in phase or differ in phase by 180°.

single-phase motor A motor energized by a single alternating voltage.

single-phase rectifier A rectifier whose input voltage is a single sinusoidal voltage, in contrast to a polyphase rectifier.

single-point grounding Grounding system that attempts to confine all return currents to a network that serves as the circuit reference; to be effective, no appreciable current is allowed to flow in the circuit reference, that is, the sum of the return currents is zero.

single-polarity pulse Pulse in which the sense of the departure from normal is in one direction only.

single-pole double-throw A three-terminal switch or relay contact arrangement that connects one terminal to either of two other terminals. Abbreviated SPDT.

single-pole single-throw A two-terminal switch or relay contact arrangement that opens or closes one circuit. Abbreviated SPST.

single-shot blocking oscillator Blocking oscillator modified to operate as a single-shot trigger circuit.

single-shot multivibrator *See* monostable multivibrator.

single-shot trigger circuit Trigger circuit in which one triggering pulse initiates one complete cycle of conditions ending with a stable condition. Also known as single-trip trigger circuit.

single-sided amplifier *See* single-end amplifier.

single-signal receiver A highly selective superheterodyne receiver for code reception, having a crystal filter in the intermediate-frequency amplifier.

single-stub transformer Shorted section of a coaxial line that is connected to a main coaxial line near a discontinuity to provide impedance matching at the discontinuity.

single-stub tuner Section of transmission line terminated by a movable short-circuiting plunger or bar, attached to a main transmission line for impedance-matching purposes.

single-throw switch A switch in which the same pair of contacts is always opened or closed.

single-trip trigger circuit *See* single-shot trigger circuit.

single-tuned amplifier An amplifier characterized by resonance at a single frequency.

single-tuned circuit A circuit whose behavior is the same as that of a circuit with a single inductance and a single capacitance, together with associated resistances.

single-tuned interstage An interstage circuit which is resonant at a single frequency.

single-unit semiconductor device Semiconductor device having one set of electrodes associated with a single carrier stream.

single-wire line 1. Transmission line that uses the ground as one side of the circuit. 2. A surface-wave transmission line that consists of a single conductor which has a dielectric coating or other treatment that confines the propagated energy close to the wire.

sink The region of a Rieke diagram where the rate of change of frequency with respect to phase of the reflection coefficient is maximum for an oscillator; operation in this region may lead to unsatisfactory performance by reason of cessation or instability of oscillations.

sinusoidal angular modulation *See* angle modulation.

sinusoidal current *See* simple harmonic current.

sinusoidal oscillator An oscillator circuit whose output voltage is a sine-wave function of time. Also known as harmonic oscillator; sine-wave oscillator.

SIP *See* single in-line package.

SIT *See* static induction transistor.

situation-display tube Large cathode-ray tube used to display tabular and vector messages pertinent to the various functions of an air defense mission.

six-phase circuit Combination of circuits energized by alternating electromotive forces which differ in phase by one-sixth of a cycle (60°).

six-phase rectifier A rectifier in which transformers are used to produce six alternating electromotive forces which differ in phase by one-sixth of a cycle, and which feed six diodes.

size control A control provided on a television receiver for changing the size of a picture either horizontally or vertically.

skew 1. The deviation of a received facsimile frame from rectangularity due to lack of synchronism between scanner and recorder; expressed numerically as the tangent of the angle of this deviation. 2. The degree of nonsynchronism of supposedly parallel bits when bit-coded characters are read from magnetic tape.

skiatron *See* dark-trace tube.

skin antenna Flush-mounted aircraft antenna made by using insulating material to isolate a portion of the metal skin of the aircraft.

skin depth The depth beneath the surface of a conductor, which is carrying current at a given frequency due to electromagnetic waves incident on its surface, at which the current density drops to one neper below the current density at the surface.

skin effect The tendency of alternating currents to flow near the surface of a conductor thus being restricted to a small part of the total sectional area and producing the effect of increasing the resistance. Also known as conductor skin effect; Kelvin skin effect.

skin resistance For alternating current of a given frequency, the direct-current resistance of a layer at the surface of a conductor whose thickness equals the skin depth.

skin tracking Tracking of an object by means of radar without using a beacon or other signal device on board the object being tracked.

skiograph An instrument used to measure the intensity of x-rays.

skip distance The minimum distance that radio waves can be transmitted between two points on the earth by reflection from the ionosphere, at a specified time and frequency.

skip-fading Fading due to fluctuations of ionization density at the place in the ionosphere where the wave is reflected which causes the skip distance to increase or decrease.

skip-keying Reduction of radar pulse repetition frequency to submultiple of that normally used, to reduce mutual interference between radar or to increase the length of radar time base.

sky noise Noise produced by radio energy from stars.

sky wave A radio wave that travels upward into space and may or may not be returned to earth by reflection from the ionosphere. Also known as ionospheric wave.

sky-wave correction The correction to be applied to the time difference readings of received sky waves to convert them to an equivalent ground-wave reading.

sky-wave transmission delay Amount by which the time of transit from transmitter to receiver of a pulse carried by sky waves reflected once from the E layer exceeds the time of transit of the same pulse carried by ground waves.

slab A relatively thick-cut crystal from which blanks are obtained by subsequent transverse cutting.

Slater's rule The ratio of the cathode radius to the anode radius of a magnetron is approximately equal to $(N - 4)/(N + 4)$, where N is the number of resonators.

slave antenna A directional antenna positioned in azimuth and elevation by a servo system; the information controlling the servo system is supplied by a tracking or positioning system.

sleeve 1. The cylindrical contact that is farthest from the tip of a phone plug. 2. Insulating tubing used over wires or components. Also known as sleeving.

sleeve antenna A single vertical half-wave radiator, the lower half of which is a metallic sleeve through which the concentric feed line runs; the upper radiating portion, one quarter-wavelength long, connects to the center of the line.

sleeve dipole antenna Dipole antenna surrounded in its central portion by a coaxial cable.

slewing motor A motor used to drive a radar antenna at high speed for slewing to pick up or track a target.

slew rate The maximum rate at which the output voltage of an operational amplifier changes for a square-wave or step-signal input; usually specified in volts per microsecond.

slicer *See* amplitude gate.

slicer amplifier *See* amplitude gate.

slicing Transmission of only those portions of a waveform lying between two amplitude values.

slide-back voltmeter An electronic voltmeter in which an unknown voltage is measured indirectly by adjusting a calibrated voltage source until its voltage equals the unknown voltage.

slider Sliding type of movable contact.

slide-wire bridge A bridge circuit in which the resistance in one or more branches is controlled by the position of a sliding contact on a length of resistance wire stretched along a linear scale.

slide-wire potentiometer A potentiometer (variable resistor) which employs a movable sliding connection on a length of resistance wire.

sliding contact *See* wiping contact.

slip 1. The difference between synchronous and operating speeds of an induction machine. Also known as slip speed. 2. Method of interconnecting multiple wiring between switching units by which trunk number 1 becomes the first choice for the first switch, trunk number 2 first choice for the second switch, and trunk number 3 first choice for the third switch, and so on. 3. Distortion produced in the recorded facsimile image which is similar to that produced by skew but is caused by slippage in the mechanical drive system.

slip ring A conductive rotating ring which, in combination with a stationary brush, provides a continuous electrical connection between rotating and stationary conductors; used in electric rotating machinery, synchros, gyroscopes, and scanning radar antennas.

slip speed *See* slip.

slot One of the conductor-holding grooves in the face of the rotor or stator of an electric rotating machine.

slot antenna An antenna formed by cutting one or more narrow slots in a large metal surface fed by a coaxial line or waveguide.

slot coupling Coupling between a coaxial cable and a waveguide by means of two coincident narrow slots, one in a waveguide wall and the other in the sheath of the coaxial cable.

slot-mask picture tube An in-line gun-type picture tube in which the shadow mask is perforated by short, vertical slots, and the screen is painted with vertical phosphor stripes.

slot radiator Primary radiating element in the form of a slot cut in the walls of a metal waveguide or cavity resonator or in a metal plate.

slotted line *See* slotted section.

slotted section A section of waveguide or shielded transmission line in which the shield is slotted to permit the use of a traveling probe for examination of standing waves. Also known as slotted line; slotted waveguide.

slotted waveguide *See* slotted section.

slot wedge The wedge that holds the windings in a slot in the rotor or stator core of an electrical machine.

slow-acting relay A time-delay relay in which an interval of several seconds may exist between energizing of the coil and pulling up of the armature; the delay can be obtained electrically by placing a solid copper ring on the core of the relay. Also known as slow-operate relay.

slow-blow fuse A fuse that can withstand up to 10 times its normal operating current for a brief period, as required for circuits and devices which draw a very heavy starting current.

slow death The gradual change of transistor characteristics with time; this change is attributed to ions which collect on the surface of the transistor.

slowed-down video Technique or method of transmitting radar data over narrow-bandwidth circuits; the procedure involves storing the radar video over the time required for the antenna to move through the beam width, and the subsequent sampling of this stored video at some periodic rate at which all of the range intervals of interest are sampled at least once each beam width or per azimuth quantum; the radar returns are quantized at the gap-filler radar site.

slow-motion video disk recorder A magnetic disk recorder that stores one field of video information per revolution, for instant replay at normal speed or any degree of slow motion down to complete stopping of action.

slow-release relay A time-delay relay in which there is an appreciable delay between deenergizing of the coil and release of the armature.

slow wave A wave having a phase velocity less than the velocity of light, as in a ridge wave guide.

slug 1. A heavy copper ring placed on the core of a relay to delay operation of the relay. 2. A movable iron core for a coil. 3. A movable piece of metal or dielectric material used in a wave guide for tuning or impedance-matching purposes.

slug tuner Waveguide tuner containing one or more longitudinally adjustable pieces of metal or dielectric.

slug tuning Means of varying the frequency of a resonant circuit by introducing a slug of material into either the electric field or magnetic field, or both.

small-scale integration Integration in which a complete major subsystem or system is fabricated on a single integrated-circuit chip that contains integrated circuits which have appreciably less complexity than for medium-scale integration. Abbreviated SSI.

small-signal parameter One of the parameters characterizing the behavior of an electronic device at small values of input, for which the device can be represented by an equivalent linear circuit.

smear A television-picture defect in which objects appear to be extended horizontally beyond their normal boundaries in a blurred or smeared manner; one cause is excessive attenuation of high video frequencies in the television receiver.

S meter *See* signal-strength meter.

Smith chart A special polar diagram containing constant-resistance circles, constant-reactance circles, circles of constant standing-wave ratio, and radius lines representing constant line-angle loci; used in solving transmission-line and waveguide problems.

smoothing choke Iron-core choke coil employed as a filter to remove fluctuations in the output current of a vacuum-tube rectifier or direct-current generator.

smoothing circuit *See* ripple filter.

smoothing filter *See* ripple filter.

S/N *See* signal-to-noise ratio.

snap-action switch A switch that responds to very small movements of its actuating button or lever and changes rapidly and positively from one contact position to the other; the trademark of one version is Micro Switch. Also known as sensitive switch.

snap-off diode Planar epitaxial passivated silicon diode that is processed so a charge is stored close to the junction when the diode is conducting; when reverse voltage is applied, the stored charge then forces the diode to snap off or switch rapidly to its blocking state.

snap-on ammeter An ac ammeter having a magnetic core in the form of hinged jaws that can be snapped around the current-carrying wire. Also known as clamp-on ammeter.

snivet Straight, jagged, or broken vertical black line appearing near the right-hand edge of a television receiver screen.

snooperscope An infrared source, an infrared image converter, and a battery-operated high-voltage direct-current source constructed in portable form to permit a foot soldier or other user to see objects in total darkness; infrared radiation sent out by the infrared source is reflected back to the snooperscope and converted into a visible image on the fluorescent screen of the image tube.

snow Small, random, white spots produced on a television or radar screen by inherent noise signals originating in the receiver.

snow static Precipitation static caused by falling snow.

SNR See signal-to-noise ratio.

socket A device designed to provide electric connections and mechanical support for an electronic or electric component requiring convenient replacement.

sodium amalgam–oxygen cell Fuel cell system in which materials functioning in the dual capacity of fuel and anode are consumed continuously; low operating temperatures and high power-to-weight ratios are significant characteristics of the system.

sodium-vapor lamp A discharge lamp containing sodium vapor, used chiefly for outdoor illumination.

soft limiting Limiting in which there is still an appreciable increase in output for increases in input signal strength up into the range at which limiting action occurs.

soft magnetic material A magnetic material which is relatively easily magnetized or demagnetized.

soft tube 1. An x-ray tube having a vacuum of about 0.000002 atmosphere (0.202650 newton per square meter), the remaining gas being left in intentionally to give less-penetrating rays than those of a more completely evacuated tube. 2. See gassy tube.

soft x-ray An x-ray having a comparatively long wavelength and poor penetrating power.

solar battery An array of solar cells, usually connected in parallel and series.

solar cell A pn-junction device which converts the radiant energy of sunlight directly and efficiently into electrical energy.

solar generator An electric generator powered by radiation from the sun and used in some satellites.

solar noise See solar radio noise.

solar radio noise Radio noise originating at the sun, and increasing greatly in intensity during sunspots and flares; it is heard as a hissing noise on shortwave radio receivers. Also known as solar noise.

solar sensor A light-sensitive diode that sends a signal to the attitude-control system of a spacecraft when it senses the sun. Also known as sun sensor.

soldering lug A stamped metal strip used as a terminal to which wires can be soldered.

sole Electrode used in magnetrons and backward-wave oscillators to carry a current that generates a magnetic field in the direction wanted.

solenoid Also known as electric solenoid. 1. An electrically energized coil of insulated wire which produces a magnetic field within the coil. 2. In particular, a coil that surrounds a movable iron core which is pulled to a central position with respect to the coil when the coil is energized by sending current through it.

solid-dielectric capacitor A capacitor whose dielectric is one of several solid materials such as ceramic, mica, glass, plastic film, or paper.

solid-electrolyte battery A primary battery whose electrolyte is either a solid crystalline salt, such as silver iodide or lead chloride, or an ion-exchange membrane; in either case, conductivity is almost entirely ionic.

solid-electrolyte fuel cell Self-contained fuel cell in which oxygen is the oxidant and hydrogen is the fuel; the oxidant and fuel are kept separated by a solid electrolyte which has a crystalline structure and a low conductivity.

solid electrolytic capacitor An electrolytic capacitor in which the dielectric is an anodized coating on one electrode, with a solid semiconductor material filling the rest of the space between the electrodes.

solid insulator An electric insulator made of a solid substance, such as sulfur, polystyrene, rubber, or porcelain.

solid logic technology A method of computer construction that makes use of miniaturized modules, resulting in faster circuitry because of the reduced distances that current must travel.

solid-state battery A battery in which both the electrodes and the electrolyte are solid-state materials.

solid-state circuit Complete circuit formed from a single block of semiconductor material.

solid-state circuit breaker A circuit breaker in which a Zener diode, silicon controlled rectifier, or solid-state device is connected to sense when load terminal voltage exceeds a safe value.

solid-state component A component whose operation depends on the control of electrical or magnetic phenomena in solids, such as a transistor, crystal diode, or ferrite device.

solid-state device A device, other than a conductor, which uses magnetic, electrical, and other properties of solid materials, as opposed to vacuum or gaseous devices.

solid-state image sensor *See* charge-coupled image sensor.

solid-state lamp *See* light-emitting diode.

solid-state relay A relay that uses only solid-state components, with no moving parts. Abbreviated SSR.

solid-state switch A microwave switch in which a semiconductor material serves as the switching element; a zero or negative potential applied to the control electrode will reverse-bias the switch and turn it off, and a slight positive voltage will turn it on.

solid-state thyratron A semiconductor device, such as a silicon controlled rectifier, that approximates the extremely fast switching speed and power-handling capability of a gaseous thyratron tube.

solid-state uninterruptible power system An uninterruptible power system in which the load operates continuously from the output of a dc-to-ac static inverter powered by a battery.

solid tantalum capacitor An electrolytic capacitor in which the anode is a porous pellet of tantalum; the dielectric is an extremely thin layer of tantalum pentoxide formed by anodization of the exterior and interior surfaces of the pellet; the cathode is a layer of semiconducting manganese dioxide that fills the pores of the anode over the dielectric.

solion An electrochemical device in which amplification is obtained by controlling and monitoring a reversible electrochemical reaction.

solution ceramic A nonbrittle, inorganic ceramic insulating coating that can be applied to wires at a low temperature; examples include ceria, chromia, titania, and zirconia.

Sommerfeld equation *See* Sommerfeld formula.

Sommerfeld formula An approximate formula for the field strength of electromagnetic radiation generated by an antenna at distances small enough so that the curvature of the earth may be neglected, in terms of radiated power, distance from the antenna, and various constants and parameters. Also known as Sommerfeld equation.

Sommerfeld model *See* free-electron theory of metals.

Sommerfeld theory *See* free-electron theory of metals.

sonar array An arrangement of several sonar transducers or sonar projectors, appropriately spaced and energized to give proper directional characteristics.

sonar detector *See* sonar receiver.

sonar receiver A receiver designed to intercept and amplify the sound signals reflected by an underwater target and display the accompanying intelligence in useful form; it may also pick up other underwater sounds. Also known as sonar detector.

sonar resolver A resolver used with echo-ranging and depth-determining sonar to calculate and record the horizontal range of a sonar target, as required for depth-bombing.

sonar self-noise Unwanted sonar signals generated in the sonar equipment itself.

sonar transmitter A transmitter that generates electrical signals of the proper frequency and form for application to a sonar transducer or sonar projector, to produce sound waves of the same frequency in water; the sound waves may carry intelligence.

sonic delay line *See* acoustic delay line.

SOS *See* silicon-on-sapphire.

sound channel The series of stages that handles only the sound signal in a television receiver.

source 1. The circuit or device that supplies signal power or electric energy or charge to a transducer or load circuit. 2. The terminal in a field-effect transistor from which majority carriers flow into the conducting channel in the semiconductor material.

source-follower amplifier *See* common-drain amplifier.

source impedance Impedance presented by a source of energy to the input terminals of a device.

source transition loss The transmission loss at the junction between an energy source and a transducer connecting that source to an energy load; measured by the ratio of the source power to the input power.

sourcing Redesign or the modification of existing equipment to eliminate a source of radio-frequency interference.

south pole The pole of a magnet at which magnetic lines of force are assumed to enter.

space charge The net electric charge within a given volume.

space-charge balanced flow A method of focusing an electron beam in the interaction region of a traveling-wave tube; there is an axial magnetic field in the interaction region which is stronger than that in the gun region; at the transition between the two values of magnetic field strength, the beam is given a rotation in such a direction as to produce an inward force that counterbalances the outward forces from space charge and from the centrifugal forces set up by rotation.

space-charge debunching A process in which the mutual interactions between electrons in a stream spread out the electrons of a bunch.

space-charge effect Repulsion of electrons emitted from the cathode of a thermionic vacuum tube by electrons accumulated in the space charge near the cathode.

space-charge grid Grid operated at a low positive potential and placed between the cathode and control grid of a vacuum tube to reduce the limiting effect of space charge on the current through the tube.

space-charge layer *See* depletion layer.

space-charge limitation The current flowing through a vacuum between a cathode and an anode cannot exceed a certain maximum value, as a result of modification of the electric field near the cathode due to space charge in this region.

space-charge polarization Polarization of a dielectric which occurs when charge carriers are present which can migrate an appreciable distance through the dielectric

sparking potential   405

but which become trapped or cannot discharge at an electrode. Also known as interfacial polarization.

space-charge region   Of a semiconductor device, a region in which the net charge density is significantly different from zero.

space current   Total current flowing between the cathode and all other electrodes in a tube; this includes the plate current, grid current, screen grid current, and any other electrode current which may be present.

spaced antenna   Antenna system consisting of a number of separate antennas spaced a considerable distance apart, used to minimize local effects of fading at short-wave receiving stations.

space diversity reception   Radio reception involving the use of two or more antennas located several wavelengths apart, feeding individual receivers whose outputs are combined; the system gives an essentially constant output signal despite fading due to variable propagation characteristics, because fading affects the spaced-out antennas at different instants of time.

space factor   1. The ratio of the space occupied by the conductors in a winding to the total cubic content or volume of the winding, or the similar ratio of cross sections. 2. The ratio of the space occupied by iron to the total cubic content of an iron core.

space permeability   Factor that expresses the ratio of magnetic induction to magnetizing force in a vacuum; in the centimeter-gram-second electromagnetic system of units, the permeability of a vacuum is arbitrarily taken as unity; in the meter-kilogram-second-ampere system, it is $4\pi \times 10^{-7}$.

space wave   The component of a ground wave that travels more or less directly through space from the transmitting antenna to the receiving antenna; one part of the space wave goes directly from one antenna to the other; another part is reflected off the earth between the antennas.

spacistor   A multiple-terminal solid-state device, similar to a transistor, that generates frequencies up to about 10,000 megahertz by injecting electrons or holes into a space-charge layer which rapidly forces these carriers to a collecting electrode.

spaghetti   Insulating tubing used over bare wires or as a sleeve for holding two or more insulated wires together; the tubing is usually made of a varnished cloth or a plastic.

spark   A short-duration electric discharge due to a sudden breakdown of air or some other dielectric material separating two terminals, accompanied by a momentary flash of light. Also known as electric spark; spark discharge; sparkover.

spark capacitor   Capacitor connected across a pair of contact points, or across the inductance which causes the spark, for the purpose of diminishing sparking at these points.

spark coil   An induction coil for producing spark discharges, as to initiate combustion in an internal combustion engine.

spark discharge   *See* spark.

spark gap   An arrangement of two electrodes between which a spark may occur; the insulation (usually air) between the electrodes is self-restoring after passage of the spark; used as a switching device, for example, to protect equipment against lightning or to switch a radar antenna from receiver to transmitter and vice versa.

spark-gap generator   A high-frequency generator in which a capacitor is repeatedly charged to a high voltage and allowed to discharge through a spark gap into an oscillatory circuit, generating successive trains of damped high-frequency oscillations.

sparking potential   *See* breakdown voltage.

406 sparking voltage

sparking voltage *See* breakdown voltage.

spark killer *See* spark suppressor.

sparkover *See* spark.

sparkover voltage *See* flashover voltage.

spark plate A metal plate insulated from the chassis of an auto radio by a thin sheet of mica, and connected to the battery lead to bypass noise signals picked up by battery wiring in the engine compartment.

spark plug A device that screws into the cylinder of an internal combustion engine to provide a pair of electrodes between which an electrical discharge is passed to ignite the explosive mixture.

spark suppressor A device used to prevent sparking between a pair of contacts when the contacts open, such as a resistor and capacitor in series between the contacts, or, in the case of an inductive circuit, a rectifier in parallel with the inductor. Also known as spark killer.

spark transmitter A radio transmitter that utilizes the oscillatory discharge of a capacitor through an inductor and a spark gap as the source of radio-frequency power.

spark voltage The voltage required to create an arc across the gap of a spark plug.

SPDT *See* single-pole double-throw.

specific charge The ratio of a particle's charge to its mass.

specific conductance *See* conductivity.

specific insulation resistance *See* volume resistivity.

specific repetition rate The pulse repetition rate of a pair of transmitting stations of an electronic navigation system using various rates differing slightly from each other, as in loran.

specific resistance *See* electrical resistivity.

spectral density *See* spectral energy distribution.

spectral energy distribution The power carried by electromagnetic radiation within some small interval of wavelength (of frequency) of fixed amount as a function of wavelength (of frequency). Also known as spectral density.

spectral sensitivity Radiant sensitivity, considered as a function of wavelength.

spectrum-selectivity characteristic Measure of the increase in the minimum input signal power over the minimum detectable signal required to produce an indication on a radar indicator, if the received signal has a spectrum different from that of the normally received signal.

spectrum signature The spectral characteristics of the transmitter, receiver, and antenna of an electronic system, including emission spectra, antenna patterns, and other characteristics.

spectrum signature analysis The evaluation of electromagnetic interference from transmitting and receiving equipment to determine operational and environment compatibility.

specular transmittance The ratio of the power carried by electromagnetic radiation which emerges from a body and is parallel to a beam entering the body, to the power carried by the beam entering the body.

speech inverter *See* scrambler.

speech scrambler *See* scrambler.

speed control A control that changes the speed of a motor or other drive mechanism, as for a phonograph or magnetic tape recorder.

speed of light The speed of propagation of electromagnetic waves in a vacuum, which is a physical constant equal to $299{,}792.4580 \pm 0.0012$ kilometers per second. Also known as electromagnetic constant; velocity of light.

speed-power product The product of the gate speed or propagation delay of an electronic circuit and its power dissipation.

speed regulator A device that maintains the speed of a motor or other device at a predetermined value or varies it in accordance with a predetermined plan.

speromagnetic state The condition of a rare-earth glass in which the spins are oriented in fixed directions which are more or less random because of electric fields which exist in the glass.

sphere gap A spark gap between two equal-diameter spherical electrodes.

spherical antenna An antenna having the shape of a sphere, used chiefly in theoretical studies.

spherical capacitor A capacitor made of two concentric metal spheres with a dielectric filling the space between the spheres.

spherical-earth attenuation Attenuation over an imperfectly conducting spherical earth in excess of that over a perfectly conducting plane.

spherical-earth factor The ratio of the electric field strength that would result from propagation over an imperfectly conducting spherical earth to that which would result from propagation over a perfectly conducting plane.

spider A structure on the shaft of an electric rotating machine that supports the core or poles of the rotor, consisting of a hub, spokes, and rim, or some similar arrangement.

spiderweb antenna All-wave receiving antenna having several different lengths of doublets connected somewhat like the web of a spider to give favorable pickup characteristics over a wide range of frequencies.

spike A sputtering event in which the process from impact of a bombarding projectile to the ejection of target atoms involves motion of a large number of particles in the target, so that collisions between particles become significant.

spike antenna *See* monopole antenna.

spin-density wave The ground state of a metal in which the conduction–electron spin density has a sinusoidal variation in space.

spin filter A device used in a Lamb-shift polarized ion source to cause those atoms having an undesired nuclear spin orientation to decay from their metastable state to the ground state, while those with the desired spin orientation are allowed to pass through without decay.

spin glass A substance in which the atomic spins are oriented in random but fixed directions.

spin-lattice interaction The state of a solid when the energy of electron spins is being shared with the thermal-vibration energy of the solid as a whole.

spin-lattice relaxation Magnetic relaxation in which the excess potential energy associated with electron spins in a magnetic field is transferred to the lattice.

spin magnetism Paramagnetism or ferromagnetism that arises from polarization of electron spins in a substance.

spin paramagnetism Paramagnetism that arises from the electron spins in a substance.

spin-polarized low-energy electron diffraction A version of low-energy electron diffraction in which electrons in the incident beam have their spins aligned in one direction; used in studies of the magnetic properties of atoms near the surface of a material.

spin-spin relaxation Magnetic relaxation, observed after application of weak magnetic fields, in which the excess potential energy associated with electron spins in a magnetic field is redistributed among the spins, resulting in heating of the spin system.

spin temperature For a system of electron spins in a lattice, a temperature such that the population of the energy levels of the spin system is given by the Boltzmann distribution with this temperature.

spinthariscope An instrument for viewing the scintillations of alpha particles on a luminescent screen, usually with the aid of a microscope.

spin wave A sinusoidal variation, propagating through a crystal lattice, of that angular momentum which is associated with magnetism (mostly spin angular momentum of the electrons).

spiral delay line A transmission line which has a helical inner conductor.

spiral four cable A quad cable in which the four conductors are twisted about a common axis, the two sets of opposite conductors being used as pairs.

splice A joint used to connect two lengths of conductor with good mechanical strength and good conductivity.

split-anode magnetron A magnetron in which the cylindrical anode is divided longitudinally into halves, between which extremely high-frequency oscillations are produced.

split-phase motor A single-phase induction motor having an auxiliary winding connected in parallel with the main winding, but displaced in magnetic position from the main winding so as to produce the required rotating magnetic field for starting; the auxiliary circuit is generally opened when the motor has attained a predetermined speed.

split-stator variable capacitor Variable capacitor having a rotor section that is common to two separate stator sections; used in grid and plate tank circuits of transmitters for balancing purposes.

splitting In the scope presentation of the standard loran (2000 kilohertz), signals the slow diminution of the leading or lagging edge of the pulse so that it resembles two pulses and eventually a single pulse, which appears to be normal but which may be displaced in time by as much as 10,000 microseconds; this phenomenon is caused by shifting of the E_1 reflections from the ionosphere, and if the deformation is that of the leading edge and is not detected, it will cause serious errors in the reading of the navigational parameter.

spoiler Rod grating monted on a parabolic reflector to change the pencil-beam pattern of the reflector to a cosecant-squared pattern; rotating the reflector and grating 90° with respect to the feed antenna changes one pattern to the other.

spontaneous magnetization Magnetization which a substance possesses in the absence of an applied magnetic field.

spontaneous polarization Electric polarization that a substance possesses in the absence of an external electric field.

spoofing Deceiving or misleading the enemy in electronic operations, as by continuing transmission on a frequency after it has been effectively jammed by the enemy, using decoy radar transmitters to lead the enemy into a useless jamming effort, or transmitting radio messages containing false information for intentional interception by the enemy.

square-law demodulator 409

sporadic reflections Sharply defined reflections of substantial intensity from the sporadic E layer at frequencies greater than the critical frequency of the layer; they are variable with respect to time of occurrence, geographic location, and range of frequencies at which they are observed.

spot In a cathode-ray tube, the area instantaneously affected by the impact of an electron beam.

spot jammer A jammer that interferes with reception of a specific channel or frequency.

spotlight 1. A strong beam of light that illuminates only a small area about an object. 2. A lamp that has a strongly focused beam.

spot noise figure Of a transducer at a selected frequency, the ratio of the output noise power per unit bandwidth to a portion thereof attributable to the thermal noise in the input termination per unit bandwidth, the noise temperature of the input termination being standard (290 K).

spot-size error The distortion of the radar returns on the radarscope presentation caused by the diameter of the electron beam which displays the returns of the scope and the lateral radiation across the scope of part of the glow produced when the electron beam strikes the phosphorescent coating of the cathode-ray tube.

spottiness Bright spots scattered irregularly over the reproduced image in a television receiver, due to man-made or static interference entering the television system at some point.

spray point One of the sharp points arranged in a row and charged to a high direct-current potential, used to charge and discharge the conveyor belt in a Van de Graaff generator.

spreader An insulating crossarm used to hold apart the wires of a transmission line or multiple-wire antenna.

spread reflection Reflection of electromagnetic radiation from a rough surface with large irregularities. Also known as mixed reflction.

spread spectrum transmission Communications technique in which many different signal waveforms are transmitted in a wide band; power is spread thinly over the band so narrow-band radios can operate within the wide-band without interference; used to achieve security and privacy, prevent jamming, and utilize signals buried in noise.

spring contact A relay or switch contact mounted on a flat spring, usually of phosphor bronze.

SPST *See* single-pole single-throw.

spurious emission *See* spurious radiation.

spurious modulation Undesired modulation occurring in an oscillator, such as frequency modulation caused by mechanical vibration.

spurious radiation Any emission from a radio transmitter at frequencies outside its frequency band. Also known as spurious emission.

spurious response 1. Response of a radio receiver to a frequency different from that to which the receiver is tuned. 2. In electronic warfare, the undesirable signal images in the intercept receiver resulting from the mixing of the intercepted signal with harmonics of the local oscillators in the receiver.

sputtering Also known as cathode sputtering. 1. The ejection of atoms or groups of atoms from the surface of the cathode of a vacuum tube as the result of heavy-ion impact. 2. The use of this process to deposit a thin layer of metal on a glass, plastic, metal, or other surface in vacuum.

square-law demodulator *See* square-law detector.

square-law detector A demodulator whose output voltage is proportional to the square of the amplitude-modulated input voltage. Also known as square-law demodulator.

square-loop ferrite A ferrite that has an approximately rectangular hysteresis loop.

squareness ratio 1. The magnetic induction at zero magnetizing force divided by the maximum magnetic induction, in a symmetric cyclic magnetization of a material. 2. The magnetic induction when the magnetizing force has changed half-way from zero toward its negative limiting value divided by the maximum magnetic induction in a symmetric cyclic magnetization of a material.

square wave An oscillation the amplitude of which shows periodic discontinuities between two values, remaining constant between jumps.

square-wave amplifier Resistance-coupled amplifier, the circuit constants of which are to amplify a square wave with the minimum amount of distortion.

square-wave generator A signal generator that generates a square-wave output voltage.

square-wave response The response of a circuit or device when a square wave is applied to the input.

squaring circuit 1. A circuit that reshapes a sine or other wave into a square wave. 2. A circuit that contains nonlinear elements proportional to the square of the input voltage.

squealing A condition in which a radio receiver produces a high-pitched note or squeal along with the desired radio program, due to interference between stations or to oscillation in some receiver circuit.

squeezable waveguide A waveguide whose dimensions can be altered periodically; used in rapid scanning.

squeeze section Length of waveguide constructed so that alteration of the critical dimension is possible with a corresponding alteration in the electrical length.

squegger *See* blocking oscillator.

squegging Condition of self-blocking in an electron-tube-oscillator circuit.

squegging oscillator *See* blocking oscillator.

squelch To automatically quiet a receiver by reducing its gain in response to a specified characteristic of the input.

squelch circuit *See* noise suppressor.

SQUID *See* superconducting quantum interference device.

squint 1. The angle between the two major lobe axes in a radar lobe-switching antenna. 2. The angular difference between the axis of radar antenna radiation and a selected geometric axis, such as the axis of the reflector. 3. The angle between the full-right and full-left positions of the beam of a conical-scan radar antenna.

squirrel-cage motor An induction motor in which the secondary circuit consists of a squirrel-cage winding arranged in slots in the iron core.

squirrel-cage rotor *See* squirrel-cage winding.

squirrel-cage winding A permanently short-circuited winding, usually uninsulated, around the periphery of the rotor and joined by continuous end rings. Also known as squirrel-cage rotor.

squitter Random firing, intentional or otherwise, of the transponder transmitter in the absence of interrogation.

SSI *See* small-scale integration.

SSR *See* solid-state relay.

stability factor A measure of a transistor amplifier's bias stability, equal to the rate of change of collector current with respect to reverse saturation current.

stabilivolt Gas tube that maintains a constant voltage drop across its terminals, essentially independent of current, over a relatively wide range.

stabilization 1. Feedback introduced into vacuum tube or transistor amplifier stages to reduce distortion by making the amplification substantially independent of electrode voltages and tube constants. 2. Treatment of a magnetic material to improve the stability of its magnetic properties.

stabilized winding Auxiliary winding used particularly in star-connected transformers to stabilize the neutral point of the fundamental frequency voltages, to protect the transformer and the system from excessive third-harmonic voltages; and to prevent telephone interference caused by third-harmonic currents and voltages in the lines and earth. Also known as tertiary winding.

stabistor A diode component having closely controlled conductance, controlled storage charge, and low leakage, as required for clippers, clamping circuits, bias regulators, and other logic circuits that require tight voltage-level tolerances.

stable strobe Series of strobes which behaves as if caused by a single jammer.

stack *See* pileup.

stacked antennas Two or more identical antennas arranged above each other on a vertical supporting structure and connected in phase to increase the gain.

stacked array An array in which the antenna elements are stacked one above the other and connected in phase to increase the gain.

stacked-dipole antenna Antenna in which directivity is increased by providing a number of identical dipole elements, excited either directly or parasitically; the resultant radiation pattern depends on the number of dipole elements used, the spacing and phase difference between the elements, and the relative magnitudes of the currents.

stacked loops Two or more loop antennas arranged above each other on a vertical supporting structure and connected in phase to increase the gain. Also known as vertically stacked loops.

stacking The placing of antennas one above the other, connecting them in phase to increase the gain.

stage A circuit containing a single section of an electron tube or equivalent device or two or more similar sections connected in parallel, push-pull, or push-push; it includes all parts connected between the control-grid input terminal of the device and the input terminal of the next adjacent stage.

stage gain The ratio of the output power of an amplifier stage to the input power, usually expressed in decibels.

staggered tuning Alignment of successive tuned circuits to slightly different frequencies in order to widen the overall amplitude-frequency response curve.

stagger-tuned amplifier An amplifier that uses staggered tuning to give a wide bandwidth.

stagger-tuned filter A filter consisting of a cascade of amplifier stages with tuned coupling networks whose resonant frequencies and bandwidths may be easily adjusted to achieve an overall transmission function of desired shape (maximally flat or equal ripple).

stake An iron peg used as a power electrode to transfer current into the ground in electrical prospecting.

stalo A highly stable local radio-frequency oscillator used for heterodyning signals to produce an intermediate frequency in radar moving-target indicators; only echoes that have changed slightly in frequency due to reflection from a moving target produce an output signal. Derived from stable local oscillator.

stamping A transformer lamination that has been cut out of a strip or sheet of metal by a punch press.

standard antenna An open single-wire antenna (including the lead-in wire) having an effective height of 4 meters.

standard capacitor A capacitor constructed in such a manner that its capacitance value is not likely to vary with temperature and is known to a high degree of accuracy. Also known as capacitance standard.

standard cell A primary cell whose voltage is accurately known and remains sufficiently constant for instrument calibration purposes; the Weston standard cell has a voltage of 1.018636 volts at 20°C.

standard conditions The allotropic form in which a substance most commonly occurs.

standard inductor An inductor (coil) having high stability of inductance value, with little variation of inductance with current or frequency and with a low temperature coefficient; it may have an air core or an iron core; used as a primary standard in laboratories and as a precise working standard for impedance measurements.

standard noise temperature The standard reference temperature for noise measurements, equal to 290 K.

standard propagation Propagation of radio waves over a smooth spherical earth of specified dielectric constant and conductivity, under conditions of standard refraction in the atmosphere.

standard refraction Refraction which would occur in an idealized atmosphere in which the index of refraction decreases uniformly with height at a rate of 39×10^{-6} per kilometer; standard refraction may be included in ground wave calculations by use of an effective earth radius of 8.5×10^6 meters, or $\frac{4}{3}$ the geometrical radius of the earth.

standard target A radar target which will produce an echo of known power under various conditions; smooth metal spheres or corner reflectors of known dimensions are such targets, and they may be used to calibrate a radar or check its performance.

standard test-tone power One milliwatt (0 decibels above one milliwatt) at 1000 hertz.

standard waveguide Any one of several rectangular waveguides whose dimensions have been specified by various agencies and which are in general use.

standby battery A storage battery held in reserve as an emergency power source in event of failure of regular power facilities at a radio station or other location.

standby power source An uninterruptible power system in which the load normally operated from the commercial power line is switched to the output of a dc-to-ac static inverter powered by a battery in the event of a power failure.

standing-wave detector An electric indicating instrument used for detecting a standing electromagnetic wave along a transmission line or in a waveguide and measuring the resulting standing-wave ratio; it can also be used to measure the wavelength, and hence the frequency, of the wave. Also known as standing-wave indicator; standing-wave meter; standing-wave-ratio meter.

standing-wave indicator See standing-wave detector.

standing-wave loss factor The ratio of the transmission loss in an unmatched waveguide to that in the same waveguide when matched.

standing-wave meter See standing-wave detector.

standing-wave method Any method of measuring the wavelength of electromagnetic waves that involves measuring the distance between successive nodes or antinodes of standing waves.

statfarad 413

standing-wave producer A movable probe inserted in a slotted waveguide to produce a desired standing-wave pattern, generally for test purposes.

standing-wave-ratio meter *See* standing-wave detector.

standoff insulator An insulator used to support a conductor at a distance from the surface on which the insulator is mounted.

star-connected circuit Polyphase circuit in which all the current paths within the region that delimits the circuit extend from each of the points of entry of the phase conductors to a common conductor (which may be the neutral conductor).

star connection *See* star network.

star-delta switching starter A type of motor starter, used with three-phase induction motors, that switches the stator windings from a star connection to a delta connection.

Stark-Lunelund effect The polarization of light emitted from a beam of moving atoms in a region where there are no electric or magnetic fields.

star lamp A high-pressure xenon arc, used in a planetarium, which produces a tiny, intense point of light focused through thousands of individual lenses and pinholes, and projected to the planetarium's dome.

star network A set of three or more branches with one terminal of each connected at a common node to give the form of a star. Also known as star connection; Y connection.

starter 1. A device used to start an electric motor and to accelerate the motor to normal speed. 2. An auxiliary control electrode used in a gas tube to establish sufficient ionization to reduce the anode breakdown voltage. Also known as trigger electrode. 3. *See* engine starter.

starting box A device for providing extra resistance in the armature of a motor while it is being started.

starting motor *See* engine starter.

starting reactor A reactor that is used to limit the starting current of electric motors, and usually consists of an iron-core inductor connected in series with the machine stator winding.

start-stop multivibrator *See* monostable multivibrator.

stat- A prefix indicating an electrical unit in the electrostatic centimeter-gram-second system of units; it is attached to the corresponding SI unit.

statΩ *See* statohm.

stat℧ *See* statmho.

statA *See* statampere.

statampere The unit of electric current in the electrostatic centimeter-gram-second system of units, equal to a flow of charge of 1 statcoulomb per second; equal to approximately 3.3356×10^{-10} ampere. Abbreviated statA.

statC *See* statcoulomb.

statcoulomb The unit of charge in the electrostatic centimeter-gram-second system of units, equal to the charge which exerts a force of 1 dyne on an equal charge at a distance of 1 centimeter in a vacuum; equal to approximately 3.3356×10^{-10} coulomb. Abbreviated statC. Also known as franklin (Fr); unit charge.

statF *See* statfarad.

statfarad Unit of capacitance in the electrostatic centimeter-gram-second system of units, equal to the capacitance of a capacitor having a charge of 1 statcoulomb, across the plates of which the charge is 1 statvolt; equal to approximately 1.1126×10^{-12} farad. Abbreviated statF. Also known as centimeter.

statH *See* stathenry.

stathenry The unit of inductance in the electrostatic centimeter-gram-second system of units, equal to the self-inductance of a circuit or the mutual inductance between two circuits if there is an induced electromotive force of 1 statvolt when the current is changing at a rate of 1 statampere per second; equal to approximately 8.9876×10^{11} henry. Abbreviated statH.

static breeze *See* convective discharge.

static characteristic A relation between a pair of variables, such as electrode voltage and electrode current, with all other operating voltages for an electron tube, transistor, or other amplifying device maintained constant.

static charge An electric charge accumulated on an object.

static discharger A rubber-covered cloth wick about 6 inches (15 centimeters) long, sometimes attached to the trailing edges of the surfaces of an aircraft to discharge static electricity in flight.

static electricity 1. The study of the effects of macroscopic charges, including the transfer of a static charge from one object to another by actual contact or by means of a spark that bridges an air gap between the objects. 2. *See* electrostatics.

static eliminator Device intended to reduce the effect of atmospheric static interference in a radio receiver.

static induction transistor A type of transistor capable of operating at high current and voltage, whose current-voltage characteristics do not saturate, and are similar in form to those of a vacuum triode. Abbreviated SIT.

static inverter A device that converts a dc voltage to a stable ac voltage for use in an uninterruptible power system.

static machine A machine for generating electric charges, usually by electric induction, sometimes used to build up high voltages for research purposes.

static regulator Transmission regulator in which the adjusting mechanism is in self-equilibrium at any setting and requires control power to change the setting.

static sensitivity In phototubes, quotient of the direct anode current divided by the incident radiant flux of constant value.

static switching Switching of circuits by means of magnetic amplifiers, semiconductors, and other devices that have no moving parts.

station 1. An assembly line or assembly machine location at which a wiring board or chassis is stopped for insertion of one or more parts. 2. A location at which radio, television, radar, or other electric equipment is installed.

stationary ergodic noise A stationary noise for which the probability that the noise voltage lies within any given interval at any time is nearly equal to the fraction of time that the noise voltage lies within this interval if a sufficiently long observation interval is recorded.

stationary noise A random noise for which the probability that the noise voltage lies within any given interval does not change with time.

statistical multiplexer A device which combines several low-speed communications channels into a single high-speed channel, and which can manage more communications traffic than a standard multiplexer by analyzing traffic and choosing different transmission patterns.

statmho The unit of conductance, admittance, and susceptance in the electrostatic centimeter-gram-second system of units, equal to the conductance between two points of a conductor when a constant potential difference of 1 statvolt applied between the points produces in this conductor a current of 1 statampere, the conductor not being the source of any electromotive force; equal to approximately 1.1126×10^{-12} mho. Abbreviated stat℧. Also known as statsiemens (statS).

statohm The unit of resistance, reactance, and impedance in the electrostatic centimeter-gram-second system of units, equal to the resistance between two points of a conductor when a constant potential difference of 1 statvolt between these points produces a current of 1 statampere; equal to approximately 8.9876×10^{11} ohm. Abbreviated statΩ.

stator The portion of a rotating machine that contains the stationary parts of the magnetic circuit and their associated windings.

stator armature A stator which includes the main current-carrying winding in which electromotive force produced by magnetic flux rotation is induced; it is found in most alternating-current machines.

stator plate One of the fixed plates in a variable capacitor; stator plates are generally insulated from the frame of the capacitor.

statS *See* statmho.

statsiemens *See* statmho.

statT *See* stattesla.

stattesla The unit of magnetic flux density in the electrostatic centimeter-gram-second system of units, equal to one statweber per square centimeter; equal to approximately 2.9979×10^6 tesla. Abbreviated statT.

statV *See* statvolt.

statvolt The unit of electric potential and electromotive force in the electrostatic centimeter-gram-second system of units, equal to the potential difference between two points such that the work required to transport 1 statcoulomb of electric charge from one to the other is equal to 1 erg; equal to approximately 2.9979×10^2 volts. Abbreviated statV.

statWb *See* statweber.

statweber The unit of magnetic flux in the electrostatic centimeter-gram-second system of units, equal to the magnetic flux which, linking a circuit of one turn, produces in it an electromotive force of 1 statvolt as it is reduced to zero at a uniform rate in 1 second; equal to approximately 2.9979×10^2 weber. Abbreviated statWb.

steady-state current An electric current that does not change with time.

steerable antenna A directional antenna whose major lobe can be readily shifted in direction.

Steinmetz coefficient The constant of proportionality in Steinmetz's law.

Steinmetz's law The energy converted into heat per unit volume per cycle during a cyclic change of magnetization is proportional to the maximum magnetic induction raised to the 1.6 power, the constant of proportionality depending only on the material.

STEM *See* scanning transmission electron microscope.

stenode circuit Superheterodyne receiving circuit in which a piezoelectric unit is used in the intermediate-frequency amplifier to balance out all frequencies except signals at the crystal frequency, thereby giving very high selectivity.

step attenuator An attenuator in which the attenuation can be varied in precisely known steps by means of switches.

step-by-step switch A bank-and-wiper switch in which the wipers are moved by electromagnet ratchet mechanisms individual to each switch.

step change The change of a variable from one value to another in a single process, taking a negligible amount of time.

step-down transformer A transformer in which the alternating-current voltages of the secondary windings are lower than those applied to the primary winding.

step-function generator A function generator whose output waveform increases and decreases suddenly in steps that may or may not be equal in amplitude.

stepped-wave static inverter A static inverter that generates several pulses in each half cycle and combines them to achieve an output voltage which needs very little filtering.

stepper motor A motor that rotates in short and essentially uniform angular movements rather than continuously; typical steps are 30, 45, and 90°; the angular steps are obtained electromagnetically rather than by the ratchet and pawl mechanisms of stepping relays. Also known as magnetic stepping motor; stepping motor; step-servo motor.

stepping *See* zoning.

stepping motor *See* stepper motor.

stepping relay A relay whose contact arm may rotate through 360° but not in one operation. Also known as rotary stepping relay; rotary stepping switch; stepping switch.

stepping switch *See* stepping relay.

step-recovery diode A varactor in which forward voltage injects carriers across the junction, but before the carriers can combine, voltage reverses and carriers return to their origin in a group; the result is abrupt cessation of reverse current and a harmonic-rich waveform.

step-servo motor *See* stepper motor.

step strobe marker Form of strobe marker in which the discontinuity is in the form of a step in the time base.

step-up transformer Transformer in which the energy transfer is from a low-voltage winding to a high-voltage winding or windings.

step voltage regulator A type of voltage regulator used on distribution feeder lines; it provides increments or steps of voltage change.

sterba curtain Type of stacked dipole antenna array consisting of one or more phased half-wave sections with a quarter-wave section at each end; the array can be oriented for either vertical or horizontal radiation, and can be either center or end fed.

stereofluoroscopy A fluoroscopic technique that gives three-dimensional images.

stiletto An advanced electronic subsystem contained in United States strike aircraft type F-4D for detection, identification, and location of ground-based radars; the location of radar targets is determined by direction finding and passive ranging techniques; it is used for the delivery of guided and unguided weapons against the target radars under all weather conditions.

stimulated emission device A device that uses the principle of amplification of electromagnetic waves by stimulated emission, namely, a maser or a laser.

stirring effect The circulation in a molten metal carrying electric current as a result of the combined forces of the pinch and motor effects.

stop band *See* rejection band.

stopping capacitor *See* coupling capacitor.

stopping potential Voltage required to stop the outward movement of electrons emitted by photoelectric or thermionic action.

storage battery A connected group of two or more storage cells or a single storage cell. Also known as accumulator; accumulator battery; rechargeable battery; secondary battery.

storage camera *See* iconoscope.

storage cell An electrolytic cell for generating electric energy, in which the cell after being discharged may be restored to a charged condition by sending a current through it in a direction opposite to that of the discharging current. Also known as secondary cell.

storage oscilloscope An oscilloscope that can retain an image for a period of time ranging from minutes to days, or until deliberately erased to make room for a new image.

storage time 1. The time required for excess minority carriers stored in a forward-biased *pn* junction to be removed after the junction is switched to reverse bias, and hence the time interval between the application of reverse bias and the cessation of forward current. 2. The time required for excess charge carriers in the collector region of a saturated transistor to be removed when the base signal is changed to cut-off level, and hence for the collector current to cease. 3. *See* retention time.

storage tube An electron tube employing cathode-ray beam scanning and charge storage for the introduction, storage, and removal of information. Also known as electrostatic storage tube; memory tube (deprecated usage).

storage-type camera tube *See* iconoscope.

store transmission bridge Transmission bridge, which consists of four identical impedance coils (the two windings of the back-bridge relay and live relay of a connector, respectively) separated by two capacitors, which couples the calling and called telephones together electrostatically for the transmission of voice-frequency (alternating) currents, but separates the two lines for the transmission of direct current for talking purposes (talking current).

straight vertical antenna An antenna consisting of a straight vertical wire.

strain insulator An insulator used between sections of a stretched wire or antenna to break up the wire into insulated sections while withstanding the total pull of the wire.

stranded conductor *See* stranded wire.

stranded wire A conductor composed of a group of wires or a combination of groups of wires, usually twisted together. Also known as stranded conductor.

strapped magnetron A multicavity magnetron in which resonator segments having the same polarity are connected together by small conducting strips to suppress undesired modes of oscillation.

strapping 1. Connecting two or more points in a circuit or device with a short piece of wire or metal. 2. Connecting together resonator segments having the same polarity in a multicavity magnetron to suppress undesired modes of oscillation.

stray capacitance Undesirable capacitance between circuit wires, between wires and the chassis, or between components and the chassis of electronic equipment.

stray current 1. A portion of a current that flows over a path other than the intended path, and may cause electrochemical corrosion of metals in contact with electrolytes. 2. An undesirable current generated by discharge of static electricity; it commonly arises in loading and unloading petroleum fuels and some chemicals, and can initiate explosions.

stray field Leakage of magnetic flux that spreads outward from a coil and does no useful work.

streaming current The electric current which is produced when a liquid is forced to flow through a diaphragm, capillary, or porous solid.

streaming potential The difference in electric potential between a diaphragm, capillary, or porous solid and a liquid that is forced to flow through it.

striation A succession of alternately luminous and dark regions sometimes observed in the positive column of a glow-discharge tube near the anode.

striking potential 1. Voltage required to start an electric arc. 2. Smallest grid-cathode potential value at which plate current begins flowing in a gas-filled triode.

strip line A strip transmission line that consists of a flat metal-strip center conductor which is separated from flat metal-strip outer conductors by dielectric strips.

strip-line circuit A circuit in which one or more strip transmission lines serve as filters or other circuit components.

strip transmission line A microwave transmission line consisting of a thin, narrow, rectangular metal strip that is supported above a ground-plane conductor or between two wide ground-plane conductors and is usually separated from them by a dielectric material.

strobe 1. Intensified spot in the sweep of a deflection-type indicator, used as a reference mark for ranging or expanding the presentation. 2. Intensified sweep on a plan-position indicator or B-scope; such a strobe may result from certain types of interference, or it may be purposely applied as a bearing or heading marker. 3. Line on a console oscilloscope representing the azimuth data generated by a jammed radar site.

strobe circuit A circuit that produces an output pulse only at certain times or under certain conditions, such as a gating circuit or a coincidence circuit.

strobe marker A small bright spot, or a short gap, or other discontinuity produced on the trace of a radar display to indicate that part of the time base which is receiving attention.

strobe pulse Pulse of duration less than the time period of a recurrent phenomenon used for making a close investigation of that phenomenon; the frequency of the strobe pulse bears a simple relation to that of the phenomenon, and the relative timing is usually adjustable.

stroboscopic lamp *See* flash lamp.

stroboscopic tube *See* strobotron.

strobotron A cold-cathode gas-filled arc-discharge tube having one or more internal or external grids to initiate current flow and produce intensely bright flashes of light for a stroboscope. Also known as stroboscopic tube.

stroke The penlike motion of a focused electron beam in cathode-ray-tube displays.

structure factor A factor which determines the amplitude of the beam reflected from a given atomic plane in the diffraction of an x-ray beam by a crystal, and is equal to the sum of the atomic scattering factors of the atoms in a unit cell, each multiplied by an appropriate phase factor.

stub 1. A short section of transmission line, open or shorted at the far end, connected in parallel with a transmission line to match the impedance of the line to that of an antenna or transmitter. 2. A solid projection one-quarter-wavelength long, used as an insulating support in a waveguide or cavity.

stub angle Right-angle elbow for a coaxial radio-frequency transmission line which has the inner conductor supported by a quarter-wave stub.

stub cable Short branch off a principal cable; the end is often sealed until it is used at a later date; pairs in the stub are referred to as stubbed-out pairs.

stub matching Use of a stub to match a transmission line to an antenna or load; matching depends on the spacing between the two wires of the stub, the position of the shorting bar, and the point at which the transmission line is connected to the stub.

stub-supported coaxial Coaxial whose inner conductor is supported by means of short-circuited coaxial stubs.

stub-supported line A transmission line that is supported by short-circuited quarter-wave sections of coaxial line; a stub exactly a quarter-wavelength long acts as an insulator because it has infinite reactance.

stub tuner Stub which is terminated by movable short-circuiting means and used for matching impedance in the line to which it is joined as a branch.

stunt box A device to control the nonprinting functions of a teletypewriter terminal.

subassembly Two or more components combined into a unit for convenience in assembling or servicing equipment; an intermediate-frequency strip for a receiver is an example.

subcarrier oscillator 1. The crystal oscillator that operates at the chrominance subcarrier or burst frequency of 3.579545 megahertz in a color television receiver; this oscillator, synchronized in frequency and phase with the transmitter master oscillator, furnishes the continuous subcarrier frequency required for demodulators in the receiver. 2. An oscillator used in a telemetering system to translate variations in an electrical quantity into variations of a frequency-modulated signal at a subcarrier frequency.

subclutter visibility A measure of the effectiveness of moving-target indicator radar, equal to the ratio of the signal from a fixed target that can be canceled to the signal from a just visible moving target.

subcycle generator Frequency-reducing device used in telephone equipment which furnishes ringing power at a submultiple of the power supply frequency.

subdivided capacitor Capacitor in which several capacitors known as sections are mounted so that they may be used individually or in combination.

subharmonic triggering A method of frequency division which makes use of a triggered multivibrator having a period of one cycle which allows triggering only by a pulse that is an exact integral number of input pulses from the last effective trigger.

submarine cable A cable designed for service under water; usually a lead-covered cable with steel armor applied between layers of jute.

submillimeter wave An electromagnetic wave whose wavelength is less than 1 millimeter, corresponding to frequencies above 300 gigahertz.

subminiature tube An extremely small electron tube designed for use in hearing aids and other miniaturized equipment; a typical subminiature tube is about 1½ inches (4 centimeters) long and 0.4 inch (1 centimeter) in diameter, with the pins emerging through the glass base.

subrefraction Atmospheric refraction which is less than standard refraction.

subscriber line A telephone line between a central office and a telephone station, private branch exchange, or other end equipment. Also known as central office line; subscriber loop.

subscriber loop *See* subscriber line.

subscriber multiple Bank of jacks in a manual switchboard providing outgoing access to subscriber lines, and usually having more than one appearance across the face of the switchboard.

substandard propagation The propagation of radio energy under conditions of substandard refraction in the atmosphere; that is, refraction by an atmosphere or section of the atmosphere in which the index of refraction decreases with height at a rate of less than 12 N units (unit of index of refraction) per 1000 feet (304.8 meters).

substation *See* electric power substation.

substitutional impurity An atom or ion which is not normally found in a solid, but which resides at the position where an atom or ion would ordinarily be located in the lattice structure, and replaces it.

substrate The physical material on which a microcircuit is fabricated; used primarily for mechanical support and insulating purposes, as with ceramic, plastic, and glass substrates; however, semiconductor and ferrite substrates may also provide useful electrical functions.

subsurface wave Electromagnetic wave propagated through water or land; operating frequencies for communications may be limited to approximately 35 kilohertz due to attenuation of high frequencies.

subsynchronous Operating at a frequency or speed that is related to a submultiple of the source frequency.

subway-type transformer Transformer of submersible construction.

Suhl amplifier A parametric microwave amplifier which utilizes the instability of certain spin waves in a ferromagnetic material subjected to intense microwave fields.

Suhl effect When a strong transverse magnetic field is applied to an n-type semiconducting filament, holes injected into the filament are deflected to the surface, where they may recombine rapidly with electrons or be withdrawn by a probe.

sulfating The formation of lead sulfate on the plates of lead-acid storage batteries reducing the energy-storing ability of the battery and eventually causing failure.

summation network *See* summing network.

summing amplifier An amplifier that delivers an output voltage which is proportional to the sum of two or more input voltages or currents.

summing network A passive electric network whose output voltage is proportional to the sum of two or more input voltages. Also known as summation network.

sun follower A photoelectric pickup and an associated servomechanism used to maintain a sun-facing orientation, as for a space vehicle. Also known as sun seeker.

S-unit meter *See* signal-strength meter.

sunlamp A mercury-vapor gas-discharge tube used to produce ultraviolet radiation for therapeutic or cosmetic purposes.

sun seeker *See* sun follower.

sun sensor *See* solar sensor.

sun strobe The signal display seen on a radar plan-position-indicator screen when the radar antenna is aimed at the sun; the pattern resembles that produced by continuous-wave interference, and is due to radio-frequency energy radiated by the sun.

superconducting material *See* superconductor.

superconducting quantum interference device A superconducting ring that couples with one or two Josephson junctions; applications include high-sensitivity magnetometers, near-magnetic-field antennas, and measurement of very small currents or voltages. Abbreviated SQUID.

superconductivity A property of many metals, alloys, and chemical compounds at temperatures near absolute zero by virtue of which their electrical resistivity vanishes and they become strongly diamagnetic.

superconductor Any material capable of exhibiting superconductivity; examples include iridium, lead, mercury, niobium, tin, tantalum, vanadium, and many alloys. Also known as cryogenic conductor; superconducting material.

supercurrent In the two-fluid model of superconductivity, the current arising from motion of superconducting electrons, in contrast to the normal current.

superemitron camera *See* image iconoscope.

superexchange A phenomenon in which two electrons from a double negative ion (such as oxygen) in a solid go to different positive ions and couple with their spins, giving rise to a strong antiferromagnetic coupling between the positive ions, which are too far apart to have a direct exchange interaction.

superhet *See* superheterodyne receiver.

superheterodyne receiver A receiver in which all incoming modulated radio-frequency carrier signals are converted to a common intermediate-frequency carrier value for additional amplification and selectivity prior to demodulation, using heterodyne action; the output of the intermediate-frequency amplifier is then demodulated in the second detector to give the desired audio-frequency signal. Also known as superhet.

superlattice 1. A structure consisting of alternating layers of two different semiconductor materials, each several nanometers thick. 2. An ordered arrangement of atoms in a solid solution which forms a lattice superimposed on the normal solid solution lattice. Also known as superstructure.

superposition theorem *See* principle of superposition.

superregeneration Regeneration in which the oscillation is broken up or quenched at a frequency slightly above the upper audibility limit of the human ear by a separate oscillator circuit connected between the grid and anode of the amplifier tube, to prevent regeneration from exceeding the maximum useful amount.

supersensitive relay A relay that operates on extremely small currents, generally below 250 microamperes.

superstandard propagation The propagation of radio waves under conditions of superstandard refraction in the atmosphere, that is, refraction by an atmosphere or section of the atmosphere in which the index of refraction decreases with height at a rate of greater than 12 N units (unit of index of refraction) per 1000 feet (304.8 meters).

superstructure *See* superlattice.

supervisory signal A signal which indicates the operating condition of a circuit or a combination of circuits in a switching apparatus or other electrical equipment to an attendant.

supervisory system A system of control, indicating, and telemetry devices which operates between the stations of an electric power distribution system, using a single common channel to transmit signals.

supervoltage A voltage in the range of 500 to 2000 kilovolts, used for some x-ray tubes.

supplementary group In wire communications, a group of trunks that directly connects local or trunk switching centers over other than a fundamental (or backbone) route.

supply voltage The voltage obtained from a power source for operation of a circuit or device.

suppression Elimination of any component of an emission, as a particular frequency or group of frequencies in an audio-frequency of a radio-frequency signal.

suppressor 1. In general, a device used to reduce or eliminate noise or other signals that interfere with the operation of a communication system, usually at the noise source. 2. Specifically, a resistor used in series with a spark plug or distributor of an automobile engine or other internal combustion engine to suppress spark noise that might otherwise interfere with radio reception. 3. *See* suppressor grid.

suppressor grid A grid placed between two positive electrodes in an electron tube primarily to reduce the flow of secondary electrons from one electrode to the other; it is usually used between the screen grid and the anode. Also known as suppressor.

suppressor pulse Pulse used to disable an ionized flow field or beacon transponder during intervals when interference would be encountered.

surface acoustic wave device Any device, such as a filter, resonator, or oscillator, which employs surface acoustic waves with frequencies in the range 10^7–10^9 hertz, traveling on the optically polished surface of a piezoelectric substrate, to process electronic signals.

surface acoustic wave filter An electric filter consisting of a piezoelectric bar with a polished surface along which surface acoustic waves can propagate, and on which are deposited metallic transducers, one of which is connected, via thermocompression-bonded leads, to the electric source, while the other drives the load.

surface barrier A potential barrier formed at a surface of a semiconductor by the trapping of carriers at the surface.

surface-barrier diode A diode utilizing thin-surface layers, formed either by deposition of metal films or by surface diffusion, to serve as a rectifying junction.

surface-barrier transistor A transistor in which the emitter and collector are formed on opposite sides of a semiconductor wafer, usually made of *n*-type germanium, by training two jets of electrolyte against its opposite surfaces to etch and then electroplate the surfaces.

surface-charge transistor An integrated-circuit transistor element based on controlling the transfer of stored electric charges along the surface of a semiconductor.

surface-controlled avalanche transistor Transistor in which avalanche breakdown voltage is controlled by an external field applied through surface-insulating layers, and which permits operation at frequencies up to the 10-gigahertz range.

surface leakage The passage of current over the surface of an insulator.

surface magnetic wave A magnetostatic wave that can be propagated on the surface of a magnetic material, as on a slab of yttrium iron garnet.

surface noise The noise component in the electric output of a phonograph pickup due to irregularities in the contact surface of the groove. Also known as needle scratch.

surface passivation A method of coating the surface of a *p*-type wafer for a diffused junction transistor with an oxide compound, such as silicon oxide, to prevent penetration of the impurity in undesired regions.

surface recombination velocity A measure of the rate of recombination between electrons and holes at the surface of a semiconductor, equal to the component of the electron or hole current density normal to the surface divided by the excess electron or hole volume charge density close to the surface.

surface resistivity The electric resistance of the surface of an insulator, measured between the opposite sides of a square on the surface; the value in ohms is independent of the size of the square and the thickness of the surface film.

surface state An electron state in a semiconductor whose wave function is restricted to a layer near the surface.

surface wave A wave that can travel along an interface between two different mediums without radiation; the interface must be essentially straight in the direction of propagation; the commonest interface used is that between air and the surface of a circular wire.

surface-wave transmission line A single conductor transmission line energized in such a way that a surface wave is propagated along the line with satisfactorily low attenuation.

surge A momentary large increase in the current or voltage in an electric circuit.

surge admittance Reciprocal of surge impedance.

surge arrester A protective device designed primarily for connection between a conductor of an electrical system and ground to limit the magnitude of transient overvoltages on equipment. Also known as arrester; lightning arrester.

surge current A short-duration, high-amperage electric current wave that may sweep through an electrical network, as a power transmission network, when some portion of it is strongly influenced by the electrical activity of a thunderstorm.

surge electrode current See fault electrode current.

surge generator A device for producing high-voltage pulses, usually by charging capacitors in parallel and discharging them in series.

surge suppressor A circuit that responds to the rate of change of a current or voltage to prevent a rise above a predetermined value; it may include resistors, capacitors, coils, gas tubes, and semiconducting disks. Also known as transient suppressor.

susceptance The imaginary component of admittance.

susceptance standard Standard that introduces calibrated small values of shunt capacitance into 50-ohm coaxial transmission arrays.

susceptibility See electric susceptibility; magnetic susceptibility.

suspension insulator A type of insulator used to support a conductor of an overhead transmission line, consisting of one or a string of insulating units suspended from a pole or tower, with the conductor attached to the end.

SW See switch.

swamping resistor Resistor placed in the emitter lead of a transistor circuit to minimize the effects of temperature on the emitter-base junction resistance.

sweep 1. The steady movement of the electron beam across the screen of a cathode-ray tube, producing a steady bright line when no signal is present; the line is straight for a linear sweep and circular for a circular sweep. 2. The steady change in the output frequency of a signal generator from one limit of its range to the other.

sweep amplifier An amplifier used with a cathode-ray tube, such as in a television receiver or cathode-ray oscilloscope, to amplify the sawtooth output voltage of the sweep oscillator, to shape the waveform for the deflection circuits of a television picture tube, or to provide balanced signals to the deflection plates.

sweep circuit The sweep oscillator, sweep amplifier, and any other stage used to produce the deflection voltage or current for a cathode-ray tube. Also known as scanning circuit.

sweep-frequency reflectometer A reflectometer that measures standing-wave ratio and insertion loss in decibels over a wide range of frequencies, in either single- or sweep-frequency operation.

sweep generator Also known as sweep oscillator. 1. An electronic circuit that generates a voltage or current, usually recurrent, as a prescribed function of time; the resulting waveform is used as a time base to be applied to the deflection system of an electron-beam device, such as a cathode-ray tube. Also known as time-base generator; timing-axis oscillator. 2. A test instrument that generates a radio-frequency voltage whose frequency varies back and forth through a given frequency range at a rapid constant rate; used to produce an input signal for circuits or devices whose frequency response is to be observed on an oscilloscope.

424 sweeping receivers

sweeping receivers Automatically and continuously tuned receivers designed to stop and lock on when a signal is found, or to continually plot band occupancy.

sweep jamming Jamming an enemy radarscope by sweeping the region of radar-beam coverage with electromagnetic waves having the same frequency as those received by the radarscope.

sweep oscillator *See* sweep generator.

sweep rate The number of times a radar radiation pattern rotates during 1 minute; sometimes expressed as the duration of one complete rotation in seconds.

sweep test Test given coaxial cable with an oscilloscope to check attenuation.

sweep-through jammer A jamming transmitter which is swept through a radio-frequency band in short steps to jam each frequency briefly, producing a sound like that of an aircraft engine.

sweep voltage Periodically varying voltage applied to the deflection plates of a cathode-ray tube to give a beam displacement that is a function of time, frequency, or other data base.

swing Variation in frequency or amplitude of an electrical quantity.

swinging choke An iron-core choke having a core that can be operated almost at magnetic saturation; the inductance is then a maximum for small currents, and swings to a lower value as current increases. Also known as swinging reactor.

switch A manual or mechanically actuated device for making, breaking, or changing the connections in an electric circuit. Also known as electric switch. Symbolized SW.

switchboard A single large panel or assembly of panels on which are mounted switches, circuit breakers, meters, fuses, and terminals essential to the operation of electric equipment. Also known as electric switchboard.

switched capacitor An integrated circuit element, consisting of a capacitor with two metal oxide semiconductor (MOS) switches, whose function is approximately equivalent to that of a resistor.

switch function A circuit having a fixed number of inputs and outputs designed such that the output information is a function of the input information, each expressed in a certain code or signal configuration or pattern.

switchgear The aggregate of switching devices for a power or transforming station, or for electric motor control.

switch hook A switch on a telephone set that operates when the receiver is placed on the hook or removed from it.

switching Making, breaking, or changing the connections in an electrical circuit.

switching circuit A constituent electric circuit of a switching or digital processing system which receives, stores, or manipulates information in coded form to accomplish the specified objectives of the system.

switching diode A crystal diode that provides essentially the same function as a switch; below a specified applied voltage it has high resistance corresponding to an open switch, while above that voltage it suddenly changes to the low resistance of a closed switch.

switching gate An electronic circuit in which an output having constant amplitude is registered if a particular combination of input signals exists; examples are the OR, AND, NOT, and INHIBIT circuits. Also known as logical gate.

switching key *See* key.

switching pad Transmission-loss pad automatically cut in and out of a toll circuit for different desired operating conditions.

switching reactor A saturable-core reactor that has several input control windings and one or more output windings that essentially duplicate the functions of a relay.

switching substation An electric power substation whose equipment is mainly for connections and interconnections, and does not include transformers.

switching theory The theory of circuits made up of ideal digital devices; included are the theory of circuits and networks for telephone switching, digital computing, digital control, and data processing.

switching-through relay Control relay of a line-finder selector, connector, or other stepping switch, which extends the loop of a calling telephone through to the succeeding switch in a switch train.

switching time 1. The time interval between the reference time and the last instant at which the instantaneous voltage response of a magnetic cell reaches a stated fraction of its peak value. 2. The time interval between the reference time and the first instant at which the instantaneous integrated voltage response of a magnetic cell reaches a stated fraction of its peak value.

switching transistor A transistor designed for on/off switching operation.

switching trunk Trunk from a long-distance office to a local exchange office used for completing a long-distance call.

switching tube A gas tube used for switching high-power radio-frequency energy in the antenna circuits of radar and other pulsed radio-frequency systems; examples are ATR tube; pre-TR tube; TR tube.

switch jack Any of the devices that provide terminals for the control circuits of the switch.

switch over–travel That movement of a switch-operating lever which takes place after the switch has been actuated either to close or open its contacts.

switch pretravel That movement of a switch-operating level that takes place before the switch is actuated either to close or to open its contacts.

switch train A series of switches in tandem.

syllabic compandor A compandor in which the effective gain variations are made at speeds allowing response to the syllables of speech but not to individual cycles of the signal wave.

symmetrical avalanche rectifier Avalanche rectifier that can be triggered in either direction, after which it has a low impedance in the triggered direction.

symmetrical band-pass filter A band-pass filter whose attenuation as a function of frequency is symmetrical about a frequency at the center of the pass band.

symmetrical band-reject filter A band-rejection filter whose attenuation as a function of frequency is symmetrical about a frequency at the center of the rejection band.

symmetrical clipper A clipper in which the upper and lower limits on the amplitude of the output signal are positive and negative values of equal magnitude.

symmetrical deflection A type of electrostatic deflection in which voltages that are equal in magnitude and opposite in sign are applied to the two deflector plates.

symmetrical H attenuator An H attenuator in which the impedance near the input terminals equals the corresponding impedance near the output terminals.

symmetrical inductive diaphragm A waveguide diaphragm which consists of two plates that leave a space at the center of the waveguide, and which introduces an inductance in the waveguide.

symmetrical O attenuator An O attenuator in which the impedance near the input terminals equals the corresponding impedance near the output terminals.

symmetrical pi attenuator A pi attenuator in which the impedance near the input terminals equals the corresponding impedance near the output terminals.

symmetrical T attenuator A T attenuator in which the impedance near the input terminals equals the corresponding impedance near the output terminals.

symmetrical transducer A transducer is symmetrical with respect to a specified pair of terminations when the interchange of that pair of terminations will not affect the transmission.

sync generator *See* synchronizing generator.

synchro Any of several devices used for transmitting and receiving angular position or angular motion over wires, such as a synchro transmitter or synchro receiver. Also known as mag-slip (British usage); self-synchronous device; self-synchronous repeater; selsyn.

synchro control transformer A transformer having its secondary winding on a rotor; when its three input leads are excited by angle-defining voltages, the two output leads deliver an alternating-current voltage that is proportional to the sine of the difference between the electrical input angle and the mechanical rotor angle.

synchro control transmitter A high-accuracy synchro transmitter, having high-impedance windings.

synchro differential motor Motor which is electrically similar to the synchro differential generator except that a damping device is added to prevent oscillations; both its rotor and stator are connected to synchro generators, and its function is to indicate the sum or difference between the two signals transmitted by the generators.

synchro differential receiver A synchro receiver that subtracts one electrical angle from another and delivers the difference as a mechanical angle. Also known as differential synchro.

synchro differential transmitter A synchro transmitter that adds a mechanical angle to an electrical angle and delivers the sum as an electrical angle. Also known as differential synchro.

synchro generator *See* synchro transmitter.

synchro motor *See* synchro receiver.

synchronism Of a synchronous motor, the condition under which the motor runs at a speed which is directly related to the frequency of the power applied to the motor and is not dependent upon variables.

synchronized blocking oscillator A blocking oscillator which is synchronized with pulses occurring at a rate slightly faster than its own natural frequency.

synchronizer The component of a radar set which generates the timing voltage for the complete set.

synchronizing generator An electronic generator that supplies synchronizing pulses to television studio and transmitter equipment. Also known as sync generator; sync-signal generator.

synchronizing reactor Current-limiting reactor for connecting momentarily across the open contacts of a circuit-interrupting device for synchronizing purposes.

synchronizing relay Relay which functions when two alternating-current sources are in agreement within predetermined limits of phase angle and frequency.

synchronous booster converter Synchronous converter having an alternating-current generator mounted on the same shaft and connected in series with it to adjust the voltage at the commutator of the converter.

synchronous capacitor A synchronous motor running without mechanical load and drawing a large leading current, like a capacitor; used to improve the power factor and voltage regulation of an alternating-current power system.

synchronous clamp circuit *See* keyed clamp circuit.

synchronous converter A converter in which motor and generator windings are combined on one armature and excited by one magnetic field; normally used to change alternating to direct current. Also known as converter; electric converter.

synchronous demodulator *See* synchronous detector.

synchronous detector 1. A detector that inserts a missing carrier signal in exact synchronism with the original carrier at the transmitter; when the input to the detector consists of two suppressed-carrier signals in phase quadrature, as in the chrominance signal of a color television receiver, the phase of the reinserted carrier can be adjusted to recover either one of the signals. Also known as synchronous demodulator. 2. *See* cross-correlator.

synchronous gate A time gate in which the output intervals are synchronized with an incoming signal.

synchronous generator A machine that generates an alternating voltage when its armature or field is rotated by a motor, an engine, or other means. The output frequency is exactly proportional to the speed at which the generator is driven.

synchronous inverter *See* dynamotor.

synchronous machine An alternating-current machine whose average speed is proportional to the frequency of the applied or generated voltage.

synchronous motor A synchronous machine that transforms alternating-current electric power into mechanical power, using field magnets excited with direct current.

synchronous phase modifier A synchronous motor that runs without mechanical load, and is provided with means for varying its power factor to simulate a capacitive or inductive reactor; used in voltage regulation of alternating-current power systems.

synchronous rectifier A rectifier in which contacts are opened and closed at correct instants of time for rectification by a synchronous vibrator or by a commutator driven by a synchronous motor.

synchronous speed The speed of rotation of a magnetic field in a synchronous machine; in revolutions per second, it is equal to twice the frequency of the alternating current in hertz, divided by the number of poles in the machine.

synchronous switch A thyratron circuit used to control the operation of ignitrons in such applications as resistance welding.

synchronous vibrator An electromagnetic vibrator that simultaneously converts a low direct voltage to a low alternating voltage and rectifies a high alternating voltage obtained from a power transformer to which the low alternating voltage is applied; in power packs, it eliminates the need for a rectifier tube.

synchro receiver A synchro that provides an angular position related to the applied angle-defining voltages; when two of its input leads are excited by an alternating-current voltage and the other three input leads are excited by the angle-defining voltages, the rotor rotates to the corresponding angular position; the torque of rotation is proportional to the sine of the difference between the mechanical and electrical angles. Also known as receiver synchro; selsyn motor; selsyn receiver; snchro motor.

synchro resolver *See* resolver.

synchroscope A cathode-ray oscilloscope designed to show a short-duration pulse by using a fast sweep that is synchronized with the pulse signal to be observed.

synchro system An electric system for transmitting angular position or motion; in the simplest form it consists of a synchro transmitter connected by wires to a synchro receiver; more complex systems include synchro control transformers and synchro differential transmitters and receivers. Also known as selsyn system.

synchro transmitter A synchro that provides voltages related to the angular position of its rotor; when its two input leads are excited by an alternating-current voltage, the magnitudes and polarities of the voltages at the three output leads define the rotor position. Also known as selsyn generator; selsyn transmitter; synchro generator; transmitter; transmitter synchro.

synchrotron process The emission of electromagnetic radiation by relativistic electrons orbiting in a magnetic field.

synchrotron radiation Electromagnetic radiation generated by the acceleration of charged relativistic particles, usually electrons, in a magnetic field.

sync separator A circuit that separates synchronizing pulses from the video signal in a television receiver.

sync-signal generator *See* synchronizing generator.

synthesizer An electronic instrument which combines simple elements to generate more complex entities; examples are frequency synthesizer and sound synthesizer.

syntony Condition in which two oscillating circuits have the same resonant frequency.

system A combination of two or more sets generally physically separated when in operation, and such other assemblies, subassemblies, and parts necessary to perform an operational function or functions.

systematic analog network testing approach An on-line minicomputer-based system with an integrated data-based and optimal human intervention, which provides computer printouts used in automatic testing of electronic systems; aimed at maximizing cost effectivity. Abbreviated SANTA.

systematic distortion Periodic or constant distortion, such as bias or characteristic distortion; the direct opposite of fortuitous distortion.

T

T *See* tesla.

TΩ *See* teraohm.

table look-up device A logic circuit in which the input signals are grouped as address digits to a memory device, and, in response to any particular combination of inputs, the memory device location that is addressed becomes the output.

tactical electronic warfare The application of electronic warfare to tactical air operations; tactical electronic warfare encompasses the three major subdivisions of electronic warfare: electronic warfare support measures, electronic countermeasures, and electronic counter-countermeasures.

Tafel slope The slope of a curve of overpotential or electrolytic polarization in volts versus the logarithm of current density.

tail 1. A small pulse that follows the main pulse of a radar set and rises in the same direction. 2. The trailing edge of a pulse.

tail clipping Method of sharpening the trailing edge of a pulse.

talking battery *See* quiet battery.

tandem Two-terminal pair networks are in tandem when the output terminals of one network are directly connected to the input terminals of the other network.

tandem connection *See* cascade connection.

tangential wave path In radio propagation over the earth, a path of propagation of a direct wave which is tangential to the surface of the earth; the tangential wave path is curved by atmospheric refraction.

tangling The reduction of motion of dislocations in a substance by increasing the total number of dislocations, so that they tangle and interfere with each other's motions.

tank 1. A unit of acoustic delay-line storage containing a set of channels, each forming a separate recirculation path. 2. The heavy metal envelope of a large mercury-arc rectifier or other gas tube having a mercury-pool cathode. 3. *See* tank circuit.

tank circuit A circuit which exhibits resonance at one or more frequencies, and which is capable of storing electric energy over a band of frequencies continuously distributed about the resonant frequency, such as a coil and capacitor in parallel. Also known as electrical resonator; tank.

tantalum capacitor An electrolytic capacitor in which the anode is some form of tantalum; examples include solid tantalum, tantalum-foil electrolytic, and tantalum-slug electrolytic capacitors.

tantalum-foil electrolytic capacitor An electrolytic capacitor that uses plain or etched tantalum foil for both electrodes, with a weak acid electrolyte.

tantalum nitride resistor A thin-film resistor consisting of tantalum nitride deposited on a substrate, such as industrial sapphire.

tantalum-slug electrolytic capacitor An electrolytic capacitor that uses a sintered slug of tantalum as the anode, in a highly conductive acid electrolyte.

T antenna An antenna consisting of one or more horizontal wires, with a lead-in connection being made at the approximate center of each wire.

tap A connection made at some point other than the ends of a resistor or coil.

tap changer A device which is used to change the ratio of the input and output voltages of a transformer over any one of a definite number of steps.

tap crystal Compound semiconductor that stores current when stimulated by light and then gives up energy as flashes of light when it is physically tapped.

taper Continuous or gradual change in electrical properties with mechanical position such as rotation or length; for example, continuous change of cross section of a waveguide, or distribution of resistance in a potentiometer.

tapered transmission line *See* tapered waveguide.

tapered waveguide A waveguide in which a physical or electrical characteristic changes continuously with distance along the axis of the waveguide. Also known as tapered transmission line.

tape-wound core A length of ferromagnetic material in tape form, wound in such a way that each turn falls directly over the preceding turn.

tapped control A rheostat or potentiometer having one or more fixed taps along the resistance element, usually to provide a fixed grid bias or for automatic bass compensation.

tapped-potentiometer function generator A device used in analog computers for representing a function of one variable, consisting of a potentiometer with a number of taps held at voltages determined by a table of values of the variable; the input variable sets the angular position of a shaft that moves a slide contact, and the output voltage is taken from the slide contact.

tapped resistor A wire-wound fixed resistor having one or more additional terminals along its length, generally for voltage-divider applications.

tap switch Multicontact switch used chiefly for connecting a load to any one of a number of taps on a resistor or coil.

target 1. In an x-ray tube, the anode or anticathode which emits x-rays when bombarded with electrons. 2. In a television camera tube, the storage surface that is scanned by an electron beam to generate an output signal current corresponding to the charge-density pattern stored there. 3. In a cathode-ray tuning indicator tube, one of the electrodes that is coated with a material that fluoresces under electron bombardment.

target acquisition 1. The first appearance of a recognizable and useful echo signal from a new target in radar and sonar. 2. *See* acquire.

target cross section *See* echo area.

target-designating system A system for designating to one instrument a target which has already been located by a second instrument; it employs electrical data transmitters and receivers which indicate on one instrument the pointing of another.

target discrimination The ability of a detection or guidance system to distinguish a target from its background or to discriminate between two or more targets that are close together.

target echo A radio signal reflected by an airborne or other target and received by the radar station which transmitted the original signal.

target glint *See* scintillation.

telegraph receiver 431

target noise Statistical variations in a radar echo signal due to the presence on the target of a number of reflecting elements randomly oriented in space; target noise can cause scintillation.

target scintillation See scintillation.

target signal The radio energy returned to a radar by a target. Also known as echo signal; video signal.

target signature Characteristic pattern of the target displayed by detection and classification equipment.

target volume The volume of that part of a precipitation-type radar target from which a target signal is received; if the precipitation completely fills the radar beam, the target volume is identical with the radar volume.

T attenuator 1. A resistive attenuator with three resistors forming a T network. 2. A power-tap type of attenuator which removes part of the power from a main line through a T connection and dissipates the power, without reflection into the main line.

Taylor connection A transformer connection for converting three-phase power to two-phase power, or vice versa.

T circulator A circulator in which three identical rectangular waveguides are joined asymmetrically to form a T-shaped structure, with a ferrite post or wedge at its center; power entering any waveguide emerges from only one adjacent waveguide.

T connector A type of electric connector that joins a through conductor to another conductor at right angles to it.

TD See transmitter-distributor.

TDR See time-domain reflectometer.

teaser transformer Transformer, of two T-connected, single-phase units for three-phase to two-phase or two-phase to three-phase operation, which is connected between the midpoint of the main transformer and the third wire of the three-phase system.

technetron High-power multichannel field-effect transistor.

technical control board Testing position in a switch center or relay station with provisions for testing switches and associated access lines and trunks.

technical load Portion of a communications-electronics facility operational power load required for primary and ancillary equipment, including necessary lighting and air conditioning or ventilation required for full continuity of operation.

telecine camera A television camera used in conjunction with film or slide projectors to televise motion pictures and still images.

telegraph cable A uniform conductive circuit consisting of twisted pairs of insulated wires or coaxially shielded wires or combinations of each, used to carry telegraph signals.

telegraph concentrator Switching arrangement by means of which a number of branch or subscriber lines or station sets may be connected to a lesser number of trunklines, operating positions, or instruments through the medium of manual or automatic switching devices to obtain more efficient use of facilities.

telegraph distributor Device which effectively associates one direct-current or carrier-telegraph channel in rapid succession with the elements of one or more sending or receiving devices.

telegraph receiver A tape reperforator, teletypewriter, or other equipment which converts telegraph signals into a pattern of holes on a tape, printed letters, or other forms of information.

telegraph repeater A repeater inserted at intervals in long telegraph lines to amplify weak code signals, with or without reshaping of pulses, and to retransmit them automatically over the next section of the line.

telegraph transmitter A device that controls an electric power source in order to form telegraph signals.

telemetering antenna A highly directional antenna, generally mounted on a servo-controlled mount for tracking purposes, used at ground stations to receive telemetering signals from a guided missile or spacecraft.

telemetering receiver A device in a telemetering system which converts electrical signals into an indication or recording of the value of the quantity being measured at a distance.

telemetering transmitter A device which converts the readings of instruments into electrical signals for transmission to a remote location by means of wires, radio waves, or other means.

telephone carrier current A carrier current used for telephone communication over power lines or to obtain more than one channel on a single pair of wires.

telephone circuit The complete circuit over which audio and signaling currents travel in a telephone system between the two telephone subscribers in communication with each other; the circuit usually consists of insulated conductors, as ground returns are now rarely used in telephony.

telephone induction coil A coil used in a telephone circuit to match the impedance of the line to that of a telephone transmitter or receiver.

telephone line The conductors extending between telephone subscriber stations and central offices.

telephone loading coil *See* loading coil.

telephone modem A piece of equipment that modulates and demodulates one or more separate telephone circuits, each containing one or more telephone channels; it may include multiplexing and demultiplexing circuits, individual amplifiers, and carrier-frequency sources.

telephone pickup A large flat coil placed under a telephone set to pick up both voices during a telephone conversation for recording purposes.

telephone plug *See* phone plug.

telephone relay A relay having a multiplicity of contacts on long spring strips mounted parallel to the coil, actuated by a lever arm or other projection of the hinged armature; used chiefly for switching in telephone circuits.

telephone repeater A repeater inserted at one or more intermediate points in a long telephone line to amplify telephone signals so as to maintain the required current strength.

telephone repeating coil A coil used in a telephone circuit for inductively coupling two sections of a line when a direct connection is undesirable.

telephone ringer An electromagnetic device that actuates a clapper which strikes one or more gongs to produce a ringing sound; used with a telephone set to signal a called party.

telering In telephony, a frequency-selector device for the production of ringing power.

telesynd Telemeter or remote-control equipment which is synchronous in both speed and position.

television antenna An antenna suitable for transmitting or receiving television broadcasts; since television transmissions in the United States are horizontally polarized, the most basic type of receiving antenna is a horizontally mounted half-wave dipole.

television camera The pickup unit used to convert a scene into corresponding electric signals; optical lenses focus the scene to be televised on the photosensitive surface of a camera tube, and the tube breaks down the visual image into small picture elements and converts the light intensity of each element in turn into a corresponding electric signal. Also known as camera.

television monitor 1. A television set connected to the transmitter at a television station, used to continuously check the image picked up by a television camera and the sound picked up by the microphones. 2. A closed-circuit television system used to provide continuous observation of such things as hazardous or remote locations, the readings of gages for process control, or microscopic or telescopic images, for greater convenience of viewing.

television picture tube *See* picture tube.

television receiver A receiver that converts incoming television signals into the original scenes along with the associated sounds. Also known as television set.

television relay system *See* television repeater.

television repeater A repeater that transmits television signals from point to point by using radio waves in free space as a medium, such transmission not being intended for direct reception by the public. Also known as television relay system.

television screen The fluorescent screen of the picture tube in a television receiver.

television set *See* television receiver.

television transmitter An electronic device that converts the audio and video signals of a television program into modulated radio-frequency energy that can be radiated from an antenna and received on a television receiver.

television tuner A component in a television receiver that selects the desired channel and converts the frequencies received to lower frequencies within the passband of the intermediate-frequency amplifier; for very-high-frequency reception there are 12 discrete positions (channels 2–13); for ultra-high-frequency reception continuous tuning is usually employed.

TEM mode *See* transverse electromagnetic mode.

TE mode *See* transverse electric mode.

temperature-compensated Zener diode Positive-temperature-coefficient reversed-bias Zener diode (*pn* junction) connected in series with one or more negative-temperature forward-biased diodes within a single package.

temperature-compensating capacitor Capacitor whose capacitance varies with temperature in a known and predictable manner; used extensively in oscillator circuits to compensate for changes in the values of other parts with temperatures.

temperature compensation The process of making some characteristic of a circuit or device independent of changes in ambient temperature.

temperature resistance coefficient The ratio of the change of electrical resistance in a wire caused by a change in its temperature of 1°C as related to its resistance at 0°C.

temperature saturation The condition in which the anode current of a thermionic vacuum tube cannot be further increased by increasing the cathode temperature at a given value of anode voltage; the effect is due to the space charge formed near the cathode. Also known as filament saturation; saturation.

TEM wave *See* transverse electromagnetic wave.

teraohm A unit of electrical resistance, equal to 10^{12} ohms. Abbreviated TΩ.

terminal 1. A screw, soldering lug, or other point to which electric connections can be made. Also known as electric terminal. 2. The equipment at the end of a microwave relay system or other communication channel. 3. One of the electric input or output points of a circuit or component.

terminal area The enlarged portion of conductor material surrounding a hole for a lead on a printed circuit. Also known as land; pad.

terminal board An insulating mounting for terminal connections. Also known as terminal strip.

terminal box An enclosure which includes, mounts, and protects one or more terminals or terminal boards; it may include a cover and such accessories as mounting hardware, brackets, locks, and conduit fittings.

terminal cutout pairs Numbered, designated pairs brought out of a cable at a terminal.

terminal pair An associated pair of accessible terminals, such as the input or output terminals of a device or network.

terminal stub Piece of cable that comes with a cable terminal for splicing into the main cable. Also known as terminal leg.

terminal voltage The voltage at the terminals connected to the source of electricity for an electric machine.

terminated line Transmission line terminated in a resistance equal to the characteristic impedance of the line, so there is no reflection and no standing waves.

terminating Closing of the circuit at either end of a line or transducer by connecting some device thereto; terminating does not imply any special condition such as the elimination of reflection.

termination 1. Load connected to a transmission line or other device; to avoid wave reflections, it must match the characteristic of the line or device. 2. In waveguide technique, the point at which energy flowing along a waveguide continues in a nonwaveguide mode of propagation.

terrain echoes *See* ground clutter.

tertiary pyroelectricity The polarization due to temperature and gradients and corresponding nonuniform stresses and strains when the crystal is heated nonuniformly; found in pyroelectric and nonpyroelectric crystals, that is, crystals which have no polar directions. Also known as false pyroelectricity.

tertiary winding *See* stabilized winding.

tesla The International System unit of magnetic flux density, equal to one weber per square meter. Symbolized T.

Tesla coil An air-core transformer used with a spark gap and capacitor to produce a high voltage at a high frequency.

testboard Switchboard equipped with testing apparatus, arranged so that connections can be made from it to telephone lines or central-office equipment for testing purposes.

test clip A spring clip used at the end of an insulated wire lead to make a temporary connection quickly for test purposes.

testing level Value of power used for reference represented by 0.001 watt working in 600 ohms.

test jack 1. Appearance of a circuit or circuit element in jacks for testing purposes. 2. In recent practice, a jack multipled with the switchboard operating jack.

test lead A flexible insulated lead, usually with a test prod at one end, used for making tests, connecting instruments to a circuit temporarily, or making other temporary connections.

test point A terminal or plug-in connector provided in a circuit to facilitate monitoring, calibration, or trouble-shooting.

test prod A metal point attached to an insulating handle and connected to a test lead for convenience in making a temporary connection to a terminal while tests are being made. Also known as prod.

test set A combination of instruments needed for servicing a particular type of electronic equipment.

tetrode A four-electrode electron tube containing an anode, a cathode, a control electrode, and one additional electrode that is ordinarily a grid.

tetrode junction transistor *See* double-base junction transistor.

tetrode thyratron A thyratron with two control electrodes. Also known as gas tetrode.

tetrode transistor A four-electrode transistor, such as a tetrode point-contact transistor or double-base junction transistor.

TE wave *See* transverse electric wave.

thallofide cell A photoconductive cell in which the active light-sensitive material is thallium oxysulfide in a vacuum; it has maximum response at the red end of the visible spectrum and in the near infrared.

theater television A large projection-type television receiver used in theaters, generally for closed-circuit showing of important sport events.

theoretical cutoff frequency Of an electric structure, a frequency at which, disregarding the effects of dissipation, the attenuation constant changes from zero to a positive value or vice versa.

thermal agitation Random movements of the free electrons in a conductor, producing noise signals that may become noticeable when they occur at the input of a high-gain amplifier. Also known as thermal effect.

thermal battery 1. A combination of thermal cells. Also known as fused-electrolyte battery; heat-activated battery. 2. A voltage source consisting of a number of bimetallic junctions connected to produce a voltage when heated by a flame.

thermal cell A reserve cell that is activated by applying heat to melt a solidified electrolyte.

thermal converter A device that converts heat energy directly into electric energy by using the Seebeck effect; it is composed of at least two dissimilar materials, one junction of which is in contact with a heat source and the other junction of which is in contact with a heat sink. Also known as thermocouple converter; thermoelectric generator; thermoelectric power generator; thermoelement.

thermal cutout A heat-sensitive switch that automatically opens the circuit of an electric motor or other device when the operating temperature exceeds a safe value.

thermal drift Drift caused by internal heating of equipment during normal operation or by changes in external ambient temperature.

thermal effect *See* thermal agitation.

thermal flasher An electric device that opens and closes a circuit automatically at regular intervals because of alternate heating and cooling of a bimetallic strip that is heated by a resistance element in series with the circuit being controlled.

thermal horsepower Electrical motor horsepower as determined by current readings from a thermal-type ammeter; will be higher than load horsepower determined from kilowatt-input methods. Also known as true motor load.

thermal imagery Imagery produced by measuring and recording electronically the thermal radiation of objects.

thermal magnon A magnon with a relatively short wavelength, on the order of 10^{-6} centimeter.

thermal noise Electric noise produced by thermal agitation of electrons in conductors and semiconductors. Also known as Johnson noise; resistance noise.

thermal noise generator A generator that uses the inherent thermal agitation of an electron tube to provide a calibrated noise source.

thermal pulse method A method of measuring properties of insulating and conducting crystals, in which a heat pulse of known duration is measured after propagating through a crystal; the pulse can be generated by directing a laser pulse at an absorbing film evaporated onto one face of the crystal, and detected by a thin-film circuit on the other face.

thermal regenerative cell Fuel-cell system in which the reactants are regenerated continuously from the products formed during the cell reaction.

thermal relay A relay operated by the heat produced by current flow.

thermal resistance *See* effective thermal resistance.

thermal resistor 1. A resistor designed so its resistance varies in a known manner with changes in ambient temperature. 2. *See* thermistor.

thermal runaway A condition that may occur in a power transistor when collector current increases collector junction temperature, reducing collector resistance and allowing a greater current to flow, which, in turn, increases the heating effect.

thermal scattering Scattering of electrons, neutrons, or x-rays passing through a solid due to thermal motion of the atoms in the crystal lattice.

thermal switch A temperature-controlled switch. Also known as thermoswitch.

thermal tuning The process of changing the operating frequency of a system by using controlled thermal expansion to alter the geometry of the system.

thermal volt *See* kelvin.

thermal wave A sound wave in a solid which has a short wavelength.

thermal x-rays The electromagnetic radiation, mainly in the soft (low-energy) x-ray region.

thermion A charged particle, either negative or positive, emitted by a heated body, as by the hot cathode of a thermionic tube.

thermionic Pertaining to the emission of electrons as a result of heat.

thermionic cathode *See* hot cathode.

thermionic converter A device in which heat energy is directly converted to electric energy; it has two electrodes, one of which is raised to a sufficiently high temperature to become a thermionic electron emitter, while the other, serving as an electron collector, is operated at a significantly lower temperature. Also known as thermionic generator; thermionic power generator; thermoelectric engine.

thermionic current Current due to directed movements of thermions, such as the flow of emitted electrons from the cathode to the plate in a thermionic vacuum tube.

thermionic detector A detector using a hot-cathode tube.

thermionic diode A diode electron tube having a heated cathode.

thermionic emission 1. The outflow of electrons into vacuum from a heated electric conductor. Also known as Edison effect; Richardson effect. 2. More broadly, the liberation of electrons or ions from a substance as a result of heat.

thermionic fuel cell A thermionic converter in which the space between the electrodes is filled with cesium or other gas, which lowers the work functions of the electrodes, and creates an ionized atmosphere, controlling the electron space charge.

thermionic generator See thermionic converter.

thermionic power generator See thermionic converter.

thermionics The study and applications of thermionic emission.

thermionic triode A three-electrode thermionic tube, containing an anode, a cathode, and a control electrode.

thermionic tube An electron tube that relies upon thermally emitted electrons from a heated cathode for tube current. Also known as hot-cathode tube.

thermionic work function Energy required to transfer an electron from the fermi energy in a given metal through the surface to the vacuum just outside the metal.

thermistor A resistive circuit component, having a high negative temperature coefficient of resistance, so that its resistance decreases as the temperature increases; it is a stable, compact, and rugged two-terminal ceramiclike semiconductor bead, rod, or disk. Derived from thermal resistor.

thermocouple converter See thermal converter.

thermoelectric converter A converter that changes solar or other heat energy to electric energy; used as a power source on spacecraft.

thermoelectric engine See thermionic converter.

thermoelectric generator See thermal converter.

thermoelectric junction See thermojunction.

thermoelectric material A material that can be used to convert thermal energy into electric energy or provide refrigeration directly from electric energy; good thermoelectric materials include lead telluride, germanium telluride, bismuth telluride, and cesium sulfide.

thermoelectric power generator See thermal converter.

thermoelectric solar cell A solar cell in which the sun's energy is first converted into heat by a sheet of metal, and the heat is converted into electricity by a semiconductor material sandwiched between the first metal sheet and a metal collector sheet.

thermoelectromotive force Voltage developed due to differences in temperature between parts of a circuit containing two or more different metals.

thermoelectron An electron liberated by heat, as from a heated filament. Also known as negative thermion.

thermoelement See thermal converter.

thermojunction One of the surfaces of contact between the two conductors of a thermocouple. Also known as thermoelectric junction.

thermojunction battery Nuclear-type battery which converts heat into electrical energy directly by the thermoelectric or Seebeck effect.

thermomigration A technique for doping semiconductors in which exact amounts of known impurities are made to migrate from the cool side of a wafer of pure semiconductor material to the hotter side when the wafer is heated in an oven.

thermopile generator An electricity source powered by the heating of an electrical resistor that can be connected to a thermopile to generate small amounts of electric current.

thermoplastic recording A recording process in which a modulated electron beam deposits charges on a thermoplastic film, and application of heat by radio-frequency heating electrodes softens the film enough to produce deformation that is proportional to the density of the stored electrostatic charges; an optical system is used for playback.

thermopower A measure of the temperature-induced voltage in a conductor.

thermostatic switch A temperature-operated switch that receives its operating energy by thermal conduction or convection from the device being controlled or operated.

theta polarization State of a wave in which the E vector is tangential to the meridian lines of some given spherical frame of reference.

Thévenin generator The voltage generator in the equivalent circuit of Thévenin's theorem.

Thévenin's theorem A valuable theorem in network problems which allows calculation of the performance of a device from its terminal properties only: the theorem states that at any given frequency the current flowing in any impedance, connected to two terminals of a linear bilateral network containing generators of the same frequency, is equal to the current flowing in the same impedance when it is connected to a voltage generator whose generated voltage is the voltage at the terminals in question with the impedance removed, and whose series impedance is the impedance of the network looking back from the terminals into the network with all generators replaced by their internal impedances. Also known as Helmholtz's theorem.

thick-film capacitor A capacitor in a thick-film circuit, made by successive screen-printing and firing processes.

thick-film circuit A microcircuit in which passive components, of a ceramic-metal composition, are formed on a ceramic substrate by successive screen-printing and firing processes, and discrete active elements are attached separately.

thick-film resistor Fixed resistor whose resistance element is a film well over one-thousandth of an inch (over 25 micrometers) thick.

thin film A film a few molecules thick deposited on a glass, ceramic, or semiconductor substrate to form a capacitor, resistor, coil, cryotron, or other circuit component.

thin-film capacitor A capacitor that can be constructed by evaporation of conductor and dielectric films in sequence on a substrate; silicon monoxide is generally used as the dielectric.

thin-film circuit A circuit in which the passive components and conductors are produced as films on a substrate by evaporation or sputtering; active components may be similarly produced or mounted separately.

thin-film cryotron A cryotron in which the transition from superconducting to normal resistivity of a thin film of tin or indium, serving as a gate, is controlled by current in a film of lead that crosses and is insulated from the gate.

thin-film ferrite coil An inductor made by depositing a thin flat spiral of gold or other conducting metal on a ferrite substrate.

thin-film integrated circuit An integrated circuit consisting entirely of thin films deposited in a patterned relationship on a substrate.

thin-film material A material that can be deposited as a thin film in a desired pattern by a variety of chemical, mechanical, or high-vacuum evaporation techniques.

thin-film resistor A fixed resistor whose resistance element is a metal, alloy, carbon, or other film having a thickness of about one-millionth inch (about 25 nanometers).

thin-film semiconductor Semiconductor produced by the deposition of an appropriate single-crystal layer on a suitable insulator.

thin-film solar cell A solar cell in which a thin film of gallium arsenide, cadmium sulfide, or other semiconductor material is evaporated on a thin, flexible metal or plastic substrate; the rather low efficiency (about 2%) is compensated by the flexibility and light weight, making these cells attractive as power sources for spacecraft.

thin-film transducer A film a few molecules thick, usually consisting of cadmium sulfide, evaporated on a crystal substrate, used to convert microwave radiation into hypersonic sound waves in the crystal.

thin-film transistor A field-effect transistor constructed entirely by thin-film techniques, for use in thin-film circuits.

thin magnetic film A data storage device consisting of a thin magnetic film of Permalloy deposited by vacuum evaporation or electrochemical deposition.

Thomson bridge *See* Kelvin bridge.

Thomson cross section The total scattering cross section for Thomson scattering, equal to $\frac{8}{3}\pi(e^2/mc^2)^2$, where e and m are the charge (in electrostatic units) and mass of the scattering particle, and c is the speed of light.

Thomson formula 1. The formula for the intensity of scattered electromagnetic radiation in Thomson scattering as a function of the scattering angle ϕ; the intensity is proportional to $1 + \cos^2 \phi$. 2. A formula for the period of oscillation of a current when a capacitor is discharged. Also known as Kelvin's formula.

Thomson parabolas A pattern of parabolas which appear on a photographic plate exposed to a beam of ions of an element which has passed through electric and magnetic fields applied in the same direction normal to the path of the ions; each parabola corresponds to a different charge-to-mass ratio, and thus to a different isotope.

Thomson scattering Scattering of electromagnetic radiation by free (or very loosely bound) charged particles, computed according to a classical nonrelativistic theory: energy is taken away from the primary radiation as the charged particles accelerated by the transverse electric field of the radiation, radiate in all directions.

thoriated emitter *See* thoriated tungsten filament.

thoriated tungsten filament A vacuum-tube filament consisting of tungsten mixed with a small quantity of thorium oxide to give improved electron emission. Also known as thoriated emitter.

three-decibel coupler Junction of two waveguides having a common H wall; the two guides are coupled together by H-type aperture coupling; the coupling is such that 50% of the power from either channel will be fed into the other. Also known as Riblet coupler; short-slot coupler.

three-dimensional display system A radar display which shows range, azimuth, and elevation; for instance, a G display.

three-input subtracter *See* full subtracter.

three-junction transistor A *pnpn* transistor having three junctions and four regions of alternating conductivity; the emitter connection may be made to the p region at the left, the base connection to the adjacent n region, and the collector connection to the n region at the right, while the remaining p region is allowed to float.

three-layer diode A junction diode with three conductivity regions.

three-phase circuit A circuit energized by alternating-current voltages that differ in phase by one-third of a cycle or 120°.

three-phase current Current delivered through three wires, with each wire serving as the return for the other two and with the three current components differing in phase successively by one-third cycle, or 120 electrical degrees.

three-phase four-wire system System of alternating-current supply comprising four conductors, three of which are connected as in a three-phase, three-wire system, the fourth being connected to the neutral point of the supply, which may be grounded.

three-phase magnetic amplifier A magnetic amplifier whose input is the sum of three alternating-current voltages that differ in phase by 120°.

three-phase rectifier A rectifier supplied by three alternating-current voltages that differ in phase by one-third of a cycle or 120°.

three-phase seven-wire system System of alternating-current supply from groups of three single-phase transformers connected in Y to obtain a three-phase, four-wire grounded neutral system of higher voltage for power, the neutral wire being common to both systems.

three-phase three-wire system System of alternating-current supply comprising three conductors between successive pairs of which are maintained alternating differences of potential successively displaced in phase by one-third cycle.

three-pulse cascaded canceler A moving-target indicator technique in which two "two-pulse cancelers" are cascaded together; this improves the velocity response.

three-way switch An electric switch with three terminals used to control a circuit from two different points.

three-wire generator Electric generator with a balance coil connected across the armature, the midpoint of the coil providing the potential of the neutral wire in a three-wire system.

three-wire system System of electric supply comprising three conductors, one of which (known as the neutral wire) is maintained at a potential midway between the potential of the other two (referred to as the outer conductors); part of the load may be connected directly between the outer conductors, the remainder being divided as evenly as possible into two parts, each of which is connected between the neutral and one outer conductor; there are thus two distinct supply voltages, one being twice the other.

threshold In a modulation system, the smallest value of carrier-to-noise ratio at the input to the demodulator for all values above which a small percentage change in the input carrier-to-noise ratio produces a substantially equal or smaller percentage change in the output signal-to-noise ratio.

threshold frequency The frequency of incident radiant energy below which there is no photoemissive effect.

threshold signal A received radio signal (or radar echo) whose power is just above the noise level of the receiver. Also known as minimum detectable signal.

threshold switch A voltage-sensitive alternating-current switch made from a semiconductor material deposited on a metal substrate; when the alternating-current voltage acting on the switch is increased above the threshold value, the number of free carriers present in the semiconductor material increases suddenly, and the switch changes from a high resistance of about 10 megohms to a low resistance of less than 1 ohm; in other versions of this switch, the threshold voltage is controlled by heat, pressure, light, or moisture.

threshold voltage 1. In general, the voltage at which a particular characteristic of an electronic device first appears. 2. The voltage at which conduction of current begins in a *pn* junction. 3. The voltage at which channel formation occurs in a metal oxide semiconductor field-effect transistor. 4. The voltage at which a solid-state lamp begins to emit light.

through repeater Microwave repeater that is not equipped to provide for connections to any local facilities other than the service channel.

thyratron A hot-cathode gas tube in which one or more control electrodes initiate but do not limit the anode current except under certain operating conditions. Also known as hot-cathode gas-filled tube.

thyratron gate In computers, an AND gate consisting of a multielement gas-filled tube in which conduction is initiated by the coincident application of two or more signals; conduction may continue after one or more of the initiating signals are removed.

time-division multiplier   441

thyratron inverter An inverter circuit that uses thyratrons to convert direct-current power to alternating-current power.

thyrector Silicon diode that acts as an insulator up to its rated voltage, and as a conductor above rated voltage; used for alternating-current surge voltage protection.

thyristor A transistor having a thyratronlike characteristic; as collector current is increased to a critical value, the alpha of the unit rises above unity to give high-speed triggering action.

tickler coil Small coil connected in series with the plate circuit of an electron tube and inductively coupled to a grid-circuit coil to establish feedback or regeneration in a radio circuit; used chiefly in regenerative detector circuits.

tie 1. Electrical connection or strap. 2. *See* tie wire.

tie cable 1. Cable between two distributing frames or distributing points. 2. Cable between two private branch exchanges. 3. Cable between a private branch exchange switchboard and main office. 4. Cable connecting two other cables.

tie point Insulated terminal to which two or more wires may be connected.

tier array Array of antenna elements, one above the other.

tie trunk Telephone line or channel directly connecting two private branch exchanges.

tie wire A short piece of wire used to tie an open-line wire to an insulator. Also known as tie.

tight binding approximation A method of calculating energy states and wave functions of electrons in a solid in which the wave function is assumed to be a sum of pure atomic wave functions centered about each of the atoms in the lattice, each multiplied by a phase factor; it is suitable for deep-lying energy levels.

tight coupling *See* close coupling.

tilt 1. Angle which an antenna forms with the horizontal. 2. In radar, the angle between the axis of radiation in the vertical plane and a reference axis which is normally the horizontal.

tilt angle The angle between the axis of radiation of a radar beam in the vertical plane and a reference axis (normally the horizontal).

time base A device which moves the fluorescent spot rhythmically across the screen of the cathode-ray tube.

time-base generator *See* sweep generator.

time-code generator A crystal-controlled pulse generator that produces a train of pulses with various predetermined widths and spacings, from which the time of day and sometimes also day of year can be determined; used in telemetry and other data-acquisition systems to provide the precise time of each event.

time-current characteristics Of a fuse, the relation between the root-mean-square alternating current or direct current and the time for the fuse to perform the whole or some specified part of its interrupting function.

time-delay circuit A circuit in which the output signal is delayed by a specified time interval with respect to the input signal. Also known as delay circuit.

time-delay fuse A fuse in which the burnout action depends on the time it takes for the overcurrent heat to build up in the fuse and melt the fuse element.

time-delay relay A relay in which there is an appreciable interval of time between energizing or deenergizing of the coil and movement of the armature, such as a slow-acting relay and a slow-release relay.

time-division multiplier *See* mark-space multiplier.

time-division switching system A type of electronic switching system in which input signals on lines and trunks are sampled periodically, and each active input is associated with the desired output for a specific phase of the period.

time-domain reflectometer An instrument that measures the electrical characteristics of wideband transmission systems, subassemblies, components, and lines by feeding in a voltage step and displaying the superimposed reflected signals on an oscilloscope equipped with a suitable time-base sweep. Abbreviated TDR.

time gate A circuit that gives an output only during chosen time intervals.

time-height section A facsimile trace of a vertically directed radar; specifically, a cloud-detection radar.

time-mark generator A signal generator that produces highly accurate clock pulses which can be superimposed as pips on a cathode-ray screen for timing the events shown on the display.

time-pulse distributor A device or circuit for allocating timing pulses or clock pulses to one or more conducting paths or control lines in specified sequence.

timer A circuit used in radar and in electronic navigation systems to start pulse transmission and synchronize it with other actions, such as the start of a cathode-ray sweep.

time-shared amplifier An amplifier used with a synchronous switch to amplify signals from different sources one after another.

timing-axis oscillator *See* sweep generator.

timing motor A motor which operates from an alternating-current power system synchronously with the alternating-current frequency, used in timing and clock mechanisms. Also known as clock motor.

timing relay Form of auxiliary relay used to introduce a definite time delay in the performance of a function.

timing signal Any signal recorded simultaneously with data on magnetic tape for use in identifying the exact time of each recorded event.

tinsel cord A highly flexible cord used for headphone leads and test leads, in which the conductors are strips of thin metal foil or tinsel wound around a strong but flexible central cord.

tip 1. The contacting part at the end of a phone plug. 2. A small protuberance on the envelope of an electron tube, resulting from the closing of the envelope after evacuation.

tip jack A small single-hole jack for a single-pin contact plug. Also known as pup jack.

tip side Conductor of a circuit which is associated with the tip of a plug or the top spring of a jack; by extension, it is common practice to designate by these terms the conductors having similar functions or arrangements in circuits where plugs or jacks may not be involved.

Tirrill regulator A device for regulating the voltage of a generator, in which the field resistance of the exciter is short-circuited temporarily when the voltage drops.

T junction A network of waveguides with three waveguide terminals arranged in the form of a letter T; in a rectangular waveguide a symmetrical T junction is arranged by having either all three broadsides in one plane or two broadsides in one plane and the third in a perpendicular plane.

T²L *See* transistor-transistor logic.

TM mode *See* transverse magnetic mode.

TM wave *See* transverse magnetic wave.

T network A network composed of three branches, with one end of each branch connected to a common junction point, and with the three remaining ends connected to an input terminal, an output terminal, and a common input and output terminal, respectively.

Toepler-Holtz machine An early type of machine for continuously producing electrical charges at high voltage by electrostatic induction, superseded by the Wimhurst machine. Also known as Holtz machine.

toggle To switch over to an alternate state, as in a flip-flop.

toggle condition Condition of a flip-flop circuit in which the internal state of the flip-flop changes from 0 to 1 or from 1 to 0.

toggle switch 1. A small switch that is operated by manipulation of a projecting lever that is combined with a spring to provide a snap action for opening or closing a circuit quickly. 2. Interconnection between stages of an amplifier which employs a transformer for connecting the plate circuit of one stage to the grid circuit of the following stage; a special case of inductive coupling. 3. An electronically operated circuit that holds either of two states until changed.

tomography *See* sectional radiography.

tone control A control used in an audio-frequency amplifier to change the frequency response so as to secure the most pleasing proportion of bass to treble; individual bass and treble controls are provided in some amplifiers.

tone dialing *See* push-button dialing.

tone generator A signal generator used to generate an audio-frequency signal suitable for signaling purposes or for testing audio-frequency equipment.

Tonotron Trademark for a type of direct-view storage tube.

top-loaded vertical antenna Vertical antenna constructed so that, because of its greater size at the top, there results modified current distribution, giving a more desirable radiation pattern in the vertical plane.

tornadotron Millimeter-wave device which generates radio-frequency power from an enclosed, orbiting electron cloud, excited by a radio-frequency field, when subjected to a strong, pulsed magnetic field.

toroidal coil *See* toroidal magnetic circuit.

toroidal core The doughnut-shaped piece of magnetic material in a toroidal magnetic circuit.

toroidal magnetic circuit Doughnut-shaped piece of magnetic material, together with one or more coils of current-carrying wire wound about the doughnut, with the permeability of the magnetic material high enough so that the magnetic flux is almost completely confined within it. Also known as toroid; toroidal coil.

torque motor A motor designed primarily to exert torque while stalled or rotating slowly.

torque-speed characteristic For electric motors, the relationship of developed torque to armature speed.

total harmonic distortion Ratio of the power at the fundamental frequency, measured at the output of the transmission system considered, to the power of all harmonics observed at the output of the system because of its nonlinearity, when a single frequency signal of specified power is applied to the input of the system; it is expressed in decibels.

touch call *See* push-button dialing.

touch control A circuit that closes a relay when two metal areas are bridged by a finger or hand.

tower A tall metal structure used as a transmitting antenna, or used with another such structure to support a transmitting antenna wire.

tower loading Load placed on a tower by its own weight, the weight of the wires with or without ice covering, the insulators, the wind pressure normal to the line acting both on the tower and the wires, and the pull from the wires.

tower radiator Metal structure used as a transmitting antenna.

Townsend avalanche *See* avalanche.

Townsend characteristic Current-voltage characteristic curve for a phototube at constant illumination and at voltages below that at which a glow discharge occurs.

Townsend coefficient The number of ionizing collisions by an electron per centimeter of path length in the direction of the applied electric field in a radiation counter.

Townsend discharge A discharge which occurs at voltages too low for it to be maintained by the electric field alone, and which must be initiated and sustained by ionization produced by other agents; it occurs at moderate pressures, above about 0.1 torr, and is free of space charges.

Townsend ionization *See* avalanche.

T pad A pad made up of resistance elements arranged as a T network (two resistors inserted in one line, with a third between their junction and the other line).

trace The visible path of a moving spot on the screen of a cathode-ray tube. Also known as line.

trace interval Interval corresponding to the direction of sweep used for delineation.

trace sensitivity The ability of an oscilloscope to produce a visible trace on the scope face for a specified input voltage.

track 1. A path for recording one channel of information on a magnetic tape, drum, or other magnetic recording medium; the location of the track is determined by the recording equipment rather than by the medium. 2. The trace of a moving target on a plan-position-indicator radar screen or an equivalent plot.

tracking 1. A leakage or fault path created across the surface of an insulating material when a high-voltage current slowly but steadily forms a carbonized path. 2. The condition in which all tuned circuits in a receiver accurately follow the frequency indicated by the tuning dial over the entire tuning range.

tracking filter Electronic device for attenuating unwanted signals while passing desired signals, by phase-lock techniques that reduce the effective bandwidth of the circuit and eliminate amplitude variations.

track in range To adjust the gate of a radar set so that it opens at the correct instant to accept the signal from a target of changing range from the radar.

track pitch The physical distance between track centers.

track-return power system A system for distributing electric power to trains or other vehicles, in which the track rails are used as an uninsulated return conductor.

track-while-scan Electronic system used to detect a radar target, compute its velocity, and predict its future position without interfering with continuous radar scanning.

trailer A bright streak at the right of a dark area or dark line in a television picture, or a dark area or streak at the right of a bright part; usually due to insufficient gain at low video frequencies.

trailing antenna An aircraft radio antenna having one end weighted and trailing free from the aircraft when in flight.

trailing edge The major portion of the decay of a pulse.

trainer A piece of equipment used for training operators of radar, sonar, and other electronic equipment by simulating signals received under operating conditions in the field.

transadmittance A specific measure of transfer admittance under a given set of conditions, as in forward transadmittance, interelectrode transadmittance, short-circuit transadmittance, small-signal forward transadmittance, and transadmittance compression ratio.

transceiver A radio transmitter and receiver combined in one unit and having switching arrangements such as to permit use of one or more tubes for both transmitting and receiving. Also known as transmitter-receiver.

transconductance An electron-tube rating, equal to the change in plate current divided by the change in control-grid voltage that causes it, when the plate voltage and all other voltages are maintained constant. Also known as grid-anode transconductance; grid-plate transconductance; mutual conductance. Symbolized G_m; g_m.

transcribe To record, as to record a radio program by means of electric transcriptions or magnetic tape for future rebroadcasting.

transducer loss The ratio of the power available to a transducer from a specified source to the power that the transducer delivers to a specified load; usually expressed in decibels.

transductor *See* magnetic amplifier; saturable reactor.

transfer admittance An admittance rating for electron tubes and other transducers or networks; it is equal to the complex alternating component of current flowing to one terminal from its external termination, divided by the complex alternating component of the voltage applied to the adjacent terminal on the cathode or reference side; all other terminals have arbitrary external terminations.

transfer characteristic 1. Relation, usually shown by a graph, between the voltage of one electrode and the current to another electrode, with all other electrode voltages being maintained constant. 2. Function which, multiplied by an input magnitude, will give a resulting output magnitude. 3. Relation between the illumination on a camera tube and the corresponding output-signal current, under specified conditions of illumination.

transfer impedance The ratio of the voltage applied at one pair of terminals of a network to the resultant current at another pair of terminals, all terminals being terminated in a specified manner.

transferred-electron amplifier A diode amplifier, which generally uses a transferred-electron diode made from doped n-type gallium arsenide, that provides amplification in the gigahertz range to well over 50 gigahertz at power outputs typically below 1 watt continuous-wave. Abbreviated TEA.

transferred-electron device A semiconductor device, usually a diode, that depends on internal negative resistance caused by transferred electrons in gallium arsenide or indium phosphide at high electric fields; transit time is minimized, permitting oscillation at frequencies up to several hundred megahertz.

transferred-electron effect The variation in the effective drift mobility of charge carriers in a semiconductor when significant numbers of electrons are transferred from ow-mobility valley of the conduction band in a zone to a high-mobility valley, or vice versa.

transfer switch A switch for transferring one or more conductor connections from one circuit to another.

transfluxor A magnetic core having two or more apertures and three or more legs for flux; used as a computer memory element, crossbar switch, channel commutator, or control element.

transformation For two networks which are equivalent as far as conditions at the terminals are concerned, a set of equations giving the admittances or impedances of the branches of one circuit in terms of the admittances or impedances of the other.

transformation matrix A two-by-two matrix which relates the amplitudes of the traveling waves on one side of a waveguide junction to those on the other.

transformer An electrical component consisting of two or more multiturn coils of wire placed in close proximity to cause the magnetic field of one to link the other; used to transfer electric energy from one or more alternating-current circuits to one or more other circuits by magnetic induction.

transformer bridge A network consisting of a transformer and two impedances, in which the input signal is applied to the transformer primary and the output is taken between the secondary center-tap and the junction of the impedances that connect to the outer leads of the secondary.

transformer-coupled amplifier Audio-frequency amplifier that uses untuned iron-core transformers to provide coupling between stages.

transformer coupling 1. Interconnection between stages of an amplifier which employs a transformer for connecting the plate circuit of one stage to the grid circuit of the following stage; a special case of inductive coupling. **2.** *See* inductive coupling.

transformer hybrid *See* hybrid set.

transformer load loss Losses in a transformer which are incident to the carrying of the load; load losses include resistance loss in the windings due to load current, stray loss due to stray fluxes in the windings, core clamps, and so on, and to circulating current, if any, in parallel windings.

transformer loss Ratio of the signal power that an ideal transformer of the same impedance ratio would deliver to the load impedance, to the signal power that the actual transformer delivers to the load impedance; this ratio is usually expressed in decibels.

transformer substation An electric power substation whose equipment includes transformers.

transformer voltage ratio Ratio of the root-mean-square primary terminal voltage to the root-mean-square secondary terminal voltage under specified conditions of load.

transforming section Length of waveguide or transmission line of modified cross section, or with a metallic or dielectric insert, used for impedance transformation.

transhybrid loss In a carrier telephone system, the transmission loss at a given frequency measured across a hybrid circuit joined to a given two-wire termination and balancing network.

transient analyzer An analyzer that generates transients in the form of a succession of equal electric surges of small amplitude and adjustable waveform, applies these transients to a circuit or device under test, and shows the resulting output waveforms on the screen of an oscilloscope.

transient distortion Distortion due to inability to amplify transients linearly.

transient phenomena Rapidly changing actions occurring in a circuit during the interval between closing of a switch and settling to a steady-state condition, or any other temporary actions occurring after some change in a circuit.

transient suppressor *See* surge suppressor.

transistance The characteristic that makes possible the control of voltages or currents so as to accomplish gain or switching action in a circuit; examples of transistance occur in transistors, diodes, and saturable reactors.

transistor An active component of an electronic circuit consisting of a small block of semiconducting material to which at least three electrical contacts are made, usually two closely spaced rectifying contacts and one ohmic (nonrectifying) contact; it may be used as an amplifier, detector, or switch.

transistor amplifier An amplifier in which one or more transistors provide amplification comparable to that of electron tubes.

transistor biasing Maintaining a direct-current voltage between the base and some other element of a transistor.

transistor characteristics The values of the impedances and gains of a transistor.

transistor chip An unencapsulated transistor of very small size used in microcircuits.

transistor circuit An electric circuit in which a transistor is connected.

transistor clipping circuit A circuit in which a transistor is used to achieve clipping action; the bias at the input is set at such a level that output current cannot flow during a portion of the amplitude excursion of the input voltage or current waveform.

transistor gain The increase in signal power produced by a transistor.

transistor input resistance The resistance across the input terminals of a transistor stage. Also known as input resistance.

transistor magnetic amplifier A magnetic amplifier together with a transistor preamplifier, the latter used to make the signal strong enough to change the flux in the core of the magnetic amplifier completely during a half-cycle of the power supply voltage.

transistor radio A radio receiver in which transistors are used in place of electron tubes.

transistor-transistor logic A logic circuit containing two transistors, for driving large output capacitances at high speed. Abbreviated T^2L; TTL.

transition element An element used to couple one type of transmission system to another, as for coupling a coaxial line to a waveguide.

transition factor *See* reflection factor.

transition loss At a junction between a source and a load, the ratio of the available power to the power delivered to the load.

transition point A point at which the constants of a circuit change in such a way as to cause reflection of a wave being propagated along the circuit.

transition region The region between two homogeneous semiconductors in which the impurity concentration changes.

transitron Thermionic-tube circuit whose action depends on the negative transconductance of the suppressor grid of a pentode with respect to the screen grid.

transitron oscillator A negative-resistance oscillator in which the screen grid is more positive than the anode, and a capacitor is connected between the screen grid and the suppressor grid; the suppressor grid periodically divides the current between the screen grid and the anode, thereby producing oscillation.

transit time The time required for an electron or other charge carrier to travel between two electrodes in an electron tube or transistor.

transit-time microwave diode A solid-state microwave diode in which the transit time of charge carriers is short enough to permit operation in microwave bands.

transit-time mode One of the three operating modes of a transferred-electron diode, in which space-charge domains are formed at the cathode and travel across the drift region to the anode.

translator A combination television receiver and low-power television transmitter, used to pick up television signals on one frequency and retransmit them on another frequency to provide reception in areas not served directly by television stations.

transmission 1. The process of transferring a signal, message, picture, or other form of intelligence from one location to another location by means of wire lines, radio, light beams, infrared beams, or other communication systems. 2. A message, signal, or other form of intelligence that is being transmitted. 3. *See* transmittance.

transmission band Frequency range above the cutoff frequency in a waveguide, or the comparable useful frequency range for any other transmission line, system, or device.

transmission electron microscope A type of electron microscope in which the specimen transmits an electron beam focused on it, image contrasts are formed by the scattering of electrons out of the beam, and various magnetic lenses perform functions analogous to those of ordinary lenses in a light microscope.

transmission electron radiography A technique used in microradiography to obtain radiographic images of very thin specimens; the photographic plate is in close contact with the specimen, over which is placed a lead foil and then a light-tight covering; hardened x-rays shoot through the light-tight covering.

transmission gain *See* gain.

transmission gate A gate circuit that delivers an output waveform that is a replica of a selected input during a specific time interval which is determined by a control signal.

transmission line A system of conductors, such as wires, waveguides, or coaxial cables, suitable for conducting electric power or signals efficiently between two or more terminals.

transmission-line admittance The complex ratio of the current flowing in a transmission line to the voltage across the line, where the current and voltage are expressed in phasor notation.

transmission-line attenuation The decrease in power of a transmission-line signal from one point to another, expressed as a ratio or in decibels.

transmission-line cable The coaxial cable, waveguide, or microstrip which forms a transmission line; a number of standard types have been designated, specified by size and materials.

transmission-line current The amount of electrical charge which passes a given point in a transmission line per unit time.

transmission-line efficiency The ratio of the power of a transmission-line signal at one end of the line to that at the other end where the signal is generated.

transmission-line impedance The complex ratio of the voltage across a transmission line to the current flowing in the line, where voltage and current are expressed in phasor notation.

transmission-line parameters The quantities which are necessary to specify the impedance per unit length of a transmission line, and the admittance per unit length between various conductors of the line. Also known as linear electrical parameters; line parameters; transmission line constants.

transmission-line power The amount of energy carried past a point in a transmission line per unit time.

transmission-line reflection coefficient The ratio of the voltage reflected from the load at the end of a transmission line to the direct voltage.

transmission-line theory The application of electrical and electromagnetic theory to the behavior of transmission lines.

transmission-line transducer loss The ratio of the power delivered by a transmission line to a load to that produced at the generator, expressed in decibels; equal to the sum of the attenuation of the line and the mismatch loss.

transmission-line voltage The work that would be required to transport a unit electrical charge between two specified conductors of a transmission line at a given instant.

transmission mode *See* mode.

transmission modulation Amplitude modulation of the reading-beam current in a charge storage tube as the beam passes through apertures in the storage surface; the degree of modulation is controlled by the stored charge pattern.

transmission regulator In electrical communications, a device that maintains substantially constant transmission levels over a system.

transmission substation An electric power substation associated with high voltage levels.

transmissivity The ratio of the transmitted radiation to the radiation arriving perpendicular to the boundary between two mediums.

transmit-receive tube A gas-filled radio-frequency switching tube used to disconnect a receiver from its antenna during the interval for pulse transmission in radar and other pulsed radio-frequency systems. Also known as TR box; TR cell (British usage); TR switch; TR tube.

transmittance The radiant power transmitted by a body divided by the total radiant power incident upon the body. Also known as transmission.

transmittancy The transmittance of a solution divided by that of the pure solvent of the same thickness.

transmitter *See* radio transmitter; synchro transmitter.

transmitter-distributor In teletypewriter opeations, a motor-driven device which translates teletypewriter code combinations form perforated tape into electrical impulses, and transmits these impulses to one or more receiving stations. Abbreviated TD.

transmitter noise *See* frying noise.

transmitter-receiver *See* transceiver.

transmitter synchro *See* synchro transmitter.

transmittivity The internal transmittance of a piece of nondiffusing substance of unit thickness.

transolver A synchro having a two-phase cylindrical rotor within a three-phase stator, for use as a transmitter or a control transformer with no degradation of accuracy or nulls.

transpolarizer An electrostatically controlled circuit impedance that can have about 30 discrete and reproducible impedance values: two capacitors, each having a crystalline ferroelectric dielectric with a nearly rectangular hysteresis loop, are connected in series and act as a single low impedance to an alternating-current sensing signal when both capacitors are polarized in the same direction; application of 1-microsecond pulses of appropriate polarity increases the impedance in steps.

transponder dead time Time interval between the start of a pulse and the earliest instant at which a new pulse can be received or produced by a transponder.

transponder set A complete electronic set which is designed to receive an interrogation signal, and which retransmits coded signals that can be interpreted by the interrogating station; it may also utilize the received signal for actuation of additional equipment such as local indicators or servo amplifiers.

transponder suppressed time delay Overall fixed time delay between reception of an interrogation and transmission of a reply to this interrogation.

transrectification Rectification that occurs in one circuit when an alternating voltage is applied to another circuit.

transrectification characteristic Graph obtained by plotting the direct-voltage values for one electrode of a vacuum tube as abscissas against the average current values in the circuit of that electrode as ordinates, for various values of alternating voltage applied to another electrode as a parameter; the alternating voltage is held constant for each curve, and the voltages on other electrodes are maintained constant.

transrectifier Device, ordinarily a vacuum tube, in which rectification occurs in one electrode circuit when an alternating voltage is applied to another electrode.

transverse Doppler effect An aspect of the optical Doppler effect, occurring when the direction of motion of the source relative to an observer is perpendicular to the direction of the light received by the observer; the observed frequency is smaller than the source frequency by the factor $[1-(v/c)^2]^{1/2}$ where v is the speed of the source and c is the speed of light.

transverse electric mode A mode in which a particular transverse electric wave is propagated in a waveguide or cavity. Abbreviated TE mode. Also known as H mode (British usage).

transverse electric wave An electromagnetic wave in which the electric field vector is everywhere perpendicular to the direction of propagation. Abbreviated TE wave. Also known as H wave (British usage).

transverse electromagnetic mode A mode in which a particular transverse electromagnetic wave is propagated in a waveguide or cavity. Abbreviated TEM mode.

transverse electromagnetic wave An electromagnetic wave in which both the electric and magnetic field vectors are everywhere perpendicular to the direction of propagation. Abbreviated TEM wave.

transverse interference Interference occurring across terminals or between signal leads.

transverse magnetic mode A mode in which a particular transverse magnetic wave is propagated in a waveguide or cavity. Abbreviated TM mode. Also known as E mode (British usage).

transverse magnetic wave An electromagnetic wave in which the magnetic field vector is everywhere perpendicular to the direction of propagation. Abbreviated TM wave. Also known as E wave (British usage).

transverse magnetoresistance One of the galvanomagnetic effects, in which a magnetic field perpendicular to an electric current gives rise to an electrical potential change in the direction of the current.

transverse recording Technique for recording television signals on magnetic tape using a four-transducer rotating head.

trap 1. A tuned circuit used in the radio-frequency or intermediate-frequency section of a receiver to reject undesired frequencies; traps in television receiver video circuits keep the sound signal out of the picture channel. Also known as rejector. **2.** Any irregularity, such as a vacancy, in a semiconductor at which an electron or hole in the conduction band can be caught and trapped until released by thermal agitation. Also known as semiconductor trap. **3.** *See* wave trap.

TRAPATT diode A *pn* junction diode, similar to the IMPATT diode, but characterized by the formation of a trapped space-charge plasma within the junction region; used in the generation and amplification of microwave power. Derived from trapped plasma avalanche transit time diode.

trapezoidal generator Electronic stage designed to produce a trapezoidal voltage wave.

trapezoidal pulse An electrical pulse in which the voltage rises linearly to some value, remains constant at this value for some time, and then drops linearly to the original value.

trapezoidal wave A wave consisting of a series of trapezoidal pulses.

traveling-wave amplifier An amplifier that uses one or more traveling-wave tubes to provide useful amplification of signals at frequencies of the order of thousands of megahertz.

traveling-wave antenna An antenna in which the current distributions are produced by waves of charges propagated in only one direction in the conductors. Also known as progressive-wave antenna.

traveling-wave magnetron A traveling-wave tube in which the electrons move in crossed static electric and magnetic fields that are substantially normal to the direction of wave propagation, as in practically all modern magnetrons.

traveling-wave magnetron oscillations Oscillations sustained by the interaction between the space-charge cloud of a magnetron and a traveling electromagnetic field whose phase velocity is approximately the same as the mean velocity of the cloud.

traveling-wave parametric amplifier Parametric amplifier which has a continuous or iterated structure incorporating nonlinear reactors and in which the signal, pump, and difference-frequency waves are propagated along the structure.

traveling-wave phototube A traveling-wave tube having a photocathode and an appropriate window to admit a modulated laser beam; the modulated laser beam causes emission of a current-modulated photoelectron beam, which in turn is accelerated by an electron gun and directed into the helical slow-wave structure of the tube.

traveling-wave tube An electron tube in which a stream of electrons interacts continuously or repeatedly with a guided electromagnetic wave moving substantially in synchronism with it, in such a way that there is a net transfer of energy from the stream to the wave; the tube is used as an amplifier or oscillator at frequencies in the microwave region.

TR box *See* transmit-receive tube.

TR cell *See* transmit-receive tube.

tree A set of connected circuit branches that includes no meshes; responds uniquely to each of the possible combinations of a number of simultaneous inputs. Also known as decoder.

TRF receiver *See* tuned-radio-frequency receiver.

triad A triangular group of three small phosphor dots, each emitting one of the three primary colors on the screen of a three-gun color picture tube.

triangular pulse An electrical pulse in which the voltage rises linearly to some value, and immediately falls linearly to the original value.

triangular wave A wave consisting of a series of triangular pulses.

triboelectricity *See* frictional electricity.

triboelectric series A list of materials that produce an electrostatic charge when rubbed together, arranged in such an order that a material has a positive charge when rubbed with a material below it in the list, and has a negative charge when rubbed with a material above it in the list.

triboelectrification The production of electrostatic charges by friction.

trickle charge A continuous charge of a storage battery at a low rate to maintain the battery in a fully charged condition.

tricolor picture tube *See* color picture tube.

triductor Arrangement of iron-core transformers and capacitors used to triple a power-line frequency.

trigatron Gas-filled, spark-gap switch used in line pulse modulators.

trigger 1. To initiate an action, which then continues for a period of time, as by applying a pulse to a trigger circuit. 2. The pulse used to initiate the action of a trigger circuit. 3. *See* trigger circuit.

trigger action Use of a weak input pulse to initiate main current flow suddenly in a circuit or device.

trigger circuit 1. A circuit or network in which the output changes abruptly with an infinitesimal change in input at a predetermined operating point. Also known as trigger. 2. A circuit in which an action is initiated by an input pulse, as in a radar modulator. 3. *See* bistable multivibrator.

trigger control Control of thyratrons, ignitrons, and other gas tubes in such a way that current flow may be started or stopped, but not regulated as to rate.

trigger diode A symmetrical three-layer avalanche diode used in activating silicon-controlled rectifiers; it has a symmetrical switching mode, and hence fires whenever the breakover voltage is exceeded in either polarity. Also known as diode alternating-current switch (diac).

triggered spark gap A fixed spark gap in which the discharge passes between two electrodes but is initiated by an auxiliary trigger electrode to which low-power pulses are applied at regular intervals by a pulse amplifier.

trigger electrode *See* starter.

triggering Phenomenon observed in some high-performance magnetic amplifiers with very low leakage rectifiers; as the input current is decreased in magnitude, the amplifier remains at cutoff for some time, and the output then suddenly shoots upward.

trigger level In a transponder, the minimum input to the receiver which is capable of causing a transmitter to emit a reply.

trigger pulse A pulse that starts a cycle of operation. Also known as tripping pulse.

trigger switch A switch that is actuated by pulling a trigger, and is usually mounted in a pistol-grip handle.

trigger tube A cold-cathode gas-filled tube in which one or more auxiliary electrodes initiate the anode current but do not control it.

trigistor A *pnpn* device with a gating control acting as a fast-acting switch similar in nature to a thyratron.

trim Fine adjustment of capacitance, inductance, or resistance of a component during manufacture or after installation in a circuit.

trimmer capacitor A relatively small variable capacitor used in parallel with a larger variable or fixed capacitor to permit exact adjustment of the capacitance of the parallel combination.

trimmer potentiometer A potentiometer which is used to provide a small-percentage adjustment and is often used with a coarse control.

triode A three-electrode electron tube containing an anode, a cathode, and a control electrode.

triode clamp A keyed clamp circuit utilizing triodes, such as a circuit which contains a complementary pair of bipolar transistors.

triode clipping circuit A clipping circuit that utilizes a transistor or vacuum triode.

triode laser Gas laser whose light output may be modulated by signal voltages applied to an integral grid.

tube   453

triode transistor   A transistor that has three terminals.

trip coil   A type of solenoid in which the moving armature opens a circuit breaker or other protective device when the coil current exceeds a predetermined value.

triple-conversion receiver   Communications receiver having three different intermediate frequencies to give higher adjacent-channel selectivity and greater image-frequency suppression.

triple-stub transformer   Microwave transformer in which three stubs are placed a quarter-wavelength apart on a coaxial line and adjusted in length to compensate for impedance mismatch.

triplexer   Dual duplexer that permits the use of two receivers simultaneously and independently in a radar system by disconnecting the receivers during the transmitted pulse.

tripping device   Mechanical or electromagnetic device used to bring a circuit breaker or starter to its off or open position, either when certain abnormal electrical conditions occur or when a catch is actuated manually.

tripping pulse   *See* trigger pulse.

trisistor   Fast-switching semiconductor consisting of an alloyed junction *pnp* device in which the collector is capable of electron injection into the base; characteristics resemble those of a thyratron electron tube, and switching time is in the nanosecond range.

tristate logic   A form of transistor-transistor logic in which the output stages or input and output stages can assume three states; two are the normal low-impedance 1 and 0 states, and the third is a high-impedance state that allows many tristate devices to time-share bus lines.

tri-tet oscillator   Crystal-controlled, electron-coupled, vacuum-tube oscillator circuit which is isolated from the output circuit through use of the screen grid electrode as the oscillator anode; used for multiband operation because it generates strong harmonics of the crystal frequency.

trolley pole   The pole which conducts electricity from the trolley wire to the trolley.

trolley wire   The means by which power is conveyed to an electric trolley locomotive; it is an overhead wire which conducts power to the locomotive by the trolley pole.

trombone   U-shaped, adjustable, coaxial-line matching assembly.

TR switch   *See* transmit-receive tube.

true-motion radar   A radar set which provides a true-motion radar presentation on the plan-position indicator, as opposed to the relative-motion, true-or-relative-bearing, presentation most commonly used.

true-motion radar presentation   A radar plan-position indicator presentation in which the center of the scope represents the same geographic position, until reset, with all moving objects, including the user's own craft, moving on the scope.

truncated paraboloid   Paraboloid antenna in which a portion of the top and bottom have been cut away to broaden the main lobe in the vertical plane.

trunk feeder   An electric power transmission line that connects two generating stations, or a generating station and an important substation, or two electrical distribution networks.

T-section filter   T network used as an electric filter.

TTL   *See* transistor-transistor logic.

tube   *See* electron tube.

tube coefficient Any of the constants that describe the characteristics of a thermionic vacuum tube, such as amplification factor, mutual conductance, or alternating-current plate resistance.

tube heating time Time required for a tube to attain operating temperature.

tube noise Noise originating in a vacuum tube, such as that due to shot effect and thermal agitation.

tube of flux *See* tube of force.

tube of force A region of space bounded by a tubular surface consisting of the lines of force which pass through a given closed curve. Also known as tube of flux.

tube tester A test instrument designed to measure and indicate the condition of electron tubes used in electronic equipment.

tube voltage drop In a gas tube, the anode voltage during the conducting period.

tubular capacitor A paper or electrolytic capacitor having the form of a cylinder, with leads usually projecting axially from the ends; the capacitor plates are long strips of metal foil separated by insulating strips, rolled into a compact tubular shape.

tunable echo box Echo box consisting of an adjustable cavity operating in a single mode; if calibrated, the setting of the plunger at resonance will indicate the wavelength.

tunable filter An electric filter in which the frequency of the passband or rejection band can be varied by adjusting its components.

tunable magnetron Magnetron which can be tuned mechanically or electronically by varying its capacitance or inductance.

tune To adjust for resonance at a desired frequency.

tuned amplifier An amplifier in which the load is a tuned circuit; load impedance and amplifier gain then vary with frequency.

tuned-anode oscillator A vacuum-tube oscillator whose frequency is determined by a tank circuit in the anode circuit, coupled to the grid to provide the required feedback. Also known as tuned-plate oscillator.

tuned-anode tuned-grid oscillator *See* tuned-grid tuned-anode oscillator.

tuned-base oscillator Transistor oscillator in which the frequency-determining resonant circuit is located in the base circuit; comparable to a tuned-grid oscillator.

tuned cavity *See* cavity resonator.

tuned circuit A circuit whose components can be adjusted to make the circuit responsive to a particular frequency in a tuning range. Also known as tuning circuit.

tuned-collector oscillator A transistor oscillator in which the frequency-determining resonant circuit is located in the collector circuit; this is comparable to a tuned-anode electron-tube oscillator.

tuned filter Filter that uses one or more tuned circuits to attenuate or pass signals at the resonant frequency.

tuned-grid oscillator Oscillator whose frequency is determined by a parallel-resonant circuit in the grid coupled to the plate to provide the required feedback.

tuned-grid tuned-anode oscillator A vacuum-tube oscillator whose frequency is determined by a tank circuit in the grid circuit, coupled to the anode to provide the required feedback. Also known as tuned-anode tuned-grid oscillator.

tuned-plate oscillator *See* tuned-anode oscillator.

tuned-radio-frequency receiver A radio receiver consisting of a number of amplifier stages that are tuned to resonance at the carrier frequency of the desired signal by a gang capacitor; the amplified signals at the original carrier frequency are fed directly into the detector for demodulation, and the resulting audio-frequency signals are amplified by an a-f amplifier and reproduced by a loudspeaker. Abbreviated TRF receiver.

tuned-radio-frequency transformer Transformer used for selective coupling in radio-frequency stages.

tuned relay A relay having mechanical or other resonating arrangements that limit response to currents at one particular frequency.

tuned resonating cavity Resonating cavity half a wavelength long or some multiple of a half wavelength, used in connection with a waveguide to produce a resultant wave with the amplitude in the cavity greatly exceeding that of the wave in the waveguide.

tuned transformer Transformer whose associated circuit elements are adjusted as a whole to be resonant at the frequency of the alternating current supplied to the primary, thereby causing the secondary voltage to build up to higher values than would otherwise be obtained.

tuner The portion of a receiver that contains circuits which can be tuned to accept the carrier frequency of the alternating current supplied to the primary, thereby causing the secondary voltage to build up to higher values than would otherwise be obtained.

tungar tube A gas tube having a heated thoriated tungsten filament serving as cathode and a graphite disk serving as anode in an argon-filled bulb at a low pressure; used chiefly as a rectifier in battery chargers.

tungsten filament A filament used in incandescent lamps, and as an incandescent cathode in many types of electron tubes, such as thermionic vacuum tubes.

tuning The process of adjusting the inductance or the capacitance or both in a tuned circuit, for example, in a radio, television, or radar receiver or transmitter, so as to obtain optimum performance at a selected frequency.

tuning capacitor A variable capacitor used for tuning purposes.

tuning circuit *See* tuned circuit.

tuning core A ferrite core that is designed to be moved in and out of a coil or transformer to vary the inductance.

tuning eye *See* cathode-ray tuning indicator.

tuning indicator A device that indicates when a radio receiver is tuned accurately to a radio station, such as a meter or a cathode-ray tuning indicator; it is connected to a circuit having a direct-current voltage that varies with the strength of the incoming carrier signal.

tuning range The frequency range over which a receiver or other piece of equipment can be adjusted by means of a tuning control.

tuning screw A screw that is inserted into the top or bottom wall of a waveguide and adjusted as to depth of penetration inside for tuning or impedance-matching purposes.

tuning stub Short length of transmission line, usually shorted at its free end, connected to a transmission line for impedance-matching purposes.

tuning susceptance Normalized susceptance of an anti-transmit-receive tube in its mount due to the deviation of its resonant frequency from the desired resonant frequency.

tuning wand Rod of insulating material having a brass plug at one end and a powered iron core at the other end; used for checking receiver alignment.

tunnel diode A heavily doped junction diode that has a negative resistance at very low voltage in the forward bias direction, due to quantum-mechanical tunneling, and a short circuit in the negative bias direction. Also known as Esaki tunnel diode.

tunneling cryotron A low-temperature current-controlled switching device that has two electrodes of superconducting material separated by an insulating film, forming a Josephson junction, and a control line whose currents generate magnetic fields that switch the device between two states characterized by the presence or absence of electrical resistance.

tunnel rectifier Tunnel diode having a relatively low peak-current rating as compared with other tunnel diodes used in memory-circuit applications.

tunnel resistor Resistor in which a thin layer of metal is plated across a tunneling junction, to give the combined characteristics of a tunnel diode and an ordinary resistor.

tunnel triode Transistorlike device in which the emitter-base junction is a tunnel diode and the collector-base junction is a conventional diode.

turbine generator An electric generator driven by a steam, hydraulic, or gas turbine.

turboalternator An alternator, such as a synchronous generator, which is driven by a steam turbine.

turn One complete loop of wire.

turn-off time The time that is takes a gate circuit to shut off a current.

turn-on time The time that it takes a gate circuit to allow a current to reach its full value.

turns ratio The ratio of the number of turns in a secondary winding of a transformer to the number of turns in the primary winding.

turnstile antenna An antenna consisting of one or more layers of crossed horizontal dipoles on a mast, usually energized so the currents in the two dipoles of a pair are equal and in quadrature; used with television, frequency modulation, and other very-high-frequency or ultra-high-frequency transmitters to obtain an essentially omnidirectional radiation pattern.

turret tuner A television tuner having one set of pretuned circuits for each channel, mounted on a drum that is rotated by the channel selector; rotation of the drum connects each set of tuned circuits in turn to the receiver antenna circuit, radio-frequency amplifier, and r-f oscillator.

twilight zone Anything resembling the twilight zone of the earth, as the narrow sector on each side of the equisignal zone of a four-course radio range station, in which one signal is barely heard above the monotone on-course signal.

twin-T filter An electric filter consisting of a parallel-T network with values of network elements chosen in such a way that the outputs due to each of the paths precisely cancel at a specified frequency.

twin-T network *See* parallel-T network.

twist A waveguide section in which there is a progressive rotation of the cross section about the longitudinal axis of the waveguide.

twisted pair A cable composed of two small insulated conductors twisted together without a common covering.

twister A piezoelectric crystal that generates a voltage when twisted.

two-carrier theory A theory of the conduction properties of a material in bulk or in a rectifying barrier which takes into account the motion of both electrons and holes.

two-input subtracter *See* half-subtracter.

two-phase alternating-current circuit A circuit in which there are two alternating currents on separate wires, the two currents being 90° out of phase.

two-phase current Current delivered through two pairs of wires or at a phase difference of one-quarter cycle (90°) between the current in the two pairs.

two-phase five-wire system System of alternating-current supply comprising five conductors, four of which are connected as in a two-phase four-wire system, the fifth being connected to the neutral points of each phase.

two-phase four-wire system System of alternating-current supply comprising two pairs of conductors, between one pair of which is maintained an alternating difference of potential displaced in phase by one-quarter of a period from an alternating difference of potential of the same frequency maintained between the other pair.

two-phase three-wire system System of alternating-current supply comprising three conductors, between one of which (known as the common return) and each of the other two are maintained alternating difference of potential displaced in phase by one-quarter of a period with relation to each other.

two-port junction A waveguide junction with two openings; it can consist either of a discontinuity or obstacle in a waveguide, or of two essentially different waveguides connected together.

two-pulse canceler A moving-target indicator canceler which compares the phase variation of two successive pulses received from a target; discriminates against signals with radial velocities which produce a Doppler frequency equal to a multiple of the pulse repetition frequency.

two-wire circuit A metallic circuit formed by two conductors insulated from each other; in contrast with a four-wire circuit, it uses only one line or channel for transmission of electric waves in both directions.

two-wire repeater Repeater that provides for transmission in both directions over a two-wire circuit; in carrier transmission, it usually operates on the principle of frequency separation for the two directions of transmission.

Twystron Very-high-power, hybrid microwave tube, combining the input section of a high-power klystron with the output section of a traveling wave tube, characterized by high operating efficiency and wide bandwidths.

type A wave *See* continuous wave.

type-M carcinotron *See* M-type backward-wave oscillator.

type-O carcinotron *See* O-type backward-wave oscillator.

U

Uda antenna *See* Yagi-Uda antenna.

UJT *See* unijunction transistor.

ultra-audion circuit Regenerative detector circuit in which a parallel resonant circuit is connected between the grid and the plate of a vacuum tube, and a variable capacitor is connected between the plate and cathode to control the amount of regeneration.

ultra-audion oscillator Variation of the Colpitts oscillator circuit; the resonant circuit employs a section of transmission line.

ultra-high-frequency tuner A tuner in a television receiver for reception of stations transmitting in the ultra-high-frequency band (channels 14–83); it usually employs continuous tuning.

ultraphotic rays Rays outside the visible part of the spectrum, including infrared and ultraviolet rays.

ultrasonic camera A device which produces a picture display of ultrasonic waves sent through a sample to be inspected or through live tissue; a piezoelectric crystal is used to convert the ultrasonic waves to voltage differences, and the voltage pattern on the crystal modulates the intensity of an electronic beam scanning the crystal; this beam in turn controls the intensity of a beam in a television tube.

ultraviolet imagery That imagery produced as a result of sensing ultraviolet radiations reflected from a given target surface.

ultraviolet lamp A lamp providing a high proportion of ultraviolet radiation, such as various forms of mercury-vapor lamps.

ultraviolet light *See* ultraviolet radiation.

ultraviolet radiation Electromagnetic radiation in the wavelength range 4–400 nanometers; this range begins at the short-wavelength limit of visible light and overlaps the wavelengths of long x-rays (some scientists place the lower limit at higher values, up to 40 nanometers). Also known as ultraviolet light.

ultraviolet spectrum **1.** The range of wavelengths of ultraviolet radiation, covering 4–400 nanometers. **2.** A display or graph of the intensity of ultraviolet radiation emitted or absorbed by a material as a function of wavelength or some related parameter.

umbrella antenna Antenna in which the wires are guyed downward in all directions from a central pole or tower to the ground, somewhat like the ribs of an open umbrella.

Umklapp process The interaction of three or more waves in a solid, such as lattice waves or electron waves, in which the sum of the wave vectors is not equal to zero but, rather, is equal to a vector in the reciprocal lattice. Also known as flip-over process.

unamplified back bias Degenerative voltage developed across a fast time constant circuit within an amplifier stage itself.

unbalanced line A transmission line in which the voltages on the two conductors are not equal with respect to ground; a coaxial line is an example.

unbalanced output An output in which one of the two input terminals is substantially at ground potential.

unbalanced wire circuit A wire circuit whose two sides are inherently electrically unlike.

unblanking pulse Voltage applied to a cathode-ray tube to overcome bias and cause trace to be visible.

uncharged Having no electric charge.

underbunching In velocity-modulated electron streams, a condition representing less than the optimum bunching.

undercurrent relay A relay designed to operate when its coil current falls below a predetermined value.

undervoltage protection An undervoltage relay which removes a motor from service when a low-voltage condition develops, so that the motor will not draw excessive current, or which prevents a large induction or synchronous motor from starting under low-voltage conditions.

undervoltage relay A relay designed to operate when its coil voltage falls below a predetermined value.

undisturbed-one output "One" output of a magnetic cell to which no partial-read pulses have been applied since that cell was last selected for writing.

undisturbed-zero output "Zero" output of a magnetic cell to which no partial-write pulses have been applied since that cell was last selected for reading.

unfired tube Condition of a TR, ATR, or pre-TR tube in which there is no radio-frequency glow discharge at either the resonant gap or resonant window.

unidirectional antenna An antenna that has a single well-defined direction of maximum gain.

unidirectional coupler Directional coupler that samples only one direction of transmission.

unidirectional log-periodic antenna Broad-band antenna in which the cut-out portions of a log-periodic antenna are mounted at an angle to each other, to give a unidirectional radiation pattern in which the major radiation is in the backward direction, off the apex of the antenna; impedance is essentially constant for all frequencies, as is the radiation pattern.

unidirectional pulses Single polarity pulses which all rise in the same direction.

unidirectional transducer Transducer that measures stimuli in only one direction from a reference zero or rest position. Also known as unilateral transducer.

uniform line Line which has substantially identical electrical properties through its length.

uniform plane wave Plane wave in which the electric and magnetic intensities have constant amplitude over the equiphase surfaces; such a wave can only be found in free space at an infinite distance from the source.

unijunction transistor An n-type bar of semiconductor with a p-type alloy region on one side; connections are made to base contacts at either end of the bar and to the p-region. Abbreviated UJT. Formely known as double-base diode; double-base junction diode.

unilateral conductivity Conductivity in only one direction, as in a perfect rectifier.

unilateralization Use of an external feedback circuit in a high-frequency transistor amplifier to prevent undesired oscillation by canceling both the resistive and reactive changes produced in the input circuit by internal voltage feedback; with neutralization, only the reactive changes are canceled.

unilateral transducer *See* unidirectional transducer.

uninterruptible power system A system that provides protection against primary alternating-current power failure and variations in power-line frequency and voltage. Abbreviated UPS.

unipolar Having but one pole, polarity, or direction; when applied to amplifiers or power supplies, it means that the output can vary in only one polarity from zero and, therefore, must always contain a direct-current component.

unipolar machine *See* homopolar generator.

unipolar transistor A transistor that utilizes charge carriers of only one polarity, such as a field-effect transistor.

unipole A hypothetical antenna that radiates or receives signals equally well in all directions. Also known as isotropic antenna.

unipotential cathode *See* indirectly heated cathode.

unipotential electrostatic lens An electrostatic lens in which the focusing is produced by application of a single potential difference; in its simplest form it consists of three apertures of which the outer two are at a common potential, and the central aperture is at a different, generally lower, potential.

unit charge *See* statcoulomb.

unit delay A network whose output is equal to the input delayed by one unit of time.

unit magnetic pole Two equal magnetic poles of the same sign have unit value when they repel each other with a force of 1 dyne if placed 1 centimeter apart in a vacuum.

unity coupling Perfect magnetic coupling between two coils, so that all magnetic flux produced by the primary winding passes through the entire secondary winding.

unity gain bandwidth Measure of the gain-frequency product of an amplifier; unity gain bandwidth is the frequency at which the open-loop gain becomes unity, based on 6 decibels per octave crossing.

unity power factor Power factor of 1.0, obtained when current and voltage are in phase, as in a circuit containing only resistance or in a reactive circuit at resonance.

universal motor A motor that may be operated at approximately the same speed and output on either direct current or single-phase alternating current. Also known as ac/dc motor.

universal receiver *See* ac/dc receiver.

universal resonance curve A plot of Y/Y_0 against $Q_0\delta$ for a series-resonant circuit, or of Z/Z_0 against $Q_0\delta$ for a parallel-resonant circuit, where Y and Z are the admittance and impedance of a circuit, Y_0 and Z_0 are the values of these quantities at resonance, Q_0 is the Q value of the circuit at resonance, and δ is the deviation of the frequency from resonance divided by the resonant frequency; it can be applied to all resonant circuits.

universal shunt *See* Ayrton shunt.

univibrator *See* monostable multivibrator.

unloaded Q The Q of a system when there is no external coupling to it.

unloading amplifer Amplifier that is capable of reproducing or amplifying a given voltage signal while drawing negligible current from the voltage source.

unsaturated standard cell One of two types of Weston standard cells (batteries); used for voltage calibration work not requiring an accuracy greater than 0.01%.

untuned Not resonant at any of the frequencies being handled.

up-converter Type of parametric amplifier which is characterized by the frequency of the output signal being greater than the frequency of the input signal.

upper half-power frequency The frequency on an amplifier response curve which is greater than the frequency for peak response and at which the output voltage is $1/\sqrt{2}$ (that is, 0.707) of its midband or other reference value.

UPS *See* uninterruptible power system.

utilization factor In electric power distribution, the maximum demand of a system or part of a system divided by its rated capacity.

V

V *See* volt.

VA *See* volt-ampere.

vacancy A defect in the form of an unoccupied lattice position in a crystal.

vacuum capacitor A capacitor with separated metal plates or cylinders mounted in an evacuated glass envelope to obtain a high breakdown voltage rating.

vacuum circuit breaker A circuit breaker in which a pair of contacts is hermetically sealed in a vacuum envelope; the contacts are separated by using a bellows to move one of them; an arc is produced by metallic vapor boiled from the electrodes, and is extinguished when the vapor particles condense on solid surfaces.

vacuum diffusion Diffusion of impurities into a semiconductor material in a continuously pumped hard vacuum.

vacuum fluorescent lamp An evacuated display tube in which the anodes are coated with a phosphor that glows when electrons from the cathode strike it, to create a display.

vacuum phototube A phototube that is evacuated to such a degree that its electrical characteristics are essentially unaffected by gaseous ionization; in a gas phototube, some gas is intentionally introduced.

vacuum relay A sensitive relay having its contacts mounted in a highly evacuated glass housing, to permit handling radio-frequency voltages as high as 20,000 volts without flashover between contacts even though contact spacing is but a few hundredths of an inch when open.

vacuum switch A switch having its contacts in an evacuated envelope to minimize sparking.

vacuum tube An electron tube evacuated to such a degree that its electrical characteristics are essentially unaffected by the presence of residual gas or vapor.

vacuum-tube amplifier An amplifier employing one or more vacuum tubes to control the power obtained from a local source.

vacuum-tube circuit An electric circuit in which a vacuum tube is connected.

vacuum-tube clipping circuit A circuit in which a vacuum tube is used to achieve clipping action; the bias at the input is set at such a level that output current cannot flow during a portion of the amplitude excursion of the input voltage or current waveform.

vacuum-tube electrometer An electrometer in which the ionization current in an ionization chamber is amplified by a special vacuum triode having an input resistance above 10,000 megohms.

vacuum-tube keying Code-transmitter keying system in which a vacuum tube is connected in series with the plate supply lead of a frequency-controlling stage of the transmitter; when the key is open, the tube blocks, interrupting the plate supply to the output stage; closing the key allows the plate current to flow through the keying tube and the output tubes.

vacuum-tube modulator A modulator employing a vacuum tube as a modulating element for impressing an intelligence signal on a carrier.

vacuum-tube oscillator A circuit utilizing a vacuum tube to convert direct-current power into alternating-current power at a desired frequency.

vacuum-tube rectifier A rectifier in which rectification is accomplished by the unidirectional passage of electrons from a heated electrode to one or more other electrodes within an evacuated space.

vacuum ultraviolet radiation Ultraviolet radiation with a wavelength of less than 200 nanometers; absorption of radiation in this region by air and other gases requires the use of evacuated apparatus for transmission. Also known as extreme ultraviolet radiation.

valence band The highest electronic energy band in a semiconductor or insulator which can be filled with electrons.

valence electron *See* conduction electron.

valley attenuation For an electric filter with an equal ripple characteristic, the maximum attenuation occurring at a frequency between two frequencies where the attenuation reaches a minimum value.

valve *See* electron tube.

valve arrester A type of lightning arrester which consists of a single gap or multiple gaps in series with current-limiting elements; gaps between spaced electrodes prevent flow of current through the arrester except when the voltage across them exceeds the critical gap flashover.

Van Atta array Antenna array in which pairs of corner reflectors or other elements equidistant from the center of the array are connected together by a low-loss transmission line in such a way that the received signal is reflected back to its source in a narrow beam to give signal enhancement without amplification.

Van de Graaff accelerator A Van de Graaff generator equipped with an evacuated tube through which charged particles may be accelerated.

Van de Graaff generator A high-voltage electrostatic generator in which electrical charge is carried from ground to a high-voltage terminal by means of an insulating belt and is discharged onto a large, hollow metal electrode.

Van der Pol oscillator A type of relaxation oscillator which has a single pentode tube and an external circuit with a capacitance that causes the device to switch between two values of the screen voltage.

vane-anode magnetron Cavity magnetron in which the walls between adjacent cavities have parallel plane surfaces.

vane attenuator *See* flap attenuator.

V antenna An antenna having a V-shaped arrangement of conductors fed by a balanced line at the apex; the included angle, length, and elevation of the conductors are proportioned to give the desired directivity. Also spelled vee antenna.

vapor lamp *See* discharge lamp.

varactor A semiconductor device characterized by a voltage-sensitive capacitance that resides in the space-charge region at the surface of a semiconductor bounded by an insulating layer. Also known as varactor diode; variable-capacitance diode; varicap; voltage-variable capacitor.

varactor diode *See* varactor.

varactor tuning A method of tuning in which varactor diodes are used to vary the capacitance of a tuned circuit.

var hour A unit of the integral of reactive power over time, equal to a reactive power of 1 var integrated over 1 hour; equal in magnitude to 3600 joules. Also known as reactive volt-ampere hour; volt-ampere-hour reactive.

variable attenuator An attenuator for reducing the strength of an alternating-current signal either continuously or in steps, without causing appreciable signal distortion, by maintaining a substantially constant impedance match.

variable-bandwidth filter An electric filter whose upper and lower cutoff frequencies may be independently selected, so that almost any bandwidth may be obtained; it usually consists of several stages of *RC* filters, each separated by buffer amplifiers; tuning is accomplished by varying the resistance and capacitance values.

variable-capacitance diode *See* varactor.

variable capacitor A capacitor whose capacitance can be varied continuously by moving one set of metal plates with respect to another.

variable coupling Inductive coupling that can be varied by moving one coil with respect to another.

variable diode function generator An improvement of a diode function generator in which fully adjustable potentiometers are used for breakpoint and slope resistances, permitting the programming of analytic, arbitrary, and empirical functions, including inflections. Abbreviated VDFG.

variable inductance *See* variable inductor.

variable inductor A coil whose effective inductance can be changed. Also known as variable inductance.

variable-mu tube An electron tube in which the amplification factor varies in a predetermined manner with control-grid voltage; this characteristic is achieved by making the spacing of the grid wires vary regularly along the length of the grid, so that a very large negative grid bias is required to block anode current completely. Also known as remote-cutoff tube.

variable-reluctance stepper motor A stepper motor having a soft iron rotor with teeth or poles so positioned that they cannot simultaneously align with all the stator poles.

variable-reluctance transducer A transducer in which a slug of magnetic material is moved between two coils by the displacement being monitored; this changes the reluctance of the coils, thereby changing their impedance.

variable resistor *See* rheostat.

variable speech control A method of removing small portions of speech from a tape recording at regular intervals and stretching the remaining sounds to fill the gaps, so that recorded speech can be played back at twice or even 2½ times the original speed without changing pitch and without significant loss of intelligibility. Abbreviated VSC.

variable-speed scanning Scanning method whereby the speed of deflection of the scanning beam in the cathode-ray tube of a television camera is governed by the optical density of the film being scanned.

variable-transconductance circuit A circuit used in four-quadrant multipliers that employs a simple differential transistor pair in which one variable input to the base of one transistor controls the device's gain or transconductance, and one transistor amplifies the other's variable input, applied to the common emitter point, in proportion to the control input.

variable transformer An iron-core transformer having provisions for varying its output voltage over a limited range or continuously from zero to maximum output voltage, generally by means of a contact arm moving along exposed turns of the secondary winding. Also known as adjustable transformer; continuously adjustable transformer.

varicap See varactor.

varindor Inductor in which the inductance varies markedly with the current in the winding.

variocoupler In radio practice, a transformer in which the self-impedance of windings remains essentially constant while the mutual impedance between the windings is adjustable.

variolosser Device in which loss can be controlled by a voltage or current.

variometer A variable inductance having two coils in series, one mounted inside the other, with provisions for rotating the inner coil in order to vary the total inductance of the unit over a wide range.

varioplex Telegraph switching system that establishes connections on a circuit-sharing basis between a multiplicity of telegraph transmitters in one locality and respective corresponding telegraph receivers in another locality over one or more intervening telegraph channels; maximum use of channel capacity is secured by momentarily storing the signals and allocating circuit time in rotation among the transmitters having information in storage.

varistor A two-electrode semiconductor device having a voltage-dependent nonlinear resistance; its resistance drops as the applied voltage is increased. Also known as voltage-dependent resistor.

Varley loop test A method of using a Wheatstone bridge to determine the distance from the test point to a fault in a telephone or telegraph line or cable.

var measurement The measurement of reactive power in a circuit.

V band A radio-frequency band of 46.0 to 56.0 gigahertz.

V-beam radar A volumetric radar system that uses two fan beams to determine the distance, bearing, and height of a target: one beam is vertical and the other inclined; the beams intersect at ground level and rotate continuously about a vertical axis; the time difference between the arrivals of the echoes of the two beams is a measure of target elevation.

VCO See voltage-controlled oscillator.

V connection See open-delta connection.

VDFG See variable diode function generator.

vector potential A vector function whose curl is equal to the magnetic induction. Symbolized **A**. Also known as magnetic vector potential.

vector power Vector quantity equal in magnitude to the square root of the sum of the squares of the active power and the reactive power.

vector-power factor Ratio of the active power to the vector power; it is the same as power factor in the case of simple sinusoidal quantities.

vector resolver See resolver.

vee antenna See V antenna.

velocity filter Storage tube device which blanks all targets that do not move more than one resolution cell in less than a predetermined number of antenna scans.

velocity-modulated oscillator Oscillator which employs velocity modulation to produce radio-frequency power. Also known as klystron oscillator.

velocity modulation 1. Modulation in which a time variation in velocity is impressed on the electrons of a stream. 2. A television system in which the intensity of the electron beam remains constant throughout a scan, and the velocity of the spot at the screen is varied to produce changes in picture brightness (not in general use).

velocity of light *See* speed of light.

velocity pickup A device that generates a voltage proportional to the relative velocity between two principal elements of the pickup, the two elements usually being a coil of wire and a source of magnetic field.

velocity shaped canceler *See* cascaded feedback canceler.

vented battery A nickel-cadmium or other battery which lacks provisions for recombination of gases produced during normal operation, so that these gases must be vented to the atmosphere to avoid rupture of the cell case.

vernier capacitor Variable capacitor placed in parallel with a larger tuning capacitor to provide a finer adjustment after the larger unit has been set approximately to the desired position.

vernitel Precision device which makes possible the transmission of data with high accuracy over standard frequency modulated-frequency modulated telemetering systems.

versatile automatic test equipment Computer-controlled tester, for missile electronic systems, that trouble-shoots faults by deductive logic and isolates them to the plug-in module or component level.

vertical antenna A vertical metal tower, rod, or suspended wire used as an antenna.

vertical blanking Blanking of a television picture tube during the vertical retrace.

vertical centering control The centering control provided in a television receiver or cathode-ray oscilloscope to shift the position of the entire image vertically in either direction on the screen.

vertical component effect *See* antenna effect.

vertical definition *See* vertical resolution.

vertical deflection oscillator The oscillator that produces, under control of the vertical synchronizing signals, the sawtooth voltage waveform that is amplified to feed the vertical deflection coils on the picture tube of a television receiver. Also known as vertical oscillator.

vertical field-strength diagram Representation of the field strength at a constant distance from an antenna and in a vertical plane passing through the antenna.

vertical hold control The hold control that changes the free-running period of the vertical deflection oscillator in a television receiver, so the picture remains steady in the vertical direction.

vertical-incidence transmission Transmission of a radio wave vertically to the ionosphere and back.

vertical interval reference A reference signal inserted into a television program signal every 1/60 second, in line 19 of the vertical blanking period between television frames, to provide references for luminance amplitude, black-level amplitude, sync amplitude, chrominance amplitude, and color-burst amplitude and phase. Abbreviated VIR.

vertical linearity control A linearity control that permits narrowing or expanding the height of the image on the upper half of the screen of a television picture tube, to give linearity in the vertical direction so circular objects appear as true circles; usually mounted at the rear of the receiver.

vertically stacked loops *See* stacked loops.

vertical metal oxide semiconductor technology For semiconductor devices, a technology that involves essentially the formation of four diffused layers in silicon and etching of a V-shaped groove to a precisely controlled depth in the layers, followed by deposition of metal over silicon dioxide in the groove to form the gate electrode. Abbreviated VMOS technology.

vertical oscillator *See* vertical deflection oscillator.

vertical resolution The number of distinct horizontal lines, alternately black and white, that can be seen in the reproduced image of a television or facsimile test pattern; it is primarily fixed by the number of horizontal lines used in scanning. Also known as vertical definition.

vertical retrace The return of the electron beam to the top of the screen at the end of each field in television.

vertical sweep The downward movement of the scanning beam from top to bottom of the picture being televised.

vertical synchronizing pulse One of the six pulses that are transmitted at the end of each field in a television system to keep the receiver in field-by-field synchronism with the transmitter. Also known as picture synchronizing pulse.

very-high-frequency oscillator An oscillator whose frequency lies in the range from a few to several hundred megahertz; it uses distributed, rather than lumped, impedances, such as parallel wire transmission lines or coaxial cables.

very-high-frequency tuner A tuner in a television receiver for reception of stations transmitting in the very-high-frequency band; it generally has 12 discrete positions corresponding to channels 2–13.

very-long-baseline interferometry A method of improving angular resolution in the observation of radio sources; these are simultaneously observed by two radio telescopes which are very far apart, and the signals are recorded on magnetic tapes which are combined electronically or on a computer. Abbreviated VLBI.

very-long-range radar Equipment whose maximum range on a reflecting target of 1 square meter normal to the signal path exceeds 800 miles (1300 kilometers), provided line-of-sight exists between the target and the radar.

very-short-range radar Equipment whose range on a reflecting target of 1 square meter normal to the signal path is less than 50 miles (80 kilometers), provided line-of-sight exists between the target and the radar.

vestigial-sideband filter A filter that is inserted between a transmitter and its antenna to suppress part of one of the sidebands.

vibrating capacitor A capacitor whose capacitance is varied in a cyclic manner to produce an alternating voltage proportional to the charge on the capacitor; used in a vibrating-reed electrometer.

vibration pickup An electromechanical transducer capable of converting mechanical vibrations into electrical voltages.

vibrator An electromechanical device used primarily to convert direct current to alternating current but also used as a synchronous rectifier; it contains a vibrating reed which has a set of contacts that alternately hit stationary contacts attached to the frame, reversing the direction of current flow; the reed is activated when a soft-iron slug at its tip is attracted to the pole piece of a driving coil.

vibrator power supply A power supply using a vibrator to produce the varying current necessary to actuate a transformer, the output of which is then rectified and filtered.

vibrator-type inverter A device that uses a vibrator and an associated transformer or other inductive device to change direct-current input power to alternating-current output power.

vibrotron A triode electron tube having an anode that can be moved or vibrated by an externally applied force.

video 1. Pertaining to picture signals or to the sections of a television system that carry these signals in either unmodulated or modulated form. 2. Pertaining to the demodulated radar receiver output that is applied to a radar indicator.

video amplifier A low-pass amplifier having a band-width on the order of 2–10 megahertz, used in television and radar transmission and reception; it is a modification of an RC-coupled amplifier, such that the high-frequency half-power limit is determined essentially by the load resistance, the internal transistor capacitances, and the shunt capacitance in the circuit.

video correlator Radar circuit that enhances automatic target detection capability, provides data for digital target plotting, and gives improved immunity to noise, interference, and jamming.

video discrimination Radar circuit used to reduce the frequency band of the video amplifier stage in which it is used.

video disk recorder A video recorder that records television visual signals and sometimes aural signals on a magnetic, optical, or other type of disk which is usually about the size of a long-playing phonograph record.

Videograph Trademark of A. B. Dick Company for a high-speed cathode-ray character generator and electrostatic printer, used for printing magazine address labels under control of magnetic tape files of computer-processed addresses; the moving electron beam applies a charge on a dielectric-coated paper to form electrostatic images of characters; powder is then attracted to the image areas and fused to give readable addresses.

video integrator 1. Electric counter-countermeasures device that is used to reduce the response to nonsynchronous signals such as noise, and is useful against random pulse signals and noise. 2. Device which uses the redundancy of repetitive signals to improve the output signal-to-noise ratio, by summing the successive video signals.

video masking Method of removing chaff echoes and other extended clutter from radar displays.

video player A player that converts a video disk, videotape, or other type of recorded television program into signals suitable for driving a home television receiver.

video recorder A magnetic tape recorder capable of storing the video signals for a television program and feeding them back later to a television transmitter or directly to a receiver.

video replay Also known as video tape replay. 1. A procedure in which the audio and video signals of a television program are recorded on magnetic tape and then the tape is run through equipment later to rebroadcast the live scene. 2. A similar procedure in which the scene is rebroadcast almost immediately after it occurs. Also known as instant replay.

video signal *See* target signal.

video tape A heavy-duty magnetic tape designed primarily for recording the video signals of television programs.

video tape recording A method of recording television video signals on magnetic tape for later rebroadcasting of television programs. Abbreviated VTR.

video tape replay *See* video replay.

video transformer A transformer designed to transfer, from one circuit to another, the signals containing picture information in television.

video transmitter *See* visual transmitter.

vidicon A camera tube in which a charge-density pattern is formed by photoconduction and stored on a photoconductor surface that is scanned by an electron beam, usually of low-velocity electrons; used chiefly in industrial television cameras.

viewfinder An auxiliary optical or electronic device attached to a television camera so the operator can see the scene as the camera sees it.

viewing screen *See* screen.

viewing storage tube *See* direct-view storage tube.

viewing time Time during which a storage tube is presenting a visible output corresponding to the stored information.

viologen display An electrochromic display based on an electrolyte consisting of an aqueous solution of a dipositively charged organic salt, containing a colorless cation that undergoes a one-electron reduction process to produce a purple radical cation, upon application of a negative potential to the electrode.

VIR *See* vertical interval reference.

virtual cathode The locus of a space-charge-potential minimum such that only some of the electrons approaching it are transmitted, the remainder being reflected back to the electron-emitting cathode.

visibility factor The ratio of the minimum signal input detectable by ideal instruments connected to the output of a receiver, to the minimum signal power detectable by a human operator through a display connected to the same receiver. Also known as display loss.

visual display unit *See* display tube.

visually coupled display *See* helmet-mounted display.

visual scanner Device that optically scans printed or written data and generates an analog or digital signal.

visual storage tube Any electrostatic storage tube that also provides a visual readout.

visual transmitter Those parts of a television transmitter that act on picture signals, including parts that act on the audio signals as well. Also known as picture transmitter; video transmitter.

vitreous state A solid state in which the atoms or molecules are not arranged in any regular order, as in a crystal, and which crystallizes only after an extremely long time. Also known as glassy state.

VLBI *See* very-long-baseline interferometry.

VMOS technology *See* vertical metal oxide semiconductor technology.

vocoder A system of electronic apparatus for synthesizing speech according to dynamic specifications derived from an analysis of that speech.

vodas A voice-operated switching device used in transoceanic radiotelephone circuits to suppress echoes and singing sounds automatically; it connects a subscriber's line automatically to the transmitting station as soon as he starts speaking and simultaneously disconnects it from the receiving station, thereby permitting the use of one radio channel for both transmitting and receiving without appreciable switching delay as the parties alternately talk. Derived from voice-operated device anti-singing.

voder An electronic system that uses electron tubes and filters, controlled through a keyboard, to produce voice sounds artificially. Derived from voice operation demonstrator.

voltage derating   **471**

vogad An automatic gain control circuit used to maintain a constant speech output level in long-distance radiotelephony. Derived from voice-operated gain-adjusted device.

voice coder Device that converts speech input into digital form prior to encipherment for secure transmission and converts the digital signals back to speech at the receiver.

voice digitization The conversion of analog voice signals to digital signals.

voice-frequency dialing Method of dialing by which the direct-current pulses from the dial are transformed into voice-frequency alternating-current pulses.

voice-operated device Any of several devices in a telephone system which are brought into operation by a sound signal, or some characteristic of such a signal.

voice-operated device anti-singing *See* vodas.

voice-operated gain-adjusted device *See* vogad.

voice-operated loss control and suppressor Voice-operated device which switches loss out of the transmitting branch and inserts loss in the receiving branch under control of the subscriber's speech.

voice operation demonstrator *See* voder.

voice synthesizer A synthesizer that simulates speech in any language by assembling a language's elements or phonemes under digital control, each with the correct inflection, duration, pause, and other speech characteristics.

volt The unit of potential difference or electromotive force in the meter-kilogram-second system, equal to the potential difference between two points for which 1 coulomb of electricity will do 1 joule of work in going from one point to the other. Symbolized V.

Volta effect *See* contact potential difference.

voltage Potential difference or electromotive force measured in volts.

voltage amplification The ratio of the magnitude of the voltage across a specified load impedance to the magnitude of the input voltage of the amplifier or other transducer feeding that load; often expressed in decibels by multiplying the common logarithm of the ratio by 20.

voltage amplifier An amplifier designed primarily to build up the voltage of a signal, without supplying appreciable power.

voltage-amplitude-controlled clamp A single diode clamp in which the diode functions as a clamp whenever the potential at point A rises above V_R; the diode is then in its forward-biased condition and acts as a very low resistance.

voltage coefficient For a resistor whose resistance varies with voltage, the ratio of the fractional change in resistance to the change in voltage.

voltage-controlled oscillator An oscillator whose frequency of oscillation can be varied by changing an applied voltage. Abbreviated VCO.

voltage corrector Active source of regulated power placed in series with an unregulated supply to sense changes in the output voltage (or current), and to correct for these changes by automatically varying its own output in the opposite direction, thereby maintaining the total output voltage (or current) constant.

voltage-current dual A pair of circuits in which the elements of one circuit are replaced by their dual elements in the other circuit according to the duality principle; for example, currents are replaced by voltages, capacitances by resistances.

voltage-dependent resistor *See* varistor.

voltage derating The reduction of a voltage rating to extend the lifetime of an electric device or to permit operation at a high ambient temperature.

voltage divider A tapped resistor, adjustable resistor, potentiometer, or a series arrangement of two or more fixed resistors connected across a voltage source; a desired fraction of the total voltage is obtained from the intermediate tap, movable contact, or resistor junction. Also known as potential divider.

voltage doubler A transformerless rectifier circuit that gives approximately double the output voltage of a conventional half-wave vacuum-tube rectifier by charging a capacitor during the normally wasted half-cycle and discharging it in series with the output voltage during the next half-cycle. Also known as doubler.

voltage drop The voltage developed across a component or conductor by the flow of current through the resistance or impedance of that component or conductor.

voltage feed Excitation of a transmitting antenna by applying voltage at a point of maximum potential (at a voltage loop or antinode).

voltage flare A higher than normal voltage purposely supplied to exposure lamps for a short period to produce full brilliance.

voltage gain The difference between the output signal voltage level in decibels and the input signal voltage level in decibels; this value is equal to 20 times the common logarithm of the ratio of the output voltage to the input voltage.

voltage generator A two-terminal circuit element in which the terminal voltage is independent of the current through the element.

voltage gradient The voltage per unit length along a resistor or other conductive path.

voltage level At any point in a transmission system, the ratio of the voltage existing t that point to an arbitrary value of voltage used as a reference.

voltage measurement Determination of the difference in electrostatic potential between two points.

voltage multiplier 1. A rectifier circuit capable of supplying a direct-current output voltage that is two or more times the peak value of the alternating-current voltage. 2. *See* instrument multiplier.

voltage-multiplier circuit A rectifier circuit capable of supplying a direct-current output voltage that is two or more times the peak value of the alternating-current input voltage; useful for high-voltage, low-current supplies.

voltage node Point having zero voltage in a stationary wave system, as in an antenna or transmission line; for example, a voltage node exists at the center of a half-wave antenna.

voltage phasor A line whose length represents the magnitude of a sinusoidally varying voltage and whose angle with the positive x-axis represents its phase.

voltage quadrupler A rectifier circuit, containing four diodes, which supplies a direct-current output voltage which is four times the peak value of the alternating-current input voltage.

voltage-range multiplier *See* instrument multiplier.

voltage rating The maximum sustained voltage that can safely be applied to an electric device without risking the possibility of electric breakdown. Also known as working voltage.

voltage ratio The root-mean-square primary terminal voltage of a transformer divided by the root-mean-square secondary terminal voltage under a specified load.

voltage reflection coefficient The ratio of the phasor representing the magnitude and phase of the electric field of the backward-traveling wave at a specified cross section of a waveguide to the phasor representing the forward-traveling wave at the same cross section.

voltage-regulating transformer Saturated-core type of transformer which holds output voltage to within a few percent (5% above or below normal) with input variations up to 20% above or below normal; considerable harmonic distortion results unless extensive filters are employed.

voltage regulation The ratio of the difference between no-load and full-load output voltage of a device to the full-load output voltage, expressed as a percentage.

voltage regulator A device that maintains the terminal voltage of a generator or other voltage source within required limits despite variations in input voltage or load. Also known as automatic voltage regulator; voltage stabilizer.

voltage-regulator diode A diode that maintains an essentially constant direct voltage in a circuit despite changes in line voltage or load.

voltage-regulator tube A glow-discharge tube in which the tube voltage drop is approximately constant over the operating range of current; used to maintain an essentially constant direct voltage in a circuit despite changes in line voltage or load. Also known as VR tube.

voltage saturation *See* anode saturation.

voltage stabilizer *See* voltage regulator.

voltage transformer An instrument transformer whose primary winding is connected in parallel with a circuit in which the voltage is to be measured or controlled. Also known as potential transformer.

voltage-tunable tube Oscillator tube whose operating frequency can be varied by changing one or more of the electrode voltages, as in a backward-wave magnetron.

voltage-variable capacitor *See* varactor.

voltaic cell A primary cell consisting of two dissimilar metal electrodes in a solution that acts chemically on one or both of them to produce a voltage.

voltaic pile An early form of primary battery, consisting of a pile of alternate pairs of dissimilar metal disks, with moistened pads between pairs.

voltammeter An instrument that may be used either as a voltmeter or ammeter.

volt-ampere The unit of apparent power in the International System; it is equal to the apparent power in a circuit when the product of the root-mean-square value of the voltage, expressed in volts, and the root-mean-square value of the current, expressed in amperes, equals 1. Abbreviated VA.

volt-ampere hour A unit for expressing the integral of apparent power over time, equal to the product of 1 volt-ampere and 1 hour, or to 3600 joules.

volt-ampere-hour reactive *See* var hour.

volt-ampere reactive The unit of reactive power in the International System; it is equal to the reactive power in a circuit carrying a sinusoidal current when the product of the root-mean-square value of the voltage, expressed in volts, by the root-mean-square value of the current, expressed in amperes, and by the sine of the phase angle between the voltage and the current, equals 1. Abbreviated var. Also known as reactive volt-ampere.

volt box A series of resistors arranged so that a desired fraction of a voltage can be measured, and the voltage thereby computed.

Volterra dislocation A model of a dislocation which is formed in a ring of crystalline material by cutting the ring, moving the cut surfaces over each other, and then rejoining them.

voltmeter-ammeter method A method of measuring resistance in which simultaneous readings of the voltmeter and ammeter are taken, and the unknown resistance is calculated from Ohm's law.

voltmeter sensitivity Ratio of the total resistance of the voltmeter to its full scale reading in volts, expressed in ohms per volt.

volume lifetime Average time interval between the generation and recombination of minority carriers in a homogeneous semiconductor.

volume-limiting amplifier Amplifier containing an automatic device that functions only when the input signal exceeds a predetermined level, and then reduces the gain so the output volume stays substantially constant despite further increases in input volume; the normal gain of the amplifier is restored when the input volume returns below the predetermined limiting level.

volume range In a transmission system, the difference, expressed in decibels, between the maximum and minimum volumes that can be satisfactorily handled by the system.

volume resistivity Electrical resistance between opposite faces of a 1-centimeter cube of insulating material, commonly expressed in ohm-centimeters. Also known as specific insulation resistance.

volume target A radar target composed of a large number of objects too close together to be resolved.

VR tube *See* voltage-regulator tube.

VSC *See* variable speech control.

VTR *See* video tape recording.

W

wafer A thin semiconductor slice on which matrices of microcircuits can be fabricated, or which can be cut into individual dice for fabricating single transistors and diodes.

wafer lever switch A lever switch in which a number of contacts are arranged on one or both sides of one or more wafers, for engaging one or more contacts on a movable wafer segment actuated by the operating lever.

wafer socket An electron-tube socket consisting of one or two wafers of insulating material having holes in which are spring metal clips that grip the terminal pins of a tube.

Wagner earth connection *See* Wagner ground.

Wagner ground A ground connection used with an alternating-current bridge to minimize stray capacitance errors when measuring high impedances; a potentiometer is connected across the bridge supply oscillator, with its movable tap grounded. Also known as Wagner earth connection.

walk down A malfunction in a magnetic core of a computer storage in which successive drive pulses or digit pulses cause charges in the magnetic flux in the core that persist after the magnetic fields associated with pulses have been removed. Also known as loss of information.

wall effect The contribution to the ionization in an ionization chamber by electrons liberated from the walls.

wall energy The energy per unit area of the boundary between two ferromagnetic domains which are oriented in different directions.

wall outlet An outlet mounted on a wall, from which electric power can be obtained by inserting the plug of a line cord.

wander *See* scintillation.

Wannier function The Fourier transform of a Bloch function defined for an entire band, regarded as a function of the wave vector.

warning-receiver system An electronic countermeasure system, carried on a tactical or transport aircraft, which is programmed to alert a pilot when his aircraft is being illuminated by a specific radar signal above predetermined power thresholds.

washer thermistor A thermistor in the shape of a washer, which may be as large as 0.75 inch (1.9 centimeters) in diameter and 0.50 inch (1.3 centimeters) thick; it is formed by pressing and sintering an oxide-binder mixture.

water-activated battery A primary battery that contains the electrolyte but requires the addition of or immersion in water before it is usable.

water-cooled tube An electron tube that is cooled by circulating water through or around the anode structure.

water cooling Cooling the electrodes of an electron tube by circulating water through or around them.

water load A matched waveguide termination in which the electromagnetic energy is absorbed in water; the resulting rise in the temperature of the water is a measure of the output power.

water rheostat *See* electrolytic rheostat.

wattage rating A rating expressing the maximum power that a device can safely handle continuously.

watt current *See* active current.

watt-hour A unit of energy used in electrical measurements, equal to the energy converted or consumed at a rate of 1 watt during a period of 1 hour, or to 3600 joules. Abbreviated Wh.

watt-hour capacity Number of watt-hours which can be delivered from a storage battery under specified conditions as to temperature, rate of discharge, and final voltage.

wattless component *See* reactive component.

wattless current *See* reactive current.

wattless power *See* reactive power.

wave angle The angle, either in bearing or elevation, at which a radio wave leaves a transmitting antenna or arrives at a receiving antenna.

wave antenna Directional antenna composed of a system of parallel, horizontal conductors, varying from a half to several wavelengths long, terminated to ground at the far end in its characteristic impedance.

wave clutter *See* sea clutter.

wave converter Device for changing a wave of a given pattern into a wave of another pattern, for example, baffle-plate converters, grating converters, and sheath-reshaping converters for waveguides.

wave duct 1. Waveguide, with tubular boundaries, capable of concentrating the propagation of waves within its boundaries. 2. Natural duct, formed in air by atmospheric conditions, through which waves of certain frequencies travel with more than average efficiency.

wave filter A transducer for separating waves on the basis of their frequency; it introduces relatively small insertion loss to waves in one or more frequency bands and relatively large insertion loss to waves of other frequencies.

waveform-amplitude distortion *See* frequency distortion.

waveguide 1. Broadly, a device which constrains or guides the propagation of electromagnetic waves along a path defined by the physical construction of the waveguide; includes ducts, a pair of parallel wires, and a coaxial cable. Also known as microwave waveguide. 2. More specifically, a metallic tube which can confine and guide the propagation of electromagnetic waves in the lengthwise direction of the tube.

waveguide assembly An item consisting of one or more definite lengths of straight or formed, flexible or rigid, prefabricated hollow tubing of conductive material; the tubing has a predetermined cross-section, and is designed to guide or conduct high-frequency electromagnetic energy through its interior; one or more ends are terminated.

waveguide attenuation The decrease from one point of a waveguide to another, in the power carried by an electromagnetic wave in the waveguide.

waveguide bend A section of waveguide in which the direction of the longitudinal axis is changed; an E-plane bend in a rectangular waveguide is bent along the narrow dimension, while an H-plane bend is bent along the wide dimension. Also known as waveguide elbow.

waveguide cavity A cavity resonator formed by enclosing a section of waveguide between a pair of waveguide windows which form shunt susceptances.

waveguide connector A mechanical device for electrically joining and locking together separable mating parts of a waveguide system. Also known as waveguide coupler.

waveguide coupler *See* waveguide connector.

waveguide critical dimension Dimension of waveguide cross section which determines the cutoff frequency.

waveguide cutoff frequency Frequency limit of propagation along a waveguide for waves of a given field configuration.

waveguide elbow *See* waveguide bend.

waveguide filter A filter made up of waveguide components, used to change the amplitude-frequency response characteristic of a waveguide system.

waveguide hybrid A waveguide circuit that has four arms so arranged that a signal entering through one arm will divide and emerge from the two adjacent arms, but will be unable to reach the opposite arm.

waveguide junction *See* junction.

waveguide plunger *See* piston.

waveguide probe *See* probe.

waveguide resonator *See* cavity resonator.

waveguide shim Thin resilient metal sheet inserted between waveguide components to ensure electrical contact.

waveguide slot A slot in a waveguide wall, either for coupling with a coaxial cable or another waveguide, or to permit the insertion of a traveling probe for examination of standing waves.

waveguide switch A switch designed for mechanically positioning a waveguide section so as to couple it to one of several other sections in a waveguide system.

waveguide window *See* iris.

wave impedance The ratio, at every point in a specified plane of a waveguide, of the transverse component of the electric field to the transverse component of the magnetic field.

wavelength constant *See* phase constant.

wavelength shifter A photofluorescent compound used with a scintillator material to increase the wavelengths of the optical photons emitted by the scintillator, thereby permitting more efficient use of the photons by the phototube or photocell.

wave-shaping circuit An electronic circuit used to create or modify a specified time-varying electrical quantity, usually voltage or current, using combinations of electronic devices, such as vacuum tubes or transistors, and circuit elements, including resistors, capacitors, and inductors.

wave tail Part of a signal-wave envelope (in time or distance) between the steady-state value (or crest) and the end of the envelope.

wave tilt Forward inclination of a radio wave due to its proximity to ground.

wave trap A resonant circuit connected to the antenna system of a receiver to suppress signals at a particular frequency, such as that of a powerful local station that is interfering with reception of other stations. Also known as trap.

478 wave-vector space

wave-vector space The space of the wave vectors of the state functions of some system; this would be used, for example, for electron wave functions in a crystal and thermal vibrations of a lattice. Also known as k-space; reciprocal space.

waxy-electrolyte battery A primary battery in which the electrolyte is a waxy material, such as polyethylene glycol, in which is dissolved a small amount of a salt, such as zinc chloride; the electrodes are frequently made of zinc and manganese dioxide, and the electrolyte is melted and painted on a paper sheet to form the separator.

Wb *See* weber.

weber The unit of magnetic flux in the meter-kilogram-second system, equal to the magnetic flux which, linking a circuit of one turn, produces in it an electromotive force of 1 volt as it is reduced to zero at a uniform rate in 1 second. Symbolized Wb.

wedge A waveguide termination consisting of a tapered length of dissipative material introduced into the guide, such as carbon.

wedge-base lamp A small indicator lamp that has wire leads folded back on opposite sides of a flat glass base.

Wehnelt cathode *See* oxide-coated cathode.

weightlessness switch *See* zero gravity switch.

Weissenberg method A method of studying crystal structure by x-ray diffraction in which the crystal is rotated in a beam of x-rays, and a photographic film is moved parallel to the axis of rotation; the crystal is surrounded by a sleeve which has a slot that passes only diffraction spots from a single layer of the reciprocal lattice, permitting positive identification of each spot in the pattern.

Weiss molecular field The effective magnetic field postulated in the Weiss theory of ferromagnetism, which acts on atomic magnetic moments within a domain, tending to align them, and is in turn generated by these magnetic moments.

Weiss theory A theory of ferromagnetism based on the hypotheses that below the Curie point a ferromagnetic substance is composed of small, spontaneously magnetized regions called domains, and that each domain is spontaneously magnetized because a strong molecular magnetic field tends to align the individual atomic magnetic moments within the domain. Also known as molecular field theory.

welding current The current that flows through a circuit while a weld is being made.

welding generator A generator used for supplying the welding current.

welding transformer A high-current, low-voltage power transformer used to supply current for welding.

Wertheim effect *See* Wiedemann effect.

Weston standard cell A standard cell used as a highly accurate voltage source for calibrating purposes; the positive electrode is mercury, the negative electrode is cadmium, and the electrolyte is a saturated cadmium sulfate solution; the Weston standard cell has a voltage of 1.018636 volts at 20°C.

wet cell A primary cell in which there is a substantial amount of free electrolyte in liquid form.

wet contact Contact through which direct current flows.

wet electrolytic capacitor An electrolytic capacitor employing a liquid electrolyte.

wet flashover voltage The voltage at which an electric discharge occurs between two electrodes that are separated by an insulator whose surface has been sprayed with water to simulate rain.

Wiegand module 479

wet-reed relay Reed-type relay containing mercury at the relay contacts to reduce arcing and contact bounce.

wetting The coating of a contact surface with an adherent film of mercury.

Wh *See* watt-hour.

Wheatstone bridge A four-arm bridge circuit, all arms of which are predominately resistive; used to measure the electrical resistance of an unknown resistor by comparing it with a known standard resistance. Also known as resistance bridge; Wheatstone network.

Wheatstone network *See* Wheatstone bridge.

wheel static Interference encountered in automobile-radio installations due to static electricity developed by friction between the tires and the street.

whiffletree switch In computers, a multiposition electronic switch composed of gate tubes and flip-flops, so named because its circuit diagram resembles a whiffletree.

whip antenna A flexible vertical rod antenna, used chiefly on vehicles. Also known as fishpole antenna.

whistling meteor Name applied to a radio meteor when a special system for detection is used in which the presence of the meteor is indicated by a rapidly changing audio-frequency radio signal.

White Alice *See* Alaska Integrated Communications Exchange.

whitening filter An electrical filter which converts a given signal to white noise. Also known as prewhitening filter.

wide band Property of a tuner, amplifier, or other device that can pass a broad range of frequencies.

wide-band amplifier An amplifier that will pass a wide range of frequencies with substantially uniform amplification.

wide-band repeater Airborne system that receives a radio-frequency signal for transmission; used in reconnaissance missions when low-altitude reconnaissance aircraft require an airborne relay platform for beyond line-of-sight data transmission to a readout station.

wide-band switching Basically, four-wire circuits using correed matrices with electronic controls capable of switching wide-band facilities up to 50 kilohertz in bandwidth.

wide-band transformer A transformer that can transfer electric energy from one circuit to another at any of a broad range of frequencies.

wide-open Refers to the untuned characteristic or lack of frequency selectivity.

width control Control that adjusts the width of the pattern on the screen of a cathode-ray tube in a television receiver or oscilloscope.

Wiedemann effect The twist produced in a current-carrying wire when placed in a longitudinal magnetic field. Also known as Wertheim effect.

Wiedemann-Franz law The law that the ratio of the thermal conductivity of a metal to its electrical conductivity is a constant, independent of the metal, times the absolute temperature. Also known as Lorentz relation.

Wiegand effect The generation of an electrical pulse in a coil wrapped around or located near a Wiegand wire subjected to a changing magnetic field.

Wiegand module The apparatus for generating an electrical pulse by means of the Wiegand effect, consisting of a Wiegand wire, two small magnets, and a pickup coil.

Wiegand wire A work-hardened wire whose magnetic permeability is much greater near its surface than at its center.

Wien bridge oscillator A phase-shift feedback oscillator that uses a Wien bridge as the frequency-determining element.

Wien capacitance bridge A four-arm alternating-current bridge used to measure capacitance in terms of resistance and frequency; two adjacent arms contain capacitors respectively in parallel and in series with resistors, while the other two arms are nonreactive resistors; bridge balance depends on frequency.

Wien-DeSauty bridge *See* DeSauty's bridge.

Wien frequency bridge A modification of the Wien capacitance bridge, used to measure frequencies.

Wien inductance bridge A four-arm alternating-current bridge used to measure inductance in terms of resistance and frequency; two adjacent arms contain inductors respectively in parallel and in series with resistors, while the other two arms are nonreactive resistors; bridge balance depends on frequency.

Wien-Maxwell bridge *See* Maxwell bridge.

Wierl equation A formula for the intensity of an electron beam scattered through a specified angle by diffraction from the molecules in a gas.

Wigner-Seitz method A method of approximating the band structure of a solid: Wigner-Seitz cells surrounding atoms in the solid are approximated by spheres, and band solutions of the Schrödinger equation for one electron are estimated by using the assumption that an electronic wave function is the product of a plane wave function and a function whose gradient has a vanishing radial component at the sphere's surface.

Williams tube A cathode-ray storage tube in which information is stored as a pattern of electric charges produced, maintained, read, and erased by suitably controlled scanning of the screen by the electron beam.

Wilson electroscope An electroscope that has a single gold leaf which, when charged, is attracted to a grounded metal plate inclined at an angle that maximizes the instrument's sensitivity.

Wilson experiment An experiment that tests the validity of electromagnetic theory; a hollow cylinder of dielectric material, having layers of metal on its outer and inner cylindrical surfaces, is rotated about its axis in a magnetic field parallel to the axis; a sensitive electrometer, connected to the metal layers, indicates a charge that has the magnitude and sign predicted by theory.

Wimshurst machine An electrostatic generator consisting of two glass disks rotating in opposite directions, having sectors of tinfoil and collecting combs so arranged that static electricity is produced for charging Leyden jars or discharging across a gap.

wind The manner in which magnetic tape is wound onto a reel; in an A wind, the coated surface faces the hub; in a B wind, the coated surface faces away from the hub.

wind charger A wind-driven direct-current generator used for charging storage batteries.

winding 1. One or more turns of wire forming a continuous coil for a transformer, relay, rotating machine, or other electric device. 2. A conductive path, usually of wire, that is inductively coupled to a magnetic storage core or cell.

window 1. A material having minimum absorption and minimum reflection of radiant energy, sealed into the vacuum envelope of a microwave or other electron tube to permit passage of the desired radiation through the envelope to the output device. 2. A hole in a partition between two cavities or waveguides, used for coupling.

wing spot generator Electronic circuit that grows wings on the video target signal of a type G indicator; these wings are inversely proportional in size to the range.

wiper That portion of the moving member of a selector, or other similar device, in communications practice, which makes contact with the terminals of a bank.

wiping contact A switch or relay contact designed to move laterally with a wiping motion after it touches a mating contact. Also known as self-cleaning contact; sliding contact.

wire A single bare or insulated metallic conductor having solid, stranded, or tinsel construction, designed to carry current in an electric circuit. Also known as electric wire.

wire bonding Lead-covered tie used to connect two cable sheaths until a splice is permanently closed and covered.

wire fusing current The electric current which will cause a wire to melt.

wiregrating A series of wires placed in a waveguide that allow one or more types of waves to pass and block all others.

wire mile Unit of measure of the length of two-conductor wire between two points; the length of the route multiplied by the number of circuits gives the number of wire miles.

wire-wound potentiometer A potentiometer which is similar to a slide-wire potentiometer, except that the resistance wire is wound on a form and contact is made by a slider which moves along an edge from turn to turn.

wire-wound resistor A resistor employing as the resistance element a length of high-resistance wire or ribbon, usually Nichrome, wound on an insulating form.

wire-wound rheostat A rheostat in which a sliding or rolling contact moves over resistance wire that has been wound on an insulating core.

wire-wrap connection A solderless connection made by wrapping several turns of bare wire around a sharp-corner rectangular terminal under tension, using either a power tool or hand tool. Also known as solderless wrapped connection; wrapped connection.

wiring The installation and utilization of a system of wire for conduction of electricity. Also known as electric wiring.

wiring diagram *See* circuit diagram.

wiring harness An array of insulated conductors bound together by lacing cord, metal bands, or other binding, in an arrangement suitable for use only in specific equipment for which the harness was designed; it may include terminations.

wobbulator A signal generator in which a motor-driven variable capacitor is used to vary the output frequency periodically between two known limits, as required for displaying a frequency-response curve on the screen of a cathode-ray oscilloscope.

work *See* load.

work function The minimum energy needed to remove an electron from the Fermi level of a metal to infinity; usually expressed in electron volts.

working voltage *See* voltage rating.

write head Device that stores digital information as coded electrical pulses on a magnetic drum, disk, or tape.

writing speed Lineal scanning rate of the electron beam across the storage surface in writing information on a cathode-ray storage tube.

Wullenweber antenna An antenna array consisting of two concentric circles of masts, connected to be electronically steerable; used for ground-to-air communication at Strategic Air Command bases.

wye Polyphase circuit whose phase differences are 120° and which when drawn resembles the letter Y.

X

xenon arc lamp An arc lamp filled with xenon giving a light intensity approaching that of the carbon arc; particularly valuable in projecting motion pictures.

xenon flash lamp A flash tube containing xenon gas, which produces an intense peak of radiant energy at a wavelength of 566 nanometers when a high direct-current pulsed voltage is applied between electrodes at opposite ends of the tube.

x-ray absorption The taking up of energy from an x-ray beam by a medium through which the beam is passing.

x-ray generator A metal from whose surface large amounts of x-rays are emitted when it is bombarded with high-velocity electrons; metals with high atomic weight are the most efficient generators.

x-ray hardness The penetrating ability of x-rays; it is an inverse function of the wavelength.

x-ray optics A title-by-analogy of those phases of x-ray physics in which x-rays demonstrate properties similar to those of light waves. Also known as roentgen optics.

x-ray target The metal body with which high-velocity electrons collide, in a vacuum tube designed to produce x-rays.

x-ray tube A vacuum tube designed to produce x-rays by accelerating electrons to a high velocity by means of an electrostatic field, then suddenly stopping them by collision with a target.

XY switching system A telephone switching system consisting of a series of flat bank and wiper switches in which the wipers move in a horizontal plane, first in one direction and then in another under the control of pulses from a subscriber's dial; the switches are stacked on frames, and are operated one after another.

Y

Yagi antenna *See* Yagi-Uda antenna.

Yagi-Uda antenna An end-fire antenna array having maximum radiation in the direction of the array line; it has one dipole connected to the transmission line and a number of equally spaced unconnected dipoles mounted parallel to the first in the same horizontal plane to serve as directors and reflectors. Also known as Uda antenna; Yagi antenna.

Y circulator Circulator in which three identical rectangular waveguides are joined to form a symmetrical Y-shaped configuration, with a ferrite post or wedge at its center; power entering any waveguide will emerge from only one adjacent waveguide.

Y connection *See* Y network.

Y-delta transformation One of two electrically equivalent networks with three terminals, one being connected internally by a Y configuration and the other being connected internally by a delta transformation. Also known as delta-Y transformation; pi-T transformation.

yig device A filter, oscillator, parametric amplifier, or other device that uses an yttrium-iron-garnet crystal in combination with a variable magnetic field to achieve wide-band tuning in microwave circuits. Derived from yttrium-iron-garnet device.

yig filter A filter consisting of an yttrium-iron-garnet crystal positioned in a magnetic field provided by a permanent magnet and a solenoid; tuning is achieved by varying the amount of direct current through the solenoid; the bias magnet serves to tune the filter to the center of the band, thus minimizing the solenoid power required to tune over wide bandwidths.

yig-tuned parametric amplifier A parametric amplifier in which tuning is achieved by varying the amount of direct current flowing through the solenoid of a yig filter.

yig-tuned tunnel-diode oscillator Microwave oscillator in which precisely controlled wide-band tuning is achieved by varying the current through a tuning solenoid that acts on a yig filter in the tunnel-diode oscillator circuit.

Y junction A waveguide in which the longitudinal axes of the waveguide form a Y.

Y network A star network having three branches. Also known as Y connection.

yoke 1. Piece of ferromagnetic material without windings, which permanently connects two or more magnet cores. 2. *See* deflection yoke.

y parameter One of a set of four transistor equivalent-circuit parameters, used especially with field-effect transistors, that conveniently specify performance for small voltage and current in an equivalent circuit; the equivalent circuit is a current source with shunt impedance at both input and output.

yttrium-iron-garnet device *See* yig device.

Zener breakdown Nondestructive breakdown in a semiconductor, occurring when the electric field across the barrier region becomes high enough to produce a form of field emission that suddenly increases the number of carriers in this region. Also known as Zener effect.

Zener diode A semiconductor breakdown diode, usually constructed of silicon, in which reverse-voltage breakdown is based on the Zener effect.

Zener diode voltage regulator *See* diode voltage regulator.

Zener effect *See* Zener breakdown.

Zener voltage *See* breakdown voltage.

Zepp antenna Horizontal antenna which is a multiple of a half-wavelength long and is fed at one end by one lead of a two-wire transmission line that is some multiple of a quarter-wavelength long.

zero beat The condition in which a circuit is oscillating at the exact frequency of an input signal, so no beat tone is produced or heard.

zero-beat reception *See* homodyne reception.

zero bias The condition in which the control grid and cathode of an electron tube are at the same direct-current voltage.

zero-bias tube Vacuum tube which is designed so that it may be operated as a class B amplifier without applying a negative bias to its control grid.

zero error Delay time occurring within the transmitter and receiver circuits of a radar system; for accurate range data, this delay time must be compensated for in the calibration of the range unit.

zero-field emission *See* field-free emission current.

zero-gravity switch A switch that closes as weightlessness or zero gravity is approached; in one version, conductive sphere of mercury encompasses two contacts at zero gravity but flattens away from the upper contact under the influence of gravity. Also known as weightlessness switch.

zero output 1. Voltage response obtained from a magnetic cell in a zero state by a reading or resetting process. 2. Integrated voltage response obtained from a magnetic cell in a zero state by a reading or resetting process; a ratio of a one output to a zero output is a one-to-zero ratio.

zero phase-sequence relay Relay which functions in conformance with the zero phase–sequence component of the current, voltage, or power of the circuit.

zero potential Expression usually applied to the potential of the earth, as a convenient reference for comparison.

zero time reference Reference point in time from which the operations of various radar circuits are measured.

zigzag reflections From a layer of the ionosphere, high-order multiple reflections which may be of abnormal intensity; they occur in waves which travel by multihop ionosphere reflections and finally turn back toward their starting point by repeated reflections from a slightly curved or sloping portion of an ionized layer.

zinc–silver chloride primary cell A reserve primary cell that is activated by adding water; it can have a high capacity, up to 40 watt-hours per pound, and long life after activation.

zirconium lamp A high-intensity point-source lamp having a zirconium oxide cathode in an argon-filled bulb, used because of its low emanation of long-wavelength light and its concentrated source.

zone blanking Method of turning off the cathode-ray tube during part of the sweep of an antenna.

zoning The displacement of various portions of the lens or surface of a microwave reflector so the resulting phase front in the near field remains unchanged. Also known as stepping.

Z parameter One of a set of four transistor equivalent-circuit parameters; they are the inverse of the Y parameters.